er Related Pergamon Titles of Interest

SANESI: Thermodynamic and Transport Properties of Organic Salts

Solubilities

ability Constants of Metal-Ion Complexes, Part A, Inorganic Ligands

LITY DATA SERIES

ilver Azide, Cyanide, Cyanamides, Cyanate, Selenocyanate and Thiocyanate

): Alkali Metal Chlorides

VINCENT: Alkali Metal, Alkaline Earth Metal and Ammonium Halides. Amide Solvents

alogenated Benzenes

etraphenylborates

ction to Chemical Equilibrium and Kinetics

ion of Laboratory Chemicals, 2nd edition

Constants of Metal-Ion Complexes, Part B, Organic Ligands

MPSEY: Ionisation Constants of Organic Acids in Aqueous Solution

l of Symbols and Terminology for Physicochemical Quantities and Units

gamon Journals*

ional

on Kinetics

Chemistry

ta, Part A, Molecular Spectroscopy

of any journal available on request

r nearest Pergamon office for further details about any of the above books or journals

DIELECTRIC
O
BINARY S

Some Oth

FRANZOSINI &
GERRARD: Ga
HOGFELDT: St
IUPAC SOLUBI
Volume 3
SALOMON: S
Volume 9
COHEN-ADA
Volume 11
SCROSATI &
Volume 16
HORVATH: H
Volume 18
POPOVYCH: T
MEITES: Introdu
PERRIN: Purifica
PERRIN: Stability
SERJEANT & DE
WHIFFEN: Manua

Related Per

Chemistry Interna
Progress in Reacti
Pure and Applied
Spectrochimica Ac
Talanta

**Free specimen copy*
Please write to you

DIELECTRIC PROPERTIES OF BINARY SOLUTIONS

by

Y. Y. Akhadov

PERGAMON PRESS

OXFORD · NEW YORK · TORONTO · SYDNEY · PARIS · FRANKFURT

U.K.	Pergamon Press Ltd., Headington Hill Hall, Oxford OX3 0BW, England
U.S.A.	Pergamon Press Inc., Maxwell House, Fairview Park, Elmsford, New York 10523, U.S.A.
CANADA	Pergamon Press Canada Ltd., Suite 104, 150 Consumers Rd., Willowdale, Ontario M2J 1P9, Canada
AUSTRALIA	Pergamon Press (Aust.) Pty. Ltd., P.O. Box 544, Potts Point, N.S.W. 2011, Australia
FRANCE	Pergamon Press SARL, 24 rue des Ecoles, 75240 Paris, Cedex 05, France
FEDERAL REPUBLIC OF GERMANY	Pergamon Press GmbH, 6242 Kronberg-Taunus, Hammerweg 6, Federal Republic of Germany

This Pergamon Press edition 1981

British Library Cataloguing in Publication Data

Akhadov, Y. Y.
Dielectric properties of binary solutions.
1. Solution (Chemistry)
2. Dielectrics
I. Title
541'.341 QD541 80-40366

ISBN 0-08-023600-6

Printed in Great Britain by A. Wheaton & Co. Ltd., Exeter

CONTENTS

PREFACE

Investigation of dielectric properties of liquids can lead to information concerning molecular interactions and the mechanism of molecular processes which go on in liquids.

The results of such investigations are of great significance both theoretically and practically.

Without a deep understanding of the mechanism of molecular movements and interactions there can be no successful solutions of problems such as the effective direction of chemico-technological processes, the discovery of the nature of physico-chemical processes in the biological sphere, synthesis of new materials satisfying specific requirements, and the contriving of rapid methods of analysis and control in chemical processes.

Dielectric radiospectroscopy is one of the most sensitive means for the physico-chemical investigation of materials.

Knowledge of the dielectric parameters of liquids is indispensable for the operation of a number of contemporary electrochemical and radiochemical enterprises, for the construction of various appliances for automatic control of chemical processes in the organic synthesis industry, and for the operation of equipment for metering costs of fuel, oil, etc.

Dielectric data for liquids are indispensable for a wide circle of scientific workers specializing in various fields in physics, chemicstry, biology and medicine. This is clear from the great number of published works which provide information on the permittivity, dielectric loss, and relaxation phenomena in particular liquids and solutions over a wide range of frequencies and temperatures.

Despite this wide field of applicability there has not been, until now, either at home or abroad, any reference book in which the numerous experimental investigations on the dielectric properties of liquids have been collected and systematized.

The general reference books [1–4] contain selective collections of dielectric parameters for liquids outside the dispersion region.

In reference [5] an attempt was made by the author to systematize dielectric data for pure liquids.

None of the reference books deals with dispersion data for solutions such as permittivity, dielectric loss, relaxation times and the distribution of relaxation times, and thermodynamic functions of dielectric relaxation.

The present book aims to fill this gap and to make data on the dielectric parameters of binary solutions accessible to readers. It collects and systematizes numerous experimental results from 1892 to 1973 on the dielectric parameters of solutions. Critical analysis forms the basis of recommendations of the most reliable values for binary solutions taken from the original works.

The book consists of four chapters. Chapter I gives the basic formulae which describe the dielectric properties of substances and relate the experimentally observed quantities to the parameters characteristic of the substance. Chapters II and III consist of tables giving the result of measurements of static permittivity, limiting high-frequency permittivity, permittivity and dielectric loss, relaxation time, coefficient of distribution of relaxation times, and also thermodynamic functions of dielectric relaxation over a wide range of temperatures and an extensive frequency variation for non-aqueous and aqueous solutions of inorganic and organic compounds. Chapter IV presents dielectric data in graphical form.

At the end of each chapter section there are additional references to figures at the end of the book and to literature sources of experimental data which are not included elsewhere in the book.

The author expresses his deep gratitude to Y. I. Gerasimov, corresponding member of the Academy of Sciences of the U.S.S.R., for his great interest in the present work and Professor M. I. Shakhparonov for valuable critical observations and useful advice.

USE OF THE TABLES

In certain cases several values are given for the dielectric quantities measured at the same temperatures and frequencies. This indicates that it is not possible to give preference to one or other of the sources.

The tables for binary solutions of organic compounds are arranged according to the summary formulae of the components. In the summary formulae the symbols for the elements are placed in the order C, H, N, O and thereafter in alphabetic order. The precedence of compounds is decided according to the number of atoms of carbon and thereafter the number of atoms of hydrogen and so on. For any binary mixture the first component is that which precedes the other according to the above rule. The various binary mixtures are then arranged according to the precedence of the first components. Under a given first component, the second components are similarly arranged according to their precedence.

Concentrations of solutions are expressed in the same units as in the original works. When a concentration refers to the first component the concentration symbol has the index 1 (ϕ_1, x_1, w_1), and when it refers to the second component it has the index 2. Concentrations are expressed as percentages.

In order to avoid repetition, values of permittivity which are shown in the tables of dielectric dispersion parameters § § 3, 4 chapter II and § § 3, 4 chapter III are not included in the tables of permittivity § § 1, 2 chapter II and § § 1, 2 chapter III. To obtain full details concerning the permittivity of systems it is essential to consult the tables accordingly.

The first literature citation indicates the source of the values given in the table.

SYMBOLS

ϵ – limiting low-frequency permittivity.

ϵ', ϵ'' – real and imaginary parts of the permittivity.

ϵ_1, ϵ_2, ϵ_3 – limiting low-frequency permittivity for the first, second and third absorption regions.

ϵ_1', ϵ_2', ϵ_3' – real part of the permittivity in the first, second and third absorption regions.

ϵ''_1, ϵ''_2, ϵ''_3 – dielectric loss (loss factor) in the first, second and third absorption regions.

tan δ – tangent of the loss angle.

ϵ_∞ – limiting high-frequency permittivity.

$\epsilon_{\infty 1}$, $\epsilon_{\infty 2}$, $\epsilon_{\infty 3}$ – limiting high-frequency permittivity for the first, second and third absorption regions.

λ – wave length.

λ_m – limiting wave length corresponding to the maximum of the dielectric loss.

ν_m – limiting frequency corresponding to the absorption maximum.

ν_{m1}, ν_{m2}, ν_{m3} – limiting frequencies corresponding to the first, second and third absorption regions.

τ – the macroscopic relaxation time.

τ_1, τ_2, τ_3 – relaxation time in the first, second and third absorption regions.

C_1, C_2 – contribution of the first and second relaxation processes to the dielectric relaxation.

g – Kirkwood's structure factor.

α – coefficient of distribution of relaxation times.

β – parameter of distribution of relaxation times according to Cole-Davidson.

ΔF – change in free energy of activation of dielectric relaxation.

ΔF_1, ΔF_2, ΔF_3 – change in free energy of activation of dielectric relaxation for the first, second and third absorption regions.

ΔH – heat of activation of the dielectric relaxation.

ΔH_1, ΔH_2, ΔH_3 – heat of activation of dielectric relaxation for the first, second and third absorption regions.

x_1 – concentration of the first component of the solution expressed as molar percentage.

x_2 – concentration of the second component of the solution expressed as molar percentage.

w_1 – concentration of the first component of the solution expressed as weight percentage.

w_2 – concentration of the second component of the solution expressed as weight percentage.

ϕ_1 – concentration of the first component of the solution expressed as volume percentage.

ϕ_2 – concentration of the second component of the solution expressed as volume percentage.

n – concentration of the solution expressed in gram-equivalents per litre of solution.

N – concentration of the solution expressed in moles per litre of solution.

σ – conductivity.

CHAPTER I

FUNDAMENTAL FORMULAE DESCRIBING THE DIELECTRIC PROPERTIES OF LIQUIDS

Results of theoretical and experimental investigations devoted to the properties of dielectrics in static and alternating electric fields have been published in various books and monographs [5–16].

For information on the methods of calculating the experimental values given in the book and of establishing relationships between them we give the fundamental equations which describe the dielectric properties of liquids. For a closer acquaintance with matters relating to these equations and to the experimental results reference may be made to the works cited in the text.

The polarization of a dielectric is represented as the sum of the electronic P_e, atomic P_a and orientational P_o polarization

$$P = P_e + P_a + P_0. \quad (1)$$

The times for completion of polarization have the orders of magnitude P_e $10^{-14} - 10^{-15}$, P_a $10^{-11} - 10^{-14}$ and P_o 10^{-10} seconds.

The Clausius–Mosotti equation for the molecular polarization P_m is

$$\frac{\varepsilon - 1}{\varepsilon + 2}\frac{M}{\rho} = \frac{4\pi N}{3} a = P_m, \quad (2)$$

where M = molecular weight, ρ = density, N = Avogadro's number, α = deformation polarizability, equal to the sum of the electronic and atomic polarizabilities.

The molecular refraction P_R:

$$P_R = \frac{n^2 - 1}{n^2 + 2}\frac{M}{\rho} = \frac{4\pi N}{3} a_e, \quad (3)$$

where a_e is the electronic polarizability.

The temperature coefficient of the permittivity

$$\frac{1}{\varepsilon}\frac{d\varepsilon}{dt} = \frac{(\varepsilon - 1)(\varepsilon + 2)}{3\varepsilon}\frac{1}{\rho}\frac{d\rho}{dt}. \quad (4)$$

The permittivity of binary solutions of non-polar liquids is given by the formula

$$P_{12}=\frac{\varepsilon_{12}-1}{\varepsilon_{12}+2}\frac{x_1M_1+x_2M_2}{\rho_{12}}=P_1x_1+P_2x_2, \tag{5}$$

where ϵ_{12} is permittivity, ρ_{12} is density and P_{12} is molecular polarization of the solution, x_1, x_2 are mole fractions, M_1, M_2 molecular weights and P_1, P_2 molecular polarizations of the components,

$$P_1=\frac{4}{3}\pi Na_1, \qquad P_2=\frac{4}{3}\pi Na_2, \tag{6}$$

where a_1, a_2 are the deformation polarizabilities of the components.

Since

$$x_1=1-x_2, \qquad P_{12}=P_1-(P_1-P_2)\,x_2. \tag{7}$$

Debye's equation [5]

$$P=\frac{\varepsilon-1}{\varepsilon+2}\frac{M}{\rho}=\frac{4\pi N}{3}\left(a+\frac{\mu^2}{3kT}\right), \tag{8}$$

where μ is the dipole moment of the molecule, k is Boltzman's constant, T is temperature. We define

$$A=\frac{4}{3}\pi Na, \qquad B=\frac{4\pi N\mu^2}{9k} \tag{9}$$

and obtain

$$PT=AT+B.$$

The numerical formula for calculating the dipole moment has the form

$$\mu=0{,}0127\cdot(10^{-8})\,B. \tag{10}$$

The effect of solvent is given in the equations [17–23]

$$\frac{\mu}{\mu_0}=1+0{,}43A\,(\varepsilon-1), \tag{11}$$

or

$$\frac{\mu}{\mu_0}=\frac{3\varepsilon\,[1-(\varepsilon_\infty-1)\,A]}{(\varepsilon+2)\,[\varepsilon+(\varepsilon_\infty-\varepsilon)\,A]}, \tag{12}$$

where μ_o is the dipole moment of the molecule in the rarefied gas, $A = 1/3$ for a spherical molecule, ϵ_∞ is the limiting high-frequency permittivity.

Onsager's equation [24]

$$\frac{(\varepsilon-\varepsilon_\infty)\,(2\varepsilon+\varepsilon_\infty)}{\varepsilon\,(\varepsilon_\infty+2)^2}=\frac{4\pi N\mu^2}{9kT}. \tag{13}$$

Kirkwood's equation [25]

$$\frac{(\varepsilon+1)(2\varepsilon+1)}{9\varepsilon}=\frac{4\pi N}{3}\left(a+\frac{g\mu^2}{3kT}\right), \tag{14}$$

where $g = 1 + z \cos\gamma$, $\cos\gamma$ is the mean value of the cosine of the angle between the dipole moment of a given molecule and the dipoles of neighbouring molecules. If $g > 1$ then the dipoles of the neighbouring molecules are parallel, and if $g < 1$ the dipoles are arranged antiparallel [26–27] and

$$\frac{\mu_l}{\mu_0}=\sqrt{g}\,, \qquad \mu_l=\frac{(2\varepsilon+1)(\varepsilon_\infty+2)}{3(2\varepsilon+\varepsilon_\infty)}\mu_0, \tag{15}$$

where μ_l is the dipole moment of the molecule in the liquid.

The Onsager–Kirkwood–Fröhlich equation has the form [28, 10]

$$\mu_l^2=\frac{(\varepsilon-\varepsilon_\infty)(2\varepsilon+\varepsilon_\infty)}{3\varepsilon}\left(\frac{3}{\varepsilon_\infty+2}\right)\frac{3VkT}{3\pi N_1}=g\mu_0^2, \tag{16}$$

where N_1 is the number of molecules contained in volume V.

The Onsager–Fröhlich equation [24, 7] for solutions of a polar liquid in a non-polar solvent is written

$$\varphi_A\frac{\varepsilon-\varepsilon_{\infty A}}{2\varepsilon+\varepsilon_{\infty A}}+\varphi_B\frac{\varepsilon-\varepsilon_{\infty B}}{2\varepsilon+\varepsilon_{\infty B}}=\frac{4\pi}{3\varepsilon}\frac{N_B\mu_B^2}{3kT}, \tag{17}$$

where ϕ_A, ϕ_B are the volume fractions of the components, $\epsilon_{\infty A}$, $\epsilon_{\infty B}$ are their high-frequency permittivities, ϵ is the permittivity of the solution, and μ_B is the dipole moment of the molecule of the polar component.

Kirkwood's equation [29] for solutions is written in the form

$$\frac{(\varepsilon-1)(2\varepsilon+1)}{9\varepsilon}\frac{M}{\rho}=P_1x_1+P_2x_2, \tag{18}$$

where P_1, P_2 are the molar polarizations of the components,

$$P_1=\frac{4\pi N}{3\varepsilon}a_1, \qquad P_2=\frac{4\pi N}{3\varepsilon}\left(a_2+\frac{g\mu^2}{3kT}\right).$$

For solutions of polar liquids in non-polar solvents the Kirkwood equation has the form [29, 30]

$$\frac{(\varepsilon-1)(2\varepsilon+1)}{9\varepsilon}-\frac{(\varepsilon_\infty-1)(2\varepsilon_\infty+1)}{9\varepsilon_\infty}=\frac{P_2x_2}{V}, \tag{19}$$

where $V = M/\rho$ is the molecular volume.

An attempt to improve the theory of polarization of liquid polar dielectrics [31, 37], taking into account the anisotropy of polarizability and the ellipsoidal form of the molecule [38–43] leads to the following equation for an ellipsoidal molecule with semi-axes a_1, a_2, a_3, and having a permanent dipole moment directed along the axis a_i [44, 45],

$$\frac{4\pi N\mu^2}{3kT}=\frac{(\varepsilon-n^2)(2\varepsilon+1)[\varepsilon-(\varepsilon-n^2)A_1]}{\varepsilon(2\varepsilon+n^2)[\varepsilon-(\varepsilon-1)A_1][1+(n^2-1)A_1]}, \tag{20}$$

where $A_1 = 1/3$ – the depolarization factor, n = refractive index.

Syrkin's equation [46] for polar liquids and their solutions in non-polar solvents:

$$\frac{\varepsilon-\varepsilon_\infty}{\varepsilon_\infty+2}\frac{\varepsilon+2}{2\varepsilon+1}=\frac{4\pi N}{9kT}\mu^2,$$

$$\frac{\dfrac{\varepsilon-1}{\varepsilon+2}\dfrac{M_1x_1+M_2x_2}{\rho}-x_1R_1-x_2R_2}{1-\left(\dfrac{\varepsilon-1}{\varepsilon+2}\right)^2}=\frac{4\pi N}{9kT}(x_1\mu_1+x_2\mu_2). \tag{21}$$

Equation for the polarization of a pure polar liquid

$$\left[\frac{(\varepsilon-1)(\varepsilon+2)}{8\varepsilon}-\frac{(n^2-1)(n^2+2)}{8n^2}\right]\frac{M}{\rho}=\frac{4\pi N}{3}\frac{\mu^2}{3kT}=P \tag{22}$$

and for binary solutions of polar liquids has the form [47, 48]

$$P_0=\frac{4}{3}\pi N\frac{x_1\mu_1^2+x_2\mu_2^2}{3kT}=\left[\frac{x_1M_1+x_2M_2}{\rho}\right]\left[\frac{(\varepsilon-1)(\varepsilon+2)}{8\varepsilon}-\frac{(n^2-1)(n^2+2)}{8n^2}\right]. \tag{23}$$

From a known value of the orientational polarization P_o it is possible to calculate the orientational polarization of the dissolved substance P_{o2} in the polar solvent:

$$P_{02}=\frac{P_0-P_{01}}{x_2}+P_{01}, \tag{24}$$

where P_{o1} is the orientational polarization of the polar solvent calculated from formula (22).

The dipole moment and the number of associated molecules in strongly polar liquids may be determined from equations [48–52]

$$\frac{\varepsilon(\varepsilon-n_1^2)}{2\varepsilon+n_1^2}\varphi_1\left(1-\gamma\frac{N_2}{N_1}\right)+\frac{\varepsilon(\varepsilon-n_2^2)}{2\varepsilon+n_2^2}\varphi_2(1-\gamma)+\frac{\varepsilon(\varepsilon-n_x^2)}{2\varepsilon+n_x^2}\gamma\left(\frac{N_2}{N_1}\varphi_1+\varphi_2\right)=$$
$$=\frac{4\pi}{9kT}\left\{(N_1-\gamma N_2)\mu_1^2\left[\frac{\varepsilon(n_1^2+2)}{2\varepsilon+n_1^2}\right]^2+N_2(1-\gamma)\mu_2^2\left[\frac{\varepsilon(n_2^2+2)}{2\varepsilon+n_2^2}\right]^2+\right.$$
$$\left.+\gamma N_2\mu_x^2\left[\frac{\varepsilon(n_x^2+2)}{2\varepsilon+n_x^2}\right]^2\right\}, \tag{25}$$

where $\gamma = N_{2x}/N_2$, N_{2x} is the number of associated molecules per cubic centimetre of solution, μ_x and n_x their dipole moment and refractive index respectively.

In equation (25) ϵ, n, ϕ_1 and ϕ_2 are measured quantities,

$$n_x = \varphi_1 n_1 + \varphi_2 n_2,$$

μ_1 and μ_2 are already determined and the remaining unknowns are μ_x and γ. The equilibrium constant K of the reaction is connected with γ by the relationship

$$K = \frac{\gamma N}{(N_1 - \gamma N_2)(1-\gamma)}. \tag{26}$$

The value of the equilibrium constant can be determined in various ways. By substitution from expression (26) into formula (25) it is possible to calculate the dipole moment of the associated molecules μ_x for varied concentrations and temperatures.

The tangent of the loss angle is

$$\tan \delta = \frac{\varepsilon''}{\varepsilon'}, \tag{27}$$

where ϵ' is the real part of the permittivity, ϵ'' is the imaginary part of the permittivity, the dielectric loss factor.

The complex permittivity is expressed in the form

$$\varepsilon^* = \varepsilon' - j\varepsilon''.$$

The mean value of the thermal energy generated during one periodic change of the field is given by the relationship

$$\langle Q \rangle = \frac{E_0^2 \varepsilon'' \omega}{8\pi}, \tag{28}$$

where E_o is the electric field intensity, w is the angular frequency.

The contribution of ionic conductivity to the loss factor is calculated by the formula

$$\varepsilon''_{cor} = \varepsilon''_{obs} - \frac{2\sigma}{\nu}, \tag{29}$$

where ϵ''_{cor} is the loss factor arising from orientation of dipoles, ϵ''_{obs} is the experimentally determined value of ϵ'', σ is the electrical conductivity of the liquid and ν the frequency at which the conductivity is measured.

The time dependence of the polarization is expressed in the form

$$P = P_0 e^{-t/\tau}, \tag{30}$$

where τ is the macroscopic relaxation time and t is the time.

Debye's equation [5, 15]

$$\varepsilon' = \varepsilon_\infty + \frac{\varepsilon - \varepsilon_\infty}{1+\omega^2\tau^2},$$

$$\varepsilon'' = \frac{\varepsilon - \varepsilon_\infty}{1+\omega^2\tau^2}\,\omega\tau, \tag{31}$$

where ϵ_∞ is the limiting high-frequency permittivity and ϵ is the limiting low frequency permittivity.

The relaxation time is given by the conditions

$$\tau = 1/\omega_m = 1/2\pi\nu_m, \tag{32}$$

where w_m is the angular frequency corresponding to the absorbtion maximum.

The wave-length corresponding to the absorbtion maximum is given by the formula

$$\lambda_m = 2\pi c/\omega_m, \tag{33}$$

where c is the velocity of light.

The maximum value of the loss factor ϵ''_m is given by

$$\varepsilon''_m = \frac{(\varepsilon - \varepsilon_\infty)}{2}. \tag{34}$$

The maximum value of the tangent of the loss angle is

$$\tan\delta_m = \frac{\varepsilon - \varepsilon_\infty}{2\sqrt{\varepsilon\varepsilon_\infty}}. \tag{35}$$

The value of ϵ_∞ can be calculated from measured values of the refractive index by the formula [53, 54]

$$\frac{\varepsilon_\infty - 1}{\varepsilon_\infty + 2} = A\frac{n^2-1}{n^2+2}, \tag{36}$$

where A has the value 1.05 – 1.15 and allows for the contribution of the atomic polarization.

Deviation of the properties of dielectrics from Debye's equation can be determined from the graph of the imaginary part ϵ'' (or the polarizability a'') against the real part ϵ' (or a') [55, 56];

$$a' = \frac{(\varepsilon'-1)(\varepsilon'+2)+\varepsilon''^2}{(\varepsilon'+2)^2+\varepsilon''^2}, \qquad a'' = \frac{3\varepsilon''}{(\varepsilon'+2)^2+\varepsilon''^2}. \tag{37}$$

The relaxation time is determined from the relationships

$$\frac{v}{u} = \frac{1-a}{1-a_\infty}\,\omega\tau, \qquad \frac{v}{u} = \omega\tau, \tag{38}$$

where

$$a = \frac{\varepsilon - 1}{\varepsilon + 2}, \qquad a_\infty = \frac{\varepsilon_\infty - 1}{\varepsilon_\infty + 2},$$

ν and u are distances from experimental points on the semicircle* to points $\epsilon' = \epsilon$ and $\epsilon' = \epsilon_\infty$ or $a' = a$ and $a' = a_\infty$ respectively,

$$v = \sqrt{(\varepsilon - \varepsilon')^2 + \varepsilon''^2}, \qquad u = \sqrt{(\varepsilon' - \varepsilon_\infty)^2 + \varepsilon''^2}. \tag{39}$$

The following expressions suitable for graphical representation of the relaxation processes described by Debye's theory [57, 58] are derived from equation (31):

$$\begin{aligned} \varepsilon' &= \varepsilon - \tau(\omega\varepsilon''), \\ \varepsilon' &= \varepsilon_\infty + \frac{1}{\tau}\frac{\varepsilon''}{\omega}, \\ \frac{\omega\tau}{\varepsilon''} &= \frac{(\omega\tau)^2}{\varepsilon - \varepsilon_\infty} + \frac{1}{\varepsilon - \varepsilon_\infty}, \\ \frac{1}{\varepsilon''\omega\tau} &= \frac{1}{(\varepsilon - \varepsilon_\infty)\,\omega^2\tau^2} + \frac{1}{\varepsilon - \varepsilon_\infty}, \\ \frac{1}{\varepsilon - \varepsilon'} &= \frac{1}{(\varepsilon - \varepsilon_\infty)\,\omega^2\tau^2} + \frac{1}{\varepsilon - \varepsilon_\infty}. \end{aligned} \tag{40}$$

The values of ϵ and ϵ_∞ can be determined as the intercepts of the straight lines derived from these equations on the ϵ' axis, and the relaxation times τ from the inclination to the ϵ' axis.

For dilute solutions of polar liquids in non-polar solvents Debye's equation may be written in the form [15]

$$\varepsilon'' = \frac{(\varepsilon' + 2)^2 \pi N \mu^2 C}{6750kT}\,\frac{\omega\tau}{1 + \omega^2\tau^2}, \tag{41}$$

where C is the concentration of the polar liquid in moles/litre, μ is the dipole moment of the polar molecule.

For $\epsilon'' < \epsilon'$ equation (41) may be written

$$\frac{\omega}{\varepsilon''} = A\left(\frac{1}{\tau} + \omega^2\tau\right), \tag{42}$$

where $A = 6750kT/(\epsilon' + 2)^2 N\pi\mu^2 C$.

From a graph of w/ϵ'' against w^2 it is possible to evaluate τ and μ.

By the method of Gopala Krishna [59–61] τ and μ may be determined from measurements at a single frequency for various concentrations and temperatures. The quantities X and Y are introduced defined by the relationships

$$X = \frac{\varepsilon' + \varepsilon''^2 + \varepsilon'' - 2}{\varepsilon''^2 + (\varepsilon' + 2)^2}, \qquad Y = \frac{3\varepsilon''}{\varepsilon''^2 + (\varepsilon' + 2)^2}. \tag{43}$$

These quantities are linearly related

$$X = P + \left(\frac{1}{\omega\tau}\right) Y, \tag{44}$$

* Translator's note: i.e. the semicircle in the plot of ϵ'' against ϵ' in the complex plane.

where

$$P=\frac{\varepsilon_\infty-1}{\varepsilon_\infty+2}.$$

From the graph of X against Y the relaxation time is determined. The dipole moment is determined from the slope of the straight line in the graph of X against C,

$$X=P+KC, \tag{45}$$

where $K=\frac{4\pi N\mu^2}{9kT}\left(\frac{1}{1+\omega^2\tau^2}\right)$ and C is the concentration in moles/cm^3.

Krishna's method also gives τ for dilute solutions:

$$\tau=\frac{1}{\omega}\frac{a''}{a'-a_\infty}. \tag{46}$$

Values a', a'' and a_∞ are determined by the slopes of the graphical lines relating concentration and the following quantities

$$\begin{aligned} \varepsilon_{12}&=\varepsilon_1+ax_2, & \varepsilon'_{12}&=\varepsilon'_1+a'x_2,\\ \varepsilon''_{12}&=a''x_2, & \varepsilon_{\infty 12}&=\varepsilon_{\infty 1}+a_\infty x_2, \end{aligned} \tag{47}$$

where the index 1 refers to the solvent and 2 to the polar component. In this case $\epsilon_\infty = n^2$, x_2 = concentration in mole fractions.

The applicability of Debye's equations to polar molecules of various forms and dimensions has been examined [62–71].

When there is a distribution of relaxation times a large number of experimental results agree well with a Gaussian distribution [72].

The equations of Cole and Cole concerning the distribution of relaxation times have the form [75]

$$\begin{aligned} \frac{\varepsilon'-\varepsilon_\infty}{\varepsilon-\varepsilon_\infty}&=\frac{1+(\omega\tau)^{1-\alpha}\sin\frac{\pi\alpha}{2}}{1+(\omega\tau)^{2(1-\alpha)}+2(\omega\tau)^{1-\alpha}\sin\frac{\pi\alpha}{2}},\\ \frac{\varepsilon''}{\varepsilon-\varepsilon_\infty}&=\frac{(\omega\tau)^{1-\alpha}\cos\frac{\pi\alpha}{2}}{1+(\omega\tau)^{2(1-\alpha)}+2(\omega\tau)^{1-\alpha}\sin\frac{\pi\alpha}{2}}, \end{aligned} \tag{48}$$

where α is called the coefficient of distribution of the relaxation times and varies from 0 to 1.

The parameter α of the distribution of relaxation times is determined from the formula

$$\psi=\alpha\pi/2, \tag{49}$$

where ψ is the angle formed between the axis of abscissae with the line joining the centre of the circle with the point $\epsilon'=\epsilon_\infty$ in the graph showing the dependence of $\frac{\varepsilon'-\varepsilon_\infty}{\varepsilon-\varepsilon_\infty}$ or $\frac{\varepsilon''}{\varepsilon-\varepsilon_\infty}$.

The relaxation time is determined from the expression

$$\frac{v}{u}=\left(\frac{\nu}{\nu_m}\right)^{1-\alpha}=(\omega\tau)^{1-\alpha}. \tag{50}$$

If $0.1 < \left(\frac{\nu}{\nu_m}\right) < 10$, then we have for ϵ'' [74]

$$\begin{aligned} \varepsilon''&=(\varepsilon-\varepsilon_\infty)\left(\frac{\nu}{\nu_m}\right)^{1-\alpha} \qquad (\nu \ll \nu_m),\\ \varepsilon''&=(\varepsilon-\varepsilon_\infty)\left(\frac{\nu}{\nu_m}\right)^{-(1-\alpha)} \qquad (\nu \gg \nu_m). \end{aligned} \tag{51}$$

It is possible to determine $(1 - \alpha)$ from experimentally measured quantities by constructing the graph of $\ln \epsilon''$ against $\ln \nu (\ln w)$. Methods of analysis of dielectric relaxation are considered in references [57, 75, 76].

The quantities τ and α for dilute solutions can be found from experimental results at a single frequency by the formula [77–79]

$$\tau=\frac{1}{\omega}\left(\frac{A^2+B^2}{C^2}\right)^{\frac{1}{2}(1-\alpha)}, \tag{52}$$

where

$$\begin{aligned} A&=a''(a-a_\infty),\\ B&=(a-a')(a'-a_\infty)-a''^2,\\ C&=(a'-a_\infty)^2+a''^2. \end{aligned}$$

The Cole–Davidson equation [80–81] has the form

$$\begin{aligned} \varepsilon'-\varepsilon_\infty&=(\varepsilon-\varepsilon_\infty)\cos^\beta\varphi\cos\beta\varphi,\\ \varepsilon''&=(\varepsilon-\varepsilon_\infty)\cos^\beta\varphi\sin\beta\varphi, \end{aligned} \tag{53}$$

where

$$\tan\varphi=\omega\tau, \qquad 0<\beta<1.$$

In this case τ is determined from the relationship

$$\frac{\tan\theta}{\beta}=\omega\tau, \tag{54}$$

where

$$\theta=\arctan\left(\frac{\varepsilon''}{\varepsilon'-\varepsilon_\infty}\right).$$

Values of β and τ can be determined from experimental quantities measured at a single frequency but at various temperatures. For this the value of ϵ'' must pass through the maximum with the variation in temperature.

Then

$$\omega\tau = \frac{\varepsilon' - \varepsilon_\infty}{\varepsilon''} = \tan \frac{\pi}{2(1+\beta)} \cdot \tag{55}$$

The macroscopic relaxation time τ and the molecular relaxation time τ^* are connected by the relationships [82–86]

$$\tau = \frac{\varepsilon + 2}{\varepsilon_\infty + 2} \tau^*, \tag{56}$$

and with $\epsilon > \epsilon_\infty$, then $\tau > \tau^*$

$$\tau = \frac{2\varepsilon_\infty + 1}{2\varepsilon + 1} \tau^*, \tag{57}$$

$$\tau = \frac{3\varepsilon}{2\varepsilon + \varepsilon_\infty} \tau^*, \tag{58}$$

$$\tau = \frac{2\varepsilon^2 + \varepsilon_\infty^3}{3\varepsilon_\infty \varepsilon (2\varepsilon + \varepsilon_\infty)} \tau^*. \tag{59}$$

The equality (58) gives a better description of the internal field than the other equations.

Equations describing the distribution of relaxation times are given also by Fröhlich and Higasi [7, 88, 89].

The Fuoss–Kirkwood equation [90]

$$\epsilon'' = \epsilon''_m \operatorname{sech}\left(\beta \ln \frac{\omega}{\omega_m}\right), \qquad 0 < \beta < 1, \tag{60}$$

where

$$\cosh^{-1}\left(\frac{\varepsilon''_m}{\varepsilon''}\right) = 2{,}3\beta\,(\lg \nu_m - \lg \nu). \tag{61}$$

A graph of the dependence on lg ν gives 2.3 β by the slope of the line and the intercept on the axis of abscissae gives the value of lg ν_m.

The parameter β in the Fuoss–Kirkwood equation is connected with the coefficient of distribution of relaxation times α of Cole and Cole by the following relationship

$$\beta\sqrt{2} = \frac{1-\alpha}{\cos \frac{1-\alpha}{4}\pi} \cdot \tag{62}$$

The dependence of α on the temperature is expressed by the formula [91]

$$\alpha = \frac{A}{T} e^{B/T}, \tag{63}$$

where A and B are constants determined from this relationship.

In the case of two independent relaxation processes, the dielectric relaxation is described by the equations [92–93]

$$\frac{\varepsilon' - \varepsilon_\infty}{\varepsilon - \varepsilon_\infty} = \frac{C_1}{1+\omega^2\tau_1^2} + \frac{C_2}{1+\omega^2\tau_2^2}, \tag{64}$$

$$\frac{\varepsilon''}{\varepsilon - \varepsilon_\infty} = \frac{C_1\omega\tau_1}{1+\omega^2\tau_1^2} + \frac{C_2\omega\tau_2}{1+\omega^2\tau_2^2}, \tag{65}$$

where C_1, C_2 are weighting factors determining the contribution of each relaxation process, and

$$C_1 + C_2 = 1. \tag{66}$$

For determination of τ_1, τ_2, C_1 and C_2 equations (64), (65) may be put in the form

$$\begin{aligned} &X = C_1X_1 + C_2X_2, && Y = C_1Y_1 + C_2Y_2, \\ &X_1 = \frac{1}{1+z_1^2}, && X_2 = \frac{1}{1+z_2^2}, \\ &Y_1 = \frac{z_1}{1+z_1^2}, && Y_2 = \frac{z_2}{1+z_2^2}, \\ &z_1 = \omega\tau_1, && z_2 = \omega\tau_2. \end{aligned} \tag{67}$$

Experimental points with co-ordinates (X, Y) lying between the points (X_1, Y_1) and (X_2, Y_2) divide the chord in the ratio C_1/C_2. The varying disposition of the chords going through the experimental points necessarily leads to the same values τ_1 and τ_2 for all experimental points. A method of solving equations (64), (65), (67) with the aid of computers and simplified methods of calculating τ_1 and τ_2 are given in articles [94–99].

In the case of pure liquids the weighting factors C_1 and C_2 in the equality (66) are related to $\mu_{\parallel}$, the component of the dipole moment directed along the rotation axis, and $\mu_{\perp}$, the component directed perpendicular to the rotation axis, by the condition [100]

$$\frac{C_1}{C_2} = \frac{\mu_{\parallel}}{\mu_{\perp}}. \tag{68}$$

For solutions this connection has the form

$$\frac{C_1}{C_2} = \frac{\mu_1^2 x_1}{\mu_2^2 x_2}, \tag{69}$$

where μ_1 and μ_2, x_1 and x_2 are the dipole moments and concentrations of the components respectively.

Application of the dipole autocorrelation function and Laplace transformation for determining the frequency dependence of ϵ^* leads to equations (64), (65) [100, 102–104].

The presence of several absorption regions can be determined from the plot of ϵ' against ϵ''/λ. In this case each absorption region corresponds to a straight line with a definite slope [105].

The frequency dependence of ϵ^* may show the presence of several overlapping dispersion regions. Assuming that the dispersion is additively made up of the contributions of each dispersion region it is possible to separate out the maxima of the different regions [106–117]. The first dispersion region is then quantitatively described by the equation

$$\varepsilon_1^* - \varepsilon_\infty = (\varepsilon_1' - \varepsilon_\infty) - j\varepsilon'' = \frac{\varepsilon - \varepsilon_\infty}{1 + j\dfrac{\nu}{\nu_{m1}}}, \tag{70}$$

where the index "1" refers to the first absorption region.

For the second absorption region ϵ' and ϵ'' are given by the formulae

$$\begin{aligned} \varepsilon_2' &= \varepsilon' - (\varepsilon - \varepsilon_\infty)\left(\frac{\nu_{m1}}{\nu}\right)^2, \\ \varepsilon_2'' &= \varepsilon'' - (\varepsilon - \varepsilon_\infty)\left(\frac{\nu_{m1}}{\nu}\right)^2. \end{aligned} \tag{71}$$

The connection of relaxation time τ with molecular dimensions is established by Debye's formula [5, 118–120]

$$\tau = 4\pi a^3 \eta / kT, \tag{72}$$

where a is the radius of the molecule, η is the viscosity of the liquid.

Experimental values agree well with the calculated if the factor $0.36\,\eta$ is used instead of η.

Several attempts have been made to improve Debye's formula [121–126].

The mean relaxation time for concentrated solutions is determined from the relationship [127, 128]

$$\tau_c = C_1\tau_1 + C_2\tau_2, \qquad \tau_c = \tau\frac{\eta_1}{\eta_2}, \tag{73}$$

where τ_1 is the relaxation time of the molecule, τ_2 the relaxation time of the clusters, C_1 and C_2 the contributions of each type of relaxation, τ the relaxation time determined from the Cole–Cole equation, η_1 the viscosity of the solution, η_2 the viscosity of the solvent.

The connection between τ and η is defined by the relationship [129–132]

$$\tau = \frac{B}{T}\eta^x, \tag{74}$$

where B is a constant, x is a parameter less than unity.

The dynamic viscosity is related to the static viscosity by the equation [131]

$$\eta_g = \eta\,[1 + \omega^2\tau^2]^{-1}. \tag{75}$$

For the connection between τ and η for pure liquids and solutions the following equation also holds [134–136]

$$\tau=\frac{3M\eta}{RT\rho}\frac{(\varepsilon-1)(2\varepsilon+1)}{9\varepsilon^2}, \tag{76}$$

where M, ρ are the molecular weight and density of the polar liquid, η and ϵ the viscosity and permittivity of the solvent and

$$\tau=\frac{4\pi a^2}{kT}(\eta+c\sqrt{\rho}), \tag{77}$$

where ρ is the density of the liquid, and c is a constant.

For dilute solutions

$$\tau=\frac{M\eta}{RT\rho}(\varepsilon^{1/2}-1). \tag{78}$$

The relaxation time τ is connected with the change of free energy ΔF, the heat ΔH and the entropy of activation ΔS of the dipole relaxation by the expression [137–139]

$$\tau=\frac{hN}{RT}\exp\left(\frac{\Delta H-T\Delta S}{RT}\right), \quad \Delta F=\Delta H-T\Delta S, \tag{79}$$

where

$$\Delta H=\frac{Rd(\ln\tau)}{d(1/T)}-RT=-RT^2\frac{1}{\tau}\frac{d\tau}{dT}-RT,$$

$$\tau=\frac{hN}{RT}\exp\left[\frac{\Delta F}{R(T-T_0)}\right],$$

where $T_0=0.25-0.35\,T_b$ (boiling point), h = Planck's constant.

The viscous relaxation process is described by the expression

$$\eta=\frac{hN}{V}\exp\left[\frac{\Delta F_\eta}{RT}\right], \tag{80}$$

where V = molar volume, ΔF_η = change of free energy of activation of the viscous process

$$\Delta F_\eta=\Delta H_\eta-T\Delta S_\eta,$$

where ΔH_η and ΔS_η are the heat and entropy of activation of the viscous process.

The processes of dielectric and viscous relaxation are connected by the relationship [140–142]

$$\ln(\tau T)=\ln B+x\ln\eta, \tag{81}$$

where

$$B=\frac{1}{k}\left[\left(\frac{V}{N}\right)^{x}h^{(1-x)}\right]\exp\left[\frac{1}{R}(x\Delta S_{\eta}-\Delta S)\right], \qquad x=\frac{\Delta H}{\Delta H_{\eta}}.$$

The graph of $\ln(\tau T)$ against $\ln\eta$ is a straight line. The slope of this gives x and the intercept on the ordinate axis gives the value of B.

The mean square fluctuation of the permittivity is equal to [30, 143–147]

$$\overline{(\Delta\varepsilon)^2}=\left(\frac{\partial\varepsilon}{\partial\varphi}\right)^2\overline{(\Delta\varphi)^2}, \tag{82}$$

where $\overline{(\Delta\phi)^2}$ is the mean square fluctuation of the concentration.

The difference between the measured value of the permittivity ϵ and the mean local permittivity $\bar{\epsilon}$ is expressed in the equation

$$\bar{\varepsilon}-\varepsilon=\frac{\overline{(\Delta\varepsilon)^2}}{\varepsilon\left(2+\dfrac{d\bar{\varepsilon}/d\varphi}{d\varepsilon/d\varphi}\right)}, \qquad \bar{\varepsilon}-\varepsilon=\varphi_1\varepsilon_1+\varphi_2\varepsilon_2. \tag{83}$$

For the concentration $\phi = 0.5$ it is permissible to take

$$\frac{d\bar{\varepsilon}}{d\varepsilon}\approx\frac{d\varepsilon}{d\varphi}, \tag{84}$$

Then

$$\bar{\varepsilon}-\varepsilon=\frac{\overline{(\Delta\varphi)^2}}{3\varepsilon}\frac{d\bar{\varepsilon}}{d\varphi},$$

where ϕ_1, ϕ_2 and ϵ_1, ϵ_2 are the volume fractions and permittivities of the first and second components.

The deviation of ϵ from additivity may also be determined from the equation [148, 149]

$$\frac{\varepsilon-1}{\varepsilon}=\sum_{i=1}^{N}\varphi_i\frac{(2\varepsilon_i+n_{i\,\infty}^2)(2\varepsilon+1)}{(2\varepsilon+n_{i\,\infty}^2)(2\varepsilon_i+1)}+ +\sum_{i=1}^{N}\left\{\varphi_i\cdot 3\left[1-\frac{(2\varepsilon_i+n_{i\,\infty}^2)(2\varepsilon+1)}{(2\varepsilon+n_{i\,\infty}^2)(2\varepsilon_i+1)}\right]\frac{n_{i\,\infty}^2-1}{2\varepsilon+n_{i\,\infty}^2}\right\}. \tag{85}$$

When the condition is satisfied that and

$$\frac{(2\varepsilon_i+n_{i\,\infty}^2)(2\varepsilon+1)}{(2\varepsilon_i+1)(2\varepsilon+n_{i\,\infty}^2)}\approx 1 \quad \text{и} \quad n_{i\,\infty}^2=\overline{n_\infty^2} \tag{86}$$

we have

$$\frac{(2\varepsilon-1)(2\varepsilon+n_{i\,\infty}^2)}{\varepsilon}=\sum_{i=1}^{N}\varphi_i\frac{(2\varepsilon_i+\overline{n_\infty^2})(\varepsilon_i-1)}{\varepsilon_i}, \tag{87}$$

where ϕ_i, ϵ_i are the volume fraction and permittivity, and $n_{i\infty}$ is the refractive index extrapolated to infinite wave length for the i-th component, ϵ is the permittivity of the solution, and $\overline{n^2}_\infty$ is the mean value of the refractive index of the solution.

The influence of fluctuations on the permittivity and the loss factor for pure liquids and solutions in the region of dispersion of electromagnetic waves is given by the equations [150, 151]

$$\overline{\varepsilon'}-\varepsilon'=\frac{\overline{(\Delta\varphi)^2}}{3(\varepsilon'^2+\varepsilon''^2)}\left\{2\varepsilon''\frac{d\varepsilon'}{d\varphi}\frac{d\varepsilon''}{d\varphi}+\varepsilon'\left[\left(\frac{d\varepsilon'}{d\varphi}\right)^2-\left(\frac{d\varepsilon''}{d\varphi}\right)^2\right]\right\},$$

$$\overline{\varepsilon''}-\varepsilon''=\frac{\overline{(\Delta\varphi)^2}}{3(\varepsilon'^2+\varepsilon''^2)}\left\{2\varepsilon'\frac{d\varepsilon'}{d\varphi}\frac{d\varepsilon''}{d\varphi}-\varepsilon''\left[\left(\frac{d\varepsilon'}{d\varphi}\right)^2-\left(\frac{d\varepsilon''}{d\varphi}\right)^2\right]\right\},$$

where

$$\frac{d\overline{\varepsilon'}}{d\varphi}\approx\frac{d\varepsilon'}{d\varphi}, \qquad \frac{d\overline{\varepsilon''}}{d\varphi}\approx\frac{d\varepsilon''}{d\varphi},$$

$$\overline{\varepsilon'}=\varphi_1\varepsilon_1'+\varphi_2\varepsilon_2', \qquad \overline{\varepsilon''}=\varphi_1\varepsilon_1''+\varphi_2\varepsilon_2''.$$

CHAPTER II

DIELECTRIC DATA FOR BINARY SYSTEMS – NON-AQUEOUS SOLUTIONS

§ 1. Permittivity of non-aqueous solutions of inorganic compounds

Aluminium bromide ($AlBr_3$) – Bromine (Br_2)
[152], $t = 20$ °C, $\lambda = 83{,}5$ cm

w_1	ε	w_1	ε	w_1	ε	w_1	ε
0,00	3,006	5,11	3,056	20,94	3,135	44,44	3,300
2,34	3,021	7,13	3,074	25,93	3,181	49,84	3,331
2,90	3,032	10,4	3,097	29,82	3,215	—	—
3,88	3,041	15,4	3,134	37,06	3,238	—	—

Aluminium bromide ($AlBr_3$) – Carbon disulphide
[152], $t = 20$ °C, $\lambda = 83{,}5$ cm

w_1	ε	w_1	ε	w_1	ε	w_1	ε
0,00	2,641	18,96	2,747	27,24	2,792	45,46	2,881
4,01	2,660	—	—	34,88	2,816	49,07	2,910
10,99	2,706	22,07	2,756	40,47	2,854	—	—

Aluminium bromide ($AlBr_3$) – Benzene
[152], $t = 20$ °C, $\lambda = 83{,}5$ cm

w_1	ε	w_1	ε	w_1	ε	w_1	ε
0,00	2,282	2,15	2,338	9,28	2,375	40,89	2,608
0,05	2,287	5,34	2,372	20,16	2,480	43,66	2,634
0,63	2,309	7,06	2,375	24,57	2,497	47,69	2,700
1,81	2,336	8,21	2,372	37,10	2,556	50,69	2,750

Iron Pentacarbonyl ($Fe(CO)_5$) – Benzene (C_6H_6)
[155], $t = 20$ °C

x_1	0,00	2,01	4,07	8,35	18,35	40,48	100,0
ε	2,292	2,305	2,317	2,331	2,368	2,437	2,602

Nitrosyl bromide (NOBr) – Carbon tetrachloride (CCl_4)
[156], $t = 12$ °C

x_1	0,0	22,5	100
ε	2,252	3,547	18,2

Nitrosyl chloride (NOCl) – Carbon tetrachloride (CCl_4)
[156], $t = 12$ °C

x_1	0,00	5,09	6,73	18,47	31,75	44,79	100
ε	2,252	2,482	2,539	3,141	4,077	5,087	18,2

Sulphuric acid (H_2SO_4) – Trifluoroacetic acid ($C_2HO_2F_3$)
[157], $t = 25$ °C, $\nu = 15$ MHz

φ_2	100	80	60	40	20	0
ε	8,25	28,6	49	68,5	84	100,5

Oxygen (O_2) – Ozone (O_3)
[158], $t = 90$ °K, $\nu = 1$ kHz

w_2	100	29,8	19,2	17,6	14,9	0
x_2	100	22,1	13,7	12,5	10,5	0
ε	4,75	2,20	1,93	1,90	1,82	1,46

Phosphorus (P_4) – Carbon disulphide (CS_2)
[159], $t = 18$ °C

x_1	0,0 (19 °C)	3,73	12,0	19,9	27,0
ε	2,65	2,76	3,15	3,35	3,4

Sulphur (S) – Carbon disulphide (CS_2)
[159], $t = 18$ °C

x_1	0,0	3,2	5,84	11,56	27,26	28,8
ε	2,65	2,55	2,37	2,15	2,08	2,05

[160]

$w_2 = 100$				$w_2 = 93,42$	
t °C	ε	t °C	ε	t °C	ε
36,05	2,6707	16,27	2,7195	30,09	2,8022
30,14	2,6841	16,22	2,7186	22,13	2,8255
29,97	2,6838	11,35	2,7272	14,91	2,8487
24,20	2,6998	11,05	2,7296	9,38	2,8648
24,05	2,6997	10,96	2,7296	8,58	2,8752
20,00	2,7118	—	—	5,29	2,8745

$w_2 = 87,25$		$w_2 = 79,17$		$w_2 = 76,70$			
t °C	ε	t °C	ε	t °C	ε	t °C	ε
35,17	2,8397	31,10	2,9375	35,33	2,9494	15,28	3,0110
35,08	2,8427	19,69	2,9747	35,26	2,9499	14,78	3,0191
25,51	2,8730	15,69	2,9834	30,04	2,9670	12,71	3,0198
17,19	2,8962	12,72	2,9919	29,84	2,9689	11,49	3,0195
12,14	2,9152	9,84	3,0004	22,31	2,9912	7,68	3,0142
9,05	2,9234	7,68	3,0044	22,16	2,9919	6,17	3,0293

[161]

$w_2 = 100$		$w_2 = 94,74$		$w_2 = 89,97$		$w_2 = 82,29$		$w_2 = 75,41$	
t °C	ε	t °C	ε	t °C	ε	t °C	ε	t °C	ε
19,2	2,654	17,7	2,697	16,8	2,732	19,1	2,790	18,0	2,850
27,0	2,630	28,8	2,663	27,0	2,701	24,9	2,770	27,2	2,820
31,0	2,618	—	—	30,6	2,690	30,5	2,752	31,2	2,807

Sulphur dioxide (SO_2) – Sulphur trioxide (SO_3)

x_1	ε at t °C, [163], ν=1 MHz					x_2	ε at t °C, [163], ν=1 MHz				
	10	15	20	25	30		10	15	20	25	30
54,3	8,13	7,93	7,73	7,54	7,35	75,6	5,76	5,64	5,33	5,41	5,30
56,4	7,87	7,68	7,49	7,31	7,12	76,7	5,61	5,50	5,39	5,28	5,17
56,7	7,76	7,58	7,40	7,22	7,03	77,1	5,57	5,46	5,35	5,25	5,14
58,0	7,67	7,50	7,32	7,14	6,96	78,5	5,43	5,33	5,22	5,12	5,02
59,6	7,51	7,34	7,16	6,98	6,81	79,2	5,35	5,25	5,15	5,06	4,95
60,5	7,39	7,22	7,05	6,88	6,71	81,5	5,14	5,04	4,95	4,85	4,76
60,9	7,26	7,10	6,94	6,78	6,62	81,8	5,11	5,02	4,92	4,82	4,73
61,0	7,13	6,98	6,82	6,66	6,50	83,0	5,00	4,90	4,81	4,72	4,62
63,0	7,03	6,87	6,71	6,56	6,40	83,6	4,94	4,84	4,75	4,65	4,56
64,5	6,98	6,72	6,56	6,40	6,24	85,0	4,71	4,63	4,54	4,51	4,36
65,0	6,82	6,67	6,51	6,35	6,19	89,1	—	4,22	4,15	4,07	3,99
66,0	6,72	6,58	6,37	6,27	6,13	91,2	—	4,02	3,96	3,88	3,81
67,1	6,62	6,47	6,32	6,18	6,03	92,6	—	3,84	3,77	3,70	3,64
67,9	6,55	6,40	6,26	6,12	5,97	94,7	—	3,66	3,60	3,54	3,48
68,3	6,52	6,36	6,23	6,09	5,95	96,5	—	3,45	3,41	3,36	3,30
68,6	6,45	6,31	6,17	6,03	5,90	97,5	—	3,36	3,31	3,26	3,20
70,0	6,33	6,20	6,06	5,93	5,70	98,9	—	3,21	3,17	3,12	3,07
70,8	6,22	6,09	5,96	5,83	5,70	99,7	—	—	3,10	3,05	3,03
72,4	6,10	5,92	5,85	5,72	5,60	100	—	—	3,06	3,02	2,97
74,2	5,91	5,79	5,67	5,55	5,43	—	—	—	—	—	—

Titanium tetrachloride ($TiCl_4$) –
Ethyl trichloroacetate ($C_4H_5Cl_3O_2$)
[164, 165], $t = 20$ °C

x_2	100	80	65	50	35	20	0
ε	8,428	8,752	8,638	7,840	6,464	5,396	2,789

Titanium tetrachloride ($TiCl_4$) –
Butyl trichloroacetate ($C_6H_9Cl_3O_2$)
[164], $t = 20$ °C

x_2	100	80	65	50	35	20	0
ε	7,480	7,591	7,593	7,231	6,420	5,283	2,789

Titanium tetrachloride ($TiCl_4$) –
Isobutyl trichloroacetate ($C_6H_9Cl_3O_2$)
[164], $t = 20$ °C

x_2	100	80	65	50	35	20	0
ε	7,667	8,079	7,930	7,572	6,754	5,467	2,789

Titanium tetrachloride ($TiCl_4$) –
Isopentyl trichloroacetate ($C_7H_{11}Cl_3O_2$)
[164], $t = 20$ °C

x_2	100	80	65	50	35	20	0
ε	7,287	7,634	7,528	7,263	6,461	5,110	2,789

Supplementary information on the permittivity of non-aqueous solutions of inorganic substances

Calcium chloride ($CaCl_2$) – Benzene (C_6H_6), fig. 1.
Chlorine (Cl_2) – Sulphur dichloride (SCl_2) [153].
Deuterium oxide (D_2O) – Dioxan ($C_4H_8O_2$) [154].
Lithium nitrate ($LiNO_3$) – Formamide (CH_3ON), fig. 2.
Lithium nitrate ($LiNO_3$) – Methanol (CH_4), fig. 2.
Lithium nitrate ($LiNO_3$) – Dimethylsulphoxide (C_2H_6SO) fig. 2.
Lithium nitrate ($LiNO_3$) – N, N–Dimethylformamide (C_3H_7NO) fig. 2.
Lithium nitrate ($LiNO_3$)
Lithium perchlorate ($LiClO_4$) – Methanol (CH_4), fig. 3.
Magnesium perchlorate ($Mg(ClO_4)_2$) – Methanol (CH_4O), fig. 3.
Sodium perchlorate ($NaClO_4$) – Methanol (CH_4O), fig. 3.
Sulphuric acid (H_2SO_4) – Benzene (C_6H_6), fig. 1.
Sulphur (S) – Ethanol (C_2H_6O) [162].

§ 2. Permittivity of Non-aqueous Solutions of Organic Compounds

Carbon tetrabromide (CBr_4) – Diethyl ether ($C_4H_{10}O$)
[166], $t = 20$ °C

x_2	100	91,577	87,170	83,651	80,214	74,603
ε	4,335	4,335	4,335	4,274	4,321	4,248

Carbon tetrabromide (CBr_4) – Di-isopropyl ether ($C_6H_{14}O$)
[166], $t = 20$ °C

x_2	100	92,647	87,130	81,613	75,699	71,116
ε	3,976	3,984	3,976	3,976	3,952	3,928

Carbon tetrachloride (CCl_4) – Tetranitromethane (CN_4O_8)
[167], $t = 25$ °C

x_2	100	30,898	26,079	18,437	14,291	8,421	4,039	0
ε	2,521	2,317	2,295	2,275	2,264	2,251	2,240	2,227

Carbon tetrachloride (CCl_4) – Carbon disulphide (CS_2) *)
[168, 169], $t = 20$ °C

x_1	ε	x_1	ε	x_1	ε	x_1	ε
0	2,63	32,49	2,447	64,09	2,333	87,29	2,264
15,07	2,538	38,50	2,422	78,80	2,285	80,20	2,261
26,12	2,476	48,97	2,382	82,75	2,275	94,66	2,242

*) See also [170, 171, 174].

[173], $t = 18$ °C

w_1	ε	w_1	ε	w_1	ε	w_1	ε
0	2,669	28,932	2,513	60,639	2,394	100	2,222
9,811	2,639	40,586	2,477	70,258	2,359	—	—
19,487	2,558	49,807	2,439	90,209	2,315	—	—

[172]

w_2	100	91,96	68,26	51,46	34,66	9,97	0
t °C	21,0	18,8	18,9	18,7	18,5	18,5	18,3
ε	2,256	2,287	2,367	2,457	2,527	2,648	2,691

Carbon tetrachloride (CCl_4) – Chloroform ($CHCl_3$) *)
[178], $t = 25$ °C, $\nu = 1$ MHz

x_1	0	25	50	75	100
ε	4,770	3,901	3,215	2,682	2,230

*) See also [168, 170, 177].

[176], $t = 20$ °C

x_2	ε	x_2	ε	x_2	ε	x_2	ε
100	4,806	33,779	2,8683	6,035	2,3398	1,308	2,2606
78,471	4,025	16,822	2,5294	3,582	2,3006	0	2,2398
54,269	2,3373	11,545	2,4367	2,441	2,2791	—	—

[172]

w_1	0	10,14	33,07	39,19	51,84	67,59	91,18	100
t °C	18,0	19,2	19,3	19,4	19,5	19,6	19,9	21,0
ε	5,220	4,803	4,055	3,820	3,460	3,059	2,441	2,256

[174]

$w_1=0$				$w_1=20$				$w_1=40$			
t °C	ε	t °C	ε	t °C	ε	t °C	ε	t °C	ε	t °C	ε
13,7	5,165	35,7	4,791	14,2	4,410	40,1	4,060	16,4	3,783	42,0	3,531
18,9	5,065	44,8	4,654	23,6	4,283	45,7	3,987	23,3	3,709	49,5	3,465
24,2	4,966	44,3	4,591	31,9	4,171	54,4	3,872	30,0	3,645	55,8	3,412
31,0	4,867	55,9	4,503								

$w_1=60$		$w_1=80$		$w_1=100$	
t °C	ε	t °C	ε	t °C	ε
14	3,145	14,5	2,660	17,7	2,197
33,5	3,070	33,5	2,591	25,3	2,184
40,5	3,024	40,1	2,565	31,0	2,176
48,1	2,976	50,4	2,525	52,3	2,135
56,6	2,933	55,5	2,508	70,2	2,121

[175], $t = 20$ °C, $\nu = 1,8$ MHz

φ_2	100	80	60	40	20	0
ε	2,20	2,66	3,13	3,64	4,20	4,80

Carbon tetrachloride (CCl_4) – Dichloromethane (CH_2Cl_2)
[179]

w_1=0		w_1=21,7		w_1=51,7		w_1=74,9		w_1=87,4	
t °C	ε	t °C	ε	t °C	ε	t °C	ε	t °C	ε
32,8	8,47	37,5	6,105	41,8	4,035	40,7	3,000	36,8	2,588
28,4	8,93	34,4	6,180	38,5	4,071	40,6	3,991	30,5	2,612
22,0	8,90	29,7	6,295	32,4	4,149	32,1	3,057	24,3	2,632
—8,6	10,27	25,1	6,412	27,6	4,210	24,6	3,106	24,2	2,638
—28,4	11,45	15,8	6,550	23,4	4,264	19,0	3,142	18,6	2,657
—28,9	11,34	7,5	6,900	11,4	4,250	7,1	3,230	11,2	2,686
—55,6	13,02	—1,2	7,106	4,2	4,424	—1,6	3,297	—1,5	2,740
—80,7	14,98	—13,7	7,560	—5,1	4,526	—15,2	3,400	—18,1	2,816
—89,1	15,9	—26,4	8,007	—16,1	4,666	—20,8	3,447	—28,1	2,867
—	—	—36,4	8,378	—29,0	4,843	—30,5	3,527	—32,7	2,885
—	—	—47,2	8,815	—40,9	5,063	—41,0	3,620	—40,8	2,930
—	—	—54,3	9,135	—52,8	—	—	—	—	—

Carbon tetrachloride (CCl_4) – Chloromethane (CH_3Cl)
[179]

w_1=82,35		w_1=82,35		w_1=88,12		w_1=94,49		w_1=97,48		w_1=100	
t °C	ε	t °C	ε	t °C	ε	t °C	ε	t °C	ε	t °C	ε
18,0	2,865	—28,4	3,150	15,1	2,668	19,2	2,425	17,3	2,329	59,6	2,161
—1,6	2,989	—36,1	3,203	3,6	2,717	14,3	2,448	5,8	2,356	44,9	2,181
—3,4	3,000	—41,8	3,244	—6,3	2,760	3,8	2,482	—4,2	2,381	38,1	2,203
—10,0	3,035	—46,0	3,272	—21,0	2,822	—6,6	2,514	—17,0	2,412	29,3	2,219
—20,0	3,086			—36,0	2,899	—24,2	2,566	—21,5	2,421	21,1	2,237
				—26,3	2,852	—32,6	2,593	—28,0	2,423	—	—

Carbon tetrachloride (CCl_4) – Iodomethane (CH_3I)
[180], t=20 °C

x_1	ε	x_1	ε	x_1	ε	x_1	ε
0	7,081	60,972	3,524	88,637	2,5605	98,724	2,2737
14,951	5,898	67,754	3,242	92,205	2,4567	99,370	2,2545
32,804	4,814	76,791	2,9365	95,718	2,3563	100,00	2,2366
50,952	3,939	81,795	2,7532	97,905	2,2942	—	—

Carbon tetrachloride (CCl_4) – Nitromethane (CH_3NO_2)

[175], $t=20$ °C, $\nu=1{,}8$ MHz

φ_2	100	80	60	40	20	0
ε	37,45	29,00	20,80	13,30	7,05	2,24

Carbon tetrachloride (CCl_4) – Methanol (CH_4O) *)

x_2	ε at t °C [181]				x_2	ε at t °C [182]		x_2	ε at t °C [184]		
	25	35	45	55		6	30		22	30	40
100	32,645	30,678	28,841	27,114	100	33,0	30,9	100	—	32,1	29,92
73,23	17,434	16,541	15,664	14,738	82,619	27,53	23,44	40,57	7,03	6,682	6,313
66,73	14,898	14,084	13,295	12,573	63,496	16,90	15,067	30,22	4,88	4,707	4,489
51,11	9,834	9,057	8,432	7,841	54,320	12,57	11,364	19,83	3,426	3,348	3,258
49,35	9,298	8,549	7,967	7,431	41,665	8,263	7,499	16,83	3,126	3,071	3,003
42,24	7,340	6,859	6,401	5,954	33,681	5,994	5,350	10,30	2,635	2,612	2,576
30,86	4,915	4,696	4,483	4,283	13,169	2,8825	2,7775	6,08	2,433	2,417	2,393
29,28	4,619	4,431	4,244	4,055	8,889	2,5921	2,5327	2,15	2,301	2,284	—
17,22	3,161	3,077	3,011	2,955	4,668	2,4049	2,3607	1,18	2,272	2,255	—
16,45	3,097	3,005	2,947	2,900	2,793	2,3428	2,2987	1,04	2,267	2,249	—
7,84	2,509	2,481	2,449	2,418	0	2,2636	2,2163	0	2,232	2,216	2,196

*) See also [183, 185].

[175], $t=20$ °C, $\nu=1{,}8$ MHz

φ_2	100	80	60	40	20	0
ε	33,60	27,05	20,35	13,20	6,26	2,24

Carbon tetrachloride (CCl_4) – Trifluoroacetic acid ($C_2HO_2F_3$)

[186], $t=25$ °C

x_2	ε	x_2	ε	x_2	ε	x_2	ε
100	8,24	70,00	3,75	41,48	2,66	10,09	2,32
89,93	6,28	60,23	3,14	29,88	2,58	0	2,24
79,33	4,81	49,60	2,85	20,13	2,40	—	—

Carbon tetrachloride (CCl_4) – 1, 1, 2, 2–Tetrachloroethane ($C_2H_2Cl_4$)

[175], $t=20$ °C, $\nu=1{,}8$ MHz

φ_2	100	80	60	40	20	0
ε	8,50	6,66	5,10	3,88	2,94	2,24

Carbon tetrachloride (CCl_4) – 1, 1, 1–Trichloroethane ($C_2H_3Cl_3$)
[27]

x_2=100		x_2=81,92		x_2=65,71		x_2=48,05		x_2=32,25	
t °C	ε	t °C	ε	t °C	ε	t °C	ε	t °C	ε
19,8	7,252	19,6	6,133	19,9	5,216	20,4	4,288	22,2	7,521
16,4	7,370	15,5	6,248	15,8	5,300	13,4	4,395	18,5	3,560
12,8	7,498	9,4	6,392	10,3	5,390	8,6	4,461	13,0	3,612
5,1	7,780	4,7	6,519	4,8	5,513	2,9	4,540	7,9	3,663
—0,1	7,975	—0,1	6,646	—1,3	5,640	—3,4	4,638	2,7	3,715
—5,6	8,191	—5,2	6,779	—5,4	5,718	—8,0	4,726	—2,6	3,770
—10,9	8,406	—12,3	6,964	—15,4	5,904	—14,3	4,806	—8,1	3,830
—15,7	8,600	—	—	—20,0	5,992	—20,2	4,897	—14,0	3,896
—20,9	8,829	—	—	—	—	—	—	—20,0	3,965
—25,0	9,013	—	—	—	—	—	—	—	—

x_2=16,01		x_2=7,67		x_2=0					
t °C	ε	t °C	ε	t °C	ε	t °C	ε	t °C	ε
20,7	2,843	18,7	2,528	21,8	2,2369	7,2	2,2664	—5,8	2,2828
14,2	2,880	15,5	2,540	21,4	2,2386	6,6	2,2679	—9,3	2,2997
13,5	2,883	9,2	2,563	19,7	2,2400	6,0	2,2689	—11,0	2,3035
10,1	2,902	4,5	2,581	18,0	2,2448	5,9	2,2688	—12,9	2,3067
1,8	2,951	—0,4	2,599	17,7	2,2446	2,7	2,2758	—14,2	2,3097
0,2	2,959	—6,7	2,623	17,4	2,2456	0,9	2,2788	—14,8	2,3099
—6,8	2,999	—10,4	2,637	13,2	2,2544	0,0	2,2811	—16,0	2,3138
—12,6	3,033	—15,7	2,657	11,1	2,2586	—4,6	2,2904	—18,0	2,3170
—20,1	3,079	—19,7	2,672	8,7	2,2628	—4,9	2,2903	—19,8	2,3212

Carbon tetrachloride (CCl_4) – Acetonitrile (C_2H_3N)
[187], t=25 °C

w_1	ε	w_1	ε	w_1	ε	w_1	ε	w_1	ε	w_1	ε
0	36,01	60,15	20,60	67,74	17,45	70,81	16,13	77,73	13,06	80,68	11,16
28,66	30,62	60,61	19,87	68,06	17,18	75,22	14,21	78,77	12,31	82,72	10,68
54,84	22,32	62,47	19,46	68,69	17,05	75,97	13,92	79,65	12,14	84,27	9,80

[189], $t=25$ °C

w_1	ε	w_1	ε	w_1	ε
0,0	36,01	68,46	16,99	75,81	13,93
63,20	18,91	68,71	17,02	78,84	12,29
63,85	18,45	74,54	14,65	80,80	11,35

[175], $t=20$ °C, ν=1,8 MHz

φ_2	100	80	60	40	20	0
ε	36,80	36,80	29,30	12,32	7,65	2,24

Carbon tetrachloride (CCl_4) – 1, 2–Dibromoethane ($C_2H_4Br_2$)

[172]

w_2	100	90,70	67,42	54,98	51,75	33,16	11,80	0
t	22,7	23	23,1	23	23	23	22,7	21
ε	4,991	4,367	3,423	3,080	3,044	2,666	2,335	2,256

Carbon tetrachloride (CCl_4) – 1, 2–Dichloroethane ($C_2H_4Cl_2$)

$t=25$ °C [190]						$t=60$ °C [190]			
x_1	ε	x_1	ε	x_1	ε	x_1	ε	x_1	ε
0	2,227	33,3	3,478	72,2	6,428	0	2,165	76,6	5,998
9,8	2,501	47,5	4,347	84,1	7,778	24,8	2,879	85,1	6,823
20,7	2,889	66,7	5,875	100	10,02	45,0	3,774	94,0	7,776
						67,9	5,282	100	8,492

[175], $t=20$ °C, ν=1,8 MHz

φ_2	100	80	60	40	20	0
ε	10,65	8,08	5,96	4,26	3,05	2,24

Carbon tetrachloride (CCl_4) – Acetic acid ($C_2H_4O_2$)

[191, 192], $t=20$ °C

φ_2	100	75	50	25	0
ε	7,1	5,9	4,5	3,3	2,2

Carbon tetrachloride (CCl_4) – Bromoethane (C_2H_5Br)

[176], $t=20$ °C

x_2	ε	x_2	ε	x_2	ε	x_2	ε
100	9,613	23,730	3,4178	4,647	2,4566	1,326	2,3004
79,674	7,501	12,218	2,8289	3,601	2,4073	0	2,2383
55,515	5,483	8,538	2,6420	2,225	2,3443	—	—

Carbon tetrachloride (CCl_4) – Iodoethane (C_2H_5I)

[180], $t=20$ °C

x_2	ε	x_2	ε	x_2	ε	x_2	ε
100	7,719	29,625	3,553	8,170	2,5758	0,807	2,2689
75,333	6,082	18,076	2,9933	4,058	2,4029	0	2,2353
47,188	4,449	11,800	2,7313	2,288	2,3296	—	—

Carbon tetrachloride (CCl_4) – Ethanol (CH_4O) *)

x_2	ε at t °C [181]				x_2	ε at t °C [181]			
	25	35	45	55		25	35	45	55
100	24,325	22,787	21,301	19,707	57,54	10,392	9,670	8,965	8,510
84,50	18,562	17,457	16,384	15,472	31,82	4,798	4,358	4,292	4,091
78,95	16,771	15,743	14,840	14,015	23,84	3,686	3,553	3,422	3,305
72,56	14,778	13,839	12,943	12,221	10,42	2,570	2,533	2,517	2,487
65,76	12,698	11,852	11,089	10,497	4,05	2,335	2,307	2,955	2,271

x_2	ε at t °C [182]		x_2	ε at t °C [182]		x_2	ε at t °C [182]	
	6	30		6	30		6	30
100	28,0	24,4	35,382	6,279	5,494	5,593	2,426	2,385
77,394	20,12	19,35	26,423	4,344	3,941	3,437	2,354	2,318
59,429	14,05	13,69	16,681	3,093	2,948	1,857	2,313	2,272
53,026	11,47	10,78	8,372	2,530	2,485	0	2,264	2,216
43,031	8,419	7,534	—	—	—	—	—	—

*) See also [196].

[193], $t=25$ °C, $\nu=0,2$ MHz

w_1	x_1	ε	w_1	x_1	ε	w_1	x_1	ε
0	0	24,91	46,82	20,86	17,80	88,91	70,60	4,45
17,59	6,01	22,69	59,81	30,82	14,60	100	100	2,25
27,57	10,23	21,20	69,27	40,30	11,76	—	—	—
37,03	14,98	19,67	79,17	53,23	8,12	—	—	—

[194], $t=25$ °C				[178], $t=25$ °C, ε=1 MHz			
x_2	ε	x_2	ε	x_2	ε	x_2	ε
100	24,69	29,26	4,45	100	25,2	25	3,74
71,27	14,71	15,48	2,939	50	7,05	10	2,56
52,45	9,40	0	2,276	40	5,45	0	2,23

[195], $t=21$ °C

w_2	100	90	80	70	60	50	40	30	20	10	0
ε	25,8	24,2	22,6	20,9	19,1	17,0	14,5	11,7	5,5	5,4	2,25

Carbon tetrachloride (CCl_4) – **Acetone** (C_3H_6O) *)

[166], $t=20$ °C, $\nu=1,8$ *Мгц*

x_2	ε	x_2	ε	x_2	ε	x_2	ε
100	21,17	75,685	14,46	20,665	4,650	0	2,242
97,431	20,30	57,880	10,66	9,496	3,265	—	—
89,866	18,06	35,814	7,069	4,186	2,675	—	—

*) See also [168, 191].

[178], $t=25$ °C, $\nu=1$ MHz

x_2	100	40	30	25	10	0
ε	21,3	6,17	5,67	4,98	3,28	2,23

[194], $t=25$ °C

w_2	100	84,04	75,45	66,39	56,84	46,75	36,07	24,78	12,76	0
ε	20,87	16,27	14,29	12,28	10,41	8,53	6,77	5,14	3,59	2,276

[175], $t=20$ °C, $\nu=1,8$ MHz

φ_2	100	80	60	40	20	0
ε	21,07	16,56	12,47	8,63	5,13	2,24

Carbon tetrachloride (CCl_4) – Methyl acetate ($C_3H_6O_2$)

[178, 241], $t=25$ °C, $\nu=1$ MHz

x_2	100	75	63,8	50	25	10	0
ε	6,68	5,40	4,880	4,225	3,192	2,610	2,230

Carbon tetrachloride (CCl_4) – *n*-Propan-1-ol (C_3H_8O)

x_2	ε at t °C [184]			x_2	ε at t °C [181]			
	22	30	40		25	35	45	55
100	20,5	19,3	18,2	100	20,000	18,565	17,227	15,999
81,12	16,0	15,0	13,8	98,18	17,328	16,072	14,833	13,662
61,78	11,3	10,5	9,7	74,96	14,004	12,804	11,675	10,592
41,71	6,37	5,978	5,526	65,77	11,837	10,764	9,795	8,762
20,26	3,152	3,094	3,02	52,48	8,860	7,930	7,199	6,506
15,57	2,811	2,782	2,743	32,63	4,557	4,300	4,072	3,867
10,87	2,569	2,552	2,530	26,26	3,717	3,572	3,441	3,320
4,99	2,374	2,361	2,346	18,64	3,017	2,969	2,901	2,843
3,50	2,336	2,320	2,302	9,39	2,490	2,464	2,448	2,424
0	2,233	2,217	2,197	4,82	2,352	2,332	2,317	2,292

Carbon tetrachloride (CCl_4) – Propan–2–ol (C_3H_8O)

[197], $t=35$ °C

x_2	ε	x_2	ε	x_2	ε	x_2	ε
100	2,205	57,52	5,09	17,49	13,10	3,77	16,27
82,79	2,84	44,44	7,04	12,52	14,30	0	17,80
68,20	3,84	37,62	8,60	6,74	15,68	—	—

x_2	ε at t °C [184]			x_2	ε at t °C [184]		
	22	30	40		22	30	40
100	20,1	18,7	17,2	18,66	2,848	2,821	2,790
44,69	5,886	5,609	5,24	14,69	2,668	2,656	2,642
37,65	4,706	4,519	4,29	13,68	2,623	2,607	2,591
36,05	4,481	4,220	4,04	9,73	2,488	2,476	2,465
30,96	3,832	3,690	3,554	9,72	2,495	2,484	2,470
27,45	3,434	3,426	3,34	5,06	2,367	2,355	2,338
19,1	2,878	2,854	2,826	4,52	2,358	2,344	2,328
—	—	—	—	0	2,233	2,217	2,197

Carbon tetrachloride (CCl_4) – Acetic anhydride ($C_4H_6O_3$)
[175], $t=20$ °C, $\nu=1,8$ MHz

φ_2	100	80	60	40	20	0
ε	22,45	16,67	13,25	8,95	5,18	2,24

Carbon tetrachloride (CCl_4) – 1, 4–Dioxan ($C_4H_8O_2$)
[166], $t=20$ °C

x_2	ε	x_2	ε	x_2	ε	x_2	ε
100	2,229	77,369	2,257	29,901	2,268	0	2,242
93,764	2,237	62,730	2,267	18,113	2,261	—	—
86,195	2,246	46,980	2,270	8,885	2,252	—	—

Carbon tetrachloride (CCl_4) – Ethyl acetate ($C_4H_8O_2$)
[199], $t=25$ °C, $\nu=4,5$ MHz

x_1	ε	x_1	ε	x_1	ε	x_1	ε
0,0	6,02	37,8	4,73	55,9	3,77	83,6	2,60
9,2	5,81	41,5	4,54	59,2	3,62	91,1	2,36
16,9	5,56	44,8	4,41	62,8	3,43	100	2,228
23,3	5,36	47,7	4,24	67,0	3,36	—	—
28,9	5,15	50,3	4,10	71,8	3,02	—	—
33,6	4,93	53,0	3,92	77,2	2,79	—	—

[178], $t = 25$ °C

x_2	100	75	50	25	0
ε	6,03	5,11	4,201	3,220	2,230

[198], $t = 25$ °C

x_2	100	81,31	74,77	61,23	48,66	35,82	23,44	11,47	0
ε	6,053	5,387	5,148	4,647	4,171	3,678	3,184	2,707	2,2364

Carbon tetrachloride (CCl_4) – *tert*–Butyl chloride (C_4H_9Cl) *)

[34, 180], ν = 1 MHz

x_2	ε at t °C					x_2	ε at t °C				
	20	10	0	−11,4	−22		20	10	0	−11,4	−22
100	10,04	10,61	11,25	12,05	12,09	11,98	2,936	3,000	3,068	2,151	3,227
77,69	7,832	8,270	8,720	9,276	9,855	8,87	2,712	2,760	2,812	2,875	2,937
64,16	6,762	7,122	7,487	7,913	8,178	4,23	2,471	2,509	2,548	2,592	2,632
51,03	5,775	6,036	6,283	6,636	6,782	1,39	2,319	2,347	2,374	2,407	2,437
37,26	4,731	4,911	5,114	5,360	5,603	0	2,240	2,258	2,276	2,300	2,324
21,05	3,305	3,612	3,723	3,855	3,978	—	—	—	—	—	—

*) In [34] the variation of ε with t is given also for −30° and −70°C.

Carbon tetrachloride (CCl_4) – Butan–1–ol ($C_4H_{10}O$)

x_2	ε at t °C [200]				x_2	ε at t °C [184, 181]		
	25	35	45	55		22	30	40
100	17,0997	15,8407	14,6638	13,6069	100	18,1	17,0	15,7
91,59	15,3527	14,0610	13,0138	12,0492	61,82	10,11	9,349	8,438
89,21	14,9138	13,6466	12,6316	11,6949	41,61	5,706	5,349	4,974
73,26	11,6742	10,5801	9,7574	9,0146	28,84	3,786	3,656	3,528
54,01	7,7084	6,9438	6,4318	5,9970	20,58	3,040	2,991	2,947
40,32	5,1016	4,7256	4,5109	4,3308	14,63	2,697	2,678	2,662
27,34	3,4995	3,3653	3,2997	3,2540	8,73	2,475	2,461	2,448
11,71	2,5535	2,5173	2,5313	5,5378	3,97	2,346	2,332	2,317
0	2,2291	2,2093	2,1887	2,1688	0	2,234	2,218	2,198

Carbon tetrachloride (CCl_4) – Isobutanol ($C_4H_{10}O$)

x_2	ε at t °C [181]				x_2	ε at t °C [184]		
	25	35	45	55		22	30	40
100	17,575	16,075	14,660	13,333	100	18,3	17,0	15,7
83,64	14,171	12,846	11,664	10,562	61,62	9,797	9,012	8,10
70,10	11,365	10,151	9,143	8,228	41,02	5,370	5,014	4,639
56,74	8,592	7,561	6,785	6,110	31,98	3,999	3,813	3,663
46,82	6,365	5,817	5,251	4,749	22,14	3,065	3,018	2,963
38,79	4,863	4,523	4,238	3,999	15,40	2,592	2,673	2,652
33,29	4,114	3,898	7,712	3,554	10,64	2,513	2,502	2,489
16,08	2,738	2,715	2,687	2,651	5,09	2,364	2,351	2,336
15,76	2,726	2,703	2,674	2,640	2,31	2,299	2,285	2,265
7,84	2,415	2,394	2,383	2,355	0	2,232	2,216	2,196
3,80	2,317	2,295	2,280	2,260	—	—	—	—

Carbon tetrachloride (CCl_4) – Butan–2–ol ($C_6H_{10}O$)

x_2	ε at t °C [184]			x_2	ε at t °C [181]			
	22	30	40		25	35	45	55
100	17,3	15,9	14,2	100	15,25	13,93	12,75	11,67
70,69	10,06	9,144	8,127	82,57	12,05	10,91	9,924	9,054
40,64	4,377	4,175	3,980	67,35	9,240	8,470	7,669	6,954
30,90	3,435	3,362	3,293	55,12	6,963	6,364	5,840	5,388
20,15	2,825	2,809	2,790	42,63	4,936	4,629	4,365	4,133
15,88	2,663	2,655	2,642	35,16	4,015	3,843	3,695	3,562
9,94	2,489	2,480	2,471	21,26	2,940	2,908	2,872	2,832
3,95	2,348	2,335	—	8,66	2,431	2,416	2,395	2,382
0	2,234	2,218	2,198	4,53	2,338	2,320	2,298	2,771

x_2	ε at t °C [200]				x_2	ε at t °C [200]			
	25	35	45	55		25	35	45	55
100	16,1760	14,5368	13,0194	11,6335	34,85	3,7698	3,5979	3,5446	3,4621
91,23	13,7895	11,9814	10,6743	9,6068	27,08	3,1366	3,0545	3,0450	3,0374
63,10	7,5672	6,6903	6,0558	5,5136	13,20	2,5628	2,5269	2,5417	2,5479
51,99	5,6712	5,1318	4,8131	4,5883	0	2,2291	2,2093	2,1887	2,1688

Carbon tetrachloride (CCl_4) – *tert*–Butyl alcohol ($C_4H_{10}O$)

[51]

$x_2=100$				$x_2=93,23$				$x_2=88,27$			
t °C	ε	*t* °C	ε	*t* °C	ε	*t* °C	ε	*t* °C	ε	*t* °C	ε
25,43	12,377	39,14	9,876	22,55	10,804	38,45	8,486	21,52	9,596	38,35	7,641
27,32	11,985	41,82	9,487	26,53	10,115	42,63	8,028	25,54	9,029	42,15	7,315
30,01	11,458	44,74	9,094	30,53	9,510	46,44	7,650	30,18	8,477	46,21	7,008
33,00	10,899	48,00	8,673	34,96	8,914	49,75	7,360	34,21	8,034	50,18	6,735
36,32	10,338	49,98	8,451								
36,17	10,365	—	—								

$x_2=78,68$		$x_2=71,37$		$x_2=60,24$		$x_2=49,23$	
t °C	ε	*t* °C	ε	*t* °C	ε	*t* °C	ε
16,62	7,797	18,81	6,070	5,44	4,967	0,05	3,724
19,93	7,439	21,89	5,883	10,50	4,804	4,91	3,690
23,62	7,097	25,76	5,684	14,69	4,687	9,80	3,664
28,08	6,732	29,83	5,496	17,74	4,617	15,10	3,641
31,94	6,469	34,01	5,334	21,86	4,530	19,91	3,625
35,94	6,217	38,31	5,185	25,24	4,468	20,01	3,624
39,74	6,007	42,23	5,072	29,73	4,394	24,78	3,612
43,55	5,821	46,18	4,962	34,05	4,329	29,88	3,599
47,00	5,667	51,10	4,866	37,92	4,277	35,17	3,589
49,93	5,547	—	—	42,03	4,230	39,88	3,573
—	—	—	—	46,08	4,182	44,86	3,558
—	—	—	—	5,09	4,141	49,47	3,543

$x_2=45,28$		$x_2=39,95$		$x_2=28,75$		$x_2=21,46$		$x_2=18,71$	
t °C	ε	*t* °C	ε	*t* °C	ε	*t* °C	ε	*t* °C	ε
—1,09	3,424	—0,60	3,146	0,64	2,760	—0,1	2,591	—0,35	2,565
+5,77	3,414	+3,61	3,153	3,92	2,773	+4,82	2,600	+5,03	2,573
9,86	3,411	7,46	3,161	8,36	2,786	10,00	2,613	9,87	2,581
15,18	3,411	12,08	3,169	12,64	2,798	11,65	2,618	14,92	2,589
20,11	3,409	16,05	3,176	15,94	2,807	14,75	2,627	20,30	2,598
25,49	3,408	16,05	3,176	20,23	2,817	17,50	2,633	24,79	2,604
30,34	3,406	20,41	3,184	24,82	2,827	19,91	2,639	31,00	2,610
35,06	3,403	24,13	3,190	28,05	2,834	24,93	2,651	35,13	2,612
40,49	3,398	28,37	3,195	32,09	2,842	27,35	2,655	40,62	2,615
45,20	3,392	32,30	3,199	36,24	2,848	29,98	2,660	45,78	2,615
—	—	36,75	3,201	40,06	2,852	33,00	2,669	49,88	2,613
—	—	40,15	3,203	44,32	2,854	35,52	2,673	—	—
—	—	44,12	3,202	—	—	39,98	2,676	—	—
—	—	47,51	3,200	—	—	43,05	2,678	—	—
—	—	49,56	3,197	—	—	46,26	2,679	—	—
—	—	—	—	—	—	49,27	2,679	—	—

x_2=13,57		x_2=7,72		x_2=6,01		x_2=2,65		x_2=1,16	
t °C	ε	t °C	ε	t °C	ε	t °C	ε	t °C	ε
−0,90	2,492	2,26	2,409	−1,21	2,388	−1,5	2,350	−2,04	2,317
+5,30	2,497	+4,19	2,408	+5,11	2,385	+4,62	2,334	+4,86	2,306
10,62	2,501	10,20	2,407	10,45	2,382	10,29	2,327	10,25	2,295
15,97	2,505	14,71	2,406	14,99	2,379	15,15	2,319	15,14	2,286
20,14	2,507	19,98	2,404	20,28	2,376	20,15	2,312	20,32	2,275
25,21	2,510	25,28	2,401	25,12	2,371	25,38	2,303	25,38	2,265
30,81	2,511	30,11	2,398	29,70	2,366	30,17	2,294	30,49	2,254
34,77	2,511	39,55	2,389	35,67	2,359	35,28	2,285	35,15	2,244
39,67	2,510	45,76	2,380	41,76	2,350	40,13	2,275	40,16	2,233
44,75	2,507	50,32	2,373	45,04	2,345	45,17	2,264	44,98	2,223
49,86	2,503	—	—	50,19	2,335	50,89	2,253	49,21	2,213

[200], ν = 1 MHz

x_2	ε at t °C				x_2	ε at t °C			
	25	35	45	55		25	35	45	55
100	11,9921	10,2043	8,7908	7,8390	19,16	2,6012	2,6136	2,6114	2,6071
80,25	7,0521	6,2741	5,7978	5,4506	12,69	2,4748	2,4845	2,4738	2,4533
53,33	3,8123	3,7693	3,7266	3,6932	11,23	2,4486	2,4450	2,4738	2,4533
34,70	2,9661	2,9861	2,9977	3,0066	4,66	2,3390	2,3210	2,3044	2,2861
24,35	2,7011	2,7238	2,7303	2,7319	0	2,2291	2,2093	2,1887	2,1688

Carbon tetrachloride (CCl_4) – Diethyl ether ($C_4H_{10}O$) *)

[166, 241], t=20 °C

x_2	ε	x_2	ε	x_2	ε	x_2	ε
100	4,335	58,144	3,453	26,413	2,761	6,053	2,364
90,979	4,179	48,548	3,232	20,399	2,639	3,862	2,323
81,740	3,985	44,406	3,134	15,809	2,550	0,00	2,242
70,024	3,705	32,161	2,879	11,003	2,463	—	—

*) See also [172, 175].

[168], t=25 °C

x_2	ε	x_2	ε	x_2	ε	x_2	ε
100	4,355	69,31	3,677	22,92	2,699	4,773	2,335
86,30	4,057	49,11	3,209	14,12	2,521	—	—
76,51	3,838	32,60	2,893	9,402	2,428	—	—

[178], $t=25$ °C, $\nu=1$ MHz

x_2	100	75	65	50	25	0
ε	4,265	3,700	3,491	3,210	2,722	2,230

[177], $t=25$ °C, $\nu=1$ MHz

φ_2	100	90	80	70	60	50	40	30	20	10	0
ε	2,20	2,36	2,56	2,75	2,94	3,15	3,36	3,56	3,78	3,99	4,21

Carbon tetrachloride (CCl_4) – 1-Pentan-1-ol ($C_5H_{12}O$)

x_2	ε при t °C [184]			x_2	ε при t °C [184]		
	22	30	40		22	30	40
100	15,5	14,6	13,5	15,20	2,669	2,654	2,637
69,24	10,35	9,50	8,64	10,44	2,491	2,484	2,470
41,16	5,144	4,827	4,528	5,33	2,370	2,361	2,347
31,32	3,750	3,631	3,508	2,75	2,811	2,299	—
20,81	2,935	2,903	2,868	0	2,231	2,215	2,195

Carbon tetrachloride (CCl_4) – Isopentyl alcohol ($C_5H_{12}O$)

[178], $t=20$ °C

x_2	100	40	25	10	0
ε	14,55	4,61	3,127	2,502	2,230

x_2	ε at t °C [184]			x_2	ε at t °C [184]		
	22	30	40		22	30	40
100	15,6	14,6	13,4	16,03	2,674	2,664	2,646
76,56	11,5	10,64	9,633	10,61	2,501	2,491	2,479
42,69	4,952	4,693	4,415	3,72	2,334	2,320	2,303
30,70	3,546	3,467	3,360	0	2,234	2,218	2,198
21,33	2,908	2,884	2,855	—	—	—	—

Carbon tetrachloride (CCl_4) – 2–Methylbutan–2–ol ($C_5H_{12}O$)

x_2	ε at *t* °C [184]			x_2	ε at *t* °C [184]		
	22	30	40		22	30	40
100	6,201	5,849	5,532	16,24	2,570	2,572	2,569
76,77	4,302	4,270	4,237	10,50	2,480	2,475	2,467
41,78	3,085	3,119	3,140	4,57	2,363	2,350	2,330
21,46	2,679	2,689	2,693	0	2,234	2,218	2,198

Carbon tetrachloride (CCl_4) – 1, 2, 4–Trichlorobenzene ($C_6H_3Cl_3$)

[175], $t = 20$ °C, $\nu = 1{,}8$ MHz

φ_2	100	80	60	40	20	0
ε	6,75	5,82	4,87	3,95	3,07	2,24

Carbon tetrachloride (CCl_4) – Bromobenzene (C_6H_5Br)

[180], $t = 20$ °C

x_2	ε	x_2	ε	x_2	ε	x_2	ε
100	5,431	30,140	3,230	5,227	2,4080	0,832	2,2654
73,379	4,627	19,572	2,8826	2,809	2,3296	0	2,2386
48,043	3,833	9,717	2,5559	1,754	2,2954	—	—

x_2	ε at *t* °C [201]					
	10	20	30	40	50	60
100,00	3,642	3,490	3,349	3,212	3,090	2,975
80,00	5,040	4,900	4,782	4,677	4,580	4,483
60,00	4,400	4,288	4,182	4,090	4,003	3,917
40,00	3,679	3,613	3,538	3,467	3,400	3,337
20,20	2,989	2,942	2,893	2,843	2,792	2,744
10,98	2,658	2,622	2,587	2,553	2,518	2,483
8,770	2,576	2,544	2,513	2,481	2,450	2,418
7,837	2,545	2,514	2,484	2,453	2,422	2,393
6,420	2,493	2,464	2,436	2,408	2,379	2,352
4,893	2,438	2,411	2,385	2,359	2,333	2,310
4,110	2,410	2,384	2,359	2,335	2,310	2,285
2,049	2,335	2,312	2,291	2,269	2,247	2,225
1,002	2,297	2,276	2,256	2,237	2,216	2,190
0	2,260	2,241	2,223	2,205	2,184	2,167

Carbon tetrachloride (CCl_4) – Chlorobenzene (C_6H_5Cl)

x_2	ε MHz t °C [201]					
	10	20	30	40	50	60
100	5,956	5,778	5,602	5,437	5,290	5,154
74,27	5,003	4,864	4,728	4,594	4,448	4,363
48,10	4,021	3,919	3,821	3,732	3,650	4,363
21,49	3,040	2,982	2,928	2,876	2,824	2,774
10,55	2,641	2,604	2,568	2,532	2,498	2,246
8,465	2,567	2,533	2,498	2,464	2,432	2,403
7,562	2,535	2,503	2,469	2,437	2,407	2,378
5,796	2,470	2,440	2,412	2,382	2,354	2,329
4,258	2,414	2,388	2,362	2,336	2,310	2,285
3,690	2,393	2,368	2,343	2,318	2,293	2,270
2,118	2,337	2,314	2,292	2,271	2,247	2,226
0,920	2,294	2,273	2,253	2,233	2,213	2,192
0	2,260	2,241	2,223	2,205	2,186	2,167

[175], $t=25$ °C, $\nu=1,8$ MHz

φ_2	100	80	60	40	20	0
ε	5,70	4,96	4,28	3,56	2,88	2,24

Carbon tetrachloride (CCl_4) – Fluorobenzene (C_6H_5F)

[180], $t=20$ °C

x_2	ε	x_2	ε	x_2	ε	x_2	ε
100	5,472	25,702	2,9571	4,220	2,3511	0,728	2,2570
75,252	4,574	14,836	2,6479	2,454	2,3026	0	2,2365
50,159	3,748	7,594	2,4431	1,443	2,2755	—	—

Carbon tetrachloride (CCl_4) – Nitrobenzene ($C_6H_5NO_2$)

[193, 202], $t=25$ °C

w_1	0	10,04	23,09	35,87	57,35	71,41	85,81	100
x_1	0	8,20	19,37	30,92	51,83	66,66	82,87	100
ε	34,60	31,11	26,79	22,23	15,30	10,86	6,25	2,25

[204, 203, 205, 206]

x_2	ε at t °C				
	10	20	30	40	50
100	39,98	37,56	35,46	33,67	32,03
89,36	34,71	32,79	31,16	29,69	28,47
80,46	30,40	28,78	27,33	26,00	24,88
71,66	26,46	25,06	23,78	22,66	21,70
59,17	21,39	20,29	19,28	18,37	17,55
50,01	17,75	16,92	16,12	15,41	14,74
40,39	14,20	13,54	12,23	12,36	11,85
31,03	10,95	10,46	10,03	9,621	9,235
20,98	7,795	7,486	7,197	6,934	6,687
15,27	6,101	5,877	5,674	5,485	5,313
9,586	4,549	4,408	4,276	4,154	4,046
7,064	3,900	3,796	3,698	3,606	3,520
4,686	3,314	3,240	3,170	3,106	3,042
3,193	2,961	2,906	2,853	2,802	2,752
2,101	2,709	2,666	2,625	2,585	2,546
0,985	2,460	2,428	2,239	2,368	2,339
0	2,245	2,224	2,203	2,183	2,163

[175], $t=20$ °C, $\nu=1,8$ MHz

φ_2	100	80	60	40	20	0
ε	35,75	26,94	19,25	12,60	6,87	2,24

Carbon tetrachloride (CCl_4) – Benzene (C_6H_6) *)

[207], $t=20$ °C, $\nu=1$ MHz

x_1	0	10	25	50	75	100
ε	2,282	2,274	2,267	2,258	2,240	2,230

*) See also [169, 178, 209].

[170], $t=20$ °C

x_1	0	26,397	42,449	76,136	100
w_1	0	41,339	59,218	86,269	100
ε	2,249	2,200	2,199	2,203	2,241

[168], $t=25$ °C

x_1	0	10,78	29,07	56,16	72,69	83,88	93,66
ε	2,280	2,276	2,268	2,256	2,248	2,244	2,240

x_1	ε at t °C [208]			x_1	ε at t °C [208]		
	18	50	65		18	50	65
100	2,248	1,874	1,691	40	2,145	1,879	1,745
80	2,031	1,764	1,650	20	2,289	1,938	1,803
60	2,111	1,823	1,693	0	2,398	2,026	1,851

[173], $t=18$ °C

w_1	ε	w_1	ε	w_1	ε	w_1	ε
0	2,28	29,427	2,270	60,376	2,329	90,348	2,239
10,852	2,278	39,330	2,258	70,843	2,240	100	2,222
20,804	2,275	49,240	2,251	81,169	2,251	—	—

Carbon tetrachloride (CCl_4) – Cyclohexane (C_6H_{12})

[166], $t=20$ °C

x_1	0	27,294	53,458	77,419	100
ε	2,032	2,084	2,137	2,194	2,242

Carbon tetrachloride (CCl_4) – Hexan–1–ol ($C_6H_{14}O$)

x_2	ε at t °C [181]				x_2	ε at t °C [181]			
	25	35	45	55		25	35	45	55
100	13,385	12,100	10,875	9,785	24,46	3,018	2,977	2,935	2,893
81,66	10,269	9,176	8,301	7,534	18,63	2,762	2,742	2,719	2,690
46,56	4,874	4,544	4,309	4,086	12,10	2,523	2,511	2,501	2,481
39,49	4,090	3,911	3,753	3,616	6,50	2,377	2,361	2,349	2,325
31,91	3,460	3,369	3,285	3,207	—	—	—	—	—

Carbon tetrachloride (CCl_4) – Diisopropyl ether ($C_4H_{10}O$)

[166], $t=20$ °C

x_2	ε	x_2	ε	x_2	ε	x_2	ε
100	3,976	65,893	3,495	26,390	2,793	0	2,242
91,995	3,877	50,208	3,243	15,477	2,585	—	—
80,056	3,700	40,649	3,059	6,916	2,394	—	—

Carbon tetrachloride (CCl_4) – Benzonitrile (C_7H_5N)

[210], $t=25$ °C

w_2	ε	w_2	ε	w_2	ε	w_2	ε
100	25,48	27,741	8,572	2,864	2,840	0	2,2436
79,492	20,69	14,645	5,480	1,577	2,564	—	—
56,660	15,39	6,286	3,575	0,703	2,387	—	—

Carbon tetrachloride (CCl_4) – Toluene (C_7H_8)

[178], $t=25$ °C, $\nu=1$ MHz

x_1	0	50	75	90	100
ε	2,378	2,310	2,271	2,246	2,230

Carbon tetrachloride (CCl_4) – Cyclohexylmethanol ($C_7H_{14}O$)

[211], $t=20$ °C

x_2	ε	x_2	ε	x_2	ε	x_2	ε
72,0	10,10	38,95	4,35	7,178	2,48	2,051	2,32
58,63	7,43	23,78	3,08	3,670	2,39	—	—
45,47	5,24	14,71	2,70	2,227	2,33	—	—

Carbon tetrachloride (CCl_4) – Heptan-1-ol ($C_7H_{16}O$)

x_2	ε at t °C [184]			x_2	ε at t °C [184]		
	22	30	40		22	30	40
100	11,5	10,7	9,74	15,14	2,619	2,609	2,595
75,68	8,76	8,12	7,31	10,05	2,480	2,437	2,460
33,92	3,544	3,455	3,369	3,91	2,340	2,328	2,308
30,14	3,278	3,225	3,169	0	2,234	2,218	2,198
21,89	2,865	2,846	2,818	—	—	—	—

Carbon tetrachloride (CCl_4) – Dipentyl ether ($C_7H_{16}O$)

[172]

w_2	100	89,21	59,38	47,22	43,02	31,30	8,93	0
t °C	23	23	23,1	23	22,8	22,7	22,5	21
ε	3,561	3,411	3,300	3,098	3,080	2,871	2,404	2,256

Carbon tetrachloride (CCl_4) – Acetophenone (C_8H_8O)

[210], $t=25$ °C

w_1	ε	w_1	ε	w_1	ε	w_1	ε
0	2,2376	2,209	2,492	19,959	4,390	87,124	15,65
0,472	2,292	4,308	2,743	44,708	8,823	100	17,73
1,068	2,357	8,458	3,273	73,426	13,44	—	—

Carbon tetrachloride (CCl_4) – m–Xylene (C_8H_{10})

[172]

w_1	0	12,32	36,96	51,91	69,86	90,48	100
t °C	17,4	20,0	20,0	20,1	20,0	20,0	21,0
ε	2,385	2,375	2,355	2,361	2,310	2,240	2,256

Carbon tetrachloride (CCl_4) – Phenetole ($C_8H_{10}O$)

[210], $t=25$ °C

w_2	ε	w_2	ε	w_2	ε	w_2	ε
100	4,224	23,845	2,800	2,156	2,298	0	2,2433
84,180	3,980	11,971	2,256	0,755	2,263	—	—
47,290	3,306	4,931	2,363	0,263	2,251	—	—

[175], $t=20$ °C, $\nu=1$ MHz

φ_2	100	80	60	40	20	0
ε	4,25	3,84	3,43	3,02	2,62	2,24

Carbon tetrachloride (CCl_4) – Octan–1–ol ($C_8H_{18}O$)

x_2	ε at t °C [184]			x_2	ε at t °C [184]		
	22	30	40		22	30	40
100	10,22	9,614	8,725	15,15	2,596	2,589	2,579
70,95	7,256	6,734	6,149	10,14	2,408	2,461	2,451
40,48	3,881	3,758	3,626	4,73	2,353	2,340	2,324
29,76	3,163	3,126	3,077	4,13	2,343	—	2,312
19,23	2,718	2,708	2,696	0	2,234	2,218	2,198

x_2	ε at t °C [181]				x_2	ε at t °C [181]			
	25	35	45	55		25	35	45	55
100	10,08	9,035	8,104	7,339	26,29	2,962	2,930	2,897	2,864
78,80	7,613	6,758	6,136	5,595	19,66	2,689	2,673	2,656	2,637
57,17	5,071	4,749	9,466	4,224	15,69	2,574	2,562	2,548	2,538
41,74	3,787	3,660	3,546	3,440	10,35	2,451	2,447	2,428	2,413
33,44	3,286	3,225	3,165	3,105	4,07	2,313	2,295	2,276	2,258

Carbon tetrachloride (CCl_4) – Quinoline (C_7H_7N)
[166], $t = 20$ °C

x_2	ε	x_2	ε	x_2	ε	x_2	ε
100	9,293	67,462	7,218	24,647	4,124	0	2,242
85,452	9,911	53,670	6,308	12,050	3,121	—	—
83,162	8,277	38,850	5,207	3,217	2,473	—	—

Carbon tetrachloride (CCl_4) – 1–Indanone (C_9H_8O)

x_2	ε at t °C [211]		x_2	ε at t °C [211]	
	20	40		20	40
61,52	—	4,990	14,25	2,775	2,695
34,72	3,870	3,650	8,64	2,540	2,510
21,45	3,110	3,005	4,51	2,400	2,375

Carbon tetrachloride (CCl_4) – Tetrahydro–2–naphthol ($C_{10}H_{12}O$)

x_2	ε at t °C [211]		x_2	ε at t °C [111]	
	20	40		20	40
100	11,70	9,570	20,50	3,170	2,795
77,10	8,815	7,450	14,28	2,832	2,565
54,03	6,115	5,430	8,92	2,600	2,505
38,12	4,495	4,210	3,32	2,380	2,360
29,15	3,727	3,568	2,19	2,338	2,308

Carbon tetrachloride (CCl_4) – Thymol ($C_{10}H_{14}O$)

[212], $t = 20$ °C, $\lambda = 66$ cm

x_2	100	80	60	40	20	0
ε	4,08	3,96	3,75	3,41	2,85	2,2

Carbon tetrachloride (CCl_4) – Dodecan-1-ol ($C_{12}H_{26}O$)

[213], $t = 20$ °C

x_2	ε	x_2	ε	x_2	ε	x_2	ε
95,17	5,75	76,36	4,66	36,73	3,16	10,31	2,44
90,39	5,50	73,11	4,60	30,94	2,97	8,77	2,41
88,62	5,42	65,56	4,23	26,20	2,81	6,05	2,36
84,82	5,19	61,35	4,16	22,49	2,70	3,31	2,28
83,67	5,12	49,17	3,57	16,77	2,56	0	2,23
78,79	4,90	43,57	3,47	12,90	2,49	—	—

Carbon tetrachloride (CCl_4) – Tributyl phosphate ($C_{12}H_{27}O_4P$)

[214], $t = 20$ °C

x_2	100	73,72	49,27	37,32	32,77	26,48	17,49	0
ε	8,13	7,27	6,38	5,61	5,23	4,75	3,77	2,238

Carbon disulphide (CS_2) – Chloroform ($CHCl_3$)

[170], $t = 20$ °C

w_2	100	79,361	65,557	28,696	0
x_2	100	70,976	51,518	20,940	0
ε	5,132	3,761	3,552	3,160	2,579

[215], $t = 18$ °C

w_2	100	79,94	70,10	48,91	35,19	18,90	10,43	0
ε	4,863	4,214	3,951	3,475	3,114	2,856	2,726	2,578

[172]

w_2	100	91,60	67,29	61,05	49,78	33,71	6,73	0
t °C	17,5	19,0	19,0	19,0	19,1	19,0	19,0	22,2
ε	5,105	4,804	4,031	3,850	3,581	3,204	2,810	2,741

[177, 241], $t=25$ °C, $\nu=1$ MHz

φ_2	100	80	60	40	20	0
ε	4,79	4,21	3,71	3,28	2,89	2,58

Carbon disulphide (CS_2) – 1, 2–Dibromoethane ($C_2H_4Br_2$)

[172]

w_2	100	90,92	71,16	68,0	50,56	34,65	12,10	0
t °C	22,7	22,7	22,7	23,8	23	23	22,5	18,3
ε	4,991	4,462	3,680	3,596	3,220	2,958	2,745	2,691

Carbon disulphide (CS_2) – 1, 2–Dichloroethane ($C_2H_4Cl_2$)

[190], $t=25$ °C

x_2	100	87,6	77,9	60,7	40,7	24,1	15,2	0
ε	10,073	8,839	7,892	6,413	4,865	3,774	3,298	2,601

Carbon disulphide (CS_2) – Ethanol (C_2H_6O)

x_2	ε at t °C [182]		x_2	ε at t °C [182]		x_2	ε at t °C [182]	
	6	30		6	30		6	30
81,505	23,47	20,25	26,714	6,965	6,002	3,841	2,8619	2,7991
53,036	15,25	12,73	21,187	5,590	4,9303	1,917	2,7569	2,7014
45,796	12,57	10,627	16,955	4,6892	4,2452	0,948	2,7177	2,6587
38,223	10,203	8,460	10,695	3,5958	3,3819	0	2,6766	2,6124
33,185	8,676	7,246	5,888	3,0257	2,9348	—	—	—

[215], $t=18$ °C

w_2	100	76,035	52,567	26,378	0
ε	26,8	22,56	17,14	9,130	2,598

Carbon disulphide (CS_2) – Acetone (C_3H_6O)

t=20 °C [198]				t=30 °C [198]			
x_1	ε	x_1	ε	x_1	ε	x_1	ε
0	20,90	51,50	11,332	0	19,96	51,55	10,91
4,67	20,11	60,80	9,563	4,34	19,26	60,86	9,186
10,09	19,13	72,55	7,324	9,63	18,35	71,37	7,320
16,08	18,036	80,78	5,812	14,73	17,469	81,06	5,683
19,71	17,405	90,93	4,097	18,37	18,827	91,41	4,043
31,02	15,257	95,30	3,391	29,88	14,759	95,41	3,308
41,54	13,18	100	2,643	41,35	12,657	100	2,619

Carbon disulphide (CS_2) – Diethyl ether ($C_4H_{10}O$)
[170], $t = 20$ °C

x_2	100	63,380	34,969	13,407	0
w_2	100	62,759	34,367	13,100	0
ε	4,261	3,697	3,334	2,842	2,579

[172]

w_2	100	91,36	67,24	49,02	31,81	12,12	0
t °C	21	17,1	17,5	17,7	17,9	17,8	18,3
ε	4,531	4,462	4,074	3,700	3,369	2,947	2,691

Carbon disulphide (CS_2) – Cyclopentanone (C_5H_8O)

t=20 °C [198]				t=30 °C [198]			
x_1	ε	x_1	ε	x_1	ε	x_1	ε
0	13,60	51,66	9,062	0	13,17	50,56	8,994
4,43	13,28	61,26	7,975	4,06	12,88	61,55	7,795
8,6	12,97	70,81	6,792	8,36	12,57	70,88	6,483
14,42	12,53	80,27	5,600	15,16	12,08	81,02	5,277
20,40	12,04	90,30	4,209	20,04	11,70	90,77	4,010
29,01	11,31	95,84	3,363	29,77	10,92	95,90	3,286
40,19	10,25	100	2,643	39,65	10,04	100	2,619

Carbon disulphide (CS_2) – Bromobenzene (C_6H_5Br)
[172]

w_2	100	89,86	65,46	50,76	48,28	32,99	13,74	0
t °C	20,0	20,5	20	20	20	20	19,9	22,4
ε	5,820	5,469	4,571	4,070	4,038	3,542	2,990	2,744

Carbon disulphide (CS_2) – Chlorobenzene (C_6H_5Cl)
[216], $t = 25$ °C, ν = 1 MHz

x_2	100	75	50	25	10	0
ε	5,610	5,094	4,449	3,627	3,087	2,633

Carbon disulphide (CS_2) – 2-Chlorophenol (C_6H_5ClO)
[177], $t = 25$ °C, ν = 1 MHz

x_2	100	80	60	40	20	0
ε	6,21	5,08	4,14	3,42	2,90	2,58

Carbon disulphide (CS_2) – Nitrobenzene ($C_6H_5NO_2$)

[216—218], $t = 25$ °C, $\nu = 1$ MHz

x_2	100	75	50	25	10	5	0
ε	36,10	24,60	17,34	9,810	6,060	4,350	2,633

Carbon disulphide (CS_2) – Benzene (C_6H_6)

[196, 216], $t = 25$ °C

x_1	0	25	50	75	100
ε	2,285	2,358	2,441	2,531	2,633

[173], $t = 18$ °C

w_1	0	40,041	48,023	59,993	70,266	79,712	89,979	100
ε	2,28	2,370	2,436	2,493	2,530	2,585	2,615	2,669

[174]

$w_1 = 0$		$w_1 = 20$		$w_1 = 40$		$w_1 = 60$		$w_1 = 80$		$w_1 = 100$	
t °C	ε	t °C	ε	t °C	ε	°C	ε	t °C	ε	t °C	ε
14	2,2927	10,5	2,3175	13,5	2,331	12,5	2,396	12,8	2,505	0,5	2,617
21,5	2,2814	16,5	2,310	22,8	2,317	26,7	2,373	21	2,494	11,5	2,610
31,2	2,2688	21,4	2,304	32,5	2,298	35	2,360	30,9	2,474	14,6	2,604
40,4	2,2555	31,4	2,285	42,3	2,283	41,7	2,348	40	2,453	21,5	2,590
50,3	2,2425	41,1	2,2632	—	—	—	—	—	—	30,0	2,573
61,2	2,2280	—	—	—	—						
69,1	2,2178	—	—	—	—						

[219], $t = 11$ °C

w_2	100	80	60	40	20	0
ε	2,6739	2,5950	2,5041	2,4396	2,3713	2,301

Carbon disulphide (CS_2) – Phenol (C_6H_6O)

[216], $t = 20$ °C, $\nu = 1$ MHz

x_1	90,2	93,1	95,5	96,9	98,4	99,6	100
ε	3,246	3,040	2,868	2,794	2,714	2,646	2,633

Carbon disulphide (CS_4) – Hexane (C_6H_{14})
[220], $t=25$ °C

x_1	0	14,41	69,31	72,70	91,22	97,25	100
ε	1,908	1,969	2,241	2,277	2,496	2,591	2,640

[216], $t=25$ °C, $\nu=1$ MHz

x_1	0	25	50	75	100
ε	1,904	2,003	2,126	2,319	2,633

[177], $t=25$ °C, $\nu=1$ MHz

φ_2	100	80	60	40	20	0
ε	1,89	2,02	2,12	2,28	2,42	2,58

Carbon disulphide (CS_2) – Toluene (C_7H_8)
[170], $t=20$ °C

x_1	0	41,301	66,395	93,918	100
w_1	0	36,758	62,008	92,731	100
ε	2,365	2,379	2,422	2,487	2,579

[172]

w_1	0	11,69	37,49	51,73	72,32	91,72	100
t °C	19,6	20,0	20,0	20,0	20,0	19,8	18,3
ε	2,385	2,357	2,465	2,544	2,558	2,618	2,691

Carbon disulphide (CS_2) – Heptane (C_7H_{16})
[221], $t=20$ °C

x_1	ε	x_1	ε	x_1	ε	x_1	ε
10,14	1,973	33,75	2,041	59,60	2,190	94,82	1,558
15,23	1,990	36,90	2,050	69,79	2,268	100	2,63
20,97	2,010	49,95	2,128	83,95	2,413	—	—

[172]

w_1	0	11,55	37,94	51,18	60,33	67,69	90,70	100
t °C	22,2	22,1	22,5	22,6	22,5	22,5	22,5	22,4
ε	1,992	2,039	2,138	2,203	2,205	2,207	2,507	2,744

Carbon disulphide (CS_2) – *m*–Xylene (C_8H_{10})
[171], $t = 15$ °C

φ_2	100	60,3	48,5	41,4	0
ε	2,61	2,54	2,50	2,48	2,37

Carbon disulphide (CS_2) – Camphor ($C_{10}H_{16}O$)
[159]

x_2	25,3	20,1	15,5	10,74	7,24	0
t °C	21,5	24	21	22	22,5	19
ε	5,82	5,0	3,65	3,0	2,85	2,65

Carbon disulphide (CS_2) – Menthol ($C_{10}H_{20}O$)
[159]

x_2	28,56	22,1	16,06	11,5	5,4	0
t °C	24	24	23,5	24	24	19
ε	4,7	4,2	3,8	3,25	2,8	2,65

Carbon disulphide (CS_2) – Diphenyl ether ($C_{12}H_{10}O$)
[177], $t = 25$ °C

φ_1	0	20	40	60	80	100
ε	4,21	3,79	3,45	3,13	2,84	2,58

Bromoform ($CHBr_3$) – Acetone (C_3H_6O)
[222], $t = 25$ °C

x_1	0	4,9	73,2	87,7	94,8	100
ε	21,17	20,30	8,8	6,52	5,34	4,39

[166], $t = 20$ °C

x_2	100	95,086	26,782	12,255	5,189	0
ε	21,17	20,30	8,802	6,519	5,321	4,385

Bromoform ($CHBr_3$) – Diethyl ether ($C_4H_{10}O$)

t=20 °C [166]				t=25 °C [222]			
x_2	ε	x_2	ε	x_1	ε	x_1	ε
100	4,335	19,056	5,016	0	4,33	80,9	5,02
95,559	4,474	10,146	4,750	4,4	4,97	89,8	4,75
85,888	4,826	3,311	4,495	14,1	4,83	96,6	4,49
66,627	5,301	0	4,385	33,4	5,30	100	4,38
44,145	5,441	—	—	65,8	5,44	—	—

Bromoform ($CHBr_3$) – Benzene (C_6H_6)
[223], ν=0,5 MHz

x_1	ε при t °C			x_1	ε при t °C		
	10	40	70		10	40	70
4,09	2,308	2,315	2,234	23,40	2,726	2,623	2,522
5,73	2,411	2,340	2,258	50,47	3,232	3,075	2,924
7,18	2,438	2,365	2,284	73,58	3,726	3,506	3,315
9,29	2,477	2,399	2,313	100	4,404	4,084	3,816

Bromoform ($CHBr_3$) – Cyclohexane (C_6H_{12})
[166], t=20 °C

x_1	0	5,026	1,145	38,967	68,373	100
ε	2,032	2,103	2,287	2,680	3,367	4,385

Bromoform ($CHBr_3$) – Diisopropyl ether ($C_6H_{14}O$)
[166], t=20 °C

x_2	ε	x_2	ε	x_2	ε	x_2	ε
100	3,976	61,709	5,215	22,803	5,133	0	4,385
88,658	4,441	46,743	5,379	12,234	4,863	—	—
77,011	4,816	36,415	5,370	6,527	4,646	—	—

Bromoform ($CHBr_3$) – Quinoline (C_9H_7N)

t=20 °C [166]				t=25 °C [222]			
x_2	ε	x_2	ε	x_2	ε	x_2	ε
100	9,293	34,593	7,738	100	4,38	48,3	8,12
93,979	9,054	20,107	6,384	96,6	4,79	33,5	8,87
80,604	8,859	9,453	5,407	90,5	5,41	19,4	8,86
66,448	8,571	3,334	4,792	79,9	6,38	6,2	9,05
51,647	8,121	0	4,385	65,4	7,74	0	9,29

Chloroform ($CHCl_4$) – Methanol (CH_4O)
[175], t=20 °C, ν=1,8 MHz

φ_2	100	80	60	40	20	0
ε	33,60	28,20	22,35	15,85	9,45	4,80

Chloroform ($CHCl_3$) – 1, 1, 2, 2–Tetrachloroethane ($C_2H_2Cl_4$)
[175], t=20 °C, ν=1,8 MHz

φ_2	100	80	60	40	20	0
ε	8,50	7,52	6,72	5,38	5,37	4,80

Chloroform ($CHCl_3$) – 1, 2–Dibromoethane ($C_2H_4Br_2$)
[166], t=20 °C

x_2	ε	x_2	ε	x_2	ε	x_2	ε
100	4,827	65,005	4,824	30,811	4,808	0	4,816
91,637	4,827	53,730	4,819	18,289	4,804	—	—
79,133	4,827	41,438	4,812	7,600	4,812	—	—

[172]

w_1	0	10,41	26,29	38,98	58,32	85,51	100
t °C	22,7	22,0	22,3	22,4	22,5	22,3	17,5
ε	4,991	4,795	4,891	4,820	4,915	4,904	5,105

Chloroform ($CHCl_3$) – 1, 2–Dichloroethane ($C_2H_4Cl_2$)
[175], t=20 °C, ν=1,8 MHz

φ_2	100	80	60	40	20	0
ε	10,65	9,20	7,90	6,72	5,68	4,80

Chloroform ($CHCl_3$) – Acetic acid ($C_2H_4O_2$)

[225], $t = 25$ °C, $\nu = 5$ MHz

w_2	ε	w_2	ε	w_2	ε	w_2	ε
100	6,21	43,41	4,68	18,30	4,47	5,41	4,60
77,41	5,49	30,78	4,51	15,08	4,48	1,89	4,68
61,46	5,06	19,06	4,47	11,92	4,50	0	4,72

[224], $t = 25$ °C, $\nu = 28$ MHz

φ_2	100	81,46	60,91	42,01	21,10	0
ε	4,67	4,48	4,49	4,79	5,36	6,20

Chloroform ($CHCl_3$) – Ethanol (C_2H_6O)

[194], $t = 25$ °C

x_2	100	92,52	80,49	67,34	52,94	37,07	19,56	0
ε	24,69	22,90	19,82	16,14	12,46	9,36	6,65	4,80

[226], $t = 18$ °C

w_2	100	66,02	39,20	23,42	0
ε	26,09	21,45	15,09	10,99	4,927

Chloroform ($CHCl_3$) – Acetone (C_3H_6O)

$t=25$ °C [222]				$t=30$ °C [229]				$t=25$ °C [166]			
x_1	ε	x_1	ε	x_1	ε	x_1	ε	x_2	ε	x_2	ε
0	21,17	64,8	11,28	0	20,37	54,17	13,07	100	21,17	31,636	11,28
14,8	18,96	82,3	8,73	21,16	17,65	57,52	12,60	97,295	20,52	17,734	8,732
27,5	17,17	96,7	5,61	40,17	15,04	70,37	10,41	94,718	20,17	7,148	6,448
46,7	14,49	100	4,81	42,35	14,67	100	4,665	85,192	18,95	3,432	5,615
								72,434	17,17	0	4,813
								53,306	14,49	—	—

[230], $t = 25$ °C

x_2	100	86,10	71,86	57,36	42,15	26,73	10,84	0
ε	20,87	18,94	17,02	15,04	12,90	10,31	7,18	4,80

w_2	x_2	ε at t °C [228]					
		20	10	0	−10	−20	−30
100	100	21,50	22,47	23,74	25,08	26,47	28,13
66,42	80,30	18,83	19,76	20,78	21,92	23,06	24,12
41,11	58,99	17,06	17,47	18,17	18,40	19,40	20,47
24,22	39,71	13,42	13,93	14,66	15,35	16,17	17,05
10,54	19,53	9,64	9,83	10,35	11,03	11,83	12,54
0	0	5,02	5,86	5,93	5,98	6,12	6,34

[175, 185], $t=20$ °C, $\nu=1,8$ MHz

φ_2	100	80	60	40	20	0
ε	21,07	18,40	15,80	13,00	9,52	4,80

[227], $t=17$ °C, $\nu=3$ MHz

φ_2	100	80	60	40	20	0
ε	20,89	18,32	15,67	12,84	9,37	4,85

Chloroform ($CHCl_3$) – Methyl acetate ($C_3H_6O_2$) *)

t=10 °C [229]		t=20 °C [229]		t=20 °C [229]		t=30 °C [229]		t=30 °C [229]	
x_2	ε	x_2	ε	x_2	ε	x_2	ε	x_2	ε
100,0	7,420	100	7,092	34,01	6,589	100	6,785	51,35	6,578
66,12	7,258	78,88	7,018	20,73	6,072	72,47	6,718	43,58	6,479
40,12	6,889	62,21	6,948	20,67	6,061	60,19	6,692	32.82	6,221
33,17	6,750	56,38	6,901	0	4,810	58,27	6,665	0	4,665
28,81	6,624								
0,00	4,989								

*) See also [165] at 10–30°C.

Chloroform ($CHCl_3$) – bis-2-Chloroethyl) ether ($C_4H_8Cl_2O$) [166], $t=20$ °C

x_2	ε	x_2	ε	x_2	ε
100	21,17	47,960	13,32	5,181	5,716
88,257	19,39	36,1 6	11,75	0	4,816
77,905	17,99	19,333	8,386	—	—
62,776	15,75	11,160	6,792	—	—

Chloroform ($CHCl_3$) – Dioxan ($C_4H_8O_2$)
[175, 166, 388], $t = 20$ °C, $\nu = 1,8$ MHz

φ_2	100	80	60	40	20	0
ε	2,22	2,72	3,16	3,58	4,08	4,08

Chloroform ($CHCl_3$) – Ethyl acetate ($C_4H_8O_2$)
[199], $t = 25$ °C, $\nu = 45$ MHz

x_1	ε	x_1	ε	x_1	ε	x_1	ε
10,9	6,16	42,2	6,325	57,6	6,26	75,3	5,92
19,8	6,23	46,1	6,325	60,5	6,22	80,3	5,77
26,8	6,29	49,3	6,32	63,6	6,17	86,0	5,58
32,8	6,32	52,3	6,32	67,1	6,12	92,3	5,23
37,9	6,325	55,0	6,29	71,0	6,02	100	4,719

Chloroform ($CHCl_3$) – Diethyl ether ($C_4H_{10}O$) *)

x_1	ε at t °C [233]			x_1	ε at t °C [233]		
	5	20	26		5	20	26
0	5,646	4,857	4,540	60	6,907	5,645	5,091
20	7,100	5,627	5,048	80	6,061	5,202	4,682
40	7,228	5,940	5,289	100	4,930	4,362	4,137

x_1	ε at t °C [234]					
	20	0	−20	−40	−60	−80
0	4,785	5,172	5,601	6,095	6,722	—
9,06	5,265	5,750	6,345	6,991	6,802	—
18,85	5,600	6,175	6,865	7,705	8,765	—
33,7	—	6,585	—	—	9,720	11,40
40,0	5,983	6,704	7,578	8,615	9,915	11,515
44,96	6,009	6,745	7,645	8,695	10,055	11,755
50,0	5,981	6,705	7,596	8,620	9,965	11,635
60,01	5,762	6,465	7,292	8,278	9,533	10,05
83,94	4,975	5,495	6,121	6,882	7,822	9,035
91,78	4,630	5,121	5,695	6,405	7,305	8,44
100	4,300	4,706	5,180	5,791	6,545	7,541

*) See also [165], $t = 10 - 30$ °C, $\lambda = 1 - 1,5$ m [172, 175, 177, 222, 232].

t=20 °C [231]		t=25 °C [166]			
x_2	ε	x_2	ε	x_2	ε
100	4,344	100	4,335	44,819	5,835
80	5,211	96,063	4,502	39,321	5,835
60	5,940	90,082	4,742	27,792	5,732
50	6,187	79,885	5,124	13,528	5,376
40	6,280	64,215	5,527	5,387	5,093
20	5,919	57,176	5,693	1,685	4,919
0	4,942	—	—	0	4,813

[215], t=18 °C

w_2	100	82,94	62,69	45,97	28,88	9,67	4,60	0
ε	4,354	4,814	5,315	5,757	5,912	5,478	5,155	4,801

Chloroform ($CHCl_3$) – Butylamine ($C_4H_{11}N$)
[175], t=20 °C, ν=1,8 MHz

φ_2	100	80	60	40	20	0
ε	4,78	5,84	6,44	6,27	5,68	4,80

Chloroform ($CHCl_3$) – Pyridine (C_5H_5N)
[235], ν=1,75 MHz

φ_2	ε at t °C				φ_2	ε at t °C			
	10	20	30	40		10	20	30	40
100	13,8	13,2	12,6	12,1	30	9,28	8,84	8,42	8,03
90	13,4	12,9	12,3	11,8	10	6,63	6,33	6,04	5,76
70	12,3	11,7	11,8	10,7	0	5,02	4,85	4,66	4,50
50	11,1	10,6	10,1	9,61	—	—	—	—	—

Chloroform ($CHCl_3$) – 1–Pentene (C_5H_{10})
[172]

w_1	0	17,65	33,55	53,70	63,01	70,12	90,61	100
t °C	21,2	21	21,2	21,3	21,2	21,2	21,2	17,5
ε	2,068	2,259	2,531	2,926	3,220	3,394	4,337	5,105

Chloroform ($CHCl_3$) – Pentane (C_5H_{12})
[172]

w_1	0	12,42	45,34	58,91	62,34	71,28	91,93	100
t °C	20,6	21,7	21,7	21,7	21,7	21,7	21,7	18,4
ε	1,836	2,044	2,528	2,873	2,960	3,281	4,532	5,352

Chloroform ($CHCl_3$) – Chlorobenzene (C_6H_5Cl)
[175], $t = 20$ °C, $\nu = 1,8$ *Мгц*

φ_2	100	80	60	40	20	0
ε	5,70	5,54	5,38	5,20	5,02	4,80

Chloroform ($CHCl_3$) – Nitrobenzene ($C_6H_5NO_2$)
[236], $t = 25$ °C

x_2	100	90,7182	85,1090	73,5052
ε	34,890	32,250	30,673	27,372

Chloroform ($CHCl_3$) – Benzene (C_6H_6)

$t=10$ °C [229]				$t=20$ °C [229]				$t=30$ °C [229]			
x_1	ε	x_1	ε	x_1	ε	x_1	ε	x_1	ε	x_1	ε
0,00	2,300	46,32	3,298	0,00	2,285	53,38	3,415	0,00	2,270	49,68	3,546
32,12	3,001	53,64	3,472	24,48	2,758	59,90	3,604	22,50	2,705	67,45	3,724
36,10	3,058	72,03	3,961	41,09	3,071	71,69	3,998	38,50	3,006	96,35	4,261
40,26	3,169	100	4,989	51,84	3,373	100	4,810	45,22	3,188	100	4,665

[209], $t = 25$ °C, $\nu = 1$ MHz

x_1	0	10	20	50	70	100
ε	2,282	2,460	2,639	3,246	3,739	4,770

[172]

w_1	0	6,91	33,96	48,45	60,47	66,92	85,20	100
t °C	18,0	19,0	19,0	19,0	18,9	18,8	18,5	17,5
ε	2,288	2,443	2,912	2,207	3,450	3,613	4,304	5,105

φ_1	ε at t °C [50, 175, 237]				
	10	20	30	40	50
0	2,308	2,284	2,262	2,240	2,218
25	2,83	2,79	2,77	2,71	2,67
50	3,45	3,37	3,32	3,21	3,18
75	4,05	4,02	3,87	3,76	3,70
100	4,97	4,81	4,64	4,48	4,34
—	—	—	—	—	—

Chloroform ($CHCl_3$) – **Aniline** (C_6H_7N)
[175], $t = 20$ °C, $\nu = 1,8$ MHz

φ_2	100	80	60	40	20	0
ε	7,06	6,38	5,83	5,38	5,04	4,80

Chloroform ($CHCl_3$) – **Cyclohexanone** ($C_6H_{10}O$)
[175], $t = 20$ °C, $\nu = 1,8$ *Мгц*

φ_2	100	80	60	40	20	0
ε	15,70	14,62	13,35	11,62	8,75	4,80

Chloroform ($CHCl_3$) – **Cyclohexane** (C_6H_{12})
[166], $t = 20$ °C

x_1	0,000	8,127	13,658	18,364	19,226	56,846	78,515	100,02
ε	2,032	2,163	2,255	2,324	2,338	3,270	3,959	4,816

Chloroform ($CHCl_3$) – **Hexane** (C_6H_{14})

x_1	ε at t °C [238], ν=0,5 MHz							
	−90	−80	−70	−60	−50	−40	−30	−20
5,17	2,201	2,177	2,155	2,133	2,113	2,092	2,072	2,053
12,77	2,409	2,371	2,335	2,302	2,271	2,241	2,212	2,186
28,62	—	2,848	2,786	2,726	2,668	2,612	2,560	2,512
49,53	—	3,710	3,584	3,469	3,363	3,265	3,174	3,090
68,56	—	4,785	4,582	4,400	4,234	4,076	3,934	3,802
78,03	—	5,465	5,210	5,984	5,778	5,586	5,406	5,256
100	—	—	—	6,770	6,402	6,100	5,825	5,580

x_1	ε at t °C [238]							
	−10	0	10	20	30	40	50	60
5,17	2,033	2,014	1,995	1,976	1,957	1,938	1,919	1,900
12,17	2,159	2,134	2,109	2,086	2,062	2,038	2,012	1,984
28,62	2,466	2,426	2,387	2,348	2,310	2,270	2,230	2,189
49,53	3,010	2,937	2,867	2,800	2,736	2,670	2,606	2,544
68,56	3,680	3,568	3,462	3,360	3,266	3,173	3,076	2,974
78,03	5,108	3,972	3,843	3,720	3,606	3,496	3,394	3,295
100	5,356	5,150	4,960	4,783	4,614	4,450	2,292	4,140

Chloroform ($CHCl_3$) – Diisopropyl ether ($C_6H_{14}O$)
[166], $t = 20$ °C

x_2	ε	x_2	ε	x_2	ε	x_2	ε
100	3,976	65,894	5,449	17,349	5,657	0	4,916
90,944	4,385	43,925	6,001	10,017	5,361	—	—
82,041	4,787	28,517	5,945	3,961	5,020	—	—

Chloroform ($CHCl_3$) – Dipropylamine (C_6H_5N)
[175], $t = 20$ °C, $\nu = 1,8$ MHz

φ_2	100	80	60	40	20	0
ε	3,24	4,59	5,34	5,46	5,25	4,80

Chloroform ($CHCl_3$) – Triethylamine ($C_6H_{15}N$)
[175], $t = 20$ °C, $\nu = 1,8$ MHz

φ_2	100	80	60	40	20	0
ε	2,64	4,10	5,05	5,31	5,16	4,80

Chloroform ($CHCl_3$) – Benzonitrile (C_7H_5N)
[175], $t = 20$ °C, $\nu = 1,8$ MHz

φ_2	100	80	60	40	20	0
ε	25,65	22,30	18,65	14,73	10,20	4,80

Chloroform ($CHCl_3$) – 3-Nitrotoluene ($C_7H_7NO_2$)
[239], $t = 25$ °C, $\nu = 10$ MHz

x_1	ε	x_1	ε	x_1	ε	x_1	ε
0	26,39	30,90	21,36	63,72	15,38	91,90	7,66
12,31	24,63	40,59	20,04	73,50	13,02	100	4,77
21,79	23,17	50,98	18,06	83,45	10,31	—	—

Chloroform ($CHCl_3$) – 4-Nitrotoluene ($C_7H_7NO_2$)
[239], $t = 25$ °C, $\nu = 10$ MHz

x_1	0	11,88	20,53	32,91	42,95	73,35	91,59	100
ε	26,52	24,47	22,91	20,65	18,70	12,09	7,30	4,77

Chloroform ($CHCl_3$) – Toluene (C_7H_8)
[236], $t = 25$ °C

x_1	0	14,0614	16,6891	22,5595
ε	2,3721	2,5997	2,6438	2,7471

Chloroform ($CHCl_3$) – Heptane (C_7H_{16})
[168], $t = 20$ °C

x_1	61,25	70,30	82,50	90,28	100
ε	2,979	3,333	4,000	4,477	5,18

[172]

w_1	0	12,92	32,38	54,38	55,24	65,73	91,25	100
t °C	22,2	19,9	20,0	20,2	20,2	22,4	20,5	17,5
ε	1,992	2,134	2,419	2,800	2,830	3,219	4,476	5,105

[240], $t = 20$ °C

φ_1	9,149	16,28	23,95	39,27	40,52	47,21	100
ε	2,055	2,140	2,253	2,450	2,480	2,606	5,18

Chloroform ($CHCl_3$) – Ethyl pentyl ether ($C_7H_{16}O$)
[172]

w_2	100	89,40	64,85	49,93	32,98	32,73	10,55	0
t °C	23,0	23,3	23,3	23,4	23,4	23,5	23,5	23,3
ε	3,655	3,889	4,636	5,053	5,313	5,313	5,244	4,931

Chloroform ($CHCl_3$) – Phenetole ($C_8H_{10}O$)
[175], $t=20$ °C, $\nu=1,8$ MHz

φ_2	100	80	60	40	20	0
ε	4,25	4,45	4,61	4,73	4,80	4,80

Chloroform ($CHCl_3$) – Quinoline (C_9H_7N)
[166], $t=20$ °C, fig. 6

x_2	ε	x_2	ε	x_2	ε	x_2	ε
100	9,293	65,313	9,090	19,265	7,054	0	4,815
93,982	9,274	49,00	8,722	9,184	6,023	—	—
82,762	9,217	31,852	7,972	3,406	5,293	—	—

Dibromomethane (CH_2Br_2) – Denzene (C_6H_6)
[242], $t=21$ °C

x_1	0	4,02	7,50	10,47	13,98	16,70	19,50
ε	2,26	2,50	2,63	2,86	3,05	3,10	3,21

x_1	ε at t °C [223]			x_1	ε at t °C [223]		
	10	40	70		10	40	70
4,96	2,460	2,383	2,293	56,18	4,420	4,088	3,841
9,47	2,596	2,502	2,401	81,14	5,884	5,335	4,848
20,38	2,957	2,815	2,676	100	7,772	6,678	—

Dichloromethane (CH_2Cl_2) – Benzene (C_6H_6)
[242], $t=21$ °C

x_1	0	6,21	15,62	22,77	29,73	41,03
ε	2,26	2,50	2,95	3,22	3,49	4,61

Di-iodomethane (CH_2I_2) – Benzene (C_6H_6)
[223, 242], ν=0,5 MHz

x_1	ε at t °C		x_1	ε at t °C	
	25	50		25	50
2,68	2,341	2,282	13,37	2,592	2,524
4,49	2,384	2,322	19,43	2,734	2,646
6,36	2,425	2,359	100	5,316	—

Bromomethane (CH_3Br) – Hexane (C_6H_{14})
[179]

x_1=10,7				x_1=19,8				x_1=26,0			
t °C	ε	t °C	ε	t °C	ε	t °C	ε	t °C	ε	t °C	ε
37,6	2,145	—35,2	2,347	24,8	2,422	—31,6	2,653	17,0	2,61	—49,3	2,983
27,0	2,170	—44,6	2,378	16,7	2,454	—41,5	2,700	5,1	2,676	—58,9	3,048
16,7	2,200	—57,2	2,420	7,2	2,484	—53,3	2,760	—0,2	2,702	—68,7	3,114
15,9	2,202	—67,7	2,460	2,7	2,503	—62,9	2,813	—4,0	2,723	—81,0	3,206
6,0	2,228	—80,0	2,506	—4,1	2,531	—73,6	2,873	—15,5	2,786	—86,0	3,246
—3,9	2,255	—84,7	2,525	—10,6	2,558	—79,4	2,908	—27,2	2,850	—90,1	3,280
—14,5	2,283	—90,1	2,550	—20,9	2,603	—85,5	2,946	—38,2	2,913	—	—
—22,7	2,309										

Chloromethane (CH_3Cl) – Hexane (C_6H_{14})
[179]

x_1=0				x_1=12,0				x_1=18,3			
t °C	ε	t °C	ε	t °C	ε	t °C	ε	t °C	ε	t °C	ε
42,2	1,880	—17,3	1,961	16,7	2,226	—39,9	2,405	—1,7	2,490	—53,6	2,740
36,9	1,884	—29,5	1,978	9,0	2,250	—53,4	2,456	—13,9	2,544	—70,1	2,830
35,1	1,891	—46,1	1,999	1,2	2,273	—62,1	2,490	—27,4	2,683	—80,5	2,896
32,0	1,891	—56,9	2,014	—5,4	2,293	—66,8	2,508	—43,1	2,683	—93,2	2,973
30,3	1,897	—58,5	2,016	—14,0	2,321	—74,7	2,541				
23,4	1,904	—66,1	2,028	—26,2	2,357	—83,0	2,588				
17,0	1,913	—70,4	2,035								
16,1	1,917	—86,6	2,058								
12,8	1,918										

Iodomethane (CH_3I) – Benzene (C_6H_{14})
[179]

$x_1=10,25$				$x_1=16,8$			
t °C	ε	t °C	ε	t °C	ε	t °C	ε
18,0	2,133	—19,4	2,234	22,2	2,277	—36,6	2,470
12,8	2,146	—31,7	2,266	17,6	2,289	—49,0	2,515
7,4	2,158	—46,2	2,305	13,4	2,303	—62,9	2,573
5,9	2,168	—77,7	2,400	5,1	2,330	—72,2	2,613
—1,9	2,181	—	—	—3,8	2,358	—81,9	2,658
—10,6	2,204	—	—	—24,1	2,426	—87,1	2,682

Formamide (CH_3NO) – Dioxan ($C_4H_8O_2$)
[243], $t=25$ °C, $\nu=1,8$ MHz

x_2	w_2	ε	x_2	w_2	ε	x_2	w_2	ε
100	100	2,209	43,44	60	33,02	17,97	30	68,17
82,14	90	6,850	34,35	50	43,17	14,56	25	74,67
67,18	80	13,41	25,42	40	55,38	11,33	20	81,42
54,39	70	21,77	21,59	35	61,54	0	0	109,5

Nitromethane (CH_3NO_2) – Methanol (CH_4O)
[175], $t=20$ °C, $\nu=1,8$ MHz

φ_2	100	80	60	40	20	0
ε	33,60	33,45	33,35	34,20	35,20	37,45

Nitromethane (CH_3NO_2) – Acetonitrile (C_2H_3N)
[175], $t=20$ °C, $\nu=1,8$ MHz

φ_1	0	20	40	60	80	100
ε	36,80	37,08	37,20	37,32	37,41	37,45

Nitromethane (CH_3NO_2) – 1, 1, 2, 2–Tetrachloroethane ($C_2H_2Cl_4$)
[175], $t=20$ °C, $\nu=1,8$ MHz

φ_1	0	20	40	60	80	100
ε	8,50	15,35	21,20	26,65	31,95	37,45

Nitromethane (CH_3NO_2) – 1, 2–Dichloroethane ($C_2H_4Cl_2$)
[175], $t=20$ °C, $\nu=1{,}8$ MHz

φ_1	0	20	40	60	80	100
ε	10,65	16,65	22,13	27,60	32,43	37,45

Nitromethane (CH_3NO_2) – Acetic acid ($C_2H_4O_2$)
[244], $t=25$ °C, $\nu=1{,}8$ MHz

φ_1	0	20,09	39,87	59,91	79,54	100
ε	6,23	11,21	16,29	22,18	28,77	37,78

Nitromethane (CH_3NO_2) – Ethanol (C_6H_6O)
[175], $t=20$ °C, $\nu=1{,}8$ MHz

φ_2	100	80	60	40	20	0
ε	25,07	26,15	27,70	36,60	33,05	37,45

Nitromethane (CH_3NO_2) – Acetone (C_3H_6O)
[175], $t=20$ °C, $\nu=1{,}8$ MHz

φ_1	0	20	40	60	80	100
ε	21,07	24,25	27,45	30,70	34,05	37,45

Nitromethane (CH_3NO_2) – Propionic acid ($C_3H_6O_2$)
[224], $t=25$ °C, $\nu=28$ MHz

φ_1	0	19,65	39,00	60,46	80,55	100
ε	3,20	8,12	13,39	20,11	27,08	37,78

Nitromethane (CH_3NO_2) – Propan–1–ol (C_3H_8O)
[175], $t=20$ °C, $\nu=1{,}8$ MHz

φ_2	100	80	60	40	20	0
ε	20,65	21,90	24,30	27,65	31,95	37,45

Nitromethane (CH_3NO_2) – Acetic anhydride ($C_4H_6O_3$)
[245], $t = 25$ °C

x_1	ε	x_1	ε	x_1	ε	x_1	ε
0	22,06	42,9	28,0	72,4	30,8	94,1	36,6
16,3	24,0	53,8	28,6	80,3	32,7	100	38,0
30,4	25,7	63,7	29,4	87,3	35,5	—	—

[175], $t = 20$ °C, $\nu = 1,8$ MHz

φ_1	0	20	40	60	80	100
ε	22,45	25,30	28,25	31,25	34,25	37,45

Нитрометан (CH_3NO_2)—n-Диоксан (C_4H_8O)
[245], $t = 25$ °C

φ_2	100	85,0	71,6	59,6	48,5	38,6	29,6	21,3	13,6	6,6	0
ε	2,2	5,6	8,5	12,0	14,8	18,9	22,8	26,8	29,1	32,0	38,0

Nitromethane (CH_3NO_2) – Diethyl ether ($C_4H_{10}O$)
[175], $t = 20$ °C

φ_1	0	20	40	60	80	100
ε	4,65	10,07	17,05	23,65	30,60	37,45

Nitromethane (CH_3NO_2) – Nitrobenzene ($C_6H_5NO_2$)
[175], $t = 20$ °C, $\nu = 1,8$ MHz

φ_2	100	80	60	40	20	0
ε	35,75	35,85	36,08	36,45	36,86	37,45

Nitromethane (CH_3NO_2) – Benzene (C_6H_6)
[175], $t = 20$ °C, $\nu = 1,8$ MHz

φ_1	0	20	40	60	80	100
ε	2,28	7,22	13,7	20,10	28,45	37,45

Nitromethane (CH_3NO_2) – Benzonitrile (C_7H_5N)
[175], $t = 25$ °C, $\nu = 1{,}8$ MHz

φ_1	0	20	40	60	80	100
ε	25,65	27,50	29,70	38,05	34,63	37,45

Nitromethane (CH_3NO_2) – Benzaldehyde (C_7H_6O)
[175], $t = 20$ °C, $\nu = 1{,}8$ MHz

φ_1	0	20	40	60	80	100
ε	17,85	21,62	25,57	29,55	33,50	37,45

Nitromethane (CH_3NO_2) – Toluene (C_7H_8)
[244], $t = 25$ °C, $\nu = 28$ MHz

φ_2	100	85,97	57,70	30,05	17,38	0
ε	2,38	5,53	23,67	23,60	29,20	37,78

Methanol (CH_4O) – 1, 1, 2, 2–Tetrachloroethane ($C_2H_2Cl_4$)
[175], $t = 20$ °C, $\nu = 1{,}8$ MHz

φ_1	0	20	40	60	80	100
ε	8,50	13,08	18,67	24,54	29,20	33,60

Methanol (CH_4O) – Acetonitrile (C_2H_3N)
[247], $t = 25$ °C

x_1	ε	x_1	ε	x_1	ε	x_1	ε
0,00	35,95	43,90	34,98	84,90	33,75	100	32,62
5,10	35,84	67,67	34,39	89,77	33,51	—	—
26,89	35,38	78,23	34,05	96,77	32,98	—	—

x_2	ε at t °C [246]		x_2	ε at t °C [246]		x_2	ε at t °C [246]		x_2	ε at t °C [246]	
	25	30		25	30		25	30		25	30
100	36,69	35,93	87,3	36,40	35,58	43,8	35,37	34,46	8,4	33,40	32,44
96,3	36,52	35,75	75,6	36,07	35,25	33,7	34,94	34,04	0	32,63	31,64
91,8	36,52	35,65	60,2	35,75	34,90	21,0	34,26	33,27	—	—	—

[175], $t = 20$ °C, $\nu = 1,8$ MHz

φ_1	0	20	40	60	80	100
ε	36,80	36,15	35,50	34,85	34,20	33,60

Methanol (CH_4O) – 1, 2-Dichloroethane ($C_2H_4Cl_2$)

[175], $t = 20$ °C, $\nu = 1,8$ MHz

φ_1	0	20	40	60	80	100
ε	10,65	14,00	18,70	24,20	29,06	33,60

Methanol (CH_4O) – Ethanol (C_2H_6O)

[248], $t = 30$ °C, $\nu = 550$ kHz

w_2	90,0	70,0	50,0	30,0	10,0
ε	24,8	26,6	28,4	30,1	32,0

[249], $t = 15$ °C

w_2	ε	w_2	ε	w_2	ε	w_2	ε
100	25,02	70	25,71	40	27,59	10	31,91
90	25,36	60	26,20	30	28,54	0	34,05
80	25,34	50	26,78	20	30,21	—	—

[175], $t = 20$ °C, $\nu = 1,8$ MHz

φ_2	100	80	60	40	20	0
ε	25,07	26,70	28,50	30,10	31,85	33,60

Methanol (CH_4O) – Acetone (C_3H_6O) *)

[175], $t = 20$ °C, $\nu = 1,8$ *Мгц*

φ_1	0	20	40	60	80	100
ε	21,07	23,60	25,80	28,37	31,04	33,60

*) See also [250].

Methanol (CH_4O) – Dimethylformamide (C_3H_7NO)

[252], $\nu = 2$ MHz

x_2	ε at t °C			x_2	ε at t °C		
	10	25	55		10	25	55
100	40,42	37,70	32,76	49,60	41,03	38,12	33,12
89,29	40,69	37,91	33,05	29,90	40,26	37,30	32,38
69,07	41,16	38,33	33,33	10,69	38,35	35,36	30,39

Methanol (CH_4O) – Propan–1–ol (C_3H_8O)

[253], $t=25$ °C

w_2	100	80,15	61,10	40,45	19,53	0
ε	20,1	22,0	24,2	26,8	29,7	32,2

[175], $t=20$ °C, $\nu=1,8$ MHz

φ_2	100	80	60	40	20	0
ε	20,65	23,00	25,45	28,20	30,95	33,60

Methanol (CH_4O) – Propan–2–ol (C_3H_8O)

[248], $t=30$ °C, $\nu=550$ kHz

w_2	90,0	70,0	50,0	30,0	10,0
ε	20,3	22,8	25,4	28,3	31,2

[175], $t=20$ °C, $\nu=1,8$ MHz

φ_2	100	80	60	40	20	0
ε	19,90	22,50	25,15	28,00	30,80	33,60

Methanol (CH_4O) – Acetic anhydride ($C_4H_6O_3$)

[175], $t=20$ °C, $\nu=1,8$ MHz

φ_1	0	20	40	60	80	100
ε	22,45	24,40	26,55	28,95	31,27	33,60

Methanol (CH_4O) – 1, 4–Dioxan ($C_4H_8O_2$)

[248], $t=25$ °C, $\nu=550$ kHz

w_1	10,0	30,0	50,0	70,0	90,0
ε	3,90	9,17	15,9	23,0	30,2

[254], $t=25$ °C

w_2	65,71	60,07	50,93	40,15	0
ε	10,50	12,32	15,40	19,05	32,66

Methanol (CH_4O) – Butan-1-ol ($C_4H_{10}O$)
[175], $t=20$ °C, $\nu=1{,}8$ MHz

φ_2	100	80	60	40	20	0
ε	17,19	20,60	23,65	27,05	30,30	33,60

Methanol (CH_4O) – *tert*-Butyl alcohol ($C_4H_{10}O$)
[248], $t=30$ °C, $\nu=550$ kHz

w_1	10,0	20,0	50,0	70,0	90,0
ε	16,0	20,0	23,4	27,2	33,7

Methanol (CH_4O) – Diethyl ether ($C_4H_{10}O$)
[175], $t=20$ °C, $\nu=1{,}8$ MHz

φ_1	0	20	40	60	80	100
ε	4,35	9,45	15,25	21,60	27,75	33,60

Methanol (CH_4O) – Nitrobenzene ($C_6H_5NO_2$)
[255], $t=25$ °C, $\nu=60$ hHz

w_1	x_1	ε	w_1	x_1	ε	w_1	x_1	ε
0	0	34,72	27,16	58,89	31,18	77,47	92,96	29,9
1,98	7,19	33,64	44,36	75,39	30,71	94,55	98,52	30,4
8,44	26,16	32,61	61,39	85,93	30,1	100	100	30,6
19,29	47,88	31,71						

[175], $t=20$ °C, $\nu=1{,}8$ MHz

φ_1	0	20	40	60	80	100
ε	37,75	33,90	33,23	33,05	33,17	33,60

Methanol (CH_4O) – Benzene (C_6H_6) *)
[258], $t=25$ °C, $\nu=0{,}5$ MHz

x_1	ε	x_1	ε	x_1	ε	x_1	ε
0	2,27	49,7	9,20	70,0	16,33	95,5	29,51
14,0	3,67	50,8	9,54	77,5	19,69	100	32,65
35,3	6,41	60,1	12,24	84,4	22,95	—	—
36,2	6,70	61,3	12,85	89,5	26,30	—	—

*) See also [262, 259].

t=20 °C [261]								t=25 °C [260]			
x_1	ε	x_1	ε	x_1	ε	x_1	ε	x_1	ε	x_1	ε
2,48	2,357	14,16	2,940	26,31	4,158	44,42	7,246	22,12	3,8	81,45	20,7
4,34	2,425	17,11	3,167	27,86	4,363	44,68	7,306	42,41	7,1	89,71	26,0
5,79	2,483	21,25	3,573	33,02	5,141	49,81	8,387	53,50	10,0	98,28	31,1
8,01	2,580	24,60	3,957	39,67	6,297	—	—	63,75	13,2	—	—
11,48	2,765			42,38	6,788			71,99	16,4		

[175, 256], $t=20$ °C, $\nu=1,8$ MHz

φ_1	0	20	40	60	80	100
ε	2,28	6,35	12,75	19,90	26,95	33,60

Methanol (CH_4O) – Aniline (C_6H_7N)

[175], $t=20$ °C, $\nu=1,8$ MHz

φ_1	0	20	40	60	80	100
ε	7,06	11,06	16,45	22,17	27,75	33,60

Methanol (CH_4O) – *o*–Xylene (C_6H_8)

[185], $t=18$ °C, $\lambda=110$ cm

x_1	0	10	30	40	50	60	70	90	100
ε	2,3	2,8	3,8	5,6	7,7	9,6	13,9	24,2	33,0

Methanol (CH_4O) - Cyclohexane (C_6H_{12})

[182, 256]

t=30 °C				t=40 °C				t=46 °C			
x_1	ε	x_1	ε	x_1	ε	x_1	ε	x_1	ε	x_1	ε
0,000	2,004	81,052	18,41	0	1,989	10,843	2,266	0	1,979	42,393	5,318
1,470	2,031	84,044	20,71	1,529	2,014	16,178	2,528	1,652	2,005	53,113	7,478
3,032	2,057	86,750	22,36	3,391	2,044	74,800	15,73	4,114	2,041	62,200	10,18
5,027	2,099	89,636	24,48	5,398	2,090	89,825	23,9	6,660	2,109	72,067	14,41
6,475	2,138	92,466	27,14	6,162	2,108	100	30,6	13,055	2,346	85,837	21,29
7,991	2,187	95,574	28,81					22,100	2,900	93,187	26,87
10,020	2,265	100,00	30,9					32,393	3,955	—	—

Methanol (CH_4O) – Benzonitrile (C_7H_5O)
[175], $t=20$ °C, $\nu=1{,}8$ MHz

φ_1	0	20	40	60	80	100
ε	25,65	27,45	29,30	31,05	32,45	32,60

Methanol (CH_4O) – Octan-1-ol ($C_8H_{18}O$)
[175], $t=20$ °C, $\nu=1{,}8$ MHz

φ_2	100	80	60	40	20	0
ε	9,95	14,50	18,85	23,70	28,60	33,60

Methanol (CH_4O) – Dibutyl sebacate ($C_{18}H_{34}O_4$)
[263], $t=20$ °C, $\nu=10$ MHz

φ_1	x_1	ε	φ_1	x_1	ε	φ_1	x_1	ε	φ_1	x_1	ε
0	0	4,56	49,0	10,41	6,71	83,8	38,5	13,95	93,1	61,8	20,89
12,5	1,70	4,85	54,9	12,85	7,27	85,9	42,5	15,19	94,6	67,9	22,75
19,0	2,75	5,07	57,0	13,8	7,49	86,4	43,5	15,45	96,3	76,1	25,28
26,8	4,24	5,34	64,3	17,9	8,55	87,7	46,3	16,31	97,9	85,1	28,16
30,3	5,00	5,51	71,5	23,3	9,92	88,7	48,6	18,95	99,1	92,9	30,56
33,9	5,84	5,66	76,5	28,7	11,44	90,1	52,6	18,09	100	100	32,68
37,9	6,86	5,85	80,3	33,0	12,66	90,9	54,6	18,74	—	—	—
43,5	8,50	6,22	80,9	33,8	12,71	91,9	57,7	19,64	—	—	—

Thiourea (CH_4N_2S) – Benzene (C_6H_6)
[212], $t=20$ °C, $\lambda=66$ cm

x_1	0	5	10	25	50	75	100
ε	2,28	4,32	6,08	11,29	14,81	10,6	6,15

Tetrachloroethylene (C_2Cl_4) – Acetone (C_3H_6O)
[198]

$t=-20$ °C				$t=30$ °C			
x_1	ε	x_1	ε	x_1	ε	x_1	ε
0	20,90	51,21	8,55	0	19,96	50,00	8,499
4,49	19,56	60,68	6,913	5,29	18,47	60,57	6,544
10,42	17,88	70,84	5,438	9,69	17,26	70,56	5,229
14,90	16,67	81,56	4,082	15,03	15,88	80,99	4,035
20,05	15,34	91,60	3,028	20,64	14,48	89,71	3,231
30,85	12,75	96,80	2,559	30,51	12,24	95,17	2,621
40,22	10,72	100	2,276	40,73	10,18	100	2,264

Hexachloroethane (C_2Cl_6) – Diethyl ether ($C_4H_{10}O$)
[166], $t=20$ °C

x_2	100	95,206	86,405	83,599
ε	4,335	4,241	4,048	3,998

Pentochloroethane (C_2HCl_5) – Diethyl ether ($C_4H_{10}O$)
[166], $t=20$ °C

x_2	ε	x_2	ε	x_2	ε	x_2	ε
100,0	4,335	76,321	4,868	30,591	4,652	0	3,833
93,847	4,503	63,244	4,980	19,518	4,398	—	—
86,063	4,697	45,035	4,883	9,078	4,120	—	—

Pentachloroethane (C_2HCl_5) – Benzene (C_6H_6)
[166], $t=20$ °C

x_1	0	5,873	14,082	24,753	46,450	100
ε	2,282	2,370	2,496	2,660	2,973	3,833

Pentachloroethane (C_2HCl_5) – Cyclohexane (C_6H_{12})
[166], $t=20$ °C

x_1	0	6,664	22,539	35,410	58,643	80,917	100
ε	2,032	2,105	2,334	2,513	2,951	3,376	3,883

Trifluoroacetic acid ($C_2HO_2F_3$) – Chlorobenzene (C_6H_5Cl)
[186], $t=25$ °C, fig. 7

x_1	0	12,73	23,19	80,31	90,75	100
ε	5,61	5,36	5,32	5,61	6,41	8,26

Trifluoroacetic acid ($C_2HO_2F_3$) – Benzene (C_6H_6)
[186], $t=25$ °C, fig. 7

w_1	ε	w_1	ε	w_1	ε	w_1	ε
0	2,28	40,05	3,62	69,49	5,28	100	8,26
9,86	2,60	49,83	4,15	79,38	5,83	—	—
19,88	2,98	59,89	4,57	89,44	6,48	—	—

cis–1, 2–dibromoethylene ($C_2H_2Br_2$) – trans–1, 2–dibromoethylene ($C_2H_2Br_2$)
[264], $t = 25$ °C, $\nu = 1,17$ MHz

w_1	0,00	17,57	25,09	49,45	63,04	75,44	85,72	100,0
ε	2,467	3,020	3,451	4,515	5,088	5,669	6,248	6,962

Dichloroacetic acid ($C_2H_2O_2Cl_2$) – Acetic acid ($C_2H_4O_2$)
[175], $t = 20$ °C, $\nu = 1,8$ MHz

φ_1	0	20	40	60	80	100
ε	6,22	9,12	10,76	11,35	10,75	8,70

Dichloroacetic acid ($C_2H_2O_2Cl_2$) – Acetone (C_3H_6O)
[175], $t = 20$ °C, $\nu = 1,8$ MHz

φ_1	0	20	40	60	80	100
ε	21,07	21,65	22,92	22,90	18,10	8,70

Dichloroacetic acid ($C_2H_2O_2Cl_2$) – 1, 4–Dioxan ($C_4H_8O_2$)
[175], $t = 20$ °C, $\nu = 1,8$ MHz

φ_1	0	20	40	60	80	100
ε	2,22	4,69	7,34	9,70	11,31	8,70

Dichloroacetic acid ($C_2H_2O_2Cl_2$) – Diethyl ether ($C_4H_{10}O$)
[175], $t = 20$ °C, $\nu = 1,8$ MHz

φ_1	0	20	40	60	80	100
ε	4,35	8,93	12,82	14,51	13,15	8,70

1, 1, 2, 2–Tetrachloroethane ($C_2H_2Cl_4$) –
1, 2–Dichloroethane ($C_2H_4Cl_2$)
[175], $t = 20$ °C, $\nu = 1,8$ MHz

φ_1	0	20	40	60	80	100
ε	10,65	10,22	9,78	9,36	8,93	8,50

1, 1, 2, 2–Tetrachloroethane ($C_2H_2Cl_4$) – Acetone (C_3H_6O)
[175], $t = 20$ °C, $\nu = 1{,}8$ MHz

φ_2	100	80	60	40	20	0
ε	21,07	20,30	18,95	16,60	13,10	8,50

1, 1, 2, 2–Tetrachloroethane ($C_2H_2Cl_4$) – Acetic anhydride ($C_4H_6O_3$)
[175], $t = 20$ °C, $\nu = 1{,}8$ MHz

φ_2	100	80	60	40	20	0
ε	22,45	21,25	19,83	17,55	13,90	8,50

1, 1, 2, 2–Tetrachloroethane ($C_2H_2Cl_4$) – Diethyl ether ($C_4H_{10}O$)
[175], $t = 20$ °C, $\nu = 1{,}8$ MHz

φ_2	100	80	60	40	20	0
ε	4,34	5,96	7,21	8,01	8,40	8,50

1, 1, 2, 2–Tetrachloroethane ($C_2H_2Cl_4$) – 1, 2, 4–Trichlorobenzene ($C_6H_3Cl_3$)
[175], $t = 20$ °C, $\nu = 1{,}8$ MHz

φ_2	100	80	60	40	20	0
ε	6,75	6,95	7,20	7,55	7,95	8,50

1, 1, 2, 2–Tetrachloroethane ($C_2H_2Cl_4$) – Nitrobenzene ($C_6H_5NO_2$)
[175], $t = 20$ °C, $\nu = 1{,}8$ MHz

φ_2	100	80	60	40	20	0
ε	37,75	30,65	25,50	20,05	14,52	8,50

1, 1, 2, 2–Tetrachloroethane ($C_2H_2Cl_4$) – Benzene (C_6H_6)
[175], $t = 20$ °C, $\nu = 1{,}8$ MHz

φ_1	0	20	40	60	80	100
ε	2,28	3,04	3,94	4,94	6,38	8,50

1, 1, 2, 2–Tetrachloroethane ($C_2H_2Cl_4$) – Aniline (C_6H_7N)
[175], $t = 20$ °C, $\nu = 1,8$ MHz

φ_2	100	80	60	40	20	0
ε	7,06	6,81	6,70	6,80	7,34	8,50

1, 1, 2, 2–Tetrachloroethane ($C_2H_2Cl_4$) – Benzonitrile (C_7H_5N)
[175], $t = 20$ °C, $\nu = 1,8$ MHz

φ_2	100	80	60	40	20	0
ε	25,65	23,80	21,25	17,90	13,70	8,50

1, 1, 1–Trichloroethane ($C_2H_3Cl_3$) – Cyclohexane (C_6H_{12})
[166], $t = 20$ °C

x_1	0,000	6,838	12,759	25,053	100
ε	2,032	2,250	2,464	2,947	7,518

1, 1, 1–Trichloroethane ($C_2H_3Cl_3$) – Diethyl ether ($C_4H_{10}O$)
[166], $t = 20$ °C

x_2	100,0	93,111	86,276	73,730	60,925	49,056	0,000
ε	4,335	4,560	4,826	5,273	5,701	6,071	7,518

1, 1, 1–Trichloroethane ($C_2H_3Cl_3$) – Dodecan–1–ol ($C_{12}H_{26}O$)
[213], $t = 20$ °C

x_2	ε	x_2	ε	x_2	ε	x_2	ε	x_2	ε
96,09	5,95	73,82	5,65	43,96	5,46	20,84	5,95	7,74	6,71
91,64	5,86	65,99	5,53	41,03	5,49	19,42	6,01	4,87	6,97
87,41	5,81	59,74	5,47	34,40	5,56	15,59	6,13	2,16	7,24
83,19	5,74	54,85	5,47	28,78	5,71	12,55	6,32	0	7,47
78,59	5,67	50,17	5,45	23,89	5,80	9,93	6,52	—	—

Acetonitrile (C_2H_3N) – 1, 2–Dichloroethane ($C_2H_4Cl_2$)
[175], $t = 20$ °C, $\nu = 1,8$ MHz

φ_1	0	20	40	60	80	100
ε	10,65	17,40	23,15	28,20	32,60	36,80

Acetonitrile (C_2H_3N) – Acetic acid ($C_2H_4O_2$)

[245], $t=25$ °C

x_1	0	10,7	21,3	41,9	52,0	61,9	71,6	81,2	90,7	100
ε	6,3	13,9	19,9	26,0	27,7	28,9	30,1	31,8	33,8	36,05

Acetonitrile (C_2H_3N) – Bromoethane (C_2H_5Br)

[175], $t=20$ °C, $\nu=1,8$ MHz

φ_1	0	20	40	60	80	100
ε	9,50	14,96	20,37	25,90	31,45	36,80

Acetonitrile (C_2H_3N) – Ethanol (C_2H_6O)

[175], $t=20$ °C, $\nu=1,8$ MHz

φ_2	100	80	60	40	20	0
ε	25,07	27,18	29,20	31,40	33,90	36,80

Acetonitrile (C_2H_3N) – Acetone (C_3H_6O)

[175], $t=20$ °C, $\nu=1,8$ MHz

φ_1	0	20	40	60	80	100
ε	21,07	24,55	27,65	30,72	33,75	36,80

Acetonitrile (C_2H_3N) – Propan-1-ol (C_3H_8O)

[175], $t=20$ °C, $\nu=1,8$ MHz

φ_2	100	80	60	40	20	0
ε	20,65	22,92	25,70	29,05	32,73	36,80

Acetonitrile (C_2H_3N) – Acetic anhydride ($C_4H_6O_3$)

[245], $t=25$ °C

x_1	0	16,6	30,9	43,4	54,4	64,2	72,9	80,7	100
ε	22,0	24,0	26,7	28,0	29,3	30,4	31,4	32,2	36,05

[175], $t=20$ °C, $\nu=1,8$ MHz

φ_1	0	20	40	60	80	100
ε	22,45	25,45	28,44	31,30	34,07	36,80

Acetonitrile (C_2H_3N) – 1, 4–Dioxan ($C_4H_8O_2$)
[265], $t=25$ °C

w_1	4,37	6,96	10,95	15,28	25,05
ε	3,84	4,84	6,34	7,90	12,11

[254], $t=25$ °C

w_2	59,92	49,49	39,68	30,31	0
ε	17,03	20,45	23,80	26,85	36,01

[175], $t=20$ °C, $\nu=1,8$ MHz

φ_1	0	20	40	60	80	100
ε	2,22	8,54	15,19	22,23	29,49	36,80

Acetonitrile (C_2H_3N) – Ethyl acetate ($C_4H_8O_2$)
[175], $t=20$ °C, $\nu=1,8$ MHz

φ_1	0	20	40	60	80	100
ε	6,19	11,90	17,70	23,85	30,25	36,80

Acetonitrile (C_2H_3N) – Diethyl ether ($C_4H_8O_2$)
[175], $t=20$ °C, $\nu=1,8$ MHz

φ_1	0	20	40	60	80	100
ε	4,35	10,95	17,53	24,34	30,55	36,80

Acetonitrile (C_2H_3N) – Butan–1–ol ($C_4H_{10}O$)
[175], $t=20$ °C, $\nu=1,8$ MHz

φ_2	100	80	60	40	20	0
ε	17,90	20,25	23,72	27,75	32,15	36,80

Acetonitrile (C_2H_3N) – 1, 2–Dichlorobenzene ($C_6H_4Cl_2$)
[245], $t=25$ °C

x_1	0	47,7	68,1	76,2	89,5	95,0	100,0
ε	10,1	19,6	25,8	28,3	30,5	33,5	36,05

Acetonitrile (C_2H_3N) – Chlorobenzene (C_6H_5Cl)
[175], $t=20$ °C, $\nu=1,8$ MHz

φ_1	0	20	40	60	80	100
ε	5,70	11,63	17,54	23,90	30,35	36,80

Acetonitrile (C_2H_3N) – Nitrobenzene ($C_6H_5NO_2$)
[175], $t=20$ °C, $\nu=1,8$ MHz

φ_2	100	80	60	40	20	0
ε	37,75	36,23	36,48	36,63	36,73	36,80

Acetonitrile (C_2H_3N) – Benzene (C_6H_6)
[175], $t=20$ °C, $\nu=1,8$ MHz

φ_1	0	20	40	60	80	100
ε	2,28	8,08	14,50	21,55	29,60	36,80

Acetonitrile (C_2H_3N) – Cyclohexanone ($C_6H_{10}O$)
[175], $t=20$ °C, $\nu=1,8$ MHz

φ_1	0	20	40	60	80	100
ε	15,70	19,95	24,15	28,40	32,60	36,80

Acetonitrile (C_2H_3N) – Triethylamine ($C_6H_{15}N$)
[175], $t=20$ °C, $\nu=1,8$ MHz

φ_2	100	80	60	40	20	0
ε	2,64	8,60	15,15	22,25	29,55	36,80

Acetonitrile (C_2H_3N) – Dipropylamine ($C_6H_{15}N$)
[175], $t=20$ °C, $\nu=1,8$ MHz

φ_2	100	80	60	40	20	0
ν	3,24	9,05	16,05	22,90	30,10	36,80

Acetonitrile (C_2H_3N) – Benzonitrile (C_7H_5N)
[175], $t=20$ °C, $\nu=1,8$ MHz

φ_2	100	80	60	40	20	0
ε	25,65	28,00	30,27	32,45	34,60	36,80

Acetonitrile (C_2H_3N) – α–Benzyl cyanide (C_8H_7N)
[175], $t = 25$ °C, $\nu = 1{,}8$ MHz

φ_2	100	80	60	40	20	0
ε	18,95	22,48	26,07	29,65	33,25	36,80

Acetonitrile (C_2H_3N) – Menthol ($C_{10}H_{20}O$)
[159]

x_2	30,23	25,2	20,2	12,7	7,55	0,0
t °C	23	23,5	23,5	23,5	23	21
ε	19,5	22,5	24,0	29,0	32,3	36,5

1–Bromo–2–chloroethane (C_2H_4BrCl) – Heptane (C_7H_{16})

x_1	ε at t °C [266], ν=0,5 MHz							
	−50	−30	−10	10	30	50	70	90
5,30	2,082	2,059	2,032	2,004	1,978	1,941	1,905	1,872
7,05	2,102	2,076	2,052	2,023	1,999	1,964	1,925	1,893
10,95	—	2,126	2,098	2,069	2,042	2,008	1,974	1,937
100	—	—	7,98	7,41	6,92	6,47	6,08	5,69

1, 2–Dibromoethane ($C_2H_4Br_2$) – Diethyl ether ($C_4H_{10}O$)
[172]

w_2	100	89,21	63,71	50,12	35,15	8,66	0
t °C	16,8	22,2	22,3	22,1	22,0	21,15	23,0
ε	4,410	4,428	4,632	4,715	4,836	5,019	5,010

1, 2–Dibromoethane ($C_2H_4Br_2$) – 1–Pentene (C_5H_{10})
[172]

w_1	0	13,44	38,47	50,79	64,38	72,28	93,00	100
t °C	21,2	21	21,45	21,4	21,3	21,5	21,5	22,8
ε	2,068	2,121	2,309	2,486	2,690	2,952	4,112	4,912

1, 2–Dibromoethane ($C_2H_4Br_2$) – Benzene (C_6H_6)
[172]

w_1	0	8,93	39,98	50,37	68,35	70,65	89,58	100
t °C	23	23	23	23	23,2	23,2	23,2	22,7
ε	2,284	2,335	2,810	2,905	3,327	3,400	4,131	4,991

1, 2–Dibromoethane ($C_2H_4Br_2$) – Hexane (C_6H_{14})
[177], $t=25$ °C, $\nu=1$ MHz

φ_1	0	20	40	60	80	100
ε	1,89	2,22	2,66	3,21	3,88	4,63

1, 2–Dibromoethane ($C_2H_4Br_2$) – Heptane (C_7H_{16})
[172]

w_1	0	11,9	39,88	52,0	65,23	66,30	88,79	100
t °C	22,2	22,6	22,7	22,8	22,9	22,9	23,0	22,8
ε	1,992	2,108	2,301	2,444	2,690	2,729	3,721	4,912

1, 2–Dichloroethane ($C_2H_4Cl_2$) – Acetic acid ($C_2H_4O_2$)
[224], $t=25$ °C, $\nu=28$ MHz

φ_1	0	20,64	40,58	60,32	79,61	100
ε	6,20	6,27	6,57	7,31	8,48	10,36

1, 2–Dichloroethane ($C_2H_4Cl_2$) – Ethanol (C_2H_6O)
[175], $t=20$ °C, $\nu=1,8$ MHz

φ_2	100	80	60	40	20	0
ε	25,07	21,85	18,30	14,73	12,10	10,65

1, 2–Dichloroethane ($C_2H_4Cl_2$) – Acetone (C_3H_6O)
[175], $t=20$ °C, $\nu=1,8$ MHz

φ_2	100	80	60	40	20	0
ε	21,07	19,80	18,12	16,05	13,50	10,65

1, 2–Dichloroethane ($C_2H_4Cl_2$) – Diethyl ether ($C_4H_{10}O$)
[267], $t=20$ °C, $\nu=3$ MHz

x_2	100	91,9	83,8	75,3	66,3	54,4	0
ε	4,35	4,69	5,05	5,50	5,98	6,61	10,61

[175], $t=20$ °C, $\nu=1,8$ MHz

φ_2	100	80	60	40	20	0
ε	4,35	5,53	6,70	7,94	9,24	10,65

1, 2–Dichloroethane ($C_2H_4Cl_2$) – Chlorobenzene (C_6H_5Cl)
[244], $t = 25$ °C, $\nu = 28$ MHz

φ_1	0	23,50	33,92	50,04	81,21	100
ε	5,61	6,75	7,30	7,96	9,47	10,36

[175], $t = 20$ °C, $\nu = 1,8$ MHz

φ_2	100	80	60	40	20	0
ε	5,70	6,31	7,10	8,04	9,20	10,65

1, 2–Dichloroethane ($C_2H_4Cl_2$) – Benzene (C_6H_6)
[255], $t = 25$ °C

x_1	11,3	21,1	34,2	44,2	60,5	71,9	85,7	100
ε	2,791	3,318	3,879	4,707	5,914	6,940	8,455	10,365

[268, 242], $t = 25$ °C						[190], $t = 25$ °C			
x_1	λ [м]	ε	x_1	λ [м]	ε	x_1	ε	x_1	ε
0	6,800	2,274	60,5	10,969	5,914	0	2,273	51,0	5,084
11,3	7,533	2,791	71,9	11,881	6,940	11,7	2,803	61,3	5,863
21,1	8,215	3,318	85,7	13,115	8,455	22,6	3,345	72,1	6,801
34,2	8,8825	3,879	100,0	14,522	10,365	32,6	3,897	81,5	7,779
44,6	9,786	4,707				41,9	4,468	100	10,07

[175], $t = 20$ °C, $\nu = 1,8$ MHz

φ_1	0	20	40	60	80	100
ε	2,28	3,43	4,70	6,28	8,44	10,65

1, 2–Dichloroethane ($C_2H_4Cl_2$) – Aniline (C_6H_7N)
[232], $t = 20$ °C

x_1	0,00	10,0	30,0	50,0	70,0	90,0	100,0
ε	7,200	7,351	7,879	8,258	8,902	9,810	10,401

1, 2–Dichloroethane ($C_2H_4Cl_2$) – Cyclohexane (C_6H_{12})
[190], $t = 25$ °C

x_1	ε	x_1	ε	x_1	ε	x_1	ε
0	2,018	35,0	3,163	64,9	5,239	88,8	8,163
10,1	2,246	47,5	3,863	74,3	6,213	100,0	10,073
23,2	2,655	54,8	4,379	82,6	7,249	—	—

1, 2–Dichloroethane ($C_2H_4Cl_2$) – Hexanoic acid ($C_6H_{12}O_2$)
[224], $t = 25$ °C, $\nu = 28$ MHz

φ_1	0	15,72	35,56	50,85	74,83	100
ε	2,82	3,46	4,67	5,69	7,88	10,36

1, 2–Dichloroethane ($C_2H_4Cl_2$) – Hexane (C_6H_{14})
[269], $t = 25$ °C, $\nu = 3$ MHz

x_1	0	6,265	21,360	42,87	61,49	79,67	100,0
ε	1,8776	1,9915	2,3412	3,1964	4,4695	6,5097	10,376

1, 2–Dichloroethane ($C_2H_4Cl_2$) – Benzonitrile (C_7H_5N)
[175], $t = 20$ °C, $\nu = 1,8$ MHz

φ_2	100	80	60	40	20	0
ε	25,65	23,52	20,85	17,80	14,40	10,65

1, 2–Dichloroethane ($C_2H_4Cl_2$) – Pyridinium acetate ($C_7H_9NO_2$)
[270], $t = 25$ °C

x_2	ε	x_2	ε	x_2	ε	x_2	ε
100	16,0	79,51	15,4	41,43	13,5	16,85	11,9
95,04	15,9	70,73	15,0	34,27	13,1	13,26	11,5
91,30	15,8	60,47	14,6	26,92	12,6	—	—
85,74	15,6	50,48	14,1	21,78	12,3	—	—

1, 2–Dichloroethane ($C_2H_4Cl_2$) – Heptane (C_7H_{16})

x_1	ε at t °C [266], $\nu = 0,5$ MHz								
	−90	−70	−50	−30	−10	10	30	50	70
0,00	2,075	2,048	2,020	1,992	1,966	1,936	1,910	1,879	1,848
3,22	—	2,099	2,072	2,050	2,021	1,991	1,963	1,926	1,888
5,08	—	2,135	2,107	2,079	2,050	2,018	1,998	1,951	1,913
6,98	—	2,172	2,146	2,116	2,087	2,054	2,021	1,982	1,938
8,78	—	—	2,178	2,148	2,117	2,084	2,050	2,008	1,971
50,24	—	—	—	—	3,824	3,613	3,415	3,241	3,092
100,0	—	—	—	14,17	12,43	11,03	9,87	8,87	8,02

1, 2–Dichloroethane ($C_2H_4Cl_2$) – Anilinium acetate ($C_8H_{11}NO_2$)
[270], $t = 25$ °C

x_2	ε	x_2	ε	x_2	ε	x_2	ε
100	8,21	68,85	8,62	35,27	9,30	12,41	9,92
90,24	8,32	55,76	8,89	27,13	9,52	5,61	10,24
82,98	8,41	45,63	9,06	19,06	9,73	0	10,36

Acetic acid ($C_2H_4O_2$) – Ethanol (C_2H_6O)
[271], $t = 25$ °C, $\nu = 1$ kHz

10^{-3} N	0	1	5	15	20	50	100
ε	24,37	24,36	24,57	25,10	24,99	25,37	26,24

Acetic acid ($C_2H_4O_2$) – Acetone (C_3H_6O)
[175], $t = 20$ °C, $\nu = 1,8$ MHz

φ_1	0	20	40	60	80	100
ε	21,07	18,25	15,82	13,25	10,30	6,22

[191], $t = 20$ °C

φ_1	0	12,5	25	37,5	50	62,5	75	87,5	100
ε	21,5	26,0	30,5	33,5	31,0	26,5	21,5	14,5	7,1

Acetic acid ($C_2H_4O_2$) – Propionic acid ($C_3H_6O_2$) *)
[224], $t = 25$ °C, $\nu = 28$ MHz

φ_2	100	80,47	60,7	39,98	19,29	0
ε	3,20	3,54	4,01	4,60	5,34	6,20

*) See also [272].

Acetic acid ($C_2H_4O_2$) – *n*–Dioxan ($C_4H_8O_2$)

$t = 25$ °C [245]								$t = 20$ °C [231]			
x_2	ε	x_2	ε	x_2	ε	x_2	ε	x_2	ε	x_2	ε
100	2,20	61,0	3,60	30,9	5,6	6,8	6,2	100	2,220	40	4,925
85,0	2,46	50,1	4,10	22,4	6,2	0	6,2	80	3,002	20	5,735
72,8	2,85	40,2	5,10	14,4	6,2	—	—	60	4,032	0	6,421
—	—	—	—	—	—	—	—	50	4,483	—	—

[272], $t = 25$ °C

x_1	ε	x_1	ε	x_1	ε	x_1	ε
0,00	2,209	59,9	4,88	81,7	6,03	93,1	6,22
27,2	2,76	73,6	5,67	85,7	6,13	96,7	6,24
44,6	3,71	78,2	5,90	89,4	6,20	100,0	6,24

Acetic acid ($C_2H_4O_2$) – Butyric acid ($C_4H_8O_2$)

[224], $t = 25$ °C, $\nu = 28$ MHz

φ_2	100	80,36	63,24	40,91	19,84	0
ε	2,88	3,08	3,47	4,19	5,05	6,20

Acetic acid ($C_2H_4O_2$) – Diethyl ether ($C_4H_{10}O$)

[273], $\nu = 0,5$ MHz

x_1	ε at t °C				x_1	ε at t °C			
	0	10	20	30		0	10	20	30
0	4,746	4,499	4,296	4,113	35,49	5,489	5,275	5,072	4,880
3,39	4,817	4,608	4,402	4,197	52,33	5,793	5,606	5,428	5,268
6,73	4,869	4,671	4,462	4,276	71,98	6,28	6,12	5,97	5,81
10,39	4,966	4,749	4,542	4,348	100,0	—	6,07	6,13	6,20
12,69	5,015	4,797	4,593	4,406	—	—	—	—	—

[175], $t = 20$ °C, $\nu = 1,8$ MHz

φ_1	0	20	40	60	80	100
ε	4,35	5,01	5,61	6,06	6,34	6,22

Acetic acid ($C_2H_4O_2$) – Pyridine (C_5H_5N)

[274], $t = 25$ °C, fig. 8 8

φ_2	100	88,0	61,7	42,8	24,3	13,1	0
ε	12,91	13,0	16,0	21,2	16,0	10,5	6,18

Acetic acid ($C_2H_4O_2$) – Chlorobenzene (C_6H_5Cl)

[224], $t = 25$ °C, $\nu = 28$ MHz

φ_2	100	79,79	59,84	49,58	20,27	0
ε	5,61	5,18	4,94	4,94	5,43	6,20

Acetic acid ($C_2H_4O_2$) – Benzene (C_6H_6), fig. 9

x_1	ε at t °C [273, 192]							
	0	10	20	30	40	50	60	70
0	2,336	2,315	2,295	2,274	2,253	2,232	2,210	2,188
9,78	2,375	2,368	2,340	2,322	2,303	2,285	2,266	2,246
23,05	2,464	2,449	2,444	2,418	2,403	2,388	2,372	2,357
43,72	2,706	2,700	2,693	2,687	2,679	2,672	2,663	2,655
61,34	3,113	3,118	3,124	3,130	3,133	3,132	3,129	3,126
79,12	3,956	3,981	4,007	4,032	4,052	4,067	4,078	4,084
100	—	6,074	6,13	6,20	6,27	6,36	6,47	6,60

[275, 276], $t = 22$ °C

x_1	ε	x_1	ε	x_1	ε	x_1	ε
13,00	2,370	22,84	2,455	50,19	3,070	93,07	7,745
16,37	2,397	28,11	2,519	66,22	3,969	—	—
19,43	2,420	37,56	2,713	87,36	6,684	—	—

Acetic acid ($C_2H_4O_2$) – Aniline (C_6H_7N)
[274], $t = 25$ °C

φ_2	100	89,3	64,6	50	39,3	11,5	0
ε	6,77	7,4	8,2	10,6	10,2	7,7	6,18

Acetic acid ($C_2H_4O_2$) – Cyclohexane (C_6H_{12})
[272], $t = 25$ °C

x_1	0,00	25,1	38,8	65,6	78,0	85,2	88,4	94,5	97,4	100
ε	2,015	2,11	2,18	2,39	2,78	3,37	3,82	4,81	5,38	6,2

Acetic acid ($C_2H_4O_2$) – *N*–Methylaniline (C_7H_9N)
[274], $t = 25$ °C

φ_2	100	88	69,5	50	48,2	28,6	9,1	0
ε	6,06	7,8	15,5	19,9	19,6	17,6	11,0	6,18

Acetic acid ($C_2H_4O_2$) – Quinoline (C_9H_7N)
[274], $t = 25$ °C

φ_2	100	90	73,2	50	34,2	20	10	0
ε	8,95	9,7	13,1	22,2	19,5	16,0	12,1	6,18

Acetic acid ($C_2H_4O_2$) – Diethylaniline ($C_{10}H_{15}N$)
[274], $t = 25$ °C

φ_2	100	87,9	75,8	60	40	12	0
ε	5,00	7,6	14,3	20,7	20,0	14,4	6,18

Bromoethane (C_2H_5Br) – Acetone (C_3H_6O)
[175], $t = 20$ °C, $\nu = 1,8$ MHz

φ_2	100	80	60	40	20	0
ε	21,07	18,75	16,43	14,11	11,80	9,50

Bromoethane (C_2H_5Br) – Acetic anhydride ($C_4H_6O_3$)
[175], $t = 20$ °C, $\nu = 1,8$ MHz

φ_2	100	80	60	40	20	0
ε	22,45	19,88	17,28	14,70	12,10	9,50

Bromoethane (C_2H_5Br) – 1, 4–Dioxan ($C_4H_8O_2$)
[175], $t = 25$ °C, $\nu = 1,8$ MHz

φ_2	100	80	60	40	20	0
ε	2,22	3,47	4,86	6,35	7,88	9,50

Bromoethane (C_2H_5Br) – Chlorobenzene (C_6H_5Cl)
[175], $t = 20$ °C, $\nu = 1,8$ MHz

φ_2	100	80	60	40	20	0
ε	5,70	6,39	7,15	7,93	8,72	9,50

Bromoethane (C_2H_5Br) – Nitrobenzene ($C_6H_5NO_2$)
[175], $t = 20$ °C, $\nu = 1,8$ MHz

φ_2	100	80	60	40	20	0
ε	35,75	29,47	23,57	18,32	13,70	9,50

Bromoethane (C_2H_5Br) – Diethyl ether ($C_6H_{12}O_2$)
[175], $t = 20$ °C, $\nu = 1,8$ MHz

φ_2	100	80	60	40	20	0
ε	6,19	7,01	7,65	8,27	8,89	9,50

Bromoethane (C_2H_5Br) – Hexane (C_6H_{14})
[238, 277], ν=0,5 MHz

x_1	ε at t °C							
	−90	−80	−70	−60	−50	−40	−30	−20
0	2,078	2,063	2,048	2,033	2,017	2,002	1,987	1,972
3,09	2,270	2,245	2,219	2,194	2,170	2,147	2,123	2,101
7,83	2,568	2,520	2,476	2,437	2,400	2,365	2,331	2,300
17,67	3,240	3,152	3,075	3,002	2,934	2,870	2,808	2,752
33,95	4,573	4,416	4,270	4,135	4,006	3,887	3,776	3,667
50,73	6,360	6,095	5,856	5,634	5,418	5,217	5,027	4,852
71,88	9,60	9,12	8,68	8,28	7,91	7,56	7,23	6,93
100,0	16,05	15,15	14,35	13,62	12,95	12,32	11,75	11,22

x_1	ε at t °C							
	−10	0	10	20	30	40	50	60
0	1,957	1,942	1,927	1,912	1,896	1,880	1,862	1,843
3,09	2,080	2,059	2,038	2,018	1,997	1,975	1,953	1,929
7,83	2,270	2,240	2,211	2,181	2,153	2,124	2,093	—
17,67	2,700	2,648	2,599	2,522	2,500	2,444	—	—
33,95	3,570	3,476	3,385	3,254	3,200	3,112	—	—
50,73	4,690	4,544	4,407	4,275	4,146	3,985	—	—
71,88	6,65	6,38	6,14	5,91	5,65	5,40	—	—
100,0	10,73	10,27	9,82	9,41	9,98	8,44	—	—

Bromoethane (C_2H_5Br) – Benzonitrile (C_7H_5N)
[175], t=20 °C, ν=1,8 MHz

φ_2	100	80	60	40	20	0
ε	25,65	22,67	19,51	16,32	12,95	9,50

Bromoethane (C_2H_5Br) – Benzaldehyde (C_7H_6O)
[175], t=20 °C, ν=1,8 MHz

φ_2	100	80	60	40	20	0
ε	17,85	16,16	14,48	12,80	11,15	9,50

Bromoethane (C_2H_5Br) – Phenetole ($C_8H_{10}O$)
[175], t=20 °C, ν=1,8 MHz

φ_2	100	80	60	40	20	0
ε	4,25	5,20	6,18	7,25	8,36	9,50

Iodoethane (C_2H_5I) – Heptane (C_7H_{17})
[278]

x_1	ε at t °C								
	−100	−90	−80	−70	−60	−50	−40	−30	−20
3,28	2,258	2,234	2,211	2,188	2,168	2,147	2,128	2,109	2,089
3,65	2,266	2,244	2,220	2,198	2,177	2,157	2,137	2,117	2,098
8,43	2,522	2,484	2,447	2,412	2,376	2,346	2,316	2,287	2,257
18,89	3,058	2,991	2,932	2,875	2,820	2,766	2,719	2,673	2,628
41,30	4,541	4,408	4,277	4,158	4,040	3,927	3,824	3,723	3,630
61,55	6,52	6,25	6,02	5,79	5,59	5,405	5,235	5,075	4,920
80,35	—	8,96	8,55	8,19	7,87	7,58	7,30	7,03	6,78
100	—	12,27	11,64	11,11	10,60	10,16	9,74	9,36	9,01

x_1	ε at t °C								
	−10	0	10	20	30	40	50	60	70
3,28	2,070	2,051	2,032	2,014	1,995	1,977	1,959	1,940	1,920
3,65	2,078	2,059	2,039	2,022	2,003	1,984	1,964	1,944	1,922
8,43	2,230	2,203	2,177	2,154	2,134	2,113	2,093	2,074	2,057
18,89	2,584	2,541	2,501	2,462	2,422	2,383	2,349	2,314	2,280
41,30	3,539	3,456	3,372	3,296	3,220	3,149	3,078	3,010	2,940
61,55	4,774	4,630	4,493	4,360	4,240	4,122	4,017	3,909	3,809
80,35	6,54	6,31	6,09	5,88	5,66	5,46	5,27	5,08	4,89
100,0	8,67	8,38	8,10	7,82	7,56	7,29	7,03	6,80	6,59

Ethanol (C_2H_6O) – Acetone (C_3H_6O)
[194], t = 25 °C

x_1	ε	x_1	ε	x_1	ε	x_1	ε
0	20,78	35,00	20,75	65,33	21,75	92,79	23,85
12,25	20,70	45,58	20,98	74,56	22,57	100	24,69
23,90	20,68	55,68	21,38	83,42	23,08	—	—

[248], t = 30 °C, ν = 0,5 MHz

w_2	90,0	70,0	50,0	30,0	10,0
ε	20,3	20,4	21,1	21,9	23,3

[175], t = 20 °C, ν = 1,8 MHz

φ_1	0	20	40	60	80	100
ε	21,07	20,85	21,30	22,13	23,40	25,07

Ethanol (C_2H_6O) – Propionic acid ($C_3H_6O_2$)
[271], $t = 25$ °C, $\nu = 1$ kHz

$10^{-3} N$	0	1	5	10	15	20	50	100
ε	24,30	24,21	24,40	24,36	24,66	24,65	25,28	25,28

Ethanol (C_2H_6O) – Propan–1–ol (C_3H_8O)
[253], $t = 25$ °C

w_2	100,0	79,75	60,64	41,00	20,46	0
ε	20,1	20,8	21,6	22,4	23,2	24,2

Ethanol (C_2H_6O) – Propan–2–ol (C_3H_8O)
[248], $t = 30$ °C, $\nu = 0,5$ MHz

w_2	90	70,0	50,0	30,0	10,0
ε	19,7	20,7	21,7	22,5	23,6

Ethanol (C_2H_6O) – Acetic anhydride ($C_4H_6O_3$)
[175], $t = 20$ °C, $\nu = 1,8$ MHz

φ_1	0	20	40	60	80	100
ε	22,45	22,15	22,30	22,95	23,90	25,07

Ethanol (C_2H_6O) – 1, 4–Dioxan ($C_4H_8O_2$)
[182], $t = 30$ °C

x_1	0	4,20	9,97	19,40	24,37	42,31	70,41	87,17	100
ε	2,383	2,383	2,692	3,342	3,773	5,996	12,71	18,14	24,4

[248], $t = 30$ °C, $\nu = 0,5$ MHz

w_1	10,0	30,0	50,0	70,0	90,0
ε	3,27	6,61	11,1	16,1	21,3

Ethanol (C_2H_6O) – Butyric acid ($C_4H_8O_2$)
[271], $t = 25$ °C

$10^{-3} N$	0	1	5	10	15	20	50	100
ε	24,20	24,39	24,28	24,08	24,46	24,36	24,57	25,12

Ethanol (C_2H_6O) – Isobutyric acid ($C_4H_8O_2$)

[271], $t=25$ °C

$10^{-3}N$	0	5	10	15	20	50	100
ε	24,30	24,40	24,58	24,74	24,56	24,71	25,53

Ethanol (C_2H_6O) – *tert*–Butyl alcohol ($C_4H_{10}O$)

[248], $t=30$ °C, $\nu=0,5$ MHz

w_1	10,0	30,0	50,0	70,0	90,0
ε	14,8	17,6	19,6	21,5	23,5

Ethanol (C_2H_6O) – Diethyl ether ($C_4H_{10}O$)

[195, 215], $t=25$ °C

w_1	ε	w_1	ε	w_1	ε	w_1	ε
0	4,235	30,82	8,712	61,98	15,26	86,20	20,96
13,08	5,885	41,24	10,76	70,05	17,20	100,0	24,28
20,86	6,998	51,47	12,89	77,40	18,90	—	—

[271], $t=25$ °C, $\nu=3$ MHz

w_1	0	11,3	21,0	31,9	57,0	100
ε	4,35	5,27	6,17	7,44	13,00	25,00

[175], $t=20$ °C, $\nu=1,8$ MHz

φ_1	0	20	40	60	80	100
ε	4,35	7,42	11,42	15,76	20,43	25,07

Ethanol (C_2H_6O) – Valeric acid ($C_5H_{10}O_2$)

[271], $t=25$ °C, $\nu=1$ kHz

$10^{-3}N$	0	5	10	15	20	50	100
ε	24,42	24,32	24,18	24,46	24,36	24,79	25,61

Ethanol (C_2H_6O) – Isovaleric acid ($C_5H_{10}O_2$)

[271], $t=25$ °C, $\nu=1$ kHz

$10^{-3}N$	0	5	10	15	20	50	100
ε	24,38	24,26	24,35	24,34	24,54	24,73	25,37

Ethanol (C_2H_6O) – 1, 4-Dichlorobenzene ($C_6H_4Cl_2$)
[279], $t = 55$ °C

x_1	ε	x_1	ε	x_1	ε	x_1	ε
0,00	2,397	20,14	3,176	58,54	7,755	89,96	16,98
5,14	2,536	34,80	4,191	69,92	10,44	100,0	19,66
9,58	2,688	50,17	5,950	79,94	12,86	—	—

Ethanol (C_2H_6O) – Chlorobenzene (C_6H_5Cl)
[175], $t = 20$ °C, $\nu = 1,8$ MHz

φ_1	0	20	40	60	80	100
ε	5,70	7,90	11,95	16,42	20,72	25,07

Ethanol (C_2H_6O) – Nitrobenzene ($C_6H_5NO_2$)
[194], $t = 25$ °C

x_1	0	16,35	42,98	63,75	84,07	94,06	100
ε	35,22	32,88	29,71	27,54	25,80	25,10	24,69

[175], $t = 20$ °C, $\nu = 1,8$ MHz

φ_1	0	20	40	60	80	100
ε	35,75	31,60	29,02	27,22	25,95	25,07

Ethanol (C_2H_6O) – Benzene (C_6H_6)
[262], $t = 25$ °C, $\lambda = 37,5$ m

x_1	ε	x_1	ε	x_1	ε	x_1	ε
0,00	2,271	18,43	3,30	42,46	7,14	70,52	15,10
3,85	2,45	24,38	3,90	50,64	9,03	76,06	16,82
6,90	2,60	29,48	4,65	54,62	10,17	85,80	19,20
11,73	2,80	36,63	5,87	59,87	11,75	89,83	21,20

[257, 195], 25 °C

x_1	ε	x_1	ε	x_1	ε	x_1	ε
0,00	2,271	30,82	4,700	59,87	11,75	89,93	21,25
9,53	2,700	39,70	6,600	69,77	14,95	100,0	24,45
20,49	3,400	49,74	8,850	80,23	18,10	—	—

[194], $t=56{,}7$ °C

x_1	0	14,45	33,74	65,02	89,66	100
ε	2,230	2,882	4,66	10,39	17,08	20,27

[215], $t=16$ °C

w_1	0	12,705	15,284	26,475	49,159	78,499	100
ε	2,244	3,583	4,165	6,813	13,09	21,30	27,1

[175], $t=20$ °C, $\nu=1{,}8$ MHz

φ_1	0	20	40	60	80	100
ε	2,28	4,66	9,30	14,70	20,00	25,07

Ethanol (C_2H_6O) – Aniline (C_6H_7N)

[175], $t=20$ °C, $\nu=1{,}8$ MHz

φ_1	0	20	40	60	80	100
ε	7,06	9,61	12,70	16,50	20,65	25,07

Ethanol (C_2H_6O) – Cyclohexane (C_6H_{12})

[280]

$t=25$ °C		$t=35$ °C		$t=45$ °C		$t=55$ °C	
N	ε	N	ε	N	ε	N	ε
0,4748	2,122	0,4691	2,110	0,4629	2,098	0,4568	2,086
0,5472	2,130	0,5405	2,123	0,5339	2,116	0,5263	2,110
0,7583	2,218	0,7484	2,203	0,7391	2,118	0,7291	2,173
0,9651	2,282	0,9535	2,264	0,9408	2,246	0,9279	2,228
1,3788	2,481	1,3625	2,446	1,3448	2,411	1,3266	2,366
2,0782	2,962	2,0332	2,896	2,0265	2,822	1,9981	2,759
3,0485	3,757	3,0116	3,607	2,9725	3,57	2,9323	3,307
4,5359	5,530	4,4808	5,195	4,4256	4,891	4,3649	4,582
7,7000	10,15	7,5985	9,200	7,5111	8,300	7,4154	7,600
11,7964	15,85	11,6626	14,757	11,5310	13,364	11,3674	12,565
17,0271	24,33	16,8549	22,787	16,6694	21,303	16,4826	19,707

x_1	ε at t °C [182]		x_1	ε at t °C [182]	
	6	30		6	30
0	2,041	2,003	36,031	5,484	4,747
2,907	2,086	2,054	43,665	—	6,178
6,328	2,158	2,177	48,551	8,631	7,295
8,389	2,226	2,185	56,526	10,953	9,572
17,505	2,773	5,620	63,822	13,33	11,54
22,722	3,307	3,308	74,134	16,99	14,92
29,777	4,227	3,756	100,0	28,0	24,4

Ethanol (C_2H_6O) – Hexanoic acid ($C_6H_{12}O_2$)

[271], $t=25$ °C, $\nu=1$ kHz

10^{-3} N	0	5	10	15	20	50	100
ε	24,27	24,34	24,44	24,44	24,44	25,61	27,43

Ethanol (C_2H_6O) – 4-Methylpentanoic acid ($C_6H_{12}O_2$)

[271], $t=25$ °C, $\nu=1$ kHz

10^{-3} N	0	5	10	15	20	50	100
ε	24,29	24,08	24,65	24,83	25,08	26,85	32,8

Ethanol (C_2H_6O) – Hexane (C_6H_{14})

[278]

x_1	ε at t °C							
	−100	−90	−80	−70	−60	−50	−40	−30
1,53	—	2,093	2,077	2,060	2,045	2,030	2,016	2,002
5,89	—	2,160	2,144	2,129	2,114	2,098	2,083	2,068
9,62	—	2,259	2,248	2,232	2,214	2,197	2,179	2,162
20,76	—	3,418	3,360	3,295	3,225	3,150	3,075	2,996
25,60	2,129	2,113	2,099	2,084	2,069	2,053	2,040	2,027
61,42	—	—	—	—	—	—	—	11,82
83,00	—	—	29,8	28,0	26,3	24,7	23,1	21,7
92,60	—	40,8	38,1	35,6	33,2	31,0	29,0	27,2
100	52,9	49,3	46,1	43,0	40,1	37,4	35,0	32,7

x_1	ε at *t* °C									
	−20	−10	0	10	20	30	40	50	60	70
1,53	1,989	1,976	1,964	1,952	1,940	1,928	1,914	1,898	1,882	—
5,89	2,054	2,040	2,026	2,014	2,002	1,990	1,978	1,968	1,958	—
9,62	2,145	2,128	2,111	2,056	2,082	2,069	2,054	2,039	2,025	—
20,76	2,913	2,830	2,755	2,687	2,622	2,563	2,507	2,453	2,403	—
25,60	2,015	2,003	1,991	1,980	1,968	1,956	1,944	1,931	1,915	1,900
61,42	10,49	9,96	9,43	8,92	8,40	7,87	7,39	6,92	6,48	6,05
83,00	20,3	19,0	17,8	16,6	15,5	14,5	13,6	12,7	11,8	11,0
92,60	25,5	24,0	22,5	21,1	19,8	18,6	17,4	16,3	15,2	14,2
100	30,7	28,7	27,0	25,3	23,8	22,4	21,0	19,8	18,7	17,6

Ethanol (C_2H_6O) – 2-Chlorobenzoic acid ($C_7H_5ClO_2$)

[271], $t = 25$ °C, $\nu = 1$ kHz

$10^{-3}N$	0	1	5	10	15	20	100
ε	24,18	24,35	25,29	24,74	25,21	25,97	34,70

Ethanol (C_2H_6O) – 3-Chlorobenzoic acid ($C_7H_5ClO_2$)

[271], $t = 25$ °C, $\nu = 1$ kHz

$10^{-3}N$	0	1	10	15	20	100
ε	24,31	27,12	50,77	78,70	111,5	110

Ethanol (C_2H_6O) – 4-Chlorobenzoic acid ($C_7H_5ClO_2$)

[271], $t = 25$ °C, $\nu = 1$ kHz

$10^{-3}N$	0	1	5	10	15	20	50	100
ε	24,43	24,33	24,72	24,36	24,51	26,9	26,97	28,02

Ethanol (C_2H_6O) – Benzonitrile (C_7H_5N)

[175], $t = 20$ °C, $\nu = 1{,}8$ MHz

φ_1	0	20	40	60	80	100
ε	25,65	25,05	24,95	24,98	25,02	25,07

Ethanol (C_2H_6O) – 2-Nitrobenzoic acid ($C_7H_5NO_4$)
[271], $t=25$ °C

$10^{-3}N$	0	1	5	10	15	20	50	100
ε	24,27	24,85	25,26	26,37	27,88	29,14	42,16	69,23

Ethanol (C_2H_6O) – 3-Nitrobenzoic acid ($C_7H_5NO_4$)
[271], $t=25$ °C

$10^{-3}N$	0	1	5	10	15	20	50	100
ε	24,21	24,77	26,34	29,10	32,72	37,37	74,22	115

Ethanol (C_2H_6O) – 4-Nitrobenzoic acid ($C_7H_5NO_4$)
[271], $t=25$ °C

$10^{-3}N$	0	1	5	10	15	20	50	100
ε	24,24	24,81	25,36	25,10	26,72	28,78	39,04	56,87

Ethanol (C_2H_6O) – 2-Hydroxybenzoic acid ($C_7H_6O_3$)
[271], $t=25$ °C

$10^{-3}N$	0	1	5	10	15	20	50	100
ε	24,24	25,73	24,46	24,56	24,76	25,09	26,70	30,35

Ethanol (C_2H_6O) – 3-Hydroxybenzoic acid ($C_7H_6O_3$)
[271], $t=25$ °C

$10^{-3}N$	0	1	5	10	15	20	50	100
ε	23,91	25,41	25,54	25,90	26,05	26,63	28,31	32,98

Ethanol (C_2H_6O) – 4-Hydroxybenzoic acid ($C_7H_6O_3$)
[271], $t=25$ °C

$10^{-3}N$	0	1	5	10	15	20	50	100
ε	24,79	24,99	27,84	33,12	38,47	44,48	96,00	115

Ethanol (C_2H_6O) – 2-Aminobenzoic acid ($C_7H_6O_3$)
[271], $t=25$ °C

$10^{-3}N$	0	1	5	10	15	20	100
ε	26,21	27,43	29,04	31,90	34,82	78,94	115

Ethanol (C_2H_6O) – 3–Aminobenzoic acid ($C_7H_6O_3$)
[271], $t = 25°$ C

$10^{-3}N$	0	1	5	10	15	20	50	100
ε	24,81	25,37	26,32	29,18	30,58	33,71	61,75	115

Ethanol (C_2H_6O) – 4–Aminobenzoic acid ($C_7H_6O_3$)
[271], $t = 25$ °C

$10^{-3}N$	0	1	5	10	15	20	50	100
ε	26,39	26,43	26,54	26,51	26,59	26,72	26,94	28,72

Ethanol (C_2H_6O) – Heptane (C_7H_{16})
[281], $t = 30$ °C

x_1	10	20	30	40	50	60	70	80	90	100
ε	2,30	2,73	3,28	4,08	5,52	7,52	10,21	13,40	17,32	22,40

Ethanol (C_2H_6O) – *o*–Toluic acid ($C_8H_8O_2$)
[271], $t = 25$ °C, $\nu = 1$ kHz

$10^{-3}N$	0	1	5	10	15	20	50	100
ε	24,36	24,22	24,50	24,83	24,83	26,06	27,21	30,83

Ethanol (C_2H_6O) – *m*–Toluic acid ($C_8H_8O_2$)
[271], $t = 25$ °C, $\nu = 1$ kHz

$10^{-3}N$	0	1	5	10	15	20	100
ε	24,43	24,68	31,32	45,82	64,25	81,44	156

Ethanol (C_2H_6O) – *p*–Toluic acid ($C_8H_8O_2$)
[271], $t = 25$ °C, $\nu = 1$ kHz

$10^{-3}N$	0	1	5	10	15	20	50	100
ε	24,09	24,37	24,6	24,79	25,05	24,81	26,54	29,38

Ethanol (C_2H_6O) – Xylene (C_8H_{10})
[282], $t = 21$ °C

w_1	0	9	17	40	50	100
ε	2,36	3,08	3,98	9,53	13,0	26,5

Glycol ($C_2H_6O_2$) – Dioxan ($C_4H_8O_2$)
[283]

x_1	ε at t °C		x_1	ε at t °C	
	15	30		15	30
0,000	2,232	2,196	14,663	3,762	3,587
2,442	2,422	2,375	26,676	5,743	5,392
4,556	2,609	2,553	38,674	8,598	8,040
6,918	2,842	2,767	58,528	15,91	15,27
7,776	2,927	2,834	79,775	28,73	27,97
10,410	2,227	3,112	100,00	46,66	—

Ethane thiol ($C_2H_6S_2$) – Benzene (C_6H_6)
[283], $t = 15$ °C

x_1	ε	x_1	ε	x_1	ε	x_1	ε
0,000	2,292	4,693	2,427	14,287	2,742	73,257	5,244
1,620	2,309	9,074	2,569	16,136	2,801	92,474	6,348
3,045	2,369	12,388	2,676	24,591	3,086	100,00	6,912

Dimethylsulphoxide (C_2H_6OS) – Benzene (C_6H_6)
[284], $t = 25$ °C, $\nu = 2,47$ MHz

x_1	ε	x_1	ε	x_1	ε	x_1	ε
0,00	2,273	20	6,90	60	23,20	100	46,4
3,00	2,903	30	10,90	70	27,44	—	—
5,00	3,356	40	15,41	80	34,04	—	—
10,00	4,45	50	18,97	90	39,11	—	—

1, 3–Dibromopropane ($C_3H_6Br_2$) – Heptane (C_7H_{16})
[285]

x_1	ε при t °C		x_1	ε при t °C	
	25	50		25	50
0,000	1,920	1,883	13,707	2,406	2,314
3,248	2,025	1,978	18,568	2,603	2,490
7,650	2,178	2,114	—	—	—

1, 3–Dichloropropane ($C_3H_6Cl_2$) – Hexane (C_6H_{14})

[269], $t = 25$ °C

x_1	0	1,891	6,723	11,095	25,03	41,80	69,10	100
ε	1,8776	1,9457	2,1306	2,3106	2,9764	3,9942	6,2919	10,085

2, 2–Dichloropropane ($C_3H_6Cl_2$) – Hexane (C_6H_{14})

[269], $t = 25$ °C

x_1	0	1,425	5,284	19,912	61,13	100,0
ε	1,8776	1,93401	2,0944	2,8098	5,9749	11,196

Acetone (C_3H_6O) – Methyl acetate ($C_3H_6O_2$)

[229], $t = 20$ °C

x_2	100	74,82	63,93	56,46	51,62	35,69	29,60	0,00
ε	7,029	9,851	11,23	12,13	12,62	14,87	15,72	21,41

Acetone (C_3H_6O) – Dimethylformamide (C_3H_7NO)

[252], ν = 2 MHz

x_2	ε при t °C			x_2	ε при t °C		
	10	25	55		15	25	55
100	40,42	37,70	32,76	49,54	32,15	30,14	27,86
90,42	38,81	36,17	33,12	29,29	28,64	26,87	24,53
70,12	35,27	32,99	30,29	10,41	24,84	23,29	21,26

Acetone (C_3H_6O) – Propan–2–ol (C_3H_8O)

[286], $t = 20$ °C, ν = 10 kHz

x_1	0	20,7	41,0	61,0	80,6	100
ε	19,13	17,62	17,34	17,95	19,26	20,54

Acetone (C_3H_6O) – Isopropylamine (C_3H_9N)

[175], $t = 20$ °C, ν = 1,8 MHz

φ_2	100	80	60	40	20	0
ε	5,38	8,45	11,55	14,60	17,70	21,07

Acetone (C_3H_6O) – Acetic anhydride ($C_4H_6O_3$)
[175], $t = 20$ °C, $\nu = 1{,}8$ MHz

φ_2	100	80	60	40	20	0
ε	22,45	22,30	22,12	21,82	21,47	21,07

Acetone (C_3H_6O) – Butan-2-one (C_4H_8O)
[253], $t = 25$ °C

w_2	100	79,88	60,28	40,69	19,68	0
ε	17,48	18,07	18,71	19,29	19,96	20,70

[175], $t = 20$ °C, $\nu = 1{,}8$ MHz

φ_2	100	80	60	40	20	0
ε	18,35	18,88	19,40	19,95	20,50	21,07

Acetone (C_3H_6O) – 1, 4-Dioxan ($C_4H_8O_2$)
[248], $t = 30$ °C, $\nu = 550$ kHz

w_1	10	30	50	70	90
ε	3,96	7,56	11,3	15,0	18,6

[175], $t = 20$ °C, $\nu = 1{,}8$ MHz

φ_1	0	20	40	60	80	100
ε	2,22	5,05	8,52	12,25	16,33	21,07

Acetone (C_3H_6O) – Ethyl acetate ($C_4H_8O_2$)
[175], $t = 20$ °C, $\nu = 1{,}8$ MHz

φ_2	100	80	60	40	20	0
ε	6,19	8,85	11,45	14,25	17,90	21,07

Acetone (C_3H_6O) – *tert*-Butyl alcohol ($C_4H_{10}O$)
[248], $t = 30$ °C, $\nu = 550$ kHz

w_1	10	30	50	70	90
ε	12,3	13,2	14,8	16,7	19,2

Acetone (C_3H_6O) – Diethyl ether ($C_4H_{10}O$)

[267], $t=20$ °C, $\nu=3$ MHz

x_1	0	10,1	19,3	29,0	36,7	64,7	100
ε	4,35	5,51	6,65	7,96	9,04	13,73	21,04

[175], $t=20$ °C, $\nu=1,8$ MHz

φ_1	0	20	40	60	80	100
ε	4,35	7,60	10,97	14,28	17,68	21,07

Acetone (C_3H_6O) – Pyridine (C_5H_5N)

φ_1	ε at t °C [287]				φ_1	ε at t °C [287]			
	10	20	30	40		10	20	30	40
0	13,8	13,2	12,6	12,0	70,6	19,6	18,6	17,7	16,9
10,0	14,7	14,0	13,4	12,8	89,9	21,1	20,0	19,0	18,1
29,9	16,4	15,6	14,9	14,2	100	22,3	21,2	20,2	19,1
50,0	17,9	17,1	16,3	15,6	—	—	—	—	—

Acetone (C_3H_6O) – Chlorobenzene (C_6H_5Cl)

[175], $t=20$ °C, $\nu=1,8$ MHz

φ_1	0	20	40	60	80	100
ε	5,70	8,45	11,35	14,37	17,55	21,07

Acetone (C_3H_6O) – Nitrobenzene ($C_6H_5NO_2$)

[194], $t=25$ °C

x_2	100	86,53	62,49	41,65	19,22	0,00
ε	35,22	33,69	30,73	27,88	24,33	20,87

[286], $t=25$ °C, $\nu=10$ kHz

x_2	100	74,2	51,9	32,4	15,2	0
ε	34,62	31,46	28,75	26,00	23,36	20,54

[175], $t=20$ °C, $\nu=1,8$ MHz

φ_2	100	80	60	40	20	0
ε	35,75	32,45	29,45	26,60	23,80	21,07

Acetone (C_3H_6O) – Benzene (C_6H_6)

$t=20$ °C

[229]				[166]			
x_1	ε	x_1	ε	x_1	ε	x_1	ε
0,00	2,285	60,33	11,22	0,000	2,032	55,818	9,333
27,09	6,010	71,07	13,23	4,595	2,397	82,135	15,56
43,15	8,391	77,67	14,90	12,911	3,128	100	21,17
53,73	10,16	100	21,41	33,234	5,529	—	—

[286, 290, 291], $t=25$ °C, $\nu=10$ kHz

x_1	0	23,2	44,6	64,4	82,8	100
ε	2,281	4,997	8,122	11,85	15,97	20,54

[182], $t=6$ °C

x_1	ε	x_1	ε	x_1	ε	x_1	ε
0,000	2,310	5,050	2,906	34,369	7,141	100	25,28
1,582	2,496	7,983	3,261	52,911	10,75	—	—
3,569	2,720	14,726	4,141	81,037	19,15	—	—

[288, 289], $t=19$ °C, $\lambda=73$ cm

w_1	0	5,3	9,8	20,0	31,0	40,0	49,5	69,4	84,7	100
ε	2,26	2,96	3,56	5,09	6,90	8,43	10,2	14,3	17,3	20,5

[292], $t=16$ °C

w_1	ε	w_1	ε	w_1	ε	w_1	ε
0	2,55	30	6,88	60	12,85	90	19,15
10	3,69	40	8,71	70	15,20	100	21,60
20	5,17	50	10,57	80	17,04	—	—

[175], $t=20$ °C, $\nu=1,8$ MHz

φ_1	0	20	40	60	80	100
ε	2,28	5,06	8,40	12,20	16,35	21,07

Acetone (C_3H_6O) – Aniline (C_6H_7N)

[175], $t=25$ °C, $\nu=1,8$ MHz

φ_2	100	80	60	40	20	0
ε	7,06	10,37	13,35	16,05	18,50	21,07

Acetone (C_3H_6O) – Cyclohexanone ($C_6H_{10}O$)

[175], $t=20$ °C, $\nu=1,8$ MHz

φ_2	100	80	60	40	20	0
ε	15,70	16,69	17,72	18,80	19,94	21,07

Acetone (C_3H_6O) – Cyclohexane (C_6H_{12}) *)

[166], $t=20$ °C

x_1	0,00	4,599	12,911	32,234	55,818	82,135	100
ε	2,032	2,397	3,128	5,520	9,333	15,56	21,17

A Acetone (C_3H_6O) – Cyclohexylamine ($C_6H_{13}N$)

[175], $t=20$ °C, $\nu=1,8$ MHz

φ_2	100	80	60	40	20	0
ε	4,72	7,78	10,95	14,27	17,65	21,07

Acetone (C_3H_6O) – Hexane (C_6H_{14})

[286], $t=25$ °C, $\nu=10$ kHz

x_1	0	30,8	54,2	72,7	57,6	100
ε	1,884	4,40	7,698	11,68	16,01	20,54

x_1	ε at t °C [182]		x_1	ε at t °C [182]	
	6	30		6	30
0,000	1,917	1,879	26,229	4,210	3,909
1,729	2,035	1,9864	36,751	5,599	5,134
4,972	2,263	2,1820	57,136	9,371	8,476
7,210	2,430	2,3296	84,968	18,128	16,21
11,661	2,771	2,6356	100,00	25,28	20,97
17,690	3,304	3,0977	—	—	—

*) See also [250], $t=25$ °C, $x_1=0-100$.

[293], $t = 15$ °C

φ_1	ε	φ_1	ε	φ_1	ε	φ_1	ε
50	1,850	59	6,649	71	13,47	91	20,44
53	3,458	63	8,424	77	16,14	100	22,06
56	4,819	67	10,22	83	17,97	—	—

Acetone (C_3H_6O) – Benzonitrile (C_7H_5N)

[175], $t = 20$ °C, $\nu = 1{,}8$ MHz

φ_2	100	80	60	40	20	0
ε	25,65	24,70	23,80	22,90	21,95	21,07

Acetone (C_3H_6O) – Benzaldehyde (C_7H_6O)

[175], $t = 20$ °C, $\nu = 1{,}8$ MHz

φ_1	0	20	40	60	80	100
ε	17,85	18,54	19,22	19,87	20,47	21,07

Acetone (C_3H_6O) – Toluene (C_7H_8)

[286], $t = 25$ °C, $\nu = 10$ kHz

x_1	0	26,5	49,0	74,2	85,2	100
ε	2,376	4,985	8,197	13,24	16,07	20,54

[294]

w_1	t °C	ε	w_1	t °C	ε	w_1	t °C	ε
0	14	3,29	38,5	17	9,13	78,5	17	16,75
10,0	17	3,97	48,5	16	11,17	89,3	16,5	19,49
19,0	16	5,92	60,0	15	12,96	100	17	22,14
29,0	17	7,11	68,5	17	14,96	—	—	—

Acetone (C_3H_6O) – Heptane (C_7H_{16})

[221], $t = 20$ °C

x_1	3,906	7,759	11,72	16,06	19,41	100
ε	2,167	2,402	2,558	2,942	3,191	21,40

Acetone (C_3H_6O) – Phenetole ($C_8H_{10}O$)

[175], $t = 20$ °C, $\nu = 1{,}8$ MHz

φ_1	0	20	40	60	80	100
ε	4,25	7,11	10,25	13,55	17,10	21,07

Methyl acetate ($C_3H_6O_2$) – Benzene (C_6H_6)

t=10 °C [229]		t=20 °C [229]		t=30 °C [229]	
x_2	ε	x_2	ε	x_2	ε
100,0	2,300	100,0	2,285	100,0	2,270
71,21	3,583	84,50	2,285	67,60	3,542
66,47	3,791	64,65	3,806	53,80	4,102
55,34	4,302	59,44	4,010	41,96	4,631
36,94	5,249	50,59	4,433	37,19	4,927
29,84	5,597	29,23	5,452	20,28	5,529
0,00	7,420	0,00	7,092	0,00	6,785

Propionic acid ($C_3H_6O_2$) – Butyric acid ($C_4H_8O_2$)
[224], t=25 °C, ν=28 MHz

φ_2	100	79,83	60,18	39,59	21,38	0
ε	2,88	2,87	2,93	3,01	3,11	3,20

Propionic acid ($C_3H_6O_2$) – Pyridine (C_5H_5N)
[295], t=25 °C

φ_2	100	90	80	70	60	40,0	25,4	10	0
ε	12,91	13,02	13,24	13,33	13,80	14,7	13,5	8,28	3,17

Propionic acid ($C_3H_6O_2$) – Chlorobenzene (C_6H_5Cl)
[224], t=25 °C, ν=28 MHz

φ_2	100	88,14	60,93	37,81	19,54	0
ε	5,61	5,13	4,53	3,98	3,57	3,20

Propionic acid ($C_3H_6O_2$) – Benzene (C_6H_6)
[275], t=22 °C

x_1	ε	x_1	ε	x_1	ε	x_1	ε
7,85	2,310	15,82	2,350	32,99	2,442	91,90	3,170
10,93	2,326	19,23	2,365	44,23	2,530	—	—
12,08	2,330	23,18	2,386	64,37	2,764	—	—
14,91	2,345	24,90	2,390	83,24	3,030	—	—

Propionic acid ($C_3H_6O_2$) – Triethylamine ($C_6H_{15}N$)
[295], $t=25$ °C

φ_2	100	90	80	66,7	54,5	45,0	26,7	10,2	0
ε	2,42	4,22	7,84	22,6	31,3	39	28	16,1	3,17

Propionic acid ($C_3H_6O_2$) – Octane (C_8H_{18})
[224], $t=25$ °C, ν=28 MHz

φ_2	100	80,84	60,29	40,23	19,65	0
ε	1,94	1,97	2,11	2,38	2,73	3,20

Propionic acid ($C_3H_6O_2$) – N, N–Diethylaniline ($C_{10}H_{15}N$)
[295], $t=25$ °C

φ_2	100	85,0	70	50	33,3	15,0	0
ε	5,26	6,63	9,85	15,9	18,6	15,0	3,17

Ethyl formate (C_3H_6O) – Benzene (C_6H_6)

x_1	ε at t °C [296]		x_1	ε at t °C [296]	
	25	50		25	50
0	2,276	2,229	11,26	2,822	2,708
3,77	2,460	2,387	16,96	3,112	2,960
5,46	2,541	2,457	20,83	3,312	3,137

1–Bromopropane (C_3H_7Br) – 1–Bromobutane (C_4H_9Br)
[244], $t=25$ °C, ν=28 MHz

φ_1	0	9,42	28,8	42,58	55,69	100
ε	6,86	6,98	7,18	7,38	7,54	8,09

Dimethylformamide (C_3H_7NO) – 1, 4–Dioxan ($C_4H_8O_2$)
[252], ν=2 MHz

x_1	ε at t °C			x_1	ε at t °C		
	10	25	55		10	25	55
10,46	4,912	4,682	4,301	65,39	26,39	24,42	22,39
31,50	11,330	10,687	9,366	89,79	35,82	33,45	29,14
55,96	20,33	19,04	16,85	100	40,42	37,70	32,76

Dimethylformamide (C_3H_7NO) – Benzene (C_6H_6)
[252], $\nu = 2$ MHz

x_1	ε at t °C			x_1	ε at t °C		
	10	25	55		10	25	55
10,04	4,682	4,463	4,071	71,34	26,83	25,12	22,09
30,02	10,652	10,232	9,440	89,67	35,31	33,05	29,31
49,99	17,73	16,88	15,13	100	40,42	37,70	32,76

***N*–Methylacetamide (C_3H_7NO) – Dioxan ($C_4H_8O_2$)**
[297], $t = 40$ °C

w_1	25,32	26,15	50,76	56,90	60,82	70,02	74,44
ε	23,2	24,8	55,8	65,7	72,4	89,5	98,1

Propan–1–ol (C_3H_8O) – Acetic anhydride ($C_4H_6O_3$)
[175], $t = 20$ °C, $\nu = 1,8$ MHz

φ_1	0	20	40	60	80	100
ε	22,45	20,80	19,87	19,55	19,75	20,65

Propan–1–ol (C_3H_8O) – Diethyl ether ($C_4H_{10}O$)
[267], $t = 20$ °C, $\nu = 3$ MHz

x_1	0	5,7	9,6	11,4	13,7	29,7	43,3	50,0	100
ε	4,35	4,68	4,93	5,07	5,22	6,50	7,96	8,79	20,0

[175], $t = 20$ °C, $\nu = 1,8$ MHz

φ_1	0	20	40	60	80	100
ε	4,35	6,26	9,30	12,80	16,65	20,65

Propan–1–ol (C_3H_8O) – 1, 4–Dichlorobenzene ($C_6H_4Cl_2$)
[279], $t = 55$ °C

x_1	ε	x_1	ε	x_1	ε	x_1	ε
0,00	2,397	19,94	3,100	49,06	5,357	78,79	11,058
5,08	2,522	30,16	3,750	59,02	6,821	88,10	13,449
10,03	2,687	39,29	4,450	68,66	8,761	100	16,774

Propan-1-ol (C_3H_8O) – Benzene (C_6H_6) *)

$t = 25$ °C

[262], λ=37,5 m				[257]			
x_1	ε	x_1	ε	x_1	ε	x_1	ε
0,00	2,271	29,09	4,25	0,00	2,271	69,82	12,76
1,06	2,33	40,84	6,16	10,12	2,823	79,84	15,64
2,08	2,375	51,58	8,40	19,49	3,603	89,78	18,08
5,53	2,515	59,75	10,35	29,93	4,800	100	20,74
9,49	2,70	68,65	13,60	39,70	6,000	—	—
12,10	2,85	77,98	15,05	49,77	8,080	—	—
20,82	3,50	91,40	18,60	59,83	10,30	—	—

[218]

t=24 °C				t=41 °C				t=70 °C			
w_1	ε	w_1	ε	w_1	ε	w_1	ε	w_1	ε	w_1	ε
0	2,261	15,51	3,381	0	2,230	15,41	3,178	0	2,185	15,52	2,931
3,15	2,412	22,09	4,295	3,19	2,362	34,50	5,251	3,30	2,300	22,15	3,482
5,17	2,540	33,90	5,860	5,20	2,469	100	18,12	5,21	2,382	34,20	4,631
9,46	2,850	100	22,15	9,50	2,740	—	—	9,52	2,593	100	12,39

*) See also [259, 215].

[175], $t = 20$ °C, $\nu = 1,8$ MHz

φ_1	0	20	40	60	80	100
ε	2,28	3,88	7,30	11,80	16,45	20,65

Propan-1-ol (C_3H_8O) – Cyclohexane (C_6H_{12})

[280]

t=25 °C		t=35 °C		t=45 °C		t=55 °C	
N_1	ε	N_1	ε	N_1	ε	N_1	ε
0,4769	2,117	0,4705	2,017	0,4646	2,095	0,4590	2,082
0,6608	2,140	0,6517	2,131	0,6438	2,122	0,6360	2,113
0,8605	2,202	0,8488	2,190	0,8386	2,178	0,8284	2,116
1,2545	2,345	1,2381	2,328	1,2231	2,311	1,2082	2,294
1,5201	2.425	1,5009	2,395	1,4830	2,370	1,4653	2,354
1,9721	2,738	1,9466	2,677	1,9248	2,623	1,9011	2,578
4,1972	5,200	4,1475	4,813	4,1030	4,480	4,0500	4,203
6,7941	8,375	6,7148	7,573	6,6527	6,835	6,5645	6,175
9,8357	13,975	9,7314	12,750	9,6421	11,650	9,5273	10,70
13,2679	20,000	13,1475	18,585	13,0361	17,257	12,8899	16,040

Propan-1-ol (C_3H_8O) – Toluene (C_7H_8)
[218], $t = 24$ °C

w_1	0	3,54	6,51	9,76	17,81	25,20	33,75	100
ε	2,373	2,531	2,703	2,921	3,722	4,704	5,431	22,15

[298], $t = 35$ °C, $\nu = 1,8$ MHz

x_2	100	88,52	73,58	61,40	49,78	28,79	6,47	0
ε	2,34	2,82	4,00	5,09	6,90	11,50	11,73	18,90

Propan-2-ol (C_3H_8O) – 1, 4-Dioxan ($C_4H_8O_2$)
[248], $t = 30$ °C, $\nu = 550$ kHz

w_1	10,0	30,0	50,0	70,0	90,0
ε	2,93	4,98	7,96	11,9	16,1

Propan-2-ol (C_3H_8O) – *tert*-Butyl alcohol ($C_4H_{10}O$)
[248], $t = 30$ °C, $\nu = 550$ kHz

w_1	10,0	30,0	50,0	70,0	90,0
ε	13,7	15,5	16,8	17,8	18,7

Propan-2-ol (C_3H_8O) – 1, 4-Dichlorobenzene ($C_6H_4Cl_2$)
[299], $t = 55$ °C

x_1	ε	x_1	ε	x_1	ε	x_1	ε
0,00	2,397	20,29	3,062	50,53	5,600	79,86	9,628
5,09	2,526	29,75	3,605	59,95	6,551	89,96	11,756
8,89	2,630	40,00	4,451	69,87	7,707	100	14,330

Propan-2-ol (C_3H_8O) – Nitrobenzene ($C_6H_5NO_2$)
[286], $t = 25$ °C, $\nu = 10$ kHz

x_2	100	74,9	52,9	33,3	15,8	0
ε	34,62	28,93	24,91	21,98	20,08	19,13

Propan-2-ol (C_3H_8O) – Benzene (C_6H_6)
[286, 300], $t = 25$ °C, $\nu = 10$ kHz

x_1	0	22,6	43,7	63,6	82,4	100
ε	2,281	3,440	6,041	9,894	14,57	19,13

[218], $t = 20$ °C

w_1	0	4,48	9,97	15,61	25,18	33,38	45,72	100
ε	2,270	2,452	2,703	3,071	3,979	5,022	6,701	19,2

Propan–2–ol (C_3H_8O) – Hexane (C_6H_{14})
[286], $t = 25$ °C, $\nu = 10$ kHz

x_1	0	30,0	53,3	72,0	87,3	100
ε	1,884	2,785	5,887	10,27	14,96	19,13

Propan–2–ol (C_3H_8O) – Toluene (C_7H_8)
[286], $t = 25$ °C, $\nu = 10$ kHz

x_1	0	25,8	48,2	77,6	84,8	100
ε	2,376	3,528	6,212	10,12	14,68	19,13

[372], $t = 25$ °C

x_2	ε	x_2	ε	x_2	ε	x_2	ε
100	2,34	66,25	4,12	36,42	8,49	13,96	13,85
89,91	2,68	55,33	5,31	27,58	10,43	6,76	15,83
77,93	3,26	45,58	6,73	22,20	11,70	0	17,80

Glycerine (C_3H_8O) – Dioxan ($C_4H_8O_2$)
[283]

x_1	ε at t °C		x_1	ε at t °C	
	15	30		15	30
0,000	2,224	2,200	3,935	2,683	2,631
1,468	2,389	2,352	4,810	2,800	2,743
2,436	2,502	2,457	—	—	—

Dimethoxymethane (C_3H_8O) – Isopentane (C_5H_{12})
[301], $\nu = 500$ kHz

$w_1 = 0$		$w_1 = 50$		$w_1 = 75$		$w_1 = 100$	
t °C	ε	t °C	ε	t °C	ε	t °C	ε
0	1,870	−1,5	2,200	−1,5	2,429	20	2,644
−33,5	1,913	−14	2,207	−23	2,426	0	2,623
−34	1,919	−35,5	2,217	−44	2,423	−48	2,575
−61	1,953	−77	2,241	−65	2,422	−75,5	2,550
−95,5	2,002	−81,5	2,247	−73,5	2,424	−102,5	2,543

Chlorotrimethylsilane (C_3H_9ClSi) – Acetophenone (C_8H_8O)
[147], $t = 25$ °C

φ_2	ε	φ_2	ε	φ_2	ε	φ_2	ε
100	17,16	70	14,63	40	11,75	10	8,87
90	16,49	60	13,57	30	10,78	0	8,01
80	15,59	50	12,60	20	9,83	—	—

Chlorotrimethylsilane (C_3H_9ClSi) – Nitrobenzene ($C_6H_5NO_2$)
[146], $t = 25$ °C

φ_2	ε	φ_2	ε	φ_2	ε	φ_2	ε
100	34,82	70	25,90	40	17,80	10	10,30
90	31,95	60	23,21	30	15,16	0	8,01
80	28,91	50	20,42	20	12,68	—	—

Propylamine (C_3H_9N) – Benzonitrile (C_7H_5N)
[175], $t = 20$ °C, $\nu = 1{,}8$ MHz

φ_1	0	20	40	60	80	100
ε	25,65	21,70	17,75	13,95	10,05	6,29

Ethyl benzoate ($C_3H_{10}O_2$) – Benzene (C_6H_6)
[227], $t = 15$ °C, $\nu = 3$ MHz

φ_1	0	20	40	60	80	100
ε	2,28	3,09	3,85	4,59	5,32	6,12

Furan (C_4H_4O) – Benzene (C_6H_6)
[302], $t = 25$ °C

x_1	0	2,86	6,64	7,08	13,38	13,55	17,31	100,0
ε	2,276	2,289	2,306	2,308	2,340	2,341	2,363	2,953

Divinyl ether (C_4H_6O) – Benzene (C_6H_6)
[302], $t = 25$ °C

x_1	0	4,185	7,059	9,017	11,617	14,231	17,611	22,112	100,00
ε	2,286	2,340	2,380	2,406	2,443	2,479	2,527	2,593	3,942

Acetic anhydride ($C_4H_6O_3$) – 1, 4–Dioxan ($C_4H_8O_2$)
[245], $t=25$ °C

x_2	100	90,9	81,6	72,1	62,4	52,7	42,5	32,1	21,7	10,8	0
ε	2,20	4,0	5,8	7,8	9,7	11,6	13,7	15,8	17,8	20,0	22,0

Acetic anhydride ($C_4H_6O_3$) – Butan–1–ol ($C_4H_{10}O$)
[175], $t=20$ °C, $\nu=1,8$ MHz

φ_2	100	80	60	40	20	0
ε	17,90	17,33	17,75	18,60	20,10	22,45

Acetic anhydride ($C_4H_6O_3$) – Diethyl ether ($C_4H_{10}O$)
[175], $t=20$ °C, $\nu=1,8$ MHz

φ_1	0	20	40	60	80	100
ε	4,35	7,87	11,50	15,25	18,85	22,45

Acetic anhydride ($C_4H_6O_3$) – 1, 2–Dichlorobenzene ($C_6H_4Cl_2$)
[245], $t=25$ °C

x_2	100	88,3	77,0	55,7	45,6	35,9	26,4	17,3	8,5	0
ε	10,1	11,8	14,0	15,7	16,8	18,0	18,9	19,5	21,0	22,0

Acetic anhydride ($C_4H_6O_3$) – Nitrobenzene ($C_6H_5NO_2$)
[175], $t=20$ °C, $\nu=1,8$ MHz

φ_2	100	80	60	40	20	0
ε	35,75	32,40	29,45	27,00	24,67	22,45

Acetic anhydride ($C_4H_6O_3$) – Benzene (C_6H_6)
[175], $t=20$ °C, $\nu=1,8$ MHz

φ_1	0	20	40	60	80	100
ε	2,28	5.36	9,00	13,10	17,50	22,45

Acetic anhydride ($C_4H_6O_3$) – Benzonitrile (C_7H_5N)
[175], $t=20$ °C, $\nu=1,8$ MHz

φ_2	100	80	60	40	20	0
ε	25,65	25,05	24,40	23,75	23,10	22,45

Acetic anhydride ($C_4H_6O_3$) – Benzaldehyde (C_7H_6O)
[175], $t = 20$ °C, $\nu = 1{,}8$ MHz

φ_1	0	20	40	60	80	100
ε	17,85	18,75	19,68	20,61	21,52	22,45

Этиленкарбонат ($C_4H_6O_6$) – Dioxan ($C_4H_8O_2$)
[303], $t = 25$ °C

w_1	ε	w_1	ε	w_1	ε	w_1	ε
0	2,209	15,00	9,00	30,00	18,33	45,00	29,75
5,00	4,15	20,00	11,79	35,00	21,95	50,00	34,12
10,00	6,42	25,00	14,90	40,00	25,75	55,00	38,88

1, 4-Dichlorobutane ($C_4H_8Cl_2$) – Hexane (C_6H_{14})
[269], $t = 25$ °C, $\nu = 3$ MHz

x_1	ε	x_1	ε	x_1	ε	x_1	ε
0	1,8776	8,832	2,2151	49,09	4,5166	100,0	9,5111
1,136	1,92237	19,389	2,6887	67,13	6,1472	—	—
4,929	2,06146	32,52	3,5860	88,75	7,8723	—	—

bis-(2-Chloroethyl) ether ($C_4H_8Cl_2O_2$) – Cyclohexane (C_6H_{12})
[166], $t = 20$ °C

x_1	0,000	7,225	17,213	31,667	48,756	64,967	81,778	100,0
ε	2,032	2,510	3,421	5,148	7,916	10,95	15,45	21,17

Tetrahydrofuran (C_4H_8O) – Benzene (C_6H_6)
[305], $t = 20$ °C

x_1	ε	x_1	ε	x_1	ε	x_1	ε
0	2,283	19,94	3,10	62,31	5,09	100,0	7,587
5,45	2,51	30,60	3,57	73,86	5,84	—	—
13,71	2,83	43,09	4,17	87,04	6,65	—	—

Butan-2-one (C_4H_8O) – Cyclohexane (C_6H_{12})
[178], $t = 20$ °C, $\nu = 1{,}8$ MHz

φ_1	0	20	40	60	80	100
ε	2,02	4,28	7,26	10,72	14,44	18,35

Dioxan 1, 4–Dioxan ($C_4H_8O_2$) – Butan–1–ol ($C_4H_{10}O$)
[306], $t = 25$ °C

x_2	100	90	80	70	60	50	40	30	20
ε	17,10	13,73	11,58	9,94	8,40	6,36	5,12	4,11	3,25

1, 4–Dioxan ($C_4H_8O_2$) – Butan–2–ol ($C_4H_{10}O$)
[306], $t = 25$ °C

x_2	100	90	80	70	60	50	40	30	20
ε	15,80	12,90	10,59	8,59	7,05	5,79	4,76	3,92	3,24

1, 4–Dioxan ($C_4H_8O_2$) – Isobutanol ($C_4H_{10}O$)
[306], $t = 25$ °C

x_2	100	90	80	70	60	50	40	30	20
ε	17,12	14,13	11,52	9,36	7,63	6,05	4,09	4,03	3,26

1, 4–Dioxan ($C_4H_8O_2$) – *tert*–Butyl alcohol ($C_4H_{10}O$)
[306], $t = 25$ °C

x_2	100	90	80	70	60	50	40	30	20
ε	11,68	9,82	8,18	7,08	6,03	5,14	4,40	3,71	3,18

[248], $t = 25$ °C, $\nu = 550$ kHz

w_1	10,0	30,0	50,0	70,0	90,0
ε	10,3	7,52	5,57	3,99	2,76

1, 4–Dioxan (C_4H_8O) – 1, 4–Butandiol ($C_4H_{10}O_2$)

x_2	ε at t °C [283]		x_2	ε at t °C [283]	
	15	30		15	30
100,00	32,90	30,16	7,106	3,018	2,938
80,401	24,14	22,00	4,759	2,714	2,659
58,721	16,37	14,95	2,749	2,493	2,455
38,936	9,400	8,678	1,687	2,383	2,351
18,891	4,804	4,516	0,000	2,228	2,209
9,099	3,290	3,169	—	—	—

Dioxan ($C_4H_8O_2$) – Pyridine (C_5H_5N)
[307], $t=25$ °C

x_1	45,32	61,13	75,27	100,00
ε	2,2555	2,2446	2,2343	2,2075

Dioxan ($C_4H_8O_2$) – 1, 2–Dichlorobenzene ($C_6H_4Cl_2$)
[245], $t=25$ °C

x_1	0,0	12,8	24,8	36,1	46,8	56,9	66,4	75,5	84,1	92,2	100,0
ε	10,1	9,3	8,2	7,5	6,3	5,9	5,0	4,3	3,6	2,70	2,20

Dioxan ($C_4H_8O_2$) – Nitrobenzene ($C_6H_5NO_2$)

w_2	ε at t °C [308]		w_2	ε at t °C [308]	
	25	65		25	65
69,270	18,820	16,211	28,583	7,630	6,785
58,410	15,352	13,470	19,70	5,711	5,169
48,924	12,651	10,982	8,666	3,656	3,401
38,613	9,940	8,716	—		

Dioxan ($C_4H_8O_2$) – Benzene (C_6H_6)

x_1	ε at t °C [309]		x_1	ε at t °C [309]	
	20	40		20	40
0	2,2825	2,2423	56,30	2,2635	2,2266
18,74	2,2751	2,2423	71,28	2,2529	2,2193
34,98	2,2751	2,2423			

Dioxan ($C_4H_8O_2$) – Nicotinamide ($C_6H_6N_2O$)
[307], $t=25$ °C

x_1	35,93	53,03	59,64	75,07	100,0
ε	2,2591	2,2499	2,2463	2,2332	2,2119

Dioxan ($C_4H_8O_2$) – 4–Nitroaniline ($C_6H_6N_2O_2$)
[265], $t=25$ °C

w_2	18,23	11,65	7,87	5,50	4,02
ε	12,09	8,28	5,98	4,76	4,01

Dioxan ($C_4H_8O_2$) – Aniline (C_6H_7N)
[231, 232], $t = 20$ °C

x_1	0,00	20,0	40,0	50,0	60,0	80,0	100,0
ε	7,20	6,22	5,25	4,79	4,30	3,25	2,22

w_2	ε at t °C [310]		w_2	ε at t °C [310]	
	25	65		25	65
100,00	6,853	6,178	50,740	4,383	4,001
88,230	6,095	5,505	39,920	3,903	3,614
79,981	5,706	5,128	30,746	3,505	3,277
70,082	5,237	4,768	20,041	3,045	2,892
60,415	4,789	4,372	8,583	2,547	2,472

Dioxan ($C_4H_8O_2$) – Cyclohexanol ($C_6H_{10}O$)
[175], $t = 20$ °C, $\nu = 1{,}8$ MHz

φ_2	100	80	60	40	20	0
ε	15,70	12,75	9,92	7,21	4,60	2,22

Dioxan ($C_4H_8O_2$) – Cyclohexane (C_6H_{12})

x_1	ε at t° C [309]		x_1	ε at t °C [309]	
	20	40		20	40
0	2,0483	2,0059	54,76	2,1437	2,0971
17,25	2,0740	2,0271	71,56	8,1872	2,1310
35,25	2,1066	2,0579	—	—	—

[166], $t = 20$ °C

x_1	0,000	26,904	52,820	73,594	100,0
ε	2,032	2,066	2,109	2,154	2,229

Dioxan ($C_4H_8O_2$) – 2–Nitrotoluene ($C_7H_7NO_2$)

w_2	ε at t °C [311]		w_2	ε at t °C [311]	
	25	65		25	65
69,957	15,230	13,420	40,594	8,640	7,723
60,196	12,770	11,130	29,638	6,615	—
51,810	10,820	9,429	19,48	4,950	4,253

Dioxan ($C_4H_8O_2$) – 3–Nitrotoluene ($C_7H_7NO_2$)

w_2	ε at t °C [311]		w_2	ε at t °C [311]	
	25	65		25	65
71,15	16,81	14,79	31,11	7,830	6,981
61,48	14,52	12,71	22,26	6,066	5,535
51,59	12,21	10,71	11,73	4,154	3,862
40,81	9,879	9,017	—		

Dioxan ($C_4H_8O_2$) – 4–Nitrotoluene ($C_7H_7NO_2$)

w_2	ε at t °C [311]		w_2	ε at t °C [311]	
	25	65		25	65
24,840	7,034	6,333	9,886	4,027	3,743
19,906	5,922	5,423	4,984	3,104	2,940
15,011	5,031	4,565			

Dioxan ($C_4H_8O_2$) – *N*–Methylnicotinamide ($C_7H_8N_2O$)

[307], $t = 25$ °C

x_1	38,88	61,31	78,40	100,00
ε	2,2602	2,2467	2,2329	2,2121

Dioxan ($C_4H_8O_2$) – *o*–Toluidine (C_7H_9N)

[231, 232], $t = 20$ °C

x_2	100	80,00	60,00	50,00	40,00	20,00	0,00
ε	6,085	5,435	4,748	4,360	3,927	3,095	2,220

w_2	ε at t °C [312]		w_2	ε at t °C [312]	
	25	65		25	65
100,00	6,138	5,635	51,124	4,238	3,922
89,395	5,741	5,218	40,551	3,800	3,492
79,114	5,384	4,844	30,727	3,398	3,186
69,480	4,591	4,501	20,196	3,003	2,845
60,074	4,591	4,184	9,462	2,558	4,484

Dioxan ($C_4H_8O_2$) – *m*–Toluidine (C_7H_9N)

w_2	ε at *t* °C [312]		w_2	ε at *t* °C [312]	
	25	65		25	65
100,0	5,816	5,246	50,314	3,884	3,602
89,582	5,304	4,793	41,351	3,534	3,343
79,520	4,965	4,530	29,871	3,174	2,998
70,064	4,587	2,193	20,421	2,867	2,743
60,195	4,224	3,894	—		

Dioxan ($C_4H_8O_2$) – *p*–Toluidine (C_7H_9N)
[312], $t = 25$ °C

w_2	19,908	15,043	9,763	4,936
ε	2,731	2,594	2,448	2,299

Dioxan ($C_4H_8O_2$) – Heptane (C_7H_{16})
[309], $t = 20$ °C

x_1	0	20,85	43,59	62,82	74,88	100
ε	1,9491	1,9780	2,0251	2,0786	2,1235	2,2326

Dioxan ($C_4H_8O_2$) – *o*–Hydroxyacetophenone ($C_8H_8O_2$)

w_2	ε at *t* °C [313]		w_2	ε at *t* °C [313]	
	25	50		25	50
24,46	5,007	4,747	8,474	3,152	3,036
16,83	4,121	3,913	0,000	2,205	2,153

Dioxan ($C_4H_8O_2$) – *N, N*–Dimethylnicotinamide ($C_8H_{10}N_2O$)
[307], $t = 25$ °C

x_1	25,49	41,12	61,04	77,47	100,0
ε	2,2679	2,2581	2,2434	2,2313	2,2133

Dioxan ($C_4H_8O_2$) – *N*–Ethylnicotinamide ($C_8H_{10}N_2O$)
[307], $t = 25$ °C

x_1	31,08	46,69	62,74	76,97	100,0
ε	2,2617	2,2561	2,2440	2,2332	2,2121

Dioxan ($C_4H_8O_2$) – 1–Nitronaphthalene ($C_{10}H_7NO_2$)

w_2	ε at t°C [311]		w_2	ε at t°C [311]	
	25	65		25	65
24,957	5,599	5,062	10,027	3,406	3,190
19,755	4,773	4,353	4,966	2,777	2,662
14,996	4,067	3,745	—		

Dioxan ($C_4H_8O_2$) – 2–Naphthylamine ($C_{10}H_9N$)

w_2	ε at t°C [308]	
	25	65
20,171	3,092	2,918
10,689	2,661	2,538
5,068	2,411	2,361

Dioxan ($C_4H_8O_2$) – *N, N*–Diethylnicotinamide ($C_{10}H_{14}N_2O$)

x_1	[307], t=25 °C		
	ε	x_1	ε
29,03	2,2658	56,00	2,2466
41,91	2,2574	100,0	2,2125

Dioxan ($C_4H_8O_2$) – Decan–1–ol ($C_{10}H_{22}O$)
[306], t=25 °C

x_1	100	90	80	70	60	50	40	30	20
ε	7,80	6,84	6,22	5,56	5,01	4,47	3,99	3,54	3,11

Dioxan ($C_4H_8O_2$) – Phenyl phenylthiosulphonate ($C_{12}H_{10}NaO_3S_2$)
[315], t=25 °C

x_2	65,1	52,8	43,4	33,9	22,1	11,1	0
ε	2,4819	2,4291	2,3889	2,3473	2,2964	2,2495	2,2027

Dioxan ($C_4H_8O_2$) – *N, N*–Dipropylnicotinamide ($C_{12}H_{18}N_2O$)
[307], t=25 °C

x_1	0,00	21,82	43,84	64,87	100,0
ε	2,2770	2,2667	2,2536	2,2380	2,2082

Dioxan ($C_4H_8O_2$) – Tributyl phosphate ($C_{12}H_{27}O_4P$)
[214], t=25 °C

x_2	100,0	72,67	69,53	38,84	21,02	10,60	0,00
ε	8,05	7,32	7,24	5,78	4,55	3,54	2,20

Butyric acid ($C_4H_8O_2$) – Pyridine (C_5H_5N)
[295], $t = 25$ °C

φ_2	ε	φ_2	ε	φ_2	ε	φ_2	ε
100	12,91	70	12,64	40	12,24	10	6,91
90	12,79	60	12,32	30	12,02	0	3,04
80	12,73	50	—	20	10,02	—	—

Butyric acid ($C_4H_8O_2$) – Valeric acid ($C_5H_{10}O_2$)
[244], $t = 25$ °C, $\nu = 28$ MHz

φ_2	100	80,76	50,04	39,29	16,00	0
ε	2,741	2,777	2,824	2,839	2,872	2,881

Butyric acid ($C_4H_8O_2$) – Benzene (C_6H_6)

x_1	ε at t °C [273]			x	ε at t °C [273]		
	10	40	70		10	40	70
0,00	2,315	2,253	2,188	18,75	2,388	2,340	2,291
3,79	2,301	2,266	2,203	47,29	2,542	2,519	2,500
6,40	2,333	2,280	2,219	71,83	2,705	2,715	2,725
8,94	2,342	2,290	2,231	100	2,932	3,001	3,074

[275], $t = 22$ °C

x_1	11,27	12,69	16,69	21,26	36,19	33,28	55,09	70,74	92,13
ε	2,334	2,338	2,360	2,384	2,430	2,446	2,590	2,700	2,850

Butyric acid ($C_4H_8O_2$) – Hexanoic acid ($C_6H_{12}O_2$)
[244], $t = 25$ °C, $\nu = 28$ MHz

φ_2	100	81,13	72,09	53,25	26,19	0
ε	2,816	2,835	2,844	2,860	2,887	2,881

Butyric acid ($C_4H_8O_2$) – Diethylaniline ($C_{10}H_{15}N$)
[295], $t = 25$ °C

φ_2	100	83,3	70,0	55,0	45,0	32,7	20,0	0
ε	5,26	6,30	7,97	10,91	12,5	12,9	11,55	3,04

Butyric acid ($C_4H_8O_2$) – Triethylamine ($C_6H_{15}N$)

φ_2	ε at t °C [295]			φ_2	ε at t °C [295]		
	25	35	45		25	35	45
85,7	4,50	4,34	4,24	25,0	23,0	—	—
66,7	14,3	14,0	13,2	11,1	13,9	13,8	13,0
50,0	22,6	24,1	25,1	0	3,04	—	—
40,0	29,5	31,8	32,2	—	—	—	—

Butyric acid ($C_4H_8O_2$) – Octane (C_8H_{18})
[244], $t = 25$ °C, $\nu = 28$ MHz

φ_1	0	4,55	12,41	35,27	56,71	73,22	100
ε	1,942	1,975	2,023	2,181	2,374	2,534	2,881

Ethyl acetate ($C_4H_8O_2$) – Dioxan ($C_4H_8O_2$)
[199], $t = 25$ °C, $\nu = 4,5$ MHz

x_2	ε	x_2	ε	x_2	ε	x_2	ε
100	2,209	69,7	3,76	53,9	4,63	36,6	5,31
92,0	2,64	65,7	3,98	50,8	4,76	31,5	5,46
85,2	3,01	62,1	4,17	47,9	4,88	25,7	5,60
79,3	3,29	59,0	4,33	44,6	5,02	18,7	5,74
74,2	3,56	56,1	4,25	40,8	5,18	10,3	5,88

Ethyl acetate ($C_4H_8O_2$) – Diethyl ether ($C_4H_{10}O$)
[175], $t = 25$ °C, $\nu = 1,8$ MHz

φ_1	0	20	40	60	80	100
ε	4,35	4,73	5,11	5,48	5,84	6,19

Ethyl acetate ($C_4H_8O_2$) – 1, 2–Dichlorobenzene ($C_6H_4Cl_2$)
[199], $t = 25$ °C, $\nu = 4,5$ MHz

x_2	ε	x_2	ε	x_2	ε	x_2	ε
100	10,12	63,5	9,37	46,6	8,86	30,3	7,90
89,7	9,93	59,2	9,24	43,9	8,73	25,9	7,61
81,4	9,78	55,4	9,12	41,1	8,58	20,7	7,22
74,3	9,64	52,1	9,03	37,9	8,39	14,8	6,89
68,5	9,51	49,1	8,95	34,3	8,18	8,0	6,45

Ethyl acetate ($C_4H_8O_2$) – Benzene (C_6H_6)

x_1	ε at t °C [296]		x_1	ε at t °C [296]	
	25	50		25	50
3,10	2,410	2,341	10,60	2,708	2,602
4,75	2,477	2,400	15,75	2,912	2,782
6,14	2,532	2,450	—	—	—

[170], $t = 20$ °C

x_1	0,000	17,395	45,597	82,750	100,0
ε	2,249	2,808	3,807	5,583	6,155

Ethyl acetate ($C_4H_8O_2$) – Cyclohexanone ($C_6H_{10}O$)
[175], $t = 20$ °C, $\nu = 1,8$ MHz

φ_1	0	20	40	60	80	100
ε	15,70	13,68	11,74	9,82	8,01	6,19

Ethyl acetate ($C_4H_8O_2$) – Cyclohexane (C_6H_{12})
[199], $t = 25$ °C, $\nu = 4,5$ MHz

x_2	ε	x_2	ε	x_2	ε	x_2	ε
100	2,015	64,3	3,14	47,4	4,14	31,1	5,00
90,0	2,14	60,1	3,42	44,8	4,28	26,6	5,19
81,8	2,42	56,3	3,62	41,9	4,42	21,3	5,39
75,0	2,71	53,0	3,80	38,7	4,61	15,4	5,61
69,2	2,93	50,0	3,98	35,1	4,77	8,3	5,82

Ethyl acetate ($C_4H_8O_2$) – Benzonitrile (C_7H_5N)
[175], $t = 20$ °C, $\nu = 1,8$ MHz

φ_2	100	80	60	40	20	0
ε	25,65	21,75	17,83	13,91	10,05	6,19

Ethyl acetate ($C_4H_8O_2$) – Octane (C_8H_{18})
[244], $t = 25$ °C, $\nu = 28$ MHz

φ_2	100	84,16	69,77	49,86	16,37	0
ε	1,91	2,45	3,08	3,83	5,20	6,00

1–Bromobutane (C_4H_9Br) – 1–Chlorobutane (C_4H_9Cl)
[244], $t = 25$ °C, $\nu = 28$ MHz

φ_2	100	87,77	48,95	34,15	15,90	0
ε	7,14	7,08	7,00	6,95	6,90	6,00

1–Bromobutane (C_4H_9Br) – Heptane (C_7H_{16})
[223], $\nu = 0,5$ MHz

x_1	ε при t °C									
	−90	−70	−50	−30	−10	10	30	50	70	90
0	2,083	2,055	2,027	1,999	1,972	1,944	1,916	1,888	1,858	—
4,56	2,349	2,289	2,239	2,193	2,149	2,107	2,064	2,022	1,979	1,938
9,30	2,588	2,513	2,441	2,373	2,310	2,251	2,197	2,146	2,097	2,049
14,09	2,854	2,751	2,653	2,568	2,491	2,417	2,346	2,279	2,217	2,158
25,79	3,548	3,379	3,226	3,086	2,957	2,843	2,737	2,640	2,547	2,455
41,54	4,687	4,389	4,119	3,886	3,679	3,499	3,336	3,187	3,050	2,914
59,59	6,280	5,810	5,405	5,044	4,732	4,454	4,203	4,966	4,761	3,577
84,13	8,970	8,220	7,565	6,984	6,474	6,039	5,640	5,290	4,969	4,662
100	11,08	10,14	9,258	8,520	7,830	7,315	6,799	6,345	5,930	5,535

1–Chlorobutane (C_4H_9Cl) – Chlorobenzene (C_6H_5Cl)
[244], $t = 25$ °C, $\nu = 28$ MHz

φ_1	0	14,42	30,18	74,09	100
ε	5,61	5,83	6,08	6,73	7,14

1–Chlorobutane (C_4H_9Cl) – Heptane (C_7H_{16})
[223], $\nu = 0,5$ MHz

x_1	ε at t °C			x_1	ε at t °C		
	−90	−70	−50		−90	−70	−50
0	2,083	2,055	2,027	26,17	3,659	3,453	3,266
2,42	2,219	2,175	2,135	56,19	6,302	5,792	5,322
8,89	2,567	2,485	2,413	86,23	10,16	9,229	8,465
15,2	2,954	2,837	2,732	100	12,24	10,98	9,940
21,91	3,359	9,184	3,038	—	—	—	—

x_1	ε at t °C					
	−30	−10	10	30	50	70
0	1,999	1,972	1,944	1,916	1,888	1,858
2,42	2,096	2,058	2,022	1,985	1,950	1,913
8,89	2,347	2,285	2,227	2,174	2,122	2,070
15,2	2,635	2,547	2,466	2,393	2,319	2,248
21,91	2,904	2,792	2,688	2,591	2,494	2,410
26,17	3,122	2,986	2,861	2,745	2,642	2,543
56,19	4,925	4,586	4,285	4,025	3,792	3,570
86,23	7,743	7,077	6,518	6,062	5,676	5,250
100	9,073	8,320	7,663	7,090	6,558	6,045

1–Chlorobutane (C_4H_9Cl) – Nitrobenzene ($C_6H_5NO_2$)
[146], $t = 25$ °C

φ_2	ε	φ_2	ε	φ_2	ε	φ_2	ε
100	34,82	70	25,22	40	16,53	10	9,37
90	31,49	60	22,21	30	13,88	0	7,19
80	28,28	50	19,15	20	11,61	—	—

1–Chlorobutane (C_4H_9Cl) – Acetophenone (C_8H_8O)
[147], $t = 25$ °C

φ_2	ε	φ_2	ε	φ_2	ε	φ_2	ε
100	17,16	70	14,00	40	10,92	10	8,15
90	16,24	60	12,94	30	10,07	0	7,19
80	15,08	50	11,89	20	9,08	—	—

***tert*–Butyl chloride (C_4H_9Cl) – Nitrobenzene ($C_6H_5NO_2$)**
[146], $t = 25$ °C

φ_2	100	80	60	50	40	20	0
ε	34,82	29,27	24,08	21,51	18,87	14,16	9,56

***tert*–Butyl chloride (C_4H_9Cl) – Acetophenone (C_8H_8O)**
[147], $t = 25$ °C

φ_2	100	80	70	50	30	20	0
ε	17,16	15,58	14,83	13,37	11,84	11,07	9,56

tert–Butyl chloride (C_4H_9Cl) – Heptane (C_7H_{16})
[316], $\nu = 0,5$ MHz

x_1	ε at t °C							
	−70	−50	−30	−10	10	30	50	70
0	2,048	2,020	1,992	1,966	1,939	1,910	1,879	1,848
3,89	2,289	2,231	2,179	2,129	2,084	2,042	2,000	1,960
5,43	2,381	2,314	2,258	2,199	2,144	2,095	2,053	2,007
13,09	2,892	2,754	2,641	2,546	2,447	2,366	2,292	2,219
21,56	3,559	3,335	3,141	2,981	2,849	2,720	2,598	2,487
100	—	—	—	11,72	10,34	9,23	—	—

1–Iodobutane (C_4H_9J) – Heptane (C_7H_{16})
[223], $\nu = 0,5$ MHz

x_1	ε at t °C							
	−60	−40	−20	0	20	40	60	80
3,86	2,201	2,157	2,115	2,076	2,038	2,001	1,964	1,925
8,31	2,386	2,325	2,271	2,219	2,169	2,119	2,073	2,027
16,98	2,760	2,668	2,585	2,506	2,432	2,360	2,295	2,229
29,43	3,351	3,195	3,063	2,939	2,820	2,718	2,630	2,556
54,62	4,729	4,445	4,188	3,971	3,783	3,614	3,465	3,318
80,90	6,543	6,082	5,673	5,304	5,005	4,755	4,507	4,254
100	8,180	7,534	7,002	6,542	6,117	5,737	5,421	5,108

Morpholine (C_4H_9NO) – Benzene (C_6H_6)
[167], $t = 25$ °C

x_1	8,588	18,193	22,967	27,621	35,608	100
ε	2,574	2,934	3,128	3,316	3,665	7,33

Ethyl ethylthiosulphonate ($C_4H_{10}O_2S_2$) – Benzene (C_6H_6)
[315], $t = 25$ °C

x_1	0	12,9	30,1	54,5	65,8	85,2
ε	2,2714	2,3054	2,3509	2,4157	2,4468	2,5000

Butan–1–ol ($C_4H_{10}O$) – Diethyl ether ($C_4H_{10}O$)
[175], $t = 20$ °C, $\nu = 1,8$ MHz

φ_1	0	20	40	60	80	100
ε	4,35	5,92	8,00	10,85	14,15	17,90

Butan-1-ol ($C_4H_{10}O$) – 1, 2-Dichlorobenzene ($C_6H_4Cl_2$)

[317], $t = 30$ °C, fig. 12

x_1	ε	x_1	ε	x_1	ε
0,00	9,875	28,90	9,8750	66,10	12,7560
5,00	9,6590	35,00	10,0670	72,20	13,8810
10,10	9,6150	36,00	10,2820	78,70	14,8720
15,10	9,6000	45,50	10,7180	88,90	15,7320
18,50	9,7950	46,30	10,8920	100,00	16,82
26,20	9,7450	57,30	11,7950	—	—

Butan-1-ol ($C_4H_{10}O$) – Nitrobenzene ($C_6H_5NO_2$)

[175], $t = 20$ ° C, $\nu = 1,8$ MHz

φ_1	0	20	40	60	80	100
ε	37,75	29,85	25,25	21,70	19,20	17,90

Butan-1-ol ($C_4H_{10}O$) – Benzene (C_6H_6)

x_1	ε at t °C [200], $\nu = 1$ MHz			
	25	35	45	55
0	2,2747	2,2530	2,2329	2,2137
14,36	2,9239	2,8818	2,8393	2,8036
17,33	3,1153	3,0621	3,0097	2,9632
27,32	3,9736	3,8289	3,7045	3,6090
37,00	4,8131	4,5718	4,3860	4,2609
43,39	6,1964	5,7289	5,3790	5,0838
68,68	10,9987	10,1353	9,3230	8,5946
78,48	13,0161	11,9381	10,9302	10,0264
100	17,0997	15,8407	14,6638	13,6069

x_1	ε at t °C [278]						
	10	20	30	40	50	60	70
0	2,315	2,294	2,274	2,253	2,231	2,209	2,186
2,11	2,400	2,378	2,354	2,330	2,304	2,274	2,243
5,78	2,548	2,526	2,502	2,473	2,438	2,401	2,358
7,98	2,655	2,631	2,600	2,565	2,525	2,480	2,435
11,17	2,835	2,797	2,756	2,713	2,665	2,612	2,555
25,25	4,066	3,899	3,746	3,604	3,472	3,350	3,238
48,99	8,29	7,58	6,94	6,37	5,96	5,43	5,03
73,85	14,03	12,84	11,72	10,68	9,74	8,91	8,11

[257, 262], $t = 25$ °C, $\lambda = 37,5$ m

x_1	ε	x_1	ε	x_1	ε	x_1	ε
0,00	2,271	14,13	2,900	39,65	5,652	66,33	10,46
0,54	2,300	19,32	3,200	45,02	6,350	69,68	11,14
1,28	2,320	20,05	3,328	50,11	7,180	70,83	11,54
3,24	2,365	24,08	3,750	50,57	7,376	79,48	13,30
4,93	2,415	29,74	4,256	54,47	8,00	84,54	13,95
10,16	2,545	32,35	4,600	59,24	8,95	91,14	15,36
10,57	2,680	38,69	5,400	60,01	9,054	100	17,38

[175], $t = 20$ °C, $\nu = 1,8$ MHz

φ_1	0	20	40	60	80	100
ε	2,28	3,32	5,80	9,70	13,90	17,90

Butan-1-ol ($C_4H_{10}O$) – Cyclohexane (C_6H_{12})

x_1	ε at t °C [319], $\nu=2$ MHz				x_1	ε at t °C [319], $\nu=2$ MHz			
	25	35	45	55		25	35	45	55
0,00	2,0151	1,9997	1,9843	1,9689	47,16	6,3492	5,4716	5,2803	4,8557
10,29	2,2487	2,2367	2,2244	2,2114	63,94	9,8075	8,9122	8,1056	7,3557
18,70	2,6024	2,5610	2,5271	2,4937	80,70	13,2138	12,0775	11,0165	10,0172
28,63	3,4776	3,2394	3,1647	3,0468	90,36	15,0804	13,9265	12,8554	11,8549
31,71	3,7495	3,5261	3,3545	3,2099	100,0	17,0997	15,8407	14,6638	13,6069
41,42	5,2377	4,7872	4,4272	4,1211					

x_1	ε at t °C [278], $\nu=0,5$ MHz						
	10	20	30	40	50	60	70
0	2,041	2,027	2,013	1,988	1,981	1,963	1,944
3,91	2,108	2,099	2,088	2,076	2,061	2,045	2,028
11,02	2,272	2,260	2,248	2,236	2,224	2,210	2,192
20,02	2,743	2,674	2,623	2,579	2,537	2,497	2,458

[306], $t = 25$ °C

x_1	20	30	40	50	60	70	80	90	100
ε	2,64	3,57	4,85	6,67	8,67	11,02	13,03	14,92	17,10

[318], $t=21,5$ °C

φ_2	ε	φ_2	ε	φ_2	ε	φ_2	ε
100	2,01	80	2,84	50	7,84	10	15,8
95	2,22	75	3,38	40	9,97	0	17,6
90	2,25	70	4,02	30	12,04	—	—
85	2,47	60	5,55	20	13,9	—	—

Butan-1-ol ($C_4H_{10}O$) – Hexane (C_6H_{14})

x_1	ε at t °C [319], ν=2 MHz				x_1	ε at t °C [319], ν=2 MHz			
	25	35	45	55		25	35	45	55
0	1,8823	1,8668	1,8513	1,8358	35,74	3,8124	3,5875	3,3931	3,2222
5,93	1,9954	1,9873	1,9769	1,9679	43,73	5,3329	4,9039	4,5324	4,2064
11,96	2,1212	2,1090	2,0954	2,0828	53,34	7,0503	6,2883	5,7918	5,2611
16,44	2,2246	2,2048	2,1851	2,1678	91,05	15,0068	13,8306	12,7473	11,7696
21,20	2,3900	2,3700	2,3500	2,3300	100,0	17,0997	15,8407	14,6638	13,6069
25,59	2,7200	2,6420	2,6235	2,5950					

Butan-1-ol ($C_4H_{10}O$) – Heptane (C_7H_{16})

[278], ν=0,5 MHz

x_1	ε at t °C								
	−90	−80	−70	−60	−50	−40	−30	−20	−10
3,12	2,113	2,098	2,082	2,068	2,052	2,041	2,028	2,016	2,004
5,25	2,131	2,118	2,105	2,092	2,080	2,066	2,054	2,042	2,032
8,05	2,196	2,168	2,148	2,134	2,122	2,109	2,096	2,084	2,073
10,42	2,223	2,208	2,194	2,179	2,164	2,149	2,136	2,123	2,111
13,83	—	2,360	2,333	2,308	2,285	2,263	2,241	2,221	2,203
26,55	4,790	4,703	4,502	4,177	9,889	3,628	3,401	3,200	3,040
44,51	9,27	10,32	10,31	9,61	8,89	8,18	7,48	6,80	6,13
61,52	—	16,85	16,20	15,40	14,44	13,40	12,40	11,44	10,49
80,42	16,3	24,01	24,4	23,0	21,6	20,0	18,6	17,3	16,0
100	—	30,0	31,3	29,8	27,9	26,2	24,6	22,9	21,4

x_1	ε at t °C				x_1	ε at t °C			
	0	10	20	30		0	10	20	30
3,12	1,993	1,982	1,972	1,961	26,55	2,910	2,798	2,696	2,608
5,25	2,021	2,011	2,001	1,991	44,51	5,55	5,16	4,82	4,48
8,05	2,063	2,053	2,044	2,034	61,52	9,57	8,78	8,00	7,30
10,42	2,110	2,088	2,077	2,068	80,42	14,9	13,8	12,6	11,6
13,83	2,186	2,167	2,153	2,141	100	20,2	18,6	17,4	16,1

x_1	ε at t °C					
	40	50	60	70	80	90
3,12	1,950	1,936	1,921	1,907	1,890	1,872
5,25	1,980	1,969	1,955	1,939	1,921	1,902
8,05	2,024	2,013	2,001	1,986	1,968	1,949
10,42	2,058	2,046	2,034	2,019	2,003	1,985
13,83	2,132	2,118	2,105	2,089	2,070	2,042
26,55	2,547	2,500	2,454	2,416	2,384	2,348
44,51	4,19	3,92	3,70	3,53	3,39	3,26
61,52	6,64	6,07	5,54	5,11	4,80	4,46
80,42	10,6	9,71	8,72	8,01	7,30	6,69
100	14,9	13,8	12,8	11,8	10,8	9,9

Butan-1-ol ($C_4H_{10}O$) – *p*-Xylene (C_8H_{10})

[306], $t = 25$ °C

x_1	20	30	40	50	60	70	80	90	100
ε	2,92	3,68	4,74	6,20	7,38	9,72	12,45	15,23	17,10

Butan-1-ol ($C_4H_{10}O$) – Mesitylene (C_9H_{12})

[306], $t = 25$ °C

x_1	20	30	40	50	60	70	80	90	100
ε	2,89	3,53	4,49	5,82	7,56	9,63	12,07	14,59	17,10

Isobutanol ($C_4H_{10}O$) – 1, 4-Dichlorobenzene ($C_6H_4Cl_2$)

[279], $t = 55$ °C, рис. 12

x_1	ε	x_1	ε	x_1	ε	x_1	ε
0,00	2,397	19,59	2,999	49,67	4,969	79,87	9,384
4,92	2,505	29,83	3,467	60,13	6,130	89,87	11,329
10,22	2,646	39,85	4,130	69,90	7,474	100	13,177

Isobutanol ($C_4H_{10}O$) – Benzene (C_6H_6)

[218], $t = 20$ °C

x_1	0	5,21	10,64	20,01	35,42	100
ε	2,270	2,494	2,742	3,410	5,083	20,2

Isobutanol ($C_4H_{10}O$) – Cyclohexane (C_6H_{12})

[306], $t = 25$ °C, fig. 10

x_1	20	30	40	50	60	70	80	90	100
ε	2,59	3,32	4,69	6,49	8,33	10,77	12,91	14,88	17,12

Isobutanol ($C_4H_{10}O$) – *m*–Xylene (C_8H_{10})
[306], $t=25$ °C

x_1	20	30	40	50	60	70	80	90	100
ε	2,98	3,58	4,55	5,94	7,74	9,75	11,96	14,67	17,12

[320]

w_1	0	10,234	18,398	40,018	62,907	100
t °C	13,5	13	13	13,5	13,5	14
ε	2,3518	2,7925	3,3742	6,5243	11,365	19,294

Isobutanol ($C_4H_{10}O$) – Mesitylene (C_9H_{12})
[306], $t=25$ °C

x_1	20	30	40	50	60	70	80	90	100
ε	2,89	3,58	4,37	5,58	7,31	9,37	11,74	14,28	17,12

Butan–2–ol ($C_4H_{10}O$) – Benzene (C_6H_6)

x_1	ε at t °C [200], ν=1 MHz				x_1	ε at t °C [200], ν=1 MHz			
	25	35	45	55		25	35	45	55
0	2,2747	2,2530	2,2329	2,2131	58,51	7,0825	6,5523	6,0606	5,5926
14,86	2,8451	2,8024	2,7608	2,7255	70,61	9,1660	8,3842	7,6150	7,0189
23,93	3,3186	3,2535	3,1969	3,1604	81,76	11,5569	10,3326	9,2650	8,3959
34,14	4,0726	3,9215	3,7847	3,6913	89,78	13,8541	12,3050	10,9525	9,9334
43,14	4,9880	4,6929	4,4422	4,2511	100	16,1760	14,5368	13,0194	11,6335
53,39	6,2523	5,8081	5,4542	5,1388					

Butan–2–ol ($C_4H_{10}O$) – Cyclohexane (C_6H_{12})

x_1	ε at t °C [319, 306], ν=2 MHz				x_1	ε at t °C [319, 306], ν=2 MHz			
	25	35	45	55		25	35	45	55
0	2,0151	1,9997	1,9843	1,9689	41,86	3,7934	3,7043	3,5089	3,3546
7,25	2,1329	2,1265	2,1191	2,1082	58,15	7,0025	6,2032	5,5528	5,0380
11,52	2,2181	2,2160	2,2108	2,2017	79,36	11,5653	10,1721	9,0075	7,9833
20,69	2,4933	2,4780	2,4615	2,4444	90,75	14,1323	12,5851	11,2137	10,0522
26,89	2,7507	2,7077	2,6706	2,6359	100,00	16,1760	14,5368	13,0194	11,6335
28,44	2,8500	2,7004	2,7630	2,7333					

Butan–2–ol ($C_4H_{10}O$) – Hexane (C_6H_{14})

x_1	ε at t °C [319], ν=2 MHz				x_1	ε at t °C [319], ν=2 MHz			
	25	35	45	55		25	35	45	55
0	1,8823	1,8668	1,8513	1,8358	44,34	4,0895	3,7448	3,4115	3,3306
6,61	1,9875	1,9801	1,9698	1,9587	56,83	6,2596	5,5443	4,9884	4,5500
11,07	2,0469	2,0423	2,0355	2,0279	76,81	10,5020	9,1472	8,0538	7,1157
21,14	2,2893	2,2801	2,2700	2,2598	90,49	13,7263	12,1739	10,8179	9,6269
35,75	3,2334	3,0848	2,9717	2,8799	100,0	16,1760	14,5369	13,0194	11,6335

Butan–2–ol ($C_4H_{10}O$) – *p*–Xylene (C_8H_{10})
[306], $t = 25$ °C

x_1	20	30	40	50	60	70	80	90	100
ε	2,86	3,31	4,00	5,02	6,50	8,72	10,68	13,20	15,80

Butan–2–ol ($C_4H_{10}O$) – Mesitylene (C_9H_{12})
[306], $t = 25$ °C

x_1	20	30	40	50	60	70	80	90	100
ε	2,80	3,22	3,86	4,79	6,26	8,11	10,28	13,00	15,80

***tert*–Butyl alcohol ($C_4H_{10}O$) – Benzene (C_6H_6)**

x_1	ε at t °C [200]				x_1	ε at t °C [200]			
	25	35	45	55		25	35	45	55
0	2,2747	2,2530	2,2329	2,2137	53,26	4,4416	4,2639	4,1919	4,2151
12,47	2,6479	2,6036	2,6080	2,5926	66,24	5,5665	5,1187	4,9411	4,8224
14,68	2,7246	2,6786	2,6786	2,6606	80,36	7,4959	6,6403	6,1742	5,8559
27,70	3,1517	3,1081	3,1053	3,0994	100	11,9921	10,2043	8,7908	7,8390

***tert*–Butyl alcohol ($C_4H_{10}O$) – Cyclohexane (C_6H_{12})**

x_1	ε at t °C [319, 306], ν=2 MHz				x_1	ε at t °C [319, 306], ν=2 MHz			
	25	35	45	55		25	35	45	55
0	2,0151	1,9997	1,9843	1,9689	52,42	3,3801	3,3267	3,2874	2,2515
5,8	2,0944	2,0903	2,0836	2,0747	60,76	4,0633	3,8822	3,7549	3,6556
10,47	2,1355	2,1384	2,1367	2,1320	71,04	5,4207	4,9388	4,6023	4,3554
17,69	2,2430	2,2551	2,2636	2,2667	89,28	9,6118	8,0891	7,0415	6,2548
23,33	2,3255	2,3436	2,3569	2,3644	100,0	11,9921	10,2043	8,7908	7,8390
43,24	2,8830	2,8922	2,8964	2,8936					

tert–**Butyl alcohol** (C_4H_{10}) – **Hexane** (C_6H_{14})

x_1	ε at t °C [319], ν=2 MHz				x_1	ε at t °C [319], ν=2 MHz			
	25	35	45	55		25	35	45	55
0	1,8823	1,8668	1,8513	1,8358	30,45	2,3378	2,3560	2,3697	2,3620
5,78	1,9511	1,9436	1,9338	1,9248	54,64	3,1995	3,1561	3,1172	3,0840
8,10	1,9735	1,9644	1,9481	1,9326	64,19	4,0897	3,8761	3,7235	3,6050
11,97	2,0210	2,0211	2,0187	2,0154	81,66	7,6417	6,5900	5,8727	5,3776
17,69	2,1092	2,1173	2,1223	2,1240	100,0	11,9921	10,2043	8,7908	7,8390
21,71	2,1498	2,1616	2,1690	2,1749					

tert–**Butyl alcohol** ($C_4H_{10}O$) – **Heptane** (C_7H_{16})

x_1	ε at t °C [316]					
	−30	−10	10	30	50	70
5,85	2,037	2,018	2,004	1,991	1,972	1,934
8,53	2,058	2,037	2,027	2,016	2,007	1,981
13,44	2,084	2,076	2,069	2,072	2,066	2,045
22,59	2,267	2,168	2,178	2,197	2,204	2,193
49,57	—	3,316	3,026	2,958	2,920	2,889
72,65	—	—	6,21	4,91	4,28	3,99
100	—	—	—	10,92	8,49	6,89

tert–**Butyl alcohol** ($C_4H_{10}O$) – *p*–**Xylene** (C_8H_{10})

[306], $t = 25$ °C

x_1	20	30	40	50	60	70	80	90	100
ε	2,71	2,98	3,32	3,83	4,52	5,71	7,11	9,32	11,68

tert–**Butyl alcohol** ($C_4H_{10}O$) – **Mesitylene** (C_9H_{12})

[306], $t = 25$ °C

x_1	20	30	40	50	60	70	80	90	100
ε	2,67	2,91	3,22	3,69	4,39	5,38	6,92	8,95	11,68

Diethyl ether ($C_4H_{10}O$) – **Butylamine** ($C_4H_{11}N$)

[175], $t = 20$ °C, $\nu = 1,8$ MHz

φ_2	100	80	60	40	20	0
ε	4,78	4,69	4,60	4,51	4,43	4,35

Diethyl ether ($C_4H_{10}O$) – Pentane (C_5H_{12})
[172]

w_1	0	14,45	36,70	50,68	54,26	69,96	89,19	100
t °C	20,6	20,6	20,8	20,8	20,8	20,9	20,7	17,2
ε	1,836	2,126	2,396	2,820	2,961	3,438	4,246	4,671

Diethyl ether ($C_4H_{10}O$) – Pentan–1–ol ($C_5H_{12}O$)
[267], $t=25$ °C, $\nu=3$ MHz

x_2	100	47,3	26,8	19,8	5,1	0
ε	15,4	7,94	6,03	5,54	4,62	4,35

Diethyl ether ($C_4H_{10}O$) – Trichlorobenzene ($C_6H_3Cl_3$)
[175], $t=20$ °C, $\nu=1,8$ MHz

φ_1	0	20	40	60	80	100
ε	6,75	6,46	6,06	5,60	5,03	4,35

Diethyl ether ($C_4H_{10}O$) – Chlorobenzene (C_6H_5Cl)
[267], $t=25$ °C, $\nu=3$ MHz

x_1	0	49,7	51,6	74,7	81,2	90,0	100
ε	5,66	5,11	5,07	4,70	4,57	4,47	4,25

x_1	ε at t °C [234]					x_1	ε at t °C [234]				
	20	0	−20	−40	−60		20	0	−20	−40	−60
0	5,605	5,983	6,401	6,872	—	60,5	4,935	5,285	5,745	6,255	6,845
20,8	5,400	5,783	6,205	6,555	—	80,35	4,615	5,015	5,469	6,005	6,655
39,0	5,205	5,582	6,005	6,481	7,050	100	4,300	4,706	5,180	5,791	6,545
50,0	5,075	5,441	5,870	6,365	6,960						

[173], $t=18$ °C

w_2	ε	w_2	ε	w_2	ε	w_2	ε
100,00	5,723	70,215	5,317	38,918	4,895	10,933	4,018
90,091	5,599	60,302	5,142	29,514	4,761	0	4,360
75,540	5,413	49,925	5,037	18,349	4,591	—	—

[175], $t = 20$ °C, $\nu = 1{,}8$ MHz

φ_1	0	20	40	60	80	100
ε	5,70	5,52	5,30	5,06	4,72	4,35

[227], $t = 15$ °C, $\nu = 3$ MHz

φ_2	0	20	40	60	80	100
ε	4,46	4,79	5,06	5,37	5,58	5,70

Diethyl ether ($C_4H_{10}O$) – 2–Chlorophenol (C_6H_5ClO)

[177], $t = 25$ °C, $\nu = 1$ MHz

φ_2	100	80	70	60	50	40	20	0
ε	6,21	8,93	9,68	9,82	9,41	8,77	6,37	4,21

Diethyl ether ($C_4H_{10}O$) – Nitrobenzene ($C_6H_5NO_2$)

[267], $t = 25$ °C, $\nu = 3$ MHz

x_2	44,2	33,5	18,2	8,5	5,5	0
ε	16,97	13,67	9,05	6,45	5,66	4,35

[227], $t = 20$ °C, $\nu = 3$ MHz

x_2	100	80	60	40	20	0
ε	33,91	27,26	20,88	15,07	9,51	4,35

[175], $t = 20$ °C, $\nu = 1{,}8$ MHz

φ_2	100	80	60	40	20	0
ε	35,75	28,60	21,85	15,40	9,70	4.35

Diethyl ether ($C_4H_{10}O$) – Benzene (C_6H_6) *)

x_1	ε at t °C [237]			x_1	ε at t °C [237]		
	5	20	26		5	20	26
0	2,527	2,372	2,278	60	3,969	3,566	3,393
20	3,008	2,770	2,650	80	4,449	3,964	3,765
40	3,488	3,168	3,622	100	4,930	4,362	4,137

*) See also [215, 291, 321].

$t = 25$ °C

[267], ν=3 MHz				[209, 322]				[170]			
x_1	ε	x_1	ε	x_1	ε	x_1	ε	x_1	ε	x_1	ε
0	2,28	89,8	4,03	0	2,282	75	3,691	0	2,249	73,498	3,884
49,7	3,19	100	4,25	25	2,713	100	4,265	25,291	2,810	100	4,261
72,6	3,66	—	—	50	3,183	—	—	44,373	3,345	—	—

$t = 20$ °C

[185], λ=120 cm		[173]				[175], ν=1,8 MHz	
x_1	ε	w_1	ε	w_1	ε	φ_1	ε
0	2,27	0	2,28	60,274	3,482	0	2,28
20	2,75	9,020	2,464	69,743	3,691	20	2,63
40	3,05	26,928	2,807	81,295	3,948	40	2,98
60	3,45	29,772	2,871	90,491	4,163	60	3,38
80	3,9	39,546	3,057	100	4,360	80	3,84
100	4,3	50,140	5,262	—	—	100	4,35

[172]

w_1	0,00	9,11	35,01	48,24	68,49	90,10	100
t °C	18,0	17,6	17,8	17,8	17,8	17,6	17,1
ε	2,288	2,55	3,073	3,399	3,870	4,300	4,50

[218], $t = 18$ °C

w_1	0	5,83	12,00	23,81	50,52	100
ε	2,270	2,375	2,483	2,708	3,281	4,400

Diethyl ether ($C_4H_{10}O$) – Aniline (C_6H_7N) *)

[231], $t = 20$ °C

x_1	0	20	40	50	60	80	100
ε	7,20	7,062	6,749	6,01	6,280	5,452	4,344

*) See also [227, 232, 165].

[175], $t = 20$ °C, ν = 1,8 MHz

φ_2	100	80	60	40	20	0
ε	7,06	7,00	6,76	6,32	5,57	4,35

Diethyl ether ($C_4H_{10}O$) – Cyclohexane (C_6H_{12})
[166], $t = 20$ °C

x_1	0,000	4,720	8,854	12,33	21,75	50,43	74,97	100
ε	2,033	2,109	2,178	2,246	3,398	3,035	3,634	4,335

Diethyl ether ($C_4H_{10}O$) – Cyclohexylamine ($C_6H_{13}N$)
[175], $t = 20$ °C, $\nu = 1,8$ MHz

φ_2	100	80	60	40	20	0
ε	4,72	4,65	4,57	4,50	4,42	4,35

Diethyl ether ($C_4H_{10}O$) – Benzonitrile (C_7H_5N)
[175], $t = 20$ °C, $\nu = 1,8$ MHz

φ_2	100	80	60	40	20	0
ε	25,65	21,65	17,58	13,35	8,85	4,35

Diethyl ether ($C_4H_{10}O$) – Benzaldehyde (C_7H_6O)
[175], $t = 20$ °C, $\nu = 1,8$ MHz

φ_2	0	20	40	60	80	100
ε	4,35	7,10	9,81	12,48	15,20	17,85

[227], $t = 18$ °C, $\nu = 3$ MHz

φ_2	100	80	60	40	20	0
ε	17,59	15,01	12,38	9,65	6,97	4,35

Diethyl ether ($C_4H_{10}O$) – Toluene (C_7H_8)
[267], $t = 25$ °C, $\nu = 3$ MHz

x_1	0	49,2	75,0	89,7	100
ε	2,38	3,12	3,65	3,98	4,35

Diethyl ether ($C_4H_{10}O$) – *m*–Cresol (C_7H_8O)
[227], $t = 17$ °C, $\nu = 3$ MHz

φ_2	100	80	60	40	20	0
ε	12,85	11,00	9,15	7,49	5,94	4,41

Diethyl ether ($C_4H_{10}O$) – Heptane (C_7H_{16})

[168], $t=21$ °C

x_1	ε	x_1	ε	x_1	ε	x_1	ε
6,330	2,026	18,87	2,196	40,03	2,527	91,59	3,962
8,474	2,058	23,67	2,241	57,55	2,896	100	4,355
13,04	2,105	25,70	2,245	76,16	3,402	—	—

[221, 167], $t=20$ °C

x_1	8,474	12,54	18,87	23,65	25,15	45,58	100
ε	2,055	2,105	2,193	2,238	2,293	2,636	4,355

φ_1	ε at t °C [323]		ε_∞ at t °C [323]	
	−22	15	−22	15
0,0	—	1,931	—	1,931
10,0	—	2,101	—	1,927
20,0	—	2,290	—	1,923
50,0	—	2,962	—	1,910
60,0	3,648	3,222	1,957	1,906
73,3	4,160	3,595	1,951	1,901
80,0	4,438	3,797	1,948	1,898
86,6	4,724	4,005	1,945	1,895
100	5,348	4,450	1,940	1,89

Diethyl ether ($C_4H_{10}O$) – Dimethylaniline ($C_8H_{11}N$)

[227], $t=17$ °C, $\nu=3$ MHz

φ_2	100	80	60	40	20	0
ε	5,04	4,90	4,80	4,65	4,56	4,40

Diethyl ether ($C_4H_{10}O$) – Quinoline (C_9H_7N)

[173], $t=18$ °C

w_2	ε	w_2	ε	w_2	ε	w_2	ε
100	9,559	59,609	7,224	29,837	5,616	0	4,360
85,072	8,758	49,787	6,578	21,593	5,242	—	—
70,408	7,665	39,911	5,089	9,687	4,729	—	—

Diethyl ether ($C_4H_{10}O$) – Bromonaphthalene ($C_{10}H_7Br$)
[172]

w_1	0	12,31	26,36	36,13	53,01	69,83	91,0	100
t °C	22,8	19,8	19,75	19,7	19,5	19,5	19,5	16,8
ε	4,912	6,006	5,050	4,993	4,789	4,614	4,400	4,410

Diethylsulphoxide ($C_4H_{10}OS$) – Benzene (C_6H_6)
[324], $t = 25$ °C

x_1	0	11,5	28,6	42,4	58,4	82,1
ε	2,2714	2,2959	2,3311	2,3594	2,3923	2,4423

Diethylsulphone ($C_4H_{10}O_2S$) – Benzene (C_6H_6)
[324], $t = 25$ °C

x_1	0	12,8	28,7	47,1	69,6	92,4
ε	2,2714	2,3069	2,3511	2,4024	2,4654	2,5297

1–Butane thiol ($C_4H_{10}S$) – Benzene (C_6H_6)

x_1	ε at t °C [285]		x_1	ε at t °C [285]	
	25	50		25	50
0	2,276	2,226	22,957	2,925	2,803
2,771	2,353	2,294	37,172	3,315	3,142
5,204	2,421	2,355	56,844	3,847	3,611
7,853	2,498	2,423	100,00	4,952	4,586
13,773	2,666	2,571	—	—	—

Diethylsulphide ($C_4H_{10}S$) – Benzene (C_6H_6)

x_1	ε at t °C [285]		x_1	ε at t °C [285]	
	25	50		25	50
0	2,276	2,226	33,000	3,358	3,183
1,815	2,334	2,278	50,070	3,946	3,705
3,496	2,389	2,326	60,390	4,313	4,002
5,603	2,458	2,386	100,00	5,723	5,236
10,292	2,615	2,523	—	—	—

Butylamine ($C_4H_{11}N$) – Benzene (C_6H_6)
[175], $t=20$ °C, $\nu=1,8$ MHz

φ_1	0	20	40	60	80	100
ε	2,28	2,69	3,14	3,66	4,20	4,78

Butylamine ($C_4H_{11}N$) – Aniline (C_6H_7N)
[175], $t=20$ °C, $\nu=1,8$ MHz

φ_2	100	80	60	40	20	0
ε	7,06	7,34	7,26	6,90	6,05	4,78

Diethylamine ($C_4H_{11}N$) – Nitrobenzene ($C_6H_5NO_2$)
[146], $t=25$ °C

φ_2	ε	φ_2	ε	φ_2	ε	φ_2	ε
100	34,82	70	21,41	40	14,94	10	6,77
90	31,41	60	20,94	30	12,10	0	4,31
80	27,81	50	18,24	20	9,45	—	—

Diethylamine ($C_4H_{11}O$) – Triethylamine ($C_6H_{15}N$)
[325]

x_1	ε at t °C		x_1	ε at t °C	
	20	30		20	30
0,00	2,423	2,349	65,98	3,178	2,955
2,66	2,466	2,366	84,47	3,523	—
6,61	2,476	2,391	92,33	3,702	3,371
12,70	2,527	2,433	96,10	3,780	3,441
24,23	2,632	2,519	97,76	3,831	3,471
47,23	2,897	2,737	100,00	3,894	3,518
56,90	3,031	2,845	—	—	—

Diethylamine ($C_4H_{11}N$) – Acetophenone (C_8H_8O)
[147], $t=25$ °C

φ_2	ε	φ_2	ε	φ_2	ε	φ_2	ε
100	17,16	70	13,41	40	9,31	10	5,42
90	15,98	60	12,04	30	7,89	0	4,31
80	14,78	50	10,73	20	6,70	—	—

Pyridine (C_5H_5N) – Valeric acid ($C_5H_{10}O_2$)
[295], $t=25$ °C

φ_1	ε	φ_1	ε	φ_1	ε	φ_1	ε
0	2,62	26,9	9,93	60	11,47	90	12,53
10	5,81	40	10,79	70	11,80	100	12,91
20	8,52	50	11,10	80	12,33	—	—

Pyridine (C_5H_5N) – Benzene (C_6H_6)
[326], $t=20$ °C, fig. 9 9

x_1	ε	x_1	ε	x_1	ε	x_1	ε
0,00	2,282	30,11	4,564	58,76	6,832	92,30	11,388
5,00	2,647	38,46	5,320	71,00	8,810	100	12,397
10,00	2,989	42,73	5,737	75,09	9,328	—	—
19,23	3,725	52,59	6,713	85,71	10,421	—	—

x_1	w_1	ε at t °C [327, 276]				
		25	30	35	40	45
18,34	18,53	3,764	3,715	3,667	3,619	3,575
36,87	37,17	5,547	5,456	5,367	5,272	5,181
51,29	51,60	7,218	7,082	6,953	6,816	6,885
67,67	67,94	9,346	9,181	9,018	8,847	8,688
83,33	83,51	11,18	10,97	10,77	10,56	10,36
100	100	13,51	—	13,04	—	12,57

[218], $t=18$ °C

w_1	0	3,41	5,12	10,57	19,70	31,59	100,0
ε	2,261	2,480	2,599	2,984	3,715	4,748	12,41

Pyridine (C_5H_5N) – Aniline (C_6H_7N) *)
[232], $t=20$ °C

x_1	0	20	40	60	80	100
ε	7,20	8,86	10,14	11,14	11,93	12,42

*) See also [231, 165, 328].

Benzonitrile (C_5H_7N) – Trichlorobenzene ($C_6H_3Cl_3$)

[175], $t=20$ °C, $\nu=1{,}8$ MHz

φ_1	0	20	40	60	80	100
ε	6,75	10,55	14,30	18,00	21,75	25,65

1, 3–Pentadiene (C_5H_8) – Benzene (C_6H_6)

[329], $t=25$ °C

x_2	100,0	95,03	89,86	87,08	77,59	66,30	49,93	22,48	0
ε	2,274	2,280	2,284	2.287	2,292	2,300	2,307	2,317	2,319

2–Methyl–1, 3–butadiene (C_5H_8) – Hexane (C_6H_{14})

x_1	ε at t °C [329]				
	–75	–50	–25	0	25
0	2,020	1,984	1,949	1,915	1,878
5,08	2,033	1,995	1,957	1,924	1,887
10,23	2,043	2,005	1,971	1,936	1,897
15,56	2,051	2,018	1,979	1,947	1,906
20,70	2,070	2,030	1,993	—	—
100	2,343	2,274	2,216	2,156	2,098

1, 5–Dichloropentane ($C_5H_{10}Cl$) – Hexane (C_6H_{14})

[269], $t=25$ °C

x_1	0	4,261	7,524	25,759	42,35	75,55	100,0
ε	1,8776	2,0820	2,2485	3,3174	4,5024	7,1266	9,9215

Valeric acid ($C_5H_{10}O_2$) – Diethylamine ($C_4H_{11}N$)

[295], $t=25$ °C

φ_2	100	87,0	66,6	50,0	41,5	14,5	0
ε	5,26	5,66	6,87	8,8	8,68	6,46	2,67

Valeric acid ($C_5H_{10}O_2$) – Triethylamine ($C_6H_{15}N$) *)

[295], $t=25$ °C

φ_2	100	75,9	61,5	44,5	40,0	37,4	34,4	28,8	17,3	0
ε	2,42	5,7	12,4	22,1	25,7	28,3	27,4	21,3	6,8	2,67

*) Values of ϵ at 35 and 45°C are given in [295].

Isovaleric acid ($C_5H_{10}O_2$) – Benzene (C_6H_6)
[275], $t = 22$ °C

x_1	ε	x_1	ε	x_1	ε	x_1	ε
7,29	2,314	15,32	2,343	34,55	2,425	92,90	2,753
8,95	2,332	18,71	2,356	53,26	2,524	—	—
11,45	2,331	26,76	2,393	74,79	2,648	—	—

Isopentyl bromide ($C_5H_{11}Br$) – 2–Methylpentane (C_6H_{14})
[330], $t = 133{,}4$ °K

x_1	0	33	50	67	95,2	100
ε	12,6	8,54	6,50	4,80	2,515	2,105

1–Chloropentane ($C_5H_{11}Cl$) – Xylene (C_8H_{10})
[320]

w_1	15,47	20,110	30,631	33,356	52,468	58,556
t °C	16,6	16,7	16,6	16	17,3	16
ε	2,8634	3,0198	3,4098	3,5094	4,1725	4,3844

Piperidine (C_5H_{11}) – Cyclohexane (C_6H_{12})

x_1	ε at t °C [304]					x_1	ε at t °C [304]				
	20	30	40	50	60		20	30	40	50	60
0	2,02	2,01	1,99	1,98	1,96	61,6	3,16	3,08	3,01	2,93	2,85
22,9	2,36	2,33	2,30	2,27	2,24	81,0	3,64	3,53	3,43	3,32	3,21
40,4	2,68	2,64	2,59	2,54	2,50	100	4,33	4,17	4,02	3,86	3,70

Pentan–1–ol ($C_5H_{12}O$) – Benzene (C_6H_6)
[331], $t = 18$ °C

x_1	0	7,951	13,73	20,22
ε	2,2951	2,3206	2,3206	2,3582

[215], $t = 16$ °C

w_1	0	3,569	6,890	11,401	26,812	31,107
ε	2,244	2,534	2,818	3,225	4,794	5,335

[259], $t = 25$ °C, $\lambda = 57$ cm

φ_1	0	10	25	50	75	100
ε	2,27	2,44	3,04	4,55	4,29	3,83

Pentan–1–ol ($C_5H_{12}O$) – Cyclohexane (C_6H_{12})
[280]

$t=25$ °C		$t=35$ °C		$t=45$ °C		$t=55$ °C	
N_1	ε	N_1	ε	N_1	ε	N_1	ε
0,440	2,101	0,434	2,096	0,430	2,081	0,423	2,068
0,690	2,131	0,682	2,128	0,673	2,125	0,664	2,121
0,929	2,371	0,917	2,192	0,905	2,153	0,894	2,144
1,212	2,253	1,197	2,250	1,181	2,247	1,167	2,244
1,404	2,371	1,387	2,362	1,369	2,353	1,352	2,344
1,848	2,535	1,825	2,517	1,802	2,498	1,780	2,480
2,864	3,218	2,830	3,100	2,797	3,008	2,763	2,932
3,701	4,152	3,657	3,880	3,617	3,663	3,575	3,505
5,525	7,560	5,462	7,780	5,408	6,130	5,350	5,610
7,366	10,50	7,294	9,375	7,221	8,300	7,148	7,325
9,206	15,25	9,122	13,925	9,038	12,750	8,953	11,675

Pentan–1–ol ($C_5H_{12}O$) – Xylene (C_8H_{10})
[320]

w_1	t °C	ε	w_1	t °C	ε
0	14,6	2,3497	10,985	15	2,713
12,048	14	2,7647	20,298	14,8	3,1540
24,148	14	3,4523	45,104	14,7	5,9024
45,196	14	5,9942	74,212	14,6	11,27
70,077	14	10,694	100	14,3	15,96
100,0	15,1	15,925	—	—	—

Isopentanol ($C_5H_{12}O$) – 1, 4–Dichlorobenzene ($C_6H_4Cl_3$)
[279], $t = 55$ °C

x_1	ε	x_1	ε	x_1	ε	x_1	ε
0,00	2,397	30,04	3,379	59,74	5,462	89,89	9,538
4,99	2,505	39,66	3,933	69,71	6,582	100,0	11,231
9,85	2,623	49,55	4,536	79,76	7,916	—	—

Isopentanol ($C_5H_{12}O$) – Benzene (C_6H_6)
[300], $t = 25$ °C

x_1	0,00	4,38	7,73	11,47	17,10	22,82	27,96
ε	2,26	2,44	2,57	2,72	2,78	3,24	3,44

[218], $t = 18$ °C

w_1	0	3,32	6,78	11,11	20,21	32,52	49,72	73,10	100
ε	2,273	2,392	2,605	2,828	3,265	4,110	7,705	14,21	14,79

[218], $t = 64$ °C

w_1	0	4,02	7,62	12,21	20,84	33,41	50,53	74,82	100
ε	2,190	2,296	2,427	2,661	3,168	3,535	5,542	10,42	11,42

1, 1–Dimethylpropan–1–ol ($C_5H_{12}O$) – Benzene (C_6H_6)
[218], $t = 24$ °C

w_1	0	3,10	5,15	10,21	35,18	50,22	100
ε	2,261	2,350	2,400	2,531	3,180	3,579	5,849

2, 2–Dimethylsulphonylpropan ($C_5H_{12}O_4S_2$) – Benzene (C_6H_6)
[324], $t = 25$ °C

x_1	0	16,5	31,9	46,2	57,7	63,7
ε	2,2714	1,2750	2,2786	2,2818	2,2845	2,2858

1–Pentanthiol ($C_5H_{12}S$) – Benzene (C_6H_6)
[285]

x_1	ε at t °C		x_1	ε at t °C		x_1	ε at t °C	
	25	50		25	50		25	50
0	2,276	2,226	6,427	2,460	2,387	40,59	3,337	3,158
2,123	2,337	2,280	13,82	2,663	2,567	51,98	3,600	3,388
2,965	2,362	2,302	21,759	2,877	2,753	100	4,547	4,230
5,362	2,427	2,361	23,52	2,914	2,789	—	—	—

Ethoxytrimethylsilane ($C_5H_{14}OSi$) – Nitrobenzene ($C_6H_5NO_2$)
[146], $t = 25°$ C

φ_2	ε	φ_2	ε	φ_2	ε	φ_2	ε
100	34,82	70	23,72	40	13,70	10	5,50
90	31,12	60	20,28	30	10,72	0	3,04
80	27,40	50	16,80	20	7,92	—	—

Ethoxytrimethylsilane ($C_5H_{14}OSi$) – Acetophenone (C_8H_8O)
[147], $t = 25$ °C

φ_2	ε	φ_2	ε	φ_2	ε	φ_2	ε
100	17,16	70	12,82	40	8,43	10	4,27
90	15,74	60	11,27	30	7,03	0	3,04
80	14,35	50	9,86	20	5,58	—	—

1, 2, 4–Trichlorobenzene ($C_6H_3Cl_3$) – Benzene (C_6H_6)
[175], $t = 20$ °C, $\nu = 1,8$ MHz

φ_1	0	20	40	60	80	100
ε	2,28	3,12	3,96	4,83	5,77	6,75

1, 2, 4–Trichlorobenzene ($C_6H_3Cl_3$) – Aniline (C_6H_7N)
[175], $t = 20$ °C, $\nu = 1,8$ MHz

φ_2	100	80	60	40	20	0
ε	7,06	6,90	6,74	6,68	6,68	6,75

Picric acid ($C_6H_3N_3O_7$) – Naphthalene ($C_{10}H_8$)
[332], $t = 21$ °C

w_1	ε	w_1	ε	w_1	ε	w_1	ε
0	2,64	30	2,69	60	2,80	9	2,99
10	2,66	40	2,70	70	2,88	100	3,05
20	2,67	50	2,73	80	2,93	—	—

Picric acid ($C_6H_3N_3O_7$) – Azobenzene ($C_{12}H_{10}N_2$)
[333, 334], $t = 21$ °C

w_1	100	89	75	62	51	38	29	18	0
ε	4,0	3,95	3,8	3,6	3,2	3,2	2,9	2,6	2,2

1, 2–Dibromobenzene ($C_6H_4Br_2$) – Benzene (C_6H_6)
[335], $t=20$ °C, $\lambda=300-1000$ m

x_1	3,33	6,66	10	20	50	100
ε	2,45	2,57	2,7	3,15	4,6	7,50

1, 3–Dibromobenzene ($C_6H_4Br_2$) – Benzene (C_6H_6)
[335], $t=20$ °C, $\lambda=300-1000$ m

x_1	3,33	6,66	10	20	50	100
ε	2,33	2,395	2,46	271	3,42	4,74

1–Chloro–2–nitrobenzene ($C_6H_4ClNO_2$) – Benzene (C_6H_6)
[321], $t=20$ °C

w_1	4,545	12,180	14,786	21,953	28,947	37,487	50,624
ε	2,910	2,055	4,483	5,750	7,120	9,055	12,477

1–Chloro–4–nitrobenzene ($C_6H_4ClNO_2$) – Benzene (C_6H_6)
[321], $t=20$ °C

w_1	1,751	3,658	4,555	7,746	9,742	12,220	13,344	14,945	22,061
ε	2,370	2,468	2,513	2,674	2,779	2,916	2,975	3,057	3,468

1, 2–Dichlorobenzene ($C_6H_4Cl_2$) – Benzene (C_6H_6)
[336], $t=25$ °C, $\nu=100$ kHz

x_1	0	10	24,82	50,03	74,81	100
ε	2,274	3,039	4,152	6,041	7,946	10,06

[337], $\nu=0,5$ MHz

x_1	ε at t °C			x_1	ε at t °C		
	0	25	50		0	25	50
0	—	2,273	2,226	20,15	4,023	3,777	3,561
2,05	—	2,453	2,373	50,10	6,520	5,988	5,547
5,4	2,873	2,724	2,607	100	11,130	9,930	8,900
11,56	3,331	3,153	2,998	—	—	—	—

[335], $t = 20$ °C, $\lambda = 300 - 1000$ m

x_1	ε	x_1	ε	x_1	ε	x_1	ε
3,33	2,49	20	3,57	50	5,64	80	8,12
6,66	2,72	30	4,23	60	6,21	90	9,15
10	2,90	40	4,97	70	7,25	100	9,82

1, 2–Dichlorobenzene ($C_6H_4Cl_2$) – Hexane (C_6H_4)

[337], $\nu = 0,5$ MHz

x_1	ε at t °C			x_1	ε at t° C		
	0	25	50		0	25	50
0	1,960	1,908	1,839	34,06	4,370	4,030	3,733
2,55	2,100	2,035	1,966	58,81	6,498	5,873	5,384
7,39	2,382	2,292	2,190	100	11,130	9,930	8,900
13,20	2,783	2,630	2,503	—	—	—	—

1, 2–Dichlorobenzene ($C_6H_4Cl_2$) – Benzonitrile (C_7H_5N)

[338], $t = 20$ °C, $\nu = 0,5 - 5$ MHz

w_2	44,99	30,04	17,51	7,53	4,99	1,51	0
ε	18,64	16,04	13,80	11,36	10,75	10,21	10,12

1, 3–Dichlorobenzene ($C_6H_4Cl_2$) – Nitrobenzene ($C_6H_5NO_2$)

[175], $t = 20$ °C, $\nu = 1,8$ MHz

φ_2	100	80	60	40	20	0
ε	35,75	28,15	21,43	15,39	10,00	5,08

1, 3–Dichlorobenzene ($C_6H_4Cl_2$) – Benzene (C_6H_6)

[335], $t = 20$ °C, $\lambda = 300 - 1000$ m

x_1	3,3	10	20	40	80	1C0
ε	2,37	2,49	2,82	3,26	4,23	4,90

[337], $\nu = 0,5$ MHz

x_1	ε at t °C			x_1	ε at t °C		
	0	25	50		0	25	50
0	—	2,273	2,226	27,03	3,283	3,130	2,989
3,72	2,497	2,414	2,342	68,37	4,552	4,254	3,997
10,08	2,719	2,629	2,533	100	5,403	5,039	4,703

1, 4-Dichlorobenzene ($C_6H_4Cl_2$) – **Benzene** (C_6H_6)
[337], $\nu=0,5$ MHz

x_1	ε at t °C			x_1	ε at t °C		
	0	25	50		0	25	50
0	—	2,273	2,226	38,60	—	2,359	2,313
13,54	2,372	2,323	2,266	67,41	—	—	2,369
14,83	2,389	2,341	2,276	—	—	—	—

1, 4-Dichlorobenzene ($C_6H_4Cl_2$) – **Heptane** (C_7H_6)
[169, 168], $t=20$ °C

x_1	4,313	5,625	9,406	13,53	18,09
ε	1,993	1,998	2,015	2,032	2,050

1, 4-Dichlorobenzene ($C_6H_4Cl_2$) – **Heptan-1-ol** ($C_7H_{16}O$)
[279], $t=55$ °C

x_2	100,0	87,00	60,00	40,00	23,00	0,00
ε	8,223	7,031	4,722	3,574	3,006	2,397

1, 4-Dichlorobenzene ($C_6H_4Cl_2$) – **Octadecan-1-ol** ($C_{18}H_{38}O$)
[279], 25 °C

x_2	100	30,17	20,33	10,50	7,74	5,07	2,63	0,00
ε	3,475	2,911	2,776	2,618	2,560	2,514	2,460	2,397

1, 2-Diiodobenzene ($C_6H_4J_2$) – **Benzene** (C_6H_6)
[335], $t=20$ °C, $\lambda=300-1000$ m

x_1	3,33	6,66	10	20	40	80	100
ε	2,38	2,54	2,64	3,07	3,74	5,1	5,7

1, 3-Diiodobenzene ($C_6H_4J_2$) – **Benzene** (C_6H_6)
[335], $t=20$ °C, $\lambda=300-1000$ m

x_1	10	20	50	100
ε	2,49	2,66	3,2	4,25

Bromobenzene (C_6H_5Br) – Chlorobenzene (C_6H_5Cl)

[232], $t = 20$ °C

x_1	0,00	20,0	40,0	50,0	60,0	80	100,0
ε	5,916	5,824	5,731	5,698	5,659	5,574	5,482

[207], $t = 25$ °C, $\nu = 1$ MHz

x_1	0	25	50	75	100
ε	5,610	5,557	5,510	5,448	5,397

Bromobenzene (C_6H_5Br) – Benzene (C_6H_6)

[339], $t = 70$ °C, $\nu = 180$ kHz

x_1	0,00	16,13	41,26	65,63	79,55	100
ε	2,193	2,648	3,335	3,962	4,300	4,783

[227], $t = 16$ °C, $\nu = 3$ MHz

x_1	0	10	20	40	60	80	100
ε	2,29	2,65	2,99	3,60	4,21	4,81	5,46

[172]

w_1	0	9,67	32,84	43,14	50,14	66,09	84,04	100
t °C	18	22	22	22,2	22,2	22,3	22,4	22
ε	2,888	2,430	2,930	3,235	3,400	3,959	4,701	5,525

Bromobenzene (C_6H_5Br) – Aniline (C_6H_7N)

[232], $t = 20$ °C

x_1	0,00	20,0	40,0	50,0	60,0	80,0	100
ε	7,200	6,719	6,351	6,180	5,938	5,709	5,432

Chlorobenzene (C_6H_5Cl) – Nitrobenzene ($C_6H_5NO_2$)

[175], $t = 20$ °C, $\nu = 1,8$ MHz

φ_2	100	80	60	40	20	0
ε	35,75	28,08	21,10	15,20	10,24	5,70

Chlorobenzene (C_6H_5Cl) – Benzene (C_6H_6)

[337], $\nu = 0,5$ MHz

x_1	ε at t °C			x_1	ε at t °C		
	0	25	50		0	25	50
0	—	2,273	2,226	60,20	4,650	4,332	5,050
7,86	2,710	2,621	2,518	70,48	5,039	4,665	5,339
19,88	3,140	2,994	2,853	100	6,088	5,628	5,226
40,07	3,879	3,659	3,443	—	—	—	—

[341], $t = 25$ °C

x_1	ε	x_1	ε	x_1	ε	x_1	ε
0	2,2823	12,3	2,8024	41,1	4,0124	83,0	5,3152
2,7	2 3873	19,9	3,1286	57,4	4,6112	86,1	5,4251
3,2	2,4133	23,2	3,2471	59,4	4,6834	89,4	5,4928
4,0	2,4798	25,5	3,3668	66,7	4,9028	100	5,6895
6,1	2,5341	28,0	3,4768	71,4	5,0479	—	—
7,2	2,6225	38,3	3,9064	73,2	5,0718	—	—

$t = 25$ °C

[340]		[322, 209, 196]		[342]		[291]	
x_1	ε	x_1	ε	x_1	ε	x_1	ε
0	2,274	0	2,282	0	2,273	0	2,16
19,60	2,944	10	2,623	5,85	2,470	17,93	2,87
39,81	3,641	25	3,131	11,80	2,679	36,82	3,46
60,46	4,299	50	3,979	14,52	2,783	56,73	4,06
77,67	4,449	100	5,610	20,86	3,014	77,76	4,58
100	5,600	—	—	25,65	3,163	100	5,22

[339], $t = 70$ °C, $\nu = 180$ kHz

x_1	0,00	28,46	56,64	87,83	100
ε	2,193	2,983	3,760	4,591	4,942

[173], $t = 18$ °C

w_1	ε	w_1	ε	w_1	ε	w_1	ε
0	2,28	28,905	3,089	58,324	4,048	89,064	5,284
10,640	2,557	38,635	3,386	69,652	4,477	100	5,723
18,677	2,795	50,849	3,800	79,517	4,885	—	—

[175], $t=20$ °C, $\nu=1,8$ MHz

φ_1	0	20	40	60	80	100
ε	2,28	2,92	3,57	4,25	4,96	5,70

[227], $t=15$ °C, $\nu=3$ kHz

φ_1	0	5	10	15	20	50	80	100
ε	2,27	2,47	2,62	2,88	3,04	4,04	4,99	5,67

Chlorobenzene (C_6H_5Cl) – Cyclohexanone ($C_6H_{10}O$)

[175], $t=20$ °C, $\nu=1,8$ MHz

φ_2	100	80	60	40	20	0
ε	15,90	13,71	11,73	9,75	7,75	5,70

Chlorobenzene (C_6H_5Cl) – Hexanoic acid ($C_6H_{12}O_2$)

[224], $t=20$ °C, $\nu=28$ MHz

φ_2	0	20,27	46,70	60,51	79,43	100
ε	2,82	3,12	3,95	4,41	4,98	5,61

Chlorobenzene (C_6H_5Cl) – Hexane (C_6H_{14})

x_1	ε at t °C [238], $\nu=0,5$ MHz							
	−80	−70	−60	−50	−40	−30	−20	−10
10,68	2,532	2,490	2,448	2,409	2,371	2,336	2,304	2,273
21,32	3,053	2,977	2,908	2,844	2,784	2,728	2,677	2,629
37,98	—	3,848	3,730	3,620	3,518	3,424	3,336	3,257
64,58	—	5,484	5,268	5,086	4,882	4,717	4,560	4,415
100	—	—	—	7,28	7,01	6,758	6,502	6,260

x_1	ε at t °C [238], $\nu=0,5$ MHz							
	0	10	20	30	40	50	60	70
10,68	2,244	2,214	2,184	2,153	2,120	2,084	2,050	—
21,32	2,584	2,540	2,497	2,456	2,416	2,377	2,347	2,300
37,98	3,180	3,108	3,040	2,977	2,913	2,846	2,777	2,706
64,58	4,280	4,156	4,038	3,928	3,827	3,723	3,627	3,538
100	6,027	5,818	5,633	5,460	5,300	5,150	5,046	4,888

[337], ν = 0,5 MHz

x_1	ε at t °C			x_1	ε at t °C		
	0	25	50		0	25	50
0	1,960	1,908	1,839	49,79	3,690	3,453	3,249
7,92	2,201	2,122	2,039	75,40	4,824	3,467	4,153
14,70	2,419	2,312	2,221	84,77	5,370	3,953	4,602
29,03	2,870	2,727	2,543	100	6,088	5,628	5,226

[216], t = 25 °C, ν = 1 MHz

x_2	100	90	75	50	25	0
ε	1,904	2,167	2,605	3,462	4,508	5,610

Chlorobenzene (C_6H_5Cl) – Benzonitrile (C_7H_5N)

[210], t = 25 °C

w_2	ε	w_2	ε	w_2	ε	w_2	ε
100,0	25,49	89,469	23,34	26,403	10,92	8,222	7,314
97,330	24,91	66,497	18,90	18,536	9,339	2,675	6,213
94,787	24,39	33,119	12,25	15,431	8,731	0,00	5,698

[175], t = 20 °C, ν = 1,8 MHz

φ_2	100	80	60	40	20	0
ε	25,65	21,55	17,60	13,70	9,73	5,70

Chlorobenzene (C_6H_5Cl) – *m*–Cresol (C_7H_8O)

[224], t = 25 °C, ν = 28 MHz

φ_1	0	22,36	42,86	62,26	79,58	100
ε	11,8	9,62	7,95	6,71	5,89	5,61

Chlorobenzene (C_6H_5Cl) – Pyridinium acetate ($C_7H_9NO_2$)

[270], t = 25 °C

x_2	ε	x_2	ε	x_2	ε	x_2	ε
100	16,0	80,42	14,3	54,54	11,6	31,69	9,2
93,39	15,4	73,24	13,5	46,00	10,7	21,96	8,2
87,40	14,9	62,52	12,5	38,14	9,8	—	—

Chlorobenzene (C_6H_5Cl) – Heptane (C_7H_{16})
[172]

w_1	0	10,41	32,62	49,96	52,91	67,67	79,75	100
t °C	22,2	22,5	22,4	22,4	22,4	22,5	22,4	22
ε	1,992	2,217	2,820	2,374	3,450	4,059	4,639	5,896

Chlorobenzene (C_6H_5Cl) – Ethylpentyl ether ($C_7H_{16}O$)
[172]

w_2	100	87,33	60,05	40,83	35,83	10,41	0
t °C	23,0	23,0	23,5	23,8	24,0	23,9	22,0
ε	3,565	3,851	4,412	4,820	4,968	5,648	5,896

Chlorobenzene (C_6H_5Cl) – *m*–Xylene (C_8H_{10})
[172]

w_1	0	8,35	30,97	50,80	51,47	67,32	90,49	100
t° C	17,4	20,1	20	20,2	20,3	20,5	20	22
ε	2,385	2,609	3,222	3,728	3,870	4,328	5,517	5,896

[171], $t=15$ °C

w_1	0	20,2	39,5	100
ε	2,37	3,18	3,76	5,98

Chlorobenzene (C_6H_5Cl) – Anilinium acetate ($C_8H_{11}NO_2$)
[270], $t=25$ °C

x_2	ε	x_2	ε	x_2	ε	x_2	ε
100	8,21	73,72	7,70	41,83	6,92	15,05	6,12
92,52	8,08	61,58	7,40	31,93	6,64	7,68	5,88
86,25	7,94	51,44	7,17	23,42	6,40	0	5,61

Chlorobenzene (C_6H_5Cl) – Quinoline (C_9H_7N)
[173], $t=18$ °C

w_2	ε	w_2	ε	w_2	ε	w_2	ε
100	9,559	71,126	8,464	40,502	7,271	10,106	6,103
91,313	9,235	60,861	8,056	30,230	6,863	0	5,723
80,184	8,788	50,710	7,715	19,678	6,464	—	—

Chlorobenzene (C_6H_5Cl) – Dipentyl ether ($C_{10}H_{12}O$)
[343], $t = 25$ °C

w_1	0,00	2,1888	4,8486	10,081	17,528	26,470
ε	2,7980	2,8354	2,8812	2,9700	3,0998	3,2822

Nitrobenzene ($C_6H_5NO_2$) – Benzene (C_6H_6)
$t = 25$ °C

[286] *)		[340] **)		[205]		[204] ***)	
x_1	ε	x_1	ε	x_1	ε	x_1	ε
0	2,281	0,0	2,274	0,00	2,273	0	2,275
17,8	6,583	9,205	4,23	5,75	3,597	6,24	3,708
36,7	11,90	20,06	7,06	7,15	3,928	14,75	5,811
56,6	18,26	29,59	9,62	8,80	4,326	29,84	10,064
77,6	25,65	36,86	11,70	9,50	4,490	40,80	13,55
100	34,62	49,62	15,20	12,44	5,224	51,05	16,92
—	—	69,69	22,15	16,95	6,372	61,32	20,56
—	—	81,97	27,1	20,78	7,374	74,78	25,60
—	—	100	34,12	22,02	7,704	84,37	29,70
						100	36,00

*) ν = 10 MHz
**) ν = 13,5 MHz
***) $t = 24$ °C, ν = 1,2 MHz

[339], $t = 70$ °C, ν = 180 kHz

x_1	0,00	18,76	39,60	60,59	78,03	100
ε	2,193	5,867	10,53	15,92	20,81	27,59

[227], $t = 15$ °C

x_1	0	20	40	60	80	100
ε	2,25	6,74	11,90	18,19	25,81	35,03

[218]

$t = 24$ °C		$t = 45$ °C		$t = 65$ °C	
w_1	ε	w_1	ε	w_1	ε
0	2,261	0	2,228	0	2,185
3,12	2,700	3,70	2,671	4,25	2,670
6,45	3,200	6,85	3,121	7,13	3,060
11,65	4,041	11,92	3,890	12,10	3,771
29,72	7,442	30,01	7,00	31,45	6,540
100	36,72	100	32,68	100	28,04

w_1	ε at t °C [344]		w_1	ε at t °C [344]	
	25	50		25	50
10,33	3,802	3,616	49,91	10,07	10,99
20,92	5,591	5,253	58,06	14,33	13,25
30,52	7,440	6,981	69,37	17,81	16,57
39,04	9,326	8,712	—	—	—

[215], $t = 16$ °C

w_1	0	24,78	57,72	73,17	100
ε	2,244	6,669	15,87	21,54	37,14

[175], $t = 20$ °C, $\nu = 1{,}8$ MHz

φ_1	0	20	40	60	80	100
ε	2,28	6,73	12,15	18,35	26,15	35,75

Nitrobenzene ($C_6H_5NO_2$) – Aniline (C_6H_7N) *)
[345], $t = 30$ °C, $\nu = 1{,}8$ MHz

w_2	100	80	60	40	20	0
ε	6,91	9,09	14,05	19,23	25,53	33,70

*) Values of ε depending on the pressure are given in [345].

[175], $t = 20$ °C, $\nu = 1{,}8$ MHz

φ_1	0	20	40	60	80	100
ε	7,06	11,30	15,78	21,85	27,75	35,75

Nitrobenzene ($C_6H_5NO_2$) – Pentyl cyanide ($C_6H_{11}N$)
[346]

w_1	0,00	28,6	37,5	44,5	50	54,5	100
t °C	18	23	23	23	23	23	22
ε	15,5	16,9	18,1	18,5	19,55	20,5	34,00

Nitrobenzene ($C_6H_5NO_2$) – Cyclohexane (C_6H_{12})
[205], $t = 25$ °C

x_1	0,00	3,23	8,14	11,17	15,45	20,21	20,67	26,77	30,18
ε	2,016	2,597	3,489	4,100	5,014	6,131	6,236	7,777	8,707

[175], $t=20$ °C, $\nu=1{,}8$ MHz

φ_1	0	20	40	60	80	100
ε	2,02	6,37	12,25	19,30	27,10	35,75

φ_1	ε at t °C [347], ν=700 kHz						
	15	17	20	25	30	35	40
0	1,890	1,890	1,890	1,890	1,890	1,890	1,890
9,5	3,825	3,805	3,778	3,632	3,688	3,642	3,598
26	—	—	8,005	7,863	7,730	7,590	7,460
45	—	—	14,196	13,910	13,615	13,325	13,035
55	—	18,182	17,915	17,485	17,055	16,620	16,190
76	26,68	26,42	26,00	25,33	24,68	24,00	23,35

Nitrobenzene ($C_6H_5NO_2$) – Hexane (C_6H_{14})

$t=25$ °C

[286] *)		[216, 350] **)		[205]	
x_1	ε	x_1	ε	x_1	ε
0	1,884	0	1,904	0,00	1,887
24,2	6,171	5	2,624	7,83	3,043
46,0	11,92	10	3,474	12,36	3,810
65,7	19,04	25	6,309	18,42	4,940
83,6	26,65	49	14,68	25,59	6,461
100	34,62	100	36,10	—	—

*) ν=10 kHz
**) ν=1 MHz

[349]

$x_1=7{,}20$		$x_1=15{,}10$		$x_1=23{,}00$		$x_1=31{,}80$	
t °C	ε	t °C	ε	t °C	ε	t °C	ε
21,0	2,820	22,00	4,180	23,00	5,500	22,00	7,670
19,0	2,825	21,00	4,190	21,00	5,520	20,50	7,720
18,0	2,830	20,00	4,200	20,00	5,530	20,00	7,745
17,5	2,835	19,00	4,210	19,00	5,540	19,50	7,770
16,8	2,835	18,50	4,215	18,00	5,550	19,10	7,780
16,6	2,835	18,50	4,220	18,00	5,560	18,70	7,895
16,2	2,835	17,60	4,225	17,60	5,565	18,30	7,815
15,8	2,835	17,40	4,230	17,40	5,565	18,10	7,840
15,7	2,840	16,90	4,230	17,20	5,570	18,00	7,855
15,5	2,840	16,80	4,235	17,10	5,570	17,90	7,860
15,0	2,840	16,00	4,240	16,70	5,575	17,70	7,855
14,5	2,840	15,50	4,245	16,50	5,575	17,40	7,840

x_1=42,70		x_1=43,50		x_1=44,50		x_1=45,57		x_1=47,08	
t °C	ε	t °C	ε	t °C	ε	t °C	ε	t °C	ε
24,00	10,670	25,00	10,670	24,00	11,950	23,00	11,240	24,00	11,880
22,00	10,690	23,00	10,760	22,00	11,040	21,00	11,310	22,00	11,950
20,80	10,715	21,00	10,860	21,00	11,070	20,00	11,350	21,00	12,000
20,00	10,725	20,50	10,880	20,50	11,100	19,80	11,380	20,50	12,030
19,90	10,735	19,80	10,950	20,20	11,120	19,60	11,430	20,00	12,060
19,75	10,745	19,70	10,960	20,00	11,140	19,52	11,520	19,80	12,090
19,65	10,735	19,60	10,980	19,75	11,190	19,50	11,570	19,75	12,150
19,62	10,70	19,50	11,100	19,70	11,240	19,45	11,680	19,70	12,250
19,60	10,840	19,45	11,140	19,65	11,340	19,42	11,700	19,64	12,400
19,50	10,810	19,40	11,12	19,60	11,440	19,30	11,660	19,60	12,380
19,40	10,790	19,30	11,100	19,50	11,380	19,20	11,620	19,50	12,350
19,20	10,760	19,10	11,070	19,40	11,350	19,00	11,570	19,30	12,330

x_1=49,20		x_1=51,19		x_1=53,84		x_1=56,49		x_1=62,00	
t °C	ε	t °C	ε	t °C	ε	t °C	ε	t °C	ε
22,00	12,150	23,00	12,95	25,00	13,46	23,00	14,290	22,00	15,870
21,00	12,200	21,00	13,05	22,00	13,64	21,00	14,330	21,00	15,940
20,50	12,200	20,00	13,09	21,00	13,70	20,00	14,380	20,00	16,000
20,20	12,240	19,40	13,13	20,00	13,76	19,50	14,420	19,50	16,030
20,00	12,260	19,25	13,21	19,00	13,85	19,10	14,450	19,00	16,080
19,80	12,280	19,20	13,37	18,70	13,90	18,80	14,470	18,80	16,120
19,60	12,350	19,10	13,48	18,60	13,96	18,50	14,500	18,75	16,150
19,55	12,450	19,04	13,52	18,50	14,05	18,10	14,590	18,70	18,130
19,52	12,530	18,90	13,50	18,40	14,12	18,00	14,620	18,50	16,100
19,40	12,500	18,70	13,48	18,30	14,21	17,80	14,600	18,40	16,100
19,30	12,490	18,30	13,45	18,00	14,19	17,50	14,590	18,30	16,120
19,10	12,460	18,10	13,42	17,60	14,16	17,00	14,580	18,00	16,135

x_1=73,64						x_2=89,00					
t °C	ε	t °C	ε	t °C	ε	t °C	ε	t °C	ε	t °C	ε
23,00	20,000	19,00	20,265	18,40	20,330	25,00	25,280	19,50	25,910	18,30	26,105
21,00	21,120	18,80	20,280	18,30	20,340	23,00	25,490	19,00	26,000	18,10	26,140
20,00	20,190	18,60	20,315	18,20	20,345	21,00	25,740	18,70	26,055	17,70	26,200
19,50	20,225	18,50	20,320	18,00	20,350	20,00	25,850	18,50	25,080	17,50	26,245

w_1	ε at t °C [348]				w_1	ε at t °C [348]			
	5	20	35	50		5	20	35	50
0,000	1,99	1,96	1,93	1,91	59,556	14,4	14,1	13,1	12,2
10,053	3,21	3,07	2,98	2,87	70,039	19,3	18,4	17,0	15,7
20,057	4,66	4,42	4,19	4,05	80,082	25,4	23,3	21,4	19,8
30,124	6,44	6,19	5,92	5,48	90,034	31,6	29,0	26,6	24,6
40,116	8,42	8,26	7,72	7,31	100,0	39,0	35,8	32,9	30,4
48,830	10,7	10,5	9,81	9,21	—	—	—	—	—

φ_1	ε at t °C [347]									
	−2	0	5	10	15	20	25	30	35	40
0	—	—	—	2,05	2,05	2,05	2,05	2,05	2,05	2,05
19	6,75	6,70	6,58	6,48	6,38	6,25	6,13	6,02	5,92	5,80
39	13,05	12,95	12,73	12,48	12,20	11,90	11,70	11,52	11,32	11,15
59	20,68	20,48	19,98	19,45	18,94	18,42	17,98	17,55	17,10	16,64
79	30,15	29,84	29,00	28,14	27,30	26,55	25,88	25,25	24,62	23,96
100	—	—	—	37,80	37,00	36,15	35,25	34,40	33,50	32,65

[175], $t=20$ °C, $\nu=1,8$ MHz

φ_1	0	20	40	60	80	100
ε	1,90	6,24	12,22	19,39	27,30	35,75

Nitrobenzene ($C_6H_5NO_2$) – *tert*–Butyl ethyl ether ($C_6H_{14}O$)
[146], $t=25$ °C

φ_1	0	20	40	50	60	80	100
ε	7,07	12,18	17,15	19,85	22,54	28,30	34,82

Nitrobenzene ($C_6H_5NO_2$) – Triethylamine ($C_6H_{13}N$)
[146], $t=25$ °C

φ_1	ε	φ_1	ε	φ_1	ε	φ_1	ε
0	2,42	30	10,28	60	20,02	90	31,00
10	4,96	40	13,28	70	23,63	100	34,82
20	7,21	50	16,53	80	27,26	—	—

Nitrobenzene ($C_6H_5NO_2$) – Benzonitrile (C_7H_5N)
[210], $t=25$ °C

w_2	ε	w_2	ε	w_2	ε	w_2	ε
100,0	25,51	90,535	26,12	27,275	32,02	7,940	34,64
97,543	25,66	79,002	26,95	22,922	32,56	5,011	35,10
94,997	25,84	57,234	28,93	10,733	34,24	0,00	35,95

[175], $t=20$ °C, $\nu=1{,}8$ MHz

φ_1	0	20	40	60	80	100
ε	25,65	27,00	28,75	30,75	33,00	37,75

Nitrobenzene ($C_6H_5NO_2$) – Benzaldehyde (C_7H_6O)

[175], $t=20$ °C, $\nu=1{,}8$ MHz

φ_1	0	20	40	60	80	100
ε	17,85	20,82	24,20	27,80	31,63	35,75

Nitrobenzene ($C_6H_5NO_2$) – 2–Nitrotoluene ($C_7H_7NO_2$)

[175], $t=20$ °C, $\nu=1{,}8$ MHz

φ_2	100	80	60	40	20	0
ε	27,20	28,92	30,68	32,33	34,34	35,75

Nitrobenzene ($C_6H_5NO_2$) – Toluene (C_7H_8)

[286], $t=25$ °C, $\nu=10$ kHz

x_1	0	20,6	40,9	60,9	80,6	100
ε	2,376	6,720	11,99	18,45	25,81	34,62

[215]

$t=16$ °C		$t=24$ °C				$t=100$ °C			
w_1	ε	w_1	ε	w_1	ε	w_1	ε	w_1	ε
0	2,342	0	2,380	19,88	5,592	0	2,204	21,42	4,745
18,211	5,432	3,41	2,880	28,35	7,261	4,21	2,609	30,20	5,985
35,718	9,199	5,37	3,160	50,02	12,33	5,80	2,780	52,86	9,830
58,104	15,91	10,73	3,940	100	36,72	11,58	3,412	100	22,20
100	37,14								

Nitrobenzene ($C_6H_5NO_2$) – Heptane (C_7H_{16})

x_1	ε at t °C [351, 352]				
	20	25	30	35	40
0	1,925	1,922	1,920	1,920	1,919
11,1	3,500	3,460	3,413	3,370	3,325
32,5	7,780	7,637	7,500	7,364	7,226
48,9	12,128	11,915	11,682	11,448	11,215
61,4	16,686	16,312	15,950	15,585	15,225
76,1	23,074	22,495	21,985	21,410	20,935

Nitrobenzene ($C_6H_5NO_2$) – Trimethylbutoxysilane ($C_7H_{18}OSi$)

[146], $t=25$ °C

φ_1	0	20	30	50	60	70	80	100
ε	7,03	11,19	13,33	18,37	21,14	24,19	27,40	34,82

Nitrobenzene ($C_6H_5NO_2$) – *p*–Xylene (C_8H_{10})

w_1	ε at *t* °C [353], ν=1 MHz					
	20	40	60	80	100	120
0	2,268	2,234	2,200	2,165	2,131	2,097
2,066	2,562	2,500	2,441	2,386	2,334	2,283
4,672	2,941	2,846	2,757	2,674	2,598	2,524
7,170	3,300	3,192	3,072	2,964	2,861	2,766
9,351	3,674	3,509	3,359	3,223	3,099	2,984
11,42	3,988	3,794	3,615	3,457	3,312	3,181
16,68	4,932	4,657	4,403	4,174	3,967	3,778
22,45	6,066	5,668	5,319	5,004	4,723	4,473
27,81	7,205	6,693	6,244	5,847	5,487	5,162

[175], $t=20$ °C, $\nu=1,8$ *Мгц*

φ_1	0	20	40	60	80	100
ε	2,27	6,68	12,22	18,80	26,65	35,75

Nitrobenzene ($C_6H_5NO_2$) – Octane (C_8H_{18})

x_1	ε at *t* °C [351]						
	15	20	25	30	35	40	45
0	1,950	1,948	1,946	1,944	1,942	1,940	1,938
9,7	3,280	3,248	3,216	3,185	3,153	3,122	3,090
28,3	6,422	6,325	6,228	6,132	6,036	5,938	5,835
51,4	—	12,263	12,063	11,822	11,585	11,345	11,070
61,1	—	15,925	15,580	15,235	14,880	14,530	14,170
80,8	25,530	24,910	24,315	23,710	23,100	22,500	21,920

Nitrobenzene ($C_6H_5NO_2$) – Tetraethoxysilane ($C_8H_{20}O_4Si$)

[146], $t=25$ °C

φ_1	0	20	40	50	60	80	100
ε	3,70	8,84	14,77	17,61	20,82	27,79	34,82

Nitrobenzene ($C_6H_5NO_2$) – Tetrapropoxysilane ($C_{12}H_{28}O_4Si$)
[146], $t=25$ °C

φ_1	0	20	40	50	60	80	100
ε	3,21	8,00	13,65	16,53	19,83	27,02	34,82

Nitrobenzene ($C_6H_5NO_2$) – Hexaethoxydisiloxan ($C_{12}H_{30}O_7Si_2$)
[146], $t=25$ °C

φ_1	0	20	40	50	60	80	100
ε	5,18	10,27	15,06	18,10	21,13	27,59	34,82

Nitrobenzene ($C_6H_5NO_2$) – Tetra(2–butoxy)silane ($C_{16}H_{36}O_4Si$)
[146], $t=25$ °C

φ_1	0	20	40	50	60	80	100
ε	2,59	7,37	13,20	16,03	19,27	26,60	34,82

Nitrobenzene ($C_6H_5NO_2$) – Tetraisobutoxysilane ($C_{16}H_{36}O_4Si$)
[146], $t=25$ °C

φ_1	0	20	40	50	60	80	100
ε	2,81	7,68	13,07	16,01	19,43	26,82	34,82

Nitrobenzene ($C_6H_5NO_2$) – Octaethoxytrisiloxan ($C_{16}H_{40}O_{10}Si_3$)
[146], $t=25$ °C

φ_1	0	20	40	50	60	80	100
ε	4,76	9,93	15,18	17,95	20,94	17,62	34,82

Nitrobenzene ($C_6H_5NO_2$) – Tetrapentoxysilane ($C_{20}H_{44}O_4Si$)
[146], $t=25$ °C

φ_1	0	20	40	50	60	80	100
ε	2,82	7,68	13,13	16,08	19,50	26,81	34,82

2–Nitrophenol ($C_6H_5NO_3$) – Azobenzene ($C_{12}H_{10}N_2$)
[334], $t=21$ °C

w_1	0	15;15	29	39	56	62	75	82	100
ε	2,2	3,9	5,6	6,8	8,8	9,6	20,2	21	24,6

Benzene (C_6H_6) – Phenol (C_6H_6O) *)

[355], $t=25$ °C, $\nu=1$ MHz, fig. 9.

x_1	65	75	90	100
ε	4,672	3,692	2,722	2,282

*) See also [178, 276].

[354], $t=25$ °C

x_2	38,88	26,66	22,27	15,39	12,10	8,83	3,88	0
ε	4,664	3,602	3,300	2,915	2,765	2,567	2,403	2,28

[339], $t=70$ °C

x_2	ε	x_2	ε	x_2	ε	x_2	ε
100	9,161	45,45	4,375	20,33	2,943	10,74	2,540
72,57	6,517	40,72	4,063	20,17	2,934	0,00	2,193
59,96	5,458	32,46	3,567	13,46	2,639	—	—

[355], $t=20$ °C, $\nu=1$ MHz

w_2	28,58	20,10	13,55	5,32	0
ε	3,805	3,252	2,871	2,493	2,29

Nitrobenzene ($C_6H_5NO_2$) – Nitroaniline ($C_6H_6N_2O_2$)

[321], $t=20$ °C, $\nu=0,6$ MHz

w_2	14,030	11,222	11,653	8,240	5,978	5,292	3,053
ε	4,977	4,328	4,277	3,707	3,275	3,148	2,758

Benzene (C_6H_6) – Aniline (C_6H_7N)

[339, 276], $t=70$ °C, $\nu=180$ MHz, fig. 9.

x_2	100	79,34	59,68	37,75	18,03	0,00
ε	5,932	5,053	4,250	3,413	2,725	2,193

w_2	ε at t °C [310]		w_2	ε at t °C [310]	
	25	50		25	50
100	6,853	—	50,08	4,165	3,960
89,48	6,183	5,878	41,37	3,711	3,583
79,05	5,568	5,309	31,17	3,255	3,141
72,63	5,248	4,964	18,75	2,858	2,799
60,08	4,673	4,438	9,274	2,525	2,466

[175], $t=20$ °C, $\nu=1{,}8$ MHz

φ_2	100	80	60	40	20	0
ε	7,06	5,96	4,94	3,94	3,06	2,28

[227], $t=14$ °C, $\nu=3$ MHz

φ_2	100	80	60	40	20	0
ε	7,20	6,07	5,02	4,05	3,16	2,28

Benzene (C_6H_6) – 2, 4–Hexadiene (C_6H_{10})

[329], $t=25$ °C, $\nu=0{,}2$ MHz, $T_{B.Pt}=80$ °C

x_1	0,00	28,62	50,47	75,66	86,90	95,07	100
ε	2,207	2,233	2,242	2,256	2,267	2,270	2,274

[329], $t=25$ °C, $\nu=0{,}2$ MHz, $T_{B.Pt}=81{,}6$ °C

x_1	0,00	28,27	50,03	75,29	89,97	95,01	100,0
ε	2,224	2,249	2,259	2,264	2,269	2,271	2,274

Benzene (C_6H_6) – 2, 3–Dimethyl–1, 3–butodiene (C_6H_{10})

[329], $t=25$ °C, $\nu=0{,}2$ MHz

x_1	0	24,46	49,82	75,38	90,16	100
ε	2,099	1,134	2,173	2,220	2,252	2,274

Benzene (C_6H_6) – 2–Methyl–1, 3–pentadiene (C_6H_{10})

[329], $t=25$ °C, $\nu=0{,}2$ MHz

x_1	0	22,19	57,19	80,03	90,15	95,04	100
ε	2,422	2,404	2,354	2,314	2,296	2,284	2,274

Benzene (C_6H_6) – 3–Methyl–1, 3–pentadiene (C_6H_{10})

[329], $t=25$ °C, $\nu=0{,}2$ MHz

x_1	0	47,12	74,93	90,05	94,76	100
ε	2,426	2,367	2,324	2,293	2,284	2,274

Benzene (C_6H_6) – Cyclohexanone ($C_6H_{10}O$)
[175], $t = 20$ °C, $\nu = 1,8$ MHz

φ_2	100	80	60	40	20	0
ε	15,70	12,72	9,83	7,22	4,67	2,28

Benzene (C_6H_6) – Diethyl oxalate ($C_6H_{10}O_4$)

x_2	ε at t °C [296]		x_2	ε at t °C [296]	
	25	50		25	50
13,81	3,326	3,177	4,92	2,675	2,583
10,74	3,104	2,966	3,18	2,537	2,463
6,29	2,784	2,680	—	—	—

Benzene (C_6H_6) – Diphenyldisulphone ($C_6H_{10}O_4S$)
[324], $t = 25$ °C

x_2	39,2	31,0	22,1	14,9	6,6	0
ε	2,3613	2,3431	2,3208	2,3058	2,2872	2,2714

Benzene (C_6H_6) – Bromocyclohexane ($C_6H_{11}Br$)
[356], $t = 60$ °C

x_2	ε	x_2	ε	x_2	ε	x_2	ε
25,19	2,402	18,47	2,356	11,53	2,311	6,161	2,277
24,96	2,401	17,30	2,348	11,43	2,309	—	—
20,12	2,368	14,78	2,333	8,088	2,289		

Benzene (C_6H_6) – Chlorocyclohexane ($C_6H_{11}Cl$)
[356], $t = 60$ °C

x_2	20,48	16,88	14,85	14,77	13,48	11,56	9,376
ε	2,357	2,335	2,325	2,323	2,315	2,304	2,292

Benzene (C_6H_6) – Iodocyclohexane ($C_6H_{11}J$)
[356], $t = 60$ °C

x_2	20,96	18,48	17,50	16,44	14,00	12,06
ε	2,378	2,354	2,350	2,340	2,327	2,312

Benzene (C_6H_6) – Cyclohexanone ($C_6H_{12}O$)
[357, 331], $t = 18$ °C

x_2	100	85	70	60	50	30	20	10	0
ε	13,4	11,65	9,99	7,70	6,78	4,21	3,36	2,75	2,29

Benzene (C_6H_6) – Hexanoic acid ($C_6H_{12}O_2$)
[358], $t = 71$ °C, $\lambda = 820$ m

N	ε	N	ε	N	ε	N	ε
0,000	2,180	1,5520	2,217	4,9516	2,369	7,6343	2,632
0,3850	2,187	2,6732	2,254	5,9880	2,444	—	—
0,7775	2,196	3,8181	2,298	6,7842	2,544	—	—

Benzene (C_6H_6) – Cyclohexylamine ($C_6H_{13}N$)
[167], $t = 25$ °C, fig. 144

x_2	27,445	20,884	12,935	0,000
ε	2,892	2,743	2,559	2,276

Benzene (C_6H_6) – Hexane (C_6H_{14})
$t = 25$ °C

[286] *)				[216] **)				[359]			
x_1	ε	x_1	ε	x_1	ε	x_1	ε	x_1	ε	x_1	ε
0	1,884	68,8	2,102	1,0	1,904	50	2,054	0	1,940	22,55	1,998
26,9	1,945	85,5	2,186	9,0	1,931	75	2,160	11,77	1,968	29,79	2,017
49,5	2,024	100	2,281	25	1,931	100	2,283	18,27	1,986	—	—

*) ν = 10 kHz
**) ν = 1 MHz

Benzene (C_6H_6) – Diisopropyl ether ($C_6H_{14}O$)
[166], $t = 20$ °C

x_2	100,0	69,922	47,866	30,333	17,152	5,982	0,000
ε	3,976	3,522	3,179	2,881	2,631	2,412	2,282

[343], $t = 25$ °C

x_2	100,0	17,733	12,468	8,560	4,153	0,000
ε	4,0370	2,6228	2,5176	2,4390	2,3512	2,2680

Benzene (C_6H_6) – Benzonitrile (C_7H_5N)

[339], $t = 70$ °C

x_2	100,0	78,18	59,68	37,78	20,68	0,00
ε	21,60	17,30	13,75	9,449	6,118	2,193

[210], $t = 25$ °C

w_2	ε	w_2	ε	w_2	ε	w_2	ε
100,0	25,51	54,281	15,12	10,845	4,788	1,359	2,588
97,173	24,91	34,732	10,60	7,414	3,982	0,758	2,452
94,878	24,41	23,770	7,910	5,502	3,536	0,000	2,2825
90,183	23,29	16,773	6,210	3,086	2,982	—	—
77,949	20,54	14,845	5,749	2,300	2,808	—	—

[175], $t = 20$ °C, $\nu = 1,8$ MHz

φ_2	100	80	60	40	20	0
ε	25,65	20,50	15,65	10,95	6,52	2,28

Benzene (C_6H_6) – 3-Nitrobenzoic acid ($C_7H_5NO_2$)

[271], $t = 25$ °C, $\nu = 1$ kHz

$10^{-3}N$	0	10	20	68,07
ε	2,273	2,288	2,305	2,377

Benzene (C_6H_6) – Benzaldehyde (C_7H_6O)

[227], $t = 15$ °C, $\nu = 3$ MHz

φ_2	100	80	60	40	20	0
ε	18,07	13,92	10,57	7,46	4,81	2,28

[175], $t = 20$ °C, $\nu = 1,8$ MHz

φ_2	100	80	60	40	20	0
ε	17,85	14,17	10,71	7,60	4,77	2,28

Benzene (C_6H_6) – 2-Nitrotoluene ($C_7H_7NO_2$)

[321], $t = 20$ °C, $\lambda = 550$ m

x_2	78,718	62,785	57,159	37,715	25,118	11,063	6,192
ε	17,917	13,217	11,839	7,790	5,669	3,645	3,020

[360], $t=25$ °C, $\nu=1$ MHz

x_2	21,20	16,50	10,50	7,54	4,20	2,92	0
ε	6,20	5,34	4,36	3,77	3,08	2,86	2,283

w_2	ε at t °C [311]		w_2	ε at t °C [311]	
	25	50		25	50
80,243	17,290	15,920	30,073	6,287	5,776
70,457	14,630	13,350	21,185	4,701	4,480
60,296	12,040	11,010	9,093	3,312	3,160
49,643	9,908	9,109	—	—	—

Benzene (C_6H_6) – 3–Nitrotoluene ($C_7H_7NO_2$)

w_2	ε at t °C [311]		w_2	ε at t °C [311]	
	25	50		25	50
70,890	15,840	14,730	32,641	7,496	6,982
60,400	13,170	12,269	21,676	5,546	5,202
50,220	10,820	10,090	13,722	4,257	4,030
40,050	8,872	8,332	—	—	—

[175], $t=20$ °C, $\nu=1,8$ MHz

φ_2	100	80	60	40	20	0
ε	26,80	20,80	15,45	10,80	6,38	2,28

Benzene (C_6H_6) – 4–Nitrotoluene ($C_7H_7NO_2$)

w_2	ε at t °C [311]	
	25	50
25,135	6,521	6,191
19,964	5,573	5,258
14,902	4,682	4,438
9,911	3,840	3,638

Benzene (C_6H_6) – Toluene (C_7H_8)

$t=25$ °C

[286], ν=10 kHz				[207, 209], ν=1 MHz			
x_2	ε	x_2	ε	x_2	ε	x_2	ε
100	2,376	35,8	2,312	100	2,378	25	2,315
77,0	2,350	17,3	2,290	75	2,362	8,5	2,304
56,6	2,329	0	2,281	50	2,337	0	2,282

Benzene (C_6H_6) – Anisole (C_7H_8O)

[361], $t=25$ °C

x_2	ε	x_2	ε	x_2	ε	x_2	ε
100	4,33	66,06	3,74	35,74	3,13	8,29	2,50
85,85	4,09	54,08	3,50	27,96	2,98	0,00	2,27
75,74	3,92	45,74	3,33	16,60	2,72	—	—

[339], $t=70$ °C, $\nu=180$ kHz

x_2	100	66,59	63,20	39,06	19,81	0,00
ε	3,887	3,348	3,306	2,892	2,547	2,193

[355], $t=20$ °C

w_2	30,89	20,81	13,90	5,67	0
ε	2,874	2,671	2,555	2,401	2,29

Benzene (C_6H_6) – Benzyl alcohol (C_7H_8O)

[339], $t=70$ °C

x_2	100	77,20	65,24	49,50	37,66	19,74	0,00
ε	9,467	7,327	6,277	4,961	4,163	3,039	2,193

[227], $t=14$ °C, $\nu=1,8$ MHz

φ_2	100	80	60	40	20	0
ε	13,63	10,58	7,46	4,86	3,37	2,28

[175], $t=20$ °C, $\nu=1,8$ MHz

φ_2	100	80	60	40	20	0
ε	13,62	10,37	7,35	4,87	3,25	2,28

Benzene (C_6H_6) – *o*–Cresol (C_7H_8O)

[355], $t=20$ °C

w_2	23,73	14,72	10,01	5,40	0
ε	3,001	2,701	2,574	2,441	2,29

Benzene (C_6H_6) – *m*–Cresol (C_7H_8O)

[355], $t=20$ °C

w_2	19,62	15,60	4,55	0
ε	3,090	2,796	2,444	2,29

[227], $t=16$ °C, $\nu=3$ MHz

φ_2	100	80	60	40	20	0
ε	12,95	9,86	6,95	4,82	3,28	2,28

Benzene (C_6H_6) – *p*–Cresol (C_7H_8O)

[355], $t=20$ °C

w_2	21,68	16,08	13,48	5,49	0
ε	3,231	2,936	2,814	2,495	2,29

Benzene (C_6H_6) – *o*–Toluidine (C_7H_9N)

w_2	ε at t °C [312]		w_2	ε at t °C [312]	
	25	50		25	50
100	6,138	—	49,42	3,919	3,748
93,30	5,867	5,528	39,67	3,527	3,416
79,18	5,207	4,998	33,65	3,316	3,240
75,90	5,047	4,796	18,27	2,784	2,718
59,10	4,297	4,129	8,961	2,531	2,470

Benzene (C_6H_6) – *m*–Toluidine (C_7H_9N)

w_2	ε at t °C [312]		w_2	ε at t °C [312]	
	25	50		25	50
100,0	5,816	—	50,19	3,714	3,624
92,21	5,500	5,192	40,94	3,400	3,298
80,71	4,991	4,714	30,24	3,052	2,973
69,36	4,480	4,301	20,23	2,788	2,713
59,36	4,072	3,922	10,05	2,533	2,473

Benzene (C_6H_6) – *p*–Toluidine (C_7H_9N)

w_2	ε at t °C [312]		w_2	ε at t °C [312]	
	25	50		25	50
38,42	3,201	3,087	21,94	2,749	2,657
30,79	3,000	2,874	15,18	2,551	2,485
25,40	2,863	2,748	9,972	2,455	2,371

Benzene (C_6H_6) – Pyridinium acetate ($C_8H_9NO_2$)

[270], $t=25$ °C

x_2	ε	x_2	ε	x_2	ε	x_2	ε
100	16,0	84,89	14,4	65,68	12,0	41,96	8,6
95,25	15,5	77,75	13,5	60,98	11,4	35,68	7,7
90,63	15,0	70,91	12,6	49,13	9,6	32,69	7,3

Benzene (C_6H_6) – Diethylmalonate ($C_7H_{12}O_4$)
[296], $\lambda = 1000$ m

x_2	ε at t °C		x_2	ε at t °C	
	25	50		25	50
16,06	3,550	3,362	3,20	2,548	2,470
10,87	3,183	3,026	2,38	2,482	2,412
5,02	2,706	2,608	—	—	—

Benzene (C_6H_6) – Cyclohexylmethanol ($C_7H_{14}O$)

x_2	ε at t °C [211]			x_2	ε at t °C [211]		
	20	40	60		20	40	60
83,59	—	9,70	8,04	21,11	3,24	3,15	3,06
51,56	6,36	5,44	4,86	12,58	2,80	2,76	2,72
40,87	5,06	4,46	—	7,006	2,59	2,54	2,48
37,00	4,85	4,14	3,99	4,256	2,52	2,44	2,42
33,55	4,22	3,88	3,86	2,138	2,35	2,32	2,26
29,35	3,85	3,60	3,43	—	—	—	—

Benzene (C_6H_6) – Methylcyclohexanol ($C_7H_{14}O$)
[331], $t = 18$ °C

x_2	23,82	17,76	12,01	0,00
ε	2,3772	2,3568	2,3370	2,2951

Benzene (C_6H_6) – Heptane (C_7H_{16})
[221, 169], $t = 20$ °C

x_1	14,98	26,82	47,12	60,32	72,51	86,49	100,0
ε	1,980	2,007	2,055	2,103	2,155	2,205	2,280

Benzene (C_6H_6) – Benzyl cyanide (C_8H_7N)
[175], $t = 20$ °C, $\nu = 1,8$ MHz

φ_2	100	80	60	40	20	0
ε	18,95	15,27	11,83	8,47	5,21	2,28

Benzene (C_6H_6) – Толуолсульфокарбонил ($C_8H_7O_3S$)
[315], $t = 25$ °C

x_2	41,4	31,0	23,1	14,5	0
ε	2,3565	2,3374	2,3210	2,3027	2,2714

Benzene (C_6H_6) – Acetophenone (C_8H_8O)
[313, 210], $t=20$ °C

w_2	ε	w_2	ε	w_2	ε	w_2	ε
100,0	17,73	90,372	16,17	9,073	3,469	1,348	2,453
97,240	17,28	82,480	14,90	6,055	3,071	0,000	2,2835
96,950	17,23	66,238	12,13	4,235	2,829	—	—
94,814	16,87	52,597	9,950	1,989	2,538	—	—
91,644	16,38	24,968	5,754	1,550	2,479	—	—

Benzene (C_6H_6) – Phenylacetic acid ($C_8H_8O_2$)
[362]

φ_2	100	77	60	50	40	25	10	0
t °C	15,0	21,0	15,0	17,0	15,0	17,0	15,0	14,2
ε	3,195	2,977	2,975	2,696	2,692	2,554	2,260	2,073

Benzene (C_6H_6) – *m*–Toluic acid ($C_8H_8O_2$)
[271], $t=25$ °C, $\nu=1$ kHz

m	0	10	20	50	100
ε	2,273	2,281	2,282	2,283	2,289

Benzene (C_6H_6) – Methyl benzoate ($C_8H_8O_2$)
[227], $t=12$ °C, $\nu=3$ MHz

φ_2	100	80	60	40	20	0
ε	6,72	5,82	4,93	4,04	3,18	2,28

Benzene (C_6H_6) – *o*–Hydroxyacetophenone (C_8H_8O)

w_2	ε at t °C [313]		w_2	ε at t °C [313]	
	25	50		25	50
26,77	4,973	4,680	9,042	3,139	2,979
18,40	4,086	3,836	0,000	2,273	2,222

Benzene (C_6H_6) – Methyl salicylate ($C_8H_8O_3$)
[363]

t=13,2 °C				t=40,2 °C			
x_2	ε	x_2	ε	x_2	ε	x_2	ε
100,00	9,35	49,939	5,16	100,00	8,64	40,232	4,20
88,551	8,39	39,811	4,49	90,442	7,77	30,312	3,67
79,172	7,45	31,156	3,95	80,341	6,87	19,951	3,12
70,623	6,72	20,762	3,32	70,123	5,94	10,512	2,77
59,092	5,84	11,132	2,79	57,989	5,39	0	2,24
—	—	0	2,29	50,102	4,87	—	—

Benzene (C_6H_6) – Ethyl phenylthiosulphonate ($C_8H_{10}O_2S_2$)
[315], t=25 °C

x_2	21,5	17,5	13,4	7,2	0
ε	2,3475	2,3331	2,3195	2,2975	2,2714

Benzene (C_6H_6) – *o*–Xylene (C_8H_{10})
[209], t=20 °C, ν=1 MHz

x_2	100	50	25	10	0
ε	2,507	2,398	2,337	2,302	2,282

Benzene (C_6H_6) – *p*–Xylene (C_8H_{10})
[209], t=25 °C, ν=1 MHz

x_2	100	50	0
ε	2,265	2,274	2,282

Benzene (C_6H_6) – 2–Methoxytoluene ($C_8H_{10}O$)
[355, 310], t=20 °C

w_2	100	25,37	17,41	11,35	5,52	0
ε	3,37	2,595	2,495	2,441	2,363	2,29

Benzene (C_6H_6) – 3–Methoxytoluene ($C_8H_{10}O$)
[355, 310], t=20 °C

w_2	23,48	16,41	10,73	5,21	0
ε	2,666	2,547	2,466	2,373	2,29

Benzene (C_6H_6) – 4–Methoxytoluene ($C_8H_{10}O$)
[355, 310], $t = 20$ °C

w_2	100	19,87	15,20	11,29	4,76	0
ε	4,03	2,590	2,523	2,467	2,357	2,29

Benzene (C_6H_6) – Phenetole ($C_8H_{10}O$)
[210], $t = 25$ °C

w_2	ε	w_2	ε	w_2	ε	w_2	ε
100,0	4,225	90,560	4,100	46,365	3,346	5,400	2,421
96,994	4,185	72,488	3,824	31,486	3,042	1,827	2,329
93,675	4,141	64,519	3,684	12,193	2,593	0,00	2,2825

[175], $t = 20$ °C, ν = 1,8 MHz

φ_2	100	80	60	40	20	0
ε	4,25	3,85	3,45	3,05	2,66	2,28

[227], $t = 15$ °C, ν = 3 MHz

φ_2	100	80	60	40	20	0
ε	4,37	3,95	3,53	3,12	2,63	2,28

Benzene (C_6H_6) – 1–Phenylethanol ($C_8H_{10}O$)

x_2	ε at t °C [211, 300]		x_2	ε at t °C [211, 300]	
	20	40		20	40
100,0	8,900	8,045	11,31	2,800	2,765
61,93	5,835	5,515	7,06	2,610	2,580
36,67	4,175	4,005	5,18	2,530	—
24,59	3,495	3,410	3,03	2,445	2,405
19,78	2,230	3,180	1,69	2,380	2,345

Benzene (C_6H_6) – 2–Phenylethanol ($C_8H_{10}O$)

x_2	ε at t °C [211]		x_2	ε at t C [211]	
	20	40		20	40
100,0	13,000	10,145	23,23	4,145	3,925
81,83	9,820	8,410	20,13	3,445	3,295
67,27	8,000	7,095	10,75	2,865	2,770
50,08	6,150	5,510	6,75	2,630	2,570
36,62	4,810	4,470	3,22	2,465	2,410

Benzene (C_6H_6) – Dimethylaniline ($C_8H_{11}N$)
[339], $t = 70$ °C, ν = 180 kHz

x_2	100	80,21	60,55	36,06	22,92	0,00
ε	4,423	4,099	3,733	3,196	2,836	2,193

[227], $t = 14$ °C, $\nu = 3$ MHz

φ_2	100	80	60	40	20	15	10	5	0
ε	5,05	4,63	4,07	3,52	3,02	2,83	2,65	2,51	2,28

Benzene (C_6H_6) – Anilinium acetate ($C_8H_{11}NO_2$)
[270], $t = 25$ °C

x_2	ε	x_2	ε	x_2	ε	x_2	ε
100	8,21	69,88	6,71	37,79	4,94	13,36	3,19
91,97	7,82	58,57	6,14	28,63	4,33	6,46	2,67
83,53	7,40	47,31	5,55	20,53	3,72	0	2,28

Benzene (C_6H_6) – Cyclohexyl acetate ($C_8H_{14}O_2$)
[331], $t = 18$ °C

x_2	37,90	24,98	16,59	12,78	0,00
ε	2,4581	2,4028	2,3682	2,3506	2,2951

Benzene (C_6H_6) – Di–isopropylxanthogendisulphide ($C_8H_{14}O_2S_4$)
[364], $t = 25$ °C

x_2	47,2	37,8	28,9	21,1	13,9	5,5	0
ε	2,3702	2,3502	2,3318	2,3159	2,3004	2,2821	2,2714

Benzene (C_6H_6) – Diethyl succinate ($C_8H_{14}O_4$)

x_2	ε at t °C [296]		x_2	ε at t °C [296]	
	25	50		26	50
15,31	3,113	3,004	7,29	2,698	2,621
10,50	2,871	2,784	5,47	2,596	2,526
9,83	2,833	2,748	3,06	2,457	2,397

Benzene (C_6H_6) – Dibutyl ether ($C_8H_{18}O$)
[343], $t = 25$ °C

x_2	100,0	73,64	57,566	29,00	5,9866	0,00
ε	3,0817	2,9315	2,8212	2,5962	2,3418	2,2705

Benzene (C_6H_6) – 2-Methylheptan-2-ol ($C_8H_{18}O$)

x_2	ε at t °C [365], ν=0,5 MHz							
	0	10	20	30	40	50	60	70
100	3,09	3,23	3,37	3,50	3,60	3,68	3,75	—
69,15	3,160	3,263	3,354	3,429	3,486	3,521	3,514	3,465
48,92	3,077	3,154	3,214	3,253	3,280	3,274	3,242	3,192
29,15	2,926	2,961	2,987	2,994	2,984	2,955	2,908	2,849
8,89	—	2,589	2,569	2,545	2,515	2,480	2,443	2,406

Benzene (C_6H_6) – Octan-2-ol ($C_8H_{18}O$)
[366], $t=20$ °C

x_2	ε	x_2	ε	x_2	ε	x_2	ε
100	8,173	25,393	3,068	3,294	2,3977	1,387	2,3308
62,086	4,815	15,966	2,790	2,938	2,3867	0	2,2813
36,130	3,455	7,302	2,5305	2,420	2,3670	—	—

Benzene (C_6H_6) – Octanoic acid ($C_8H_{16}O_2$)
[358], $t=71$ °C

N	ε	*N*	ε	*N*	ε	*N*	ε
0,000	2,180	1,2048	2,209	3,9002	2,338	6,0524	2,544
0,3013	2,185	2,0701	2,238	4,8330	2,449	—	—
0,6225	2,193	3,0212	2,283	5,5637	2,501	—	—

Benzene (C_6H_6) – Quinoline (C_9H_7N)
[173], $t=18$ °C

w_2	ε	w_2	ε	w_2	ε	w_2	ε
100	9,559	71,803	6,949	40,613	4,463	10,397	2,759
90,108	8,755	61,228	5,986	30,714	3,850	0	2,28
80,733	7,692	50,764	5,193	20,550	3,283	—	—

Benzene (C_6H_6) – 1–Indanone (C_9H_8O)
[211]

t=10 °C		t=20 °C		t=40 °C		t=60 °C	
x_2	ε	x_2	ε	x_2	ε	x_2	ε
28,82	3,71	72,15	—	72,15	6,150	100	7,820
15,45	3,14	51,39	—	51,39	4,770	78,35	6,295
9,16	2,79	38,65	4,35	38,65	4,215	52,95	4,700
6,25	2,63	26,95	3,63	26,95	3,405	35,47	3,800
3,32	2,47	14,55	3,03	14,55	2,900	21,84	3,165
1,68	2,28	11,83	2,89	11,83	2,775	13,53	2,780
—	—	7,40	2,66	7,40	2,57	8,66	2,555
—	—	4,32	2,48	4,32	2,415	5,28	2,410
—	—	2,11	2,36	2,11	2,31	2,12	2,28

Benzene (C_6H_6) – 2–Indanone (C_9H_8O)
[211]

t=20 °C		t=60 °C			
x_2	ε	x_2	ε	x_2	ε
20,52	3,415	71,28	5,800	6,66	2,500
12,37	2,975	48,04	4,410	3,51	2,375
7,91	2,730	27,75	3,413	1,46	2,300
4,86	2,560	17,75	2,990	—	—
1,73	2,400	11,21	2,695	—	—

Benzene (C_6H_6) – Ethyl salicylate ($C_9H_{10}O_3$)
[363], t=40,2 °C

x_2	ε	x_2	ε	x_2	ε	x_2	ε
100	8,02	69,309	5,94	39,272	4,20	11,067	2,80
89,000	7,22	60,276	5,41	31,716	3,79	0	2,24
80,173	6,66	51,315	4,87	22,181	3,30	—	—

Benzene (C_6H_6) – Ethyl tolylthiosulphonate ($C_9H_{12}O_2S_2$)
[315], t=25 °C

x_2	37,3	25,5	17,8	3,9	0
ε	2,4672	2,3716	2,3409	2,2868	2,2714

Benzene (C_6H_6) – Ethylphenyldithiocarbamate ($C_9H_{11}NS_2$)
[364], t=25 °C

x_2	85,6	68,4	53,2	41,2	25,1	14,1	0
ε	2,3770	2,3570	2,3371	2,3226	2,3032	2,2880	2,2714

Benzene (C_6H_6) – Phenylpropyl alcohol ($C_9H_{12}O$)
[300], $t = 25$ °C

x_2	12,27	8,96	6,85	5,20	3,74	2,38	1,0	0,00
ε	2,95	2,80	2,71	2,62	2,53	2,50	2,32	2,26

Benzene (C_6H_6) – 1, 3, 5–Trimethylcyclohexanol ($C_9H_{18}O$)
[331], $t = 18$ °C

x_2	20,82	13,33	7,000	0,000
ε	2,3855	2,3512	2,3246	2,2951

Benzene (C_6H_6) – 1–Nitronaphthalene ($C_{10}H_7NO_2$)
[314]

w_2	ε at t °C		w_2	ε at t °C	
	25	50		25	50
24,315	5,065	4,685	9,850	3,267	3,126
15,18	3,861	3,696	5,017	2,749	2,672

Benzene (C_6H_6) – Naphthylamine ($C_{10}H_9N$)
[308]

w_2	ε at t °C		w_2	ε at t °C	
	25	50		25	50
37,42	3,328	3,185	22,25	2,832	2,761
28,43	3,045	2,922	7,896	2,463	2,402

Benzene (C_6H_6) – Tetrahydro–2–naphthol ($C_{10}H_{12}O$)
[211]

x_2	ε at t °C		x_2	ε at t °C	
	20	40		20	40
100	11,70	9,570	14,41	3,075	2,935
80,92	9,375	8,025	9,23	2,765	2,660
62,46	7,165	6,345	6,08	2,630	2,535
49,60	5,880	5,315	4,22	2,535	2,455
39,75	4,932	4,545	2,52	2,442	2,375
29,11	4,040	3,800	1,23	2,365	2,320
21,62	3,540	3,360	—	—	—

Benzene (C_6H_6) – Ethylphenyl acetate ($C_{10}H_{12}O_2$)
[362]

w_2	t °C	ε	w_2	t °C	ε	w_2	t °C	ε	w_2	t °C	ε
100	15	4,279	50,0	15	3,195	33,3	17	2,696	9,1	15	2,260
73,0	21	3,639	43,5	21	2,977	28,6	15	2,692	0	14,2	2,073
68,8	21	3,508	37,5	15	2,975	20,0	17	2,554	—	—	—

Benzene (C_6H_6) – Thymol ($C_{10}H_{14}O$)
[212], $t=20$ °C, $\lambda=66$ cm

x_2	100	80	60	40	20	0
ε	4,03	4,13	4,03	3,58	2,98	2,28

Benzene (C_6H_6) – *N, N*–Diethylaniline ($C_{10}H_{15}N$)
[212], $t=20$ °C, $\lambda=66$ cm

x_2	100	80	60	40	20	10	5	0
ε	5,28	4,97	4,50	3,93	3,22	2,80	5,52	2,28

Benzene (C_6H_6) – Camphor ($C_{10}H_{16}O$)

w_2	x_2	ε at t °C [367]					
		25	30	35	40	45	50
66,04	49,95	—	7,328	7,214	7,106	7,018	6,953
47,71	31,88	5,775	5,714	5,652	5,584	5,520	5,458
39,50	25,11	5,142	5,139	5,007	4,933	4,861	4,793
29,60	17,75	4,302	4,728	4,250	4,228	4,199	4,171
23,52	13,63	3,888	3,868	3,841	3,819	3,793	3,771

Benzene (C_6H_6) – 1, 4–Cyclohexandiol diacetate ($C_{10}H_{16}O$)
[331], $t=25$ °C

x_2	32,32	26,26	18,33	13,09	10,76
ε	2,3832	2,3670	2,3419	2,3283	2,3229

Benzene (C_6H_6) – Menthol ($C_{10}H_{20}O$)
[159]

x_2	31,7	29,3	19,0	17,8	8,2	0,0
t °C	24	23	22,5	22,5	23	24
ε	4,9	5,0	3,82	3,7	3,05	2,25

Benzene (C_6H_6) – 4–Propylheptan–4–ol ($C_{10}H_{22}O$)
[331], $t=18$ °C

x_2	19,77	14,39	7,894	0,00
ε	2,3539	2,3386	2,3191	2,2951

Benzene (C_6H_6) – Dipentyl ether ($C_{10}H_{22}O$)
[343], $t=25$ °C

x_2	ε	x_2	ε	x_2	ε	x_2	ε
100,00	2,7980	41,682	2,5919	7,9138	2,3512	0,00	2,2668
73,526	2,7315	24,410	2,4902	3,434	2,3049	—	—
64,046	2,6957	13,458	2,4042	1,6642	2,2845	—	—

Benzene (C_6H_6) – Di–isopentylsulphone ($C_{10}H_{22}O_2S$)
[324], $t=25$ °C

x_2	28,5	22,9	17,1	10,7	5,3	0
ε	2,3520	2,3364	2,3198	2,3018	2,2865	2,2714

Benzene (C_6H_6) – Dipentyl sulphide ($C_{10}H_{22}S$)
[285]

x_2	ε at t °C		x_2	ε at t °C	
	25	50		25	50
100,00	3,826	3,594	3,627	2,384	2,322
46,309	3,244	3,088	2,397	2,349	2,291
28,299	2,956	3,832	1,269	2,315	2,261
14,166	2,665	2,568	0	2,276	2,226
9,009	2,531	2,453	—	—	—

Benzene (C_6H_6) – Ethyl cinnamate ($C_{11}H_{12}O_2$)
[227], $t=15$ °C, $\nu=3$ MHz

φ_2	100	80	60	40	20	0
ε	5,83	5,14	4,42	3,74	3,03	2,28

Benzene (C_6H_6) – *S*–Ethyl–dibutyldithiocarbamate ($C_{11}H_{23}NS_2$)
[364], $t=25$ °C

x_2	64,5	50,1	39,6	28,3	17,0	0
ε	2,3683	2,3501	2,3366	2,3155	2,2980	2,2714

Benzene (C_6H_6) – Undecan–1–ol ($C_{11}H_{24}O$)
[300], $t = 25$ °C

x_2	12,47	9,62	7,19	5,29	3,34	2,17	1,15	0,00
ε	2,74	2,65	2,53	2,44	2,38	2,32	2,27	2,26

Benzene (C_6H_6) – Bromophenyl bromophenylthiosulphonate ($C_{12}H_8O_2Br_2S_2$)

$t=25$ °C [315]				$t=40$ °C [315]			
x_2	ε	x_2	ε	x_2	ε	x_2	ε
30,1	2,3228	6,4	2,2811	49,4	2,3150	19,6	2,2670
22,9	2,3104	0	2,2714	38,1	2,2967	8,4	2,2511
14,4	2,2960	—	—	26,2	2,2774	0	2,2386

Benzene (C_6H_6) – Chlorophenyl chlorophenylthiosulphonate ($C_{12}H_8O_2Cl_2S_2$)

$t=25$ °C [315]				$t=40$ °C [315]			
x_2	ε	x_2	ε	x_2	ε	x_2	ε
42,60	2,3411	17,5	2,2997	60,8	2,3327	23,1	2,2734
34,4	2,3282	8,8	2,2850	46,8	2,3107	10,6	2,2531
25,9	2,3142	0	2,2714	34,1	2,2903	0	2,2386

Benzene (C_6H_6) – 4, 4′–Dibromophenyldisulphide ($C_{12}H_8Br_2S_2$)
[324], $t = 25$ °C

x_2	32,9	28,8	21,4	13,3	6,8	0
ε	2,2781	2,2770	2,2757	2,2743	2,2730	2,2714

Benzene (C_6H_6) – Azobenzene ($C_{12}H_{10}N_2$)
[334], $t = 21$ °C

w_2	100	88	73	68	62	56	48	38	29	20	10	0
ε	2,2	2,4	3,2	3,2	3,2	2,95	2,75	2,2	2,3	2,3	2,2	2,3

Benzene (C_6H_6) – Phenyl phenylthiosulphonate ($C_{12}H_{10}O_2S_2$)
[315], $t = 25$ °C

x_2	51,2	40,6	29,9	24,0	11,6	0
ε	2,4784	2,4353	2,3919	2,3680	2,3178	2,2714

Benzene (C_6H_6) – Diphenylsulphoxide ($C_{12}H_{10}OS$)
[324], $t = 25$ °C

x_2	31,4	29,8	25,1	17,6	9,5	0
ε	2,3457	2,3419	2,3307	2,3129	2,2938	2,2714

Benzene (C_6H_6) – Diphenylsulphone ($C_{12}H_{10}O_2S$)
[324], $t = 25$ °C

x_2	46,8	36,9	27,6	20,1	11,3	0
ε	2,4397	2,4041	2,3704	2,3434	2,3119	2,2714

Benzene (C_6H_6) – Diphenyldisulphide ($C_{12}H_{10}S_2$)
[324], $t = 25$ °C

x_2	37,2	29,8	23,4	16,5	7,3	0
ε	2,2936	2,2893	2,2855	2,2811	2,2759	2,2714

Benzene (C_6H_6) – Tolyl tolylthiosulphonate ($C_{14}H_{16}O_2S_2$)
[315]

$t = 25$ °C				$t = 40$ °C			
x_2	ε	x_2	ε	x_2	ε	x_2	ε
52,8	2,5326	18,7	2,3641	47,3	2,4553	18,3	2,3216
45,3	2,4957	9,0	2,3110	37,6	2,4089	7,9	2,2745
30,1	2,4202	0	2,2714	27,6	2,3644	0	2,2386

Benzene (C_6H_6) – Dodecan-1-ol ($C_{12}H_{26}O$)
[300], $t = 25$ °C

x_2	13,90	9,98	7,56	4,90	2,90	1,32	0,63	0,00
ε	2,80	2,62	2,53	2,41	2,35	2,32	2,23	2,28

Benzene (C_6H_6) – Tributyl phosphate ($C_{12}H_{27}O_4P$)
[214], $t = 20$ °C

x_2	100	65,46	61,45	49,52	28,62	26,73	5,23	0,00
ε	8,13	6,97	6,86	6,31	5,25	4,99	3,01	2,283

Benzene (C_6H_6) – Phenyl salicylate ($C_{13}H_{10}O_3$)

[363], $t = 40,2$ °C

x_2	ε	x_2	ε	x_2	ε	x_2	ε
100	6,20	69,992	4,87	38,878	3,63	9,838	2,72
90,031	5,76	59,992	4,32	29,855	3,36	0	2,24
80,008	5,25	49,993	3,97	16,853	2,94	—	—

Benzene (C_6H_6) – Hexadecanoic acid ($C_{16}H_{32}O_2$)

[358], $t = 71$ °C, $\lambda = 820$ m

N	ε	N	ε	N	ε	N	ε
3,3160	2,348	2,1480	2,258	0,6660	2,196	0,000	2,180
2,9770	2,321	1,6650	2,236	0,3410	2,187	—	—
2,6330	2,295	1,1540	2,212	0,1659	2,183	—	—

Benzene (C_6H_6) – Dioctylsulphone ($C_{16}H_{34}O_2S$)

[324], $t = 25$ °C

x_2	36,5	29,4	22,2	15,1	6,7	0
ε	2,3739	2,3538	2,3337	2,3141	2,2902	2,2714

Benzene (C_6H_6) – 2, 2–Di(butylsulphonylmethylene)–3–butylsulphonylpropane ($C_{17}H_{36}O_6S_3$)

[324], $t = 25$ °C

x_2	30,7	25,7	20,8	13,19	8,4	0
ε	2,3475	2,3328	2,3228	2,3055	2,2920	2,2714

Benzene (C_6H_6) – Linoleic acid ($C_{18}H_{32}O_2$)

[368], $t = 20$ °C

φ_2	100	50	20	10	5	2	0
ε	2,742	2,530	2,417	2,360	2,325	2,301	2,280

Benzene (C_6H_6) – Oleic acid ($C_{18}H_{34}O_2$)

[368], $t = 20$ °C

φ_2	100	50	20	10	5	2	0
ε	2,465	2,393	2,367	2,336	2,307	2,299	2,280

Benzene (C_6H_6) – Octadecanoic acid ($C_{18}H_{36}O_2$)

[358], $t=71$ °C, $\lambda=820$ m

N	ε	N	ε	N	ε	N	ε
0,00	2,180	0,6050	2,191	1,9280	2,249	2,9780	2,318
0,1496	2,183	1,0350	2,208	2,3820	2,278	—	—
0,300	2,186	1,4760	2,225	2,7100	2,299	—	—

Benzene (C_6H_6) – Glycerol trioleate ($C_{57}H_{104}O_6$)

[368], $t=20$ °C

φ_2	100	50	20	10	5	2	0
ε	3,202	2,757	2,576	2,430	2,351	2,305	2,280

Phenol (C_6H_6O) – Resorcinol ($C_6H_6O_2$)

[334], $t=21$ °C

w_2	100	91	76	60,6	53	44	25	10	0
ε	12,2	11,1	12,5	12,0	11,8	12,9	11,3	11,2	9,8

Phenol (C_6H_6O) – Aniline (C_6H_7N)

[369, 276], $t=50$ °C, $\nu=1$ MHz

w_1	x_1	ε	w_1	x_1	ε	w_1	x_1	ε
0	0	6,30	57,50	57,24	8,88	77,51	77,32	9,50
10,00	9,90	6,78	59,95	59,69	8,95	79,98	79,81	9,54
20,01	19,84	7,28	62,47	62,22	9,06	82,50	82,35	9,61
30,00	29,78	7,73	64,94	64,73	9,13	85,01	84,87	9,74
40,00	39,74	8,22	67,00	66,76	9,17	90,00	89,90	9,87
45,01	44,75	8,43	70,00	69,77	9,28	95,00	94,95	10,08
50,02	49,75	8,66	72,50	72,29	9,33	100,0	100,0	10,28
55,00	54,74	8,79	75,01	74,81	9,46	—	—	—

Phenol (C_6H_6O) – Toluene (C_7H_8)

[361], $t=50$ °C

x_1	ε	x_1	ε	x_1	ε	x_1	ε
0,00	2,38	32,46	4,20	63,36	7,35	91,02	11,05
10,48	2,74	41,98	4,92	67,46	7,82	100	12,35
22,24	3,50	54,65	6,30	83,58	9,99	—	—

Phenol (C_6H_6O) – *m*–Cresol (C_7H_8O)
[369], $t = 50$ °C, $\nu = 1$ MHz

w_1	x_1	ε	w_1	x_1	ε	w_1	x_1	ε	w_1	x_1	ε
0	0	9,32	35,00	38,22	9,60	59,98	63,26	9,86	80,00	82,13	10,03
9,98	11,30	9,41	40,00	43,37	9,64	65,00	68,09	9,87	84,99	86,68	10,07
19,99	22,30	9,51	44,98	48,43	9,72	70,02	72,85	9,95	90,01	91,19	10,17
25,00	27,69	9,53	50,00	53,47	9,76	75,01	77,52	9,97	100	100,0	10,28
30,00	32,99	9,59	55,00	58,41	9,80	79,97	82,10	10,00	—	—	—

Phenol (C_6H_6O) – Toluidine (C_7H_9N)
[369], $t = 50$ °C, $\nu = 1$ MHz

w_1	x_1	ε	w_1	x_1	ε	w_1	x_1	ε	w_1	x_1	ε
0	0	5,07	47,50	50,74	8,04	60,00	63,06	8,73	72,50	75,01	9,13
10,00	11,23	5,68	50,00	53,23	8,14	62,50	65,48	8,89	75,00	77,35	9,20
20,00	22,15	6,29	52,50	55,72	8,31	65,00	67,89	8,96	80,01	82,00	9,43
30,00	32,79	6,91	55,00	58,18	8,47	67,50	70,27	9,01	90,00	91,11	9,83
40,00	43,15	7,57	57,52	60,65	8,63	70,00	72,65	9,07	100,0	100,0	10,28

[334, 333], $t = 21$ °C

w_1	ε	w_1	ε	w_1	ε	w_1	ε
0	3,8	25	7,0	50	8,95	82	10,05
7	4,6	34	7,75	66	10,2	90	9,9
20	5,7	40	8,0	80	10,1	100	9,8

Phenol (C_6H_6O) – *m*–Xylene (C_8H_{10})
[355], $t = 20$ °C

w_1	0	5,59	12,40	26,28
ε	2,375	2,576	2,890	3,201

Phenol (C_6H_6O) – Naphthylamine ($C_{10}H_9N$)
[334], $t = 21$ °C

w_1	ε	w_1	ε	w_1	ε	w_1	ε
0	4,0	32	7	67	8,95	83	10,2
10	4,8	50	7,95	72	8,3	90	10,7
20	6,1	56	8,5	78	9,7	100	9,8

Phenol (C_6H_6O) – Benzophenone ($C_{13}H_{10}O$)
[334], $t=21$ °C

w_1	ε	w_1	ε	w_1	ε	w_1	ε
100	9,7	69	12,5	45	13,2	16	11,45
90	10,3	63	12,5	42	13,0	10	11,3
81	11,3	56	12,95	39	12,3	0	9,55
73	11,5	50	13,45	25	11,8	—	—

Resorcinol (C_6H_6O) – Anisidine (C_7H_9NO)
[334], $t=21$ °C

w_1	0	15	31	44	58	68	75	83	100
ε	13	16	12,5	9,8	9,6	10,8	12,2	14,2	7,0

Aniline (C_6H_7N) – Benzonitrile (C_7H_5N)
[175], $t=20$ °C, $\nu=18$ MHz

φ_1	0	20	40	60	80	100
ε	25,65	22,33	18,90	15,17	11,28	7,06

Aniline (C_6H_7N) – *m*–Cresol (C_7H_8O)
[227], $t=17$ °C, $\nu=3$ MHz

φ_2	100	80	60	50	40	20	0
ε	12,98	10,45	9,47	9,03	8,62	7,87	7,04

Aniline (C_6H_7N) – Quinoline (C_9H_7N) *)
[328], $t=20$ °C, $\nu=3$ MHz

x_2	100,0	80,00	60,00	40,00	20,00	0,00
ε	9,12	9,57	9,64	9,18	8,42	7,20

*) See also [231, 232, 165].

2, 3–Dimethyl–1, 3–butadiene (C_6H_{10}) – Hexane (C_6H_{14})

x_1	ε at t °C					
	−75	−50	−25	0	25	50
0	2,020	1,948	1,949	1,915	1,878	1,839
11,75	2,046	2,010	1,976	1,940	1,906	—
20,57	2,064	2,030	1,997	1,961	1,925	—
33,23	2,096	2,064	2,020	1,982	1,947	—
100	—	2,229	2,188	2,148	2,102	2,059

Cyclohexanone ($C_6H_{10}O$) – Cyclohexane (C_6H_{12})
[175], $t=20$ °C, $\nu=1,8$ MHz

φ_1	0	20	40	60	80	100
ε	2,02	4,17	6,78	9,67	12,67	15,70

Cyclohexanone ($C_6H_{10}O$) – Benzonitrile (C_7H_5N)
[175], $t=20$ °C, $\nu=1,8$ MHz

φ_2	100	80	60	40	20	0
ε	25,65	23,62	21,65	19,65	17,70	15,70

Cyclohexanone ($C_6H_{10}O$) – Benzaldehyde (C_7H_6O)
[175], $t=20$ °C, $\nu=1,8$ MHz

φ_1	0	20	40	60	80	100
ε	17,85	17,40	17,00	16,55	16,12	15,70

Cyclohexanone ($C_6H_{10}O$) – Phenetole ($C_8H_{10}O$)
[175], $t=20$ °C, $\nu=1,8$ MHz

φ_1	0	20	40	60	80	100
ε	4,25	6,21	8,80	11,10	13,38	15,70

Cyclohexane (C_6H_{12}) – Cyclohexanol ($C_6H_{12}O$)
[357], $t=18$ °C

x_2	100	85	60	50	30	20	10	0
ε	13,40	10,35	7,66	6,37	3,40	2,64	2,28	2,05

Cyclohexane (C_6H_{12}) – Hexane (C_6H_{14})
[198], $t=20$ °C

x_2	ε	x_2	ε	x_2	ε
100	1,8893	64,02	1,9337	26,11	1,9844
83,66	1,9100	51,61	1,9490	10,93	2,0079
76,81	1,9184	38,34	1,9672	0	2,0230

Cyclohexane (C_6H_{12}) – Di–isopropyl ether ($C_6H_{14}O$)
[166], $t = 20$ °C

x_2	ε	x_2	ε	x_2	ε
100,0	3,976	35,236	2,692	8,219	2,176
69,424	3,372	20,642	2,408	6,900	2,147
52,276	3,027	19,207	2,373	0,00	2,032

Cyclohexane (C_6H_{12}) – 2–Nitrotoluene ($C_7H_7NO_2$)
[239], $t = 25$ °C, $\nu = 10$ MHz

x_1	ε	x_1	ε	x_1	ε	x_1	ε
0	26,52	42,25	14,69	49,98	17,29	100	2,02
11,23	23,28	53,21	11,81	60,59	15,04	—	—
21,63	23,36	62,68	9,44	81,58	5,23	—	—
32,43	17,39	72,50	7,16	89,99	3,66	—	—

Cyclohexane (C_6H_{12}) – 3–Nitrotoluene ($C_7H_7NO_2$)
[239], $t = 25$ °C

x_1	ε	x_1	ε	x_1	ε	x_1	ε
0	26,39	31,27	18,27	61,35	10,43	90,0	3,90
10,89	23,57	40,25	15,88	71,01	8,08	100	2,02
19,99	21,23	50,95	13,09	80,91	5,79	—	—

Cyclohexane (C_6H_{12}) – Cyclohexylmethanol ($C_7H_{14}O$)
[211], $t = 60$ °C

x_2	ε	x_2	ε	x_2	ε	x_2	ε
100,0	9,700	35,41	3,666	6,048	2,122	3,471	2,080
66,89	8,920	17,26	2,410	5,318	2,115	2,509	2,052
54,99	6,770	12,73	2,270	5,287	2,115	—	—
45,14	5,008	8,799	2,180	4,265	2,098	—	—

Cyclohexane (C_6H_{12}) – Quinoline (C_9H_7N)
[166], $t = 20$ °C

x_2	100,0	76,571	62,017	39,563	23,971	11,034	0,000
ε	9,293	7,709	6,541	4,746	3,583	2,669	2,032

Cyclohexane (C_6H_{12}) – Tetrahydro–2–naphthol ($C_{10}H_{12}O$)
[211]

x_2	ε at t °C		x_2	ε at t °C		x_2	ε at t °C	
	20	40		20	40		20	40
100,0	11,70	9,570	52,79	5,800	5,060	15,78	2,695	2,570
82,50	9,325	7,935	37,09	4,515	3,975	6,17	2,225	2,210
61,97	6,895	5,920	27,35	3,420	3,190	2,15	2,100	2,098

Cyclohexane (C_6H_{12}) – Decan–1–ol ($C_{10}H_{22}O$)
[306], $t=25$ °C

x_2	100	90	80	70	60	50	40	30	20
ε	7,80	7,07	6,37	5,53	4,80	3,94	3,28	2,83	2,46

1, 6–Dichlorohexane ($C_6H_{12}Cl_2$) – Hexane (C_6H_{14})
[269], $t=25$ °C

x_1	0	9,23	18,912	38,59	55,08	75,62	100,0
ε	1,8776	2,3500	2,9166	4,1803	5,3642	6,9103	8,7885

Cyclohexanol ($C_6H_{12}O$) – Decahydronaphthalene ($C_{10}H_8$)
[357], $t=18$ °C

x_1	0	10	20	30	50	60	85	100
ε	2,20	2,36	2,59	3,13	5,04	6,82	9,92	13,4

Hexanoic acid ($C_6H_{12}O$) – Octane (C_8H_{18})
[244], $t=25$ °C

φ_1	0	27,56	49,56	68,06	84,70	100
ε	1,942	2,156	2,313	2,481	2,654	2,816

Chlorohexane ($C_6H_{13}Cl$) – Hexane (C_6H_{14})
[269], $t=25$ °C, $\nu=3$ MHz

x_1	0	10,735	19,196	39,95	54,41	81,69	100,0
ε	1,8778	2,2395	2,5344	3,3102	4,0879	5,0340	5,8205

Cyclohexylamine ($C_6H_{13}N$) – Benzonitrile (C_7H_5N)
[175], $t=20$ °C, $\nu=1,8$ MHz

φ_1	0	20	40	60	80	100
ε	25,65	21,45	17,30	13,15	8,98	4,72

Hexane (C_6H_{14}) – Benzonitrile (C_7H_5N)
[175], $t=20$ °C, $\nu=1,8$ MHz

φ_2	100	80	60	40	20	0
ε	25,65	20,68	15,60	10,61	5,85	1,90

Hexane (C_6H_{14}) – 3–Nitrotoluene ($C_7H_7NO_2$)
[175], $t=20$ °C, $\nu=1,8$ MHz

φ_2	100	80	60	40	20	0
ε	26,80	21,10	15,37	10,30	5,67	1,90

Hexane (C_6H_{14}) – Toluene (C_7H_8)
[359], $t=20$ °C

x_2	37,11	28,37	21,78	16,06	0
ε	2,092	2,055	2,033	2,010	1,940

[286], $t=20$ °C, $\nu=10$ kHz

x_2	100	83,1	64,8	45,1	23,5	0
ε	2,376	2,266	2,161	2,063	1,967	1,884

Hexane (C_6H_{14}) – 1, 7–Dichloroheptane ($C_7H_{17}Cl_2$)
[269], $t=25$ °C, $\nu=3$ MHz

x_2	100,0	76,91	60,62	38,69	16,684	5,906	0
ε	8,3372	6,8132	5,7125	4,2372	2,8307	2,2023	1,8776

Hexane (C_6H_{14}) – Heptane (C_7H_{16})
[198], $t=20$ °C

x_1	ε	x_1	ε	x_1	ε
0	1,9246	37,98	1,9133	76,23	1,8992
11,67	1,9218	50,95	1,9086	83,31	1,8961
25,37	1,9175	63,76	1,9044	100	1,8893

Hexane (C_6H_{14}) – 1, 8–Dichloro–octane ($C_6H_{16}Cl_2$)
[269], $t = 25$ °C, $\nu = 3$ MHz

x_2	100,0	45,61	25,376	12,637	4,906	1,111	0
ε	7,6372	4,5808	3,3543	2,5971	2,1518	1,9395	1,8776

Hexane (C_6H_{14}) – Dibutyl ether ($C_8H_{18}O$)
[343], $t = 25$ °C

x_2	ε	x_2	ε	x_2	ε
100	3,0817	38,624	2,3480	4,440	1,9390
84,250	2,8946	19,716	2,1215	2,0544	1,9102
63,757	2,6528	10,019	2,0063	0	1,8863

Hexane (C_6H_{14}) – 1, 10–Dichlorodecane ($C_{10}H_{20}Cl_2$)
[269], $t = 25$ °C, $\nu = 3$ *Мгц*

x_2	100,0	76,14	43,46	26,183	10,911	4,369	0
ε	6,8128	5,8839	4,3343	3,4024	2,5212	2,1352	1,8776

Hexane (C_6H_{14}) – Dipentyl ether ($C_{10}H_{22}O$)
[343], $t = 25$ °C

x_2	ε	x_2	ε
100,00	2,8220	9,1623	1,9855
79,597	2,6619	6,5344	1,9582
68,072	2,5698	3,8187	1,9273
32,879	2,2439	1,989	1,9076
18,150	2,0863	0	1,8863

Hexan–1–ol ($C_6H_{14}O$) – Heptane (C_7H_{16})
[371]

φ_1	x_1	ε at t °C		φ_1	x_1	ε at t °C	
		25	50			25	50
0	0	1,917	1,879	30	33,47	2,746	2,633
1	1,17	1,941	1,8896	50	53,99	4,40	3,71
5	5,82	2,008	1,983	70	73,25	7,45	5,84
10	11,54	2,090	2,069	100	100	12,50	9,76

***tert*–Butyl ethyl ether ($C_6H_{14}O$) – Acetophenone (C_8H_8O)**
[146], $t = 25$ °C

φ_2	100	80	60	50	40	20	0
ε	17,16	15,37	13,36	12,36	11,37	9,38	7,07

Dipropylamine ($C_6H_{15}N$) – Benzaldehyde (C_7H_6O)
[175], $t=20$ °C, $\nu=1,8$ MHz

φ_1	0	20	40	60	80	100
ε	17,85	14,90	11,89	8,72	5,73	3,24

Triethylamine ($C_6H_{15}N$) – Benzonitrile (C_7H_5N)
[175], $t=20$ °C, $\nu=1,8$ MHz

φ_1	0	20	40	60	80	100
ε	25,65	21,05	16,45	11,80	7,20	2,64

Triethylamine ($C_6H_{15}N$) – Acetophenone (C_8H_8O)
[147], $t=25$ °C

φ_2	100	90	80	70	60	50	40	30	15	0
ε	17,16	15,65	14,12	12,51	10,85	9,22	7,64	6,23	4,19	2,42

Benzonitrile (C_7H_5N) – Benzaldehyde (C_7H_6O)
[175], $t=20$ °C, $\nu=1,8$ MHz

φ_1	0	20	40	60	80	100
ε	17,85	19,38	20,93	22,51	24,10	25,65

Benzonitrile (C_7H_5N) – – Толунитрил (C_8H_7N)

φ_2	100	80	60	40	20	0
ε	18,95	20,26	21,57	22,98	24,29	25,65

Benzonitrile (C_7H_5N) – Phenetole ($C_8H_{10}O$)
[175], $t=20$ °C, $\nu=1,8$ MHz

φ_1	0	20	40	60	80	100
ε	4,25	8,52	12,80	17,03	21,29	25,65

Benzaldehyde (C_7H_6O) – Phenetole ($C_8H_{10}O$)
[175], $t=20$ °C, $\nu=1,8$ MHz

φ_1	0	20	40	60	80	100
ε	4,25	6,71	9,40	12,11	14,95	17,85

Benzaldehyde (C_7H_6O) – Octane (C_8H_{18})
[244], $t = 25$ °C, $\nu = 28$ MHz

φ_2	100	85,58	70,21	50,56	25,08	0
ε	1,94	3,48	5,43	8,25	12,57	17,01

Benzaldehyde (C_7H_6O) – Tributylamine ($C_{12}H_{27}N$)
[175], $t = 20$ °C, $\nu = 1,8$ MHz

φ_2	100	80	60	40	20	0
ε	2,49	5,05	8,28	11,85	15,40	17,85

2–Nitrotoluene ($C_7H_7NO_2$) – 3–Nitrotoluene ($C_7H_7NO_2$)
[239], $t = 25$ °C, $\nu = 10$ MHz

x_2	ε	x_2	ε	x_2	ε	x_2	ε
100	26,39	69,48	26,69	38,97	26,80	10,14	26,64
89,63	26,61	59,69	26,77	29,67	26,78	0	26,52
79,59	26,61	49,67	26,80	19,52	26,73	—	—

2–Nitrotoluene ($C_7H_7NO_2$) – *o*–Xylene (C_8H_{10})
$\nu = 10$ MHz

x_2	ε at t °C [239]		x_2	ε at t °C [239]	
	25	35		25	35
100	2,56	2,53	40,31	14,41	13,80
90,37	4,03	3,91	30,21	17,10	16,34
80,09	5,77	5,59	20,06	20,04	19,10
70,74	7,53	7,27	10,60	22,96	21,86
59,95	9,79	9,41	0	26,52	25,19
50,23	11,97	11,48	—	—	—

2–Nitrotoluene ($C_7H_7NO_2$) – *m*–Xylene (C_8H_{10})
[239], $\nu = 10$ MHz

x_2	ε at t °C		x_2	ε at t °C		x_2	ε at t °C	
	25	35		25	35		25	35
100	2,35	2,34	59,27	9,71	9,34	20,18	19,90	18,96
89,68	3,92	2,82	47,40	12,44	11,94	10,64	22,87	21,83
80,06	5,56	5,39	39,50	14,47	13,82	0	26,52	25,19
70,80	7,30	7,04	30,43	16,86	16,15	—	—	—

2–Nitrotoluene ($C_7H_7NO_2$) – *p*–**Xylene** (C_8H_{10})
[239], ν = 10 MHz

x_2	ε at t °C		x_2	ε at t °C		x_2	ε at t °C	
	25	35		25	35		25	35
100	2,27	2,25	59,40	9,60	9,20	19,75	19,87	18,94
89,85	3,81	3,68	49,44	11,89	11,36	9,72	23,17	22,05
79,59	5,56	5,37	39,48	14,38	13,75	0	26,52	25,19
69,45	7,50	7,20	29,74	17,02	16,24	—	—	—

3–Nitrotoluene ($C_7H_7NO_2$) – *o*–**Xylene** (C_8H_{10})
[239], ν = 10 MHz

x_2	ε at t °C		x_2	ε at t °C		x_2	ε at t °C	
	25	35		25	35		25	35
100	2,56	2,53	59,94	10,73	—	19,93	20,88	19,99
90,38	4,32	4,24	50,70	12,89	12,46	10,43	23,38	22,48
80,36	6,33	6,13	39,91	15,49	14,99	0	26,39	25,35
69,95	8,56	8,34	30,35	18,03	17,32	—	—	—

3–Nitrotoluene ($C_7H_7NO_2$) – *m*–**Xylene** (C_8H_{10})
[239], ν = 10 MHz

x_2	ε at t °C		x_2	ε at t °C		x_2	ε at t °C	
	25	35		25	35		25	35
100	2,35	2,34	60,33	10,44	10,09	19,69	20,75	19,94
89,19	4,33	4,22	50,25	12,82	12,37	10,45	23,35	22,44
79,53	6,26	6,09	39,76	15,44	14,85	0	26,39	25,35
69,32	8,43	8,17	29,98	17,96	17,26	—	—	—

3–Nitrotoluene ($C_7H_7NO_2$) – *p*–**Xylene** (C_8H_{10})
[239], ν = 10 MHz

x_2	ε at t °C		x_2	ε at t °C		x_2	ε at t °C	
	25	35		25	35		25	35
100	2,27	2,25	58,70	10,76	10,37	19,22	20,89	20,05
89,56	4,19	4,07	48,51	13,21	12,69	9,69	23,54	22,64
79,20	6,26	6,04	38,61	15,68	15,07	0	26,39	25,35
68,89	8,47	8,15	28,97	18,22	17,50	—	—	—

4–Nitrotoluene ($C_7H_7NO_2$) – *o*–Xylene (C_8H_{10})
[239], $\nu = 10$ MHz

x_2	ε at t °C		x_2	ε at t °C		x_2	ε at t °C	
	25	35		25	35		25	35
100	2,56	2,53	80,39	6,70	6,48	60,97	11,19	10,82
88,58	4,86	4,77	72,29	8,55	8,27	53,63	13,05	12,58

4–Nitrotoluene ($C_7H_7NO_2$) – *m*–Xylene (C_8H_{10})
[239], $\nu = 10$ MHz

x_2	ε at t °C		x_2	ε at t °C		x_2	ε at t °C	
	25	35		25	35		25	35
100	2,35	2,34	79,20	6,66	6,51	62,56	10,58	10,22
89,31	4,37	4,30	70,92	8,61	8,34	51,64	13,28	12,81

4–Nitrotoluene ($C_7H_7NO_2$) – *p*–Xylene (C_8H_{10})
[239], $\nu = 10$ MHz

x_2	ε at t °C		x_2	ε at t °C		x_2	ε at t °C	
	25	35		25	35		25	35
100	2,27	2,25	72,41	8,19	7,89	47,02	14,26	13,82
89,80	4,32	4,19	63,32	10,31	9,92	—	—	—
81,51	6,12	5,91	51,41	13,26	12,73	—	—	—

Toluene (C_7H_8) – Heptane (C_7H_{16})
[172, 168]

w_1	0	11,50	36,57	49,94	60,40	88,73	100
t °C	22,5	21,0	21,2	21,2	21,1	21,0	19,55
ε	1,992	2,042	2,109	2,183	2,228	2,315	2,385

Toluene (C_7H_8) – Ethyl benzoate ($C_9H_{10}O_2$)
[170], $t = 20$ °C

x_2	100	77,274	40,540	16,082	0,000
ε	4,852	4,108	3,345	2,653	2,356

Toluene (C_7H_8) – Turpentine ($C_{10}H_{16}$)
[170], $t=20$ °C

x_1	0	9,724	24,464	63,162	89,884	100
w_1	0	6,791	20,654	53,701	75,091	100
ε	2,158	2,169	2,157	2,188	2,199	2,356

Toluene (C_7H_8) – Linoleic acid ($C_{18}H_{32}O_2$)
[368], $t=20$ °C

φ_2	100	75	50	33	20	10	5	2	0
ε	2,742	2,653	2,589	2,525	2,495	2,442	2,422	2,391	2,375

Anisole (C_7H_8O) – *m*–Xylene (C_8H_{10})
[355], $t=20$ °C

w_1	0	4,67	11,09	19,41	25,53	100
ε	2,375	2,466	2,566	2,696	2,801	4,39

***m*–Cresol (C_7H_8O) – Octane (C_8H_{18})**
[224], $t=25$ °C, $\nu=28$ MHz

φ_2	100	80,12	60,00	40,05	20,02	0
ε	1,94	2,73	4,43	6,68	9,15	11,8

***o*–Cresol (C_7H_8O) – Quinoline (C_9H_7N)**
[373], $t=25$ °C

x_2	100	89	70	48,2	29,7	11,2	0,0
ε	9,22	10,21	11,51	12,66	13,39	13,27	12,29

1–Bromoheptane ($C_7H_{15}Br$) – Heptane (C_7H_{16})

x_1	ε at t °C [223], $\nu=0,5$ MHz							
	−70	−50	−30	−10	10	30	50	70
4,88	2,297	2,247	2,198	2,152	2,109	2,067	2,024	1,980
8,89	2,491	2,422	2,356	2,294	2,237	2,184	2,133	2,083
16,26	2,862	2,755	2,656	2,567	2,489	2,417	2,349	2,278
33,30	3,704	3,515	3,348	3,200	3,067	2,950	2,840	2,734
53,77	4,809	4,498	4,214	3,969	3,767	3,595	3,437	3,287
100	7,385	6,875	6,392	5,957	5,582	5,255	4,970	4,711

Heptane (C_7H_{16}) – *p*–Xylene (C_8H_{10})
[168], $t = 20$ °C

x_2	18,64	13,751	9,621	7,750	4,740
ε	2,026	2,015	2,002	1,997	1,989

Heptane (C_7H_{16}) – Octan–1–ol ($C_8H_{18}O$)

x_2	ε at t °C [278]					x_2	ε at t °C				
	−30	−20	−10	0	10		20	30	40	50	60
4,47	2,046	2,033	2,021	2,010	2,000	4,47	1,991	1,981	1,971	1,960	1,948
6,71	2,070	2,059	2,049	2,039	2,029	6,71	2,020	2,012	2,004	1,994	1,982
12,60	2,162	2,152	2,141	2,131	2,122	12,60	2,114	2,107	2,100	2,091	2,082
23,47	2,489	2,457	2,426	2,399	2,376	23,47	2,357	2,341	2,326	2,311	2,295
43,74	4,928	4,456	4,045	3,770	3,554	43,74	3,384	3,233	3,130	3,030	2,943
73,60	11,22	10,21	9,25	8,36	7,52	73,60	6,78	6,14	5,60	5,14	4,75
100	—	—	13,31	12,26	11,26	100	10,34	9,45	8,62	7,84	7,09

Heptane (C_7H_{16}) – Bromonaphthalene ($C_{10}H_7Br$)
[172]

w_2	100	87,54	67,40	65,69	45,93	33,84	15,06	0
t °C	20,8	20,7	20,8	20,8	20,8	20,8	21	22,2
ε	4,912	4,289	3,350	3,212	2,762	2,498	2,183	1,992

Trimethylbutoxysilane ($C_7H_{18}OSi$) – Acetophenone (C_8H_8O)
[147], $t = 25$ °C

φ_2	100	80	60	50	40	20	0
ε	17,16	14,84	12,68	11,67	10,65	8,73	7,03

Acetophenone (C_8H_8O) – Phenetole ($C_8H_{10}O$)
[175], $t = 20$ °C, $\nu = 1,8$ MHz

φ_1	0	20	40	60	80	100
ε	4,25	6,36	8,80	11,43	14,28	17,65

Acetophenone (C_8H_8O) – Tetraethoxysilane ($C_8H_{20}O_4Si$)
[147], $t = 25$ °C

φ_1	0	20	40	50	60	80	100
ε	3,70	6,37	9,13	10,54	11,91	14,72	17,16

Acetophenone (C_8H_8O) – Hexaethyldisiloxane ($C_{12}H_{30}OSi_2$)
[147], $t = 20$ °C

φ_1	0	23	40	50	60	80	100
ε	2,26	4,74	7,07	8,70	10,27	13,57	17,16

Acetophenone (C_8H_8O) – Tetrapropoxysilane ($C_{12}H_{28}O_4Si$)
[147], $t = 25$ °C

φ_1	0	20	40	50	70	80	100
ε	3,21	6,06	8,77	10,20	12,02	15,28	17,16

Acetophenone (C_8H_8O) – Hexaethoxydisiloxane ($C_{12}H_{30}O_7Si_2$)
[147], $t = 25$ °C

φ_1	0	23	40	50	60	80	100
ε	5,18	8,10	10,13	11,20	12,52	14,90	17,16

Acetophenone (C_8H_8O) – Tetraisobutoxysilane ($C_{16}H_{36}O_4Si$)
[147], $t = 25$ °C

φ_1	0	23	40	50	60	80	100
ε	2,81	5,67	7,89	9,50	11,00	14,27	17,16

Acetophenone (C_8H_8O) – Tetra(2–butoxy)silane ($C_{16}H_{36}O_4Si$)
[147], $t = 25$ °C

φ_1	0	23	40	50	60	80	100
ε	2,59	5,32	7,67	9,40	10,84	14,11	17,16

Acetophenone (C_8H_8O) – Octaethoxytrisiloxane ($C_{16}H_{40}O_{10}Si_2$)
[147], $t = 25$ °C

φ_1	0	23	40	50	60	80	100
ε	4,76	7,73	9,83	10,9	12,28	14,74	17,16

Acetophenone (C_8H_8O) – Tetrapentoxysilane ($C_{20}H_{44}O_4Si$)
[147], $t = 25$ °C

φ_1	0	23	42	50	55	70	85	100
ε	2,82	5,64	8,38	9,60	10,21	12,51	14,98	17,16

p–Xylene (C_8H_{10}) – Decan–1–ol ($C_{10}H_{22}O$)
[306], $t = 25$ °C

x_2	100	90	80	70	60	50	40	30	20
ε	7,80	7,03	6,22	5,42	4,62	3,92	3,45	3,07	2,78

Mesitylene (C_9H_{12}) – Decan–1–ol ($C_{10}H_{22}O$)
[306], $t = 25$ °C

x_2	100	90	80	70	60	50	40	30	20
ε	7,80	6,94	6,05	5,22	4,46	3,81	3,34	2,98	2,72

Naphthalene ($C_{10}H_8$) – 2–Naphthol ($C_{10}H_8O$)
[172], $t = 21$ °C

w_2	100	90	70	60	50	40	30	10	0
x_2	100	88,9	67,4	57,1	47,0	37,3	27,6	9,0	0
ε	3,15	3,06	2,96	2,95	2,91	2,83	2,76	2,68	2,62

Naphthalene ($C_{10}H_8$) – 2–Naphthylamine ($C_{10}H_9N$)
[172], $t = 21$ °C

w_2	100	90	70	60	50	30	20	10	0
x_2	100	89,0	67,6	57,3	47,2	27,7	18,3	9,0	0
ε	3,13	3,10	3,04	2,99	2,96	2,89	2,85	2,78	2,67

Naphthalene ($C_{10}H_8$) – 1–Naphthylamine ($C_{10}H_9N$)
[172], $t = 21$ °C

w_2	100	90	80	70	50	40	20	10	0
x_2	100	89,0	78,2	67,6	47,2	37,4	18,3	9,0	0
ε	3,19	3,11	3,02	2,99	2,91	2,88	2,82	2,70	2,65

Naphthalene ($C_{10}H_8$) – Bromocamphor ($C_{10}H_{15}Br$)
[172], $t = 21$ °C

x_2	100	83,3	68,9	56,4	45,4	35,7	19,2	5,8	0
ε	2,67	2,68	2,67	2,66	2,66	2,66	2,64	2,63	2,62

Naphthalene ($C_{10}H_8$) – Anthracene ($C_{14}H_{10}$)
[172], $t = 21$ °C

w_2	100	90	80	70	50	40	30	20	0
x_2	100	86,6	74,2	62,7	41,8	32,4	23,6	15,2	0
ε	2,85	2,84	2,82	2,79	2,74	2,71	2,68	2,65	2,62

Naphthalene ($C_{10}H_8$) – Phenanthrene ($C_{14}H_{10}$)
[172], $t = 21$ °C

x_2	100	86,6	62,7	51,9	41,8	23,6	15,2	7,4	0
ε	2,83	2,81	2,80	2,77	2,75	2,73	2,72	2,68	2,65

2–Naphthol ($C_{10}H_8O$) – Antracene ($C_{14}H_{10}$)
[172], $t = 21$ °C

x_2	100	87,9	76,4	54,8	44,7	25,7	16,8	8,3	0
ε	2,86	2,89	2,95	2,96	2,98	3,04	3,09	3,12	3,14

2–Naphthlyamine ($C_{10}H_9N$) – Anthracene ($C_{14}H_{10}$)
[172], $t = 21$ °C

x_2	100	87,9	76,3	54,7	44,6	25,9	16,6	8,2	0
ε	2,85	2,88	2,90	2,92	2,96	3,04	3,09	3,11	3,14

Camphor ($C_{10}H_{16}O$) – Fenchone ($C_{10}H_{16}O$)

x_2	ε at t °C [374]											
	−173	−123	−113	−103	−98	−73	−66	−53	−38	−28	−23	−13
0	2	2	2	2	2	2	2	2	2	2	7	7
6	—	2,1	2,1	2,2	2,2	2,2	2,2	2,3	2,5	7,5	8,5	8,5
9	—	2,2	2,2	2,3	2,3	2,3	2,5	8	7	8,5	9	9
20	2,4	4,5	7	7	10,5	10,5	10	10	19,9	9,8	9,7	—
100	11,5	5,0	10	11,5	12	12,5	12,5	13	11,5	11	10,5	—

Supplementary information on the permittivity of non-aqueous solutions of organic substances

Carbon tetrachloride (CCl_4) – Hexamethyldisiloxane ($C_6H_{18}Si_2O$), fig. 4.
Chloroform ($CHCl_3$) – *n*–Propylamine (C_3H_9N) [175].
Chloroform ($CHCl_3$) – Pentamethylcyclopentasiloxane ($C_5H_{20}Si_5O_5$) [241].

Chloroform ($CHCl_3$) – Hexamethyldisiloxane ($C_6H_{18}Si_2O$), fig. 5.
Chloroform ($CHCl_3$) – Dibutyl ether ($C_8H_{18}O$), fig. 6.
Chloroform ($CHCl_3$) – Octamethyltrisiloxane ($C_8H_{24}O_3Si_3$), fig. 5.
Chloroform ($CHCl_3$) – Octamethylcyclotetrasiloxane ($C_8H_{24}O_4Si_4$), fig. 5.
Chloroform ($CHCl_3$) – Benzyl acetate ($C_9H_{10}O_2$) [165].
Methanol (CH_4O) – *N*–Methylacetamide (C_3H_7NO) [251].
Ethanol (C_2H_6O) – *N*–Methylacetamide (C_3H_7NO) [251].
Ethanol (C_2H_6O) – Benzoic acid ($C_7H_6O_2$) [271].
Acetone (C_3H_6O) – Isobutanol ($C_4H_{10}O$), fig. 10.
Methyl acetate ($C_3H_6O_2$) – Piperidine ($C_5H_{11}N$) [165].
Methylacetamide (C_3H_7NO) – Propan–1–ol (C_3H_8O) [251].
N–Methylacetamide (C_3H_7NO) – Butan–1–ol ($C_4H_{10}O$) [251].
N–Methylacetamide (C_3H_7NO) – Pentan–1–ol ($C_5H_{12}O$) [251].
Propan–1–ol (C_3H_8O) – Butan–1–ol ($C_4H_{10}O$) [165].
Butan–2–one (C_4H_8O) – Benzaldehyde (C_7H_6O), fig. 11.
Tetrahydrofuran (C_4H_8O) – Cyclohexane (C_6H_{12}) [304].
Tetrahydrofuran (C_4H_8O) – Piperidine ($C_5H_{11}N$) [304].
Isobutanol ($C_4H_{10}O$) – Nitrobenzene ($C_6H_5NO_2$), fig. 10.
Isobutanol ($C_4H_{10}O$) – Aniline (C_6H_7N), fig. 10.
Isobutanol ($C_4H_{10}O$) – Cyclohexanone ($C_6H_{10}O$), fig. 13.
Butan–2–ol ($C_4H_{10}O$) – 1, 2–Dichlorobenzene ($C_6H_4Cl_2$), fig. 12.
tert–Butyl alcohol ($C_4H_{10}O$) – 1, 2–Dichlorobenzene ($C_6H_4Cl_2$), fig. 12.
Pyridine (C_5H_5N) – Quinoline (C_9H_7N), fig. 11.
Pentamethylcyclopentasiloxane ($C_5H_{20}O_5Si_5$) – Cyclohexane (C_6H_{12}), fig. 4.
Benzene (C_6H_6) – 2, 6–Dimethyl–4–pentanol ($C_9H_{20}O$), fig. 15.
Cyclohexane (C_6H_{12}) – 2, 6–Dimethyl–4–pentanol ($C_9H_{20}O$), fig. 15.
Cyclohexane (C_6H_{12}) – Hexadecane ($C_{16}H_{34}$), fig. 4.
±–3, 4–Dibromohexane ($C_6H_{12}Br_2$) – meso–3, 4–Dibromohexane ($C_6H_{12}Br_2$) [370], $t = 25°C, x_1 = 60{,}7\%, \epsilon = 5.922$.
Toluene (C_7H_8) – *m*–Xylene (C_8H_{10}) [320].
Toluene (C_7H_8) – 2, 6–Dimethyl–4–pentanol ($C_9H_{20}O$), fig. 15.
p–Toluidine (C_7H_9N) – Naphthalene ($C_{10}H_8$) [296].
Isopentyl acetate ($C_7H_{14}O_2$) – Methyl hexanoate ($C_7H_{14}O_2$), fig. 16.
Methyl hexanoate ($C_7H_{14}O_2$) – *N, N*–Diethylaniline ($C_{10}H_{15}N$), fig. 11.
±–2, 3–Diacetoxybutane ($C_8H_4O_4$) – meso–2, 3–Diacetoxybutane ($C_8H_4O_4$) [370], $t = 25°C, x_1 = 58{,}5\%, \epsilon = 5{,}740$.
Xylene (C_8H_{18}) – 2, 6–Dimethyl–4–pentanol ($C_9H_{20}O$), fig. 15.
N, N–Dimethylaniline ($C_8H_{11}N$) – *N, N*–Diethylaniline ($C_{10}H_{15}N$), fig. 17.

§ 3. Dielectric dispersion parameters of non-aqueous solutions of inorganic compounds

Deuterium oxide (D_2O) – Dioxan ($C_4H_8O_2$)
[375], $t = 25$ °C

x_1	λ=10 cm		λ=3,22 cm		λ=1,25 cm		λ=0,429 cm	
	ε′	ε″	ε′	ε″	ε′	ε″	ε′	ε″
4,96	2,50	0,028	2,48	0,066	2,41	0,056	—	—
8,15	2,72	0,054	2,66	0,11	2,54	0,095	—	—
20,3	3,79	0,28	3,43	0,53	3,02	0,52	2,54	0,31
30,3	4,98	0,67	4,11	1,08	3,35	0,90	2,73	0,47
30,5 *)	—	—	4,34	0,75	3,63	0,94	—	—
40,2	6,69	1,27	4,92	1,79	3,75	1,46	2,83	0,68
50,0	—	—	6,01	2,95	4,28	3,03	—	—

x_1	ε	ε_∞	$\varepsilon_{\infty 2}$	$10^{-12}\tau_1$	$10^{-12}\tau_2$	C_1	$10^{-12}\tau$
4,96	2,51	—	2,37	—	—	—	10,0
8,15	2,73	—	2,50	—	—	—	12,4
20,3	3,89	3,14	2,50	20,0	5,2	0,54	—
30,3	5,25	3,61	2,60	22,5	5,7	0,62	—
30,5	4,85	4,02	2,60	16,0	4,5	0,37	—
40,2	7,10	4,14	2,74	25,0	6,0	0,68	—
50,0	9,98	4,67	2,90	27,0	5,4	0,75	—

*) Measurements of all values for x_1 = 30,5 were made at t = 50°C.

[154], ν = 1800 *кгц*

x_1	ΔH	$10^{-11}\tau$ at t °C			x_1	ΔH	$10^{-11}\tau$ at t °C		
		30	40	50			30	40	50
2,8091	3,55	1,21	1,00	0,84	52,203	4,54	1,79	1,41	1,12
8,2352	3,64	1,22	1,01	0,84	59,766	4,79	1,95	1,52	1,19
13,217	3,64	1,25	1,03	0,86	65,342	4,93	2,06	1,58	1,24
15,607	3,73	1,27	1,03	0,86	74,309	5,26	2,22	1,67	1,29
18,306	3,82	1,30	1,06	0,88	81,257	5,29	2,21	1,66	1,28
34,669	4,13	1,48	1,19	0,97	100	4,57	1,10	0,85	0,68

Barium chloride ($BaCl_2$) – Methanol (CH_4O)
[377], t=20°C, λ=3, 22 cm

N	ε	ε′	ε″	λ_m	α
0,05	29,7	7,88	7,47	9,26	0,01

Cadmium nitrate ($Cd(NO_3)_2$) – **Ethanol** (C_2H_6O)
[378], $t = 28$ °C, $\lambda = 3,14$ cm

N	ε'	ε''	λ_m	ε
0,25	4,68	2,61	17,07	18,87
0,5	4,65	2,35	16,40	16,93
0,75	4,61	2,04	15,63	14,76
1,0	4,57	1,78	15,10	13,13

Cobalt nitrate ($Co(NO_3)_2$) – **Ethanol** (C_2H_6O)
[378], $t = 28$ °C, $\lambda = 3,14$ cm

N	ε'	ε''	λ_m	ε
0,25	4,65	2,71	18,91	20,97
0,5	4,61	2,44	18,68	19,13
0,75	4,57	2,11	17,91	16,61
1,0	4,53	1,70	16,17	12,89

Copper(II)nitrate ($Cu(NO_3)_2$) – **Ethanol** (C_2H_6O)
[378], $t = 28$ °C, $\lambda = 3,14$ cm

N	ε'	ε''	λ_m	ε
0,25	4,67	2,65	17,7	17,6
0,50	4,64	2,19	15,99	16,05
0,75	4,61	1,93	14,78	13,286
1,0	4,58	1,80	13,22	11,314

Potassium iodide (KJ) – **Methanol** (CH_4O) *)
[377], $t = 20$ °C, $\lambda = 3,22$ cm, fig. 18

N	ε	ε'	ε''	λ_m	α
0,1	28,6	7,76	7,14	9,26	0,01
0,2	25,8	7,64	6,60	8,84	—
0,3	23,4	7,67	6,08	8,17	—
0,4	21,7	7,56	5,68	7,89	—

*) See also [376] for λ = 0,62; 1,24; 3,22 cm.

Lanthanum chloride ($LaCl_3$) – **Methanol** (CH_4O)
[377], $t = 20$ °C, $\lambda = 3,22$ cm

N	ε	ε'	ε''	λ_m	α
0,1	21	4,63	2,49	21,8	0,1
0,33	19,2	4,63	2,34	20,1	—

Lithium chloride (LiCl) – **Methanol** (CH_4O)
[377, 376], $t = 20$ °C, $\lambda = 3,22$ cm, fig. 18

N	ε	ε'	ε''	λ_m	α
0,5	20,3	7,26	5,06	8,13	0,01
1,0	16,1	7,00	3,50	6,97	—

Lithium chloride (LiCl) – **Ethanol** (C_2H_6O)
[377], $t = 20$ °C, $\lambda = 3,22$ cm, fig. 18

N	ε	ε'	ε''	λ_m	α
1,7	18,6	4,5	2,22	20,8	0,01

Lithium perchlorate ($LiClO_4$) – **Ethanol** (C_2H_6O)

[381], $t=25$ °C, $\nu=300-3300$ MHz

N	0	0,125	0,31	0,48	0,94
ε_∞	44	4,5	4,55	4,6	4,75
ε	24,3	21,6	17,2	15,9	11,5
λ_m	31	34	29	29	27

Lithium perchlorate ($LiClO_4$) – **Tetrahydrofuran** (C_4H_8O)

[382], $t=30$ °C

ν, *Ггц*	$N=0$		$N=0,25$		$N=0,4$		$N=0,6$	
	ε'	ε''	ε'	ε''	ε'	ε''	ε'	ε''
0,13742	7,255	0,0	12,09	5,43	14,10	11,40	16,31	23,85
0,13742 *)	7,223	0,012	12,09	5,64	14,11	11,50	16,36	23,67
0,2604	7,221	0,026	11,91	3,54	13,78	6,95	15,82	13,71
0,2604 *)	7,248	0,022	11,82	3,49	13,75	7,06	16,08	13,46
0,37500	7,251	0,036	11,73	3,04	13,41	5,46	15,60	10,23
0,38375	7,222	0,030	11,65	2,98	13,43	5,60	15,58	10,36
0,38375 *)	7,227	0,032	11,59	2,96	13,37	5,57	15,47	10,33
0,62500	7,244	0,051	11,26	2,72	12,68	4,61	14,48	8,08
1,06060	7,245	0,075	10,45	2,77	11,50	4,28	12,67	7,00
1,66667	7,252	0,124	9,59	2,74	10,30	3,93	10,97	6,01
2,8250	7,224	0,203	8,53	2,44	8,90	3,36	9,27	4,67
9,342	7,105	0,690	7,07	1,53	7,13	2,14	7,14	2,65
34,719	6,27	2,117	6,04	2,34	5,92	2,36	5,7	2,47

*) Measurements made in another measuring cell.

Lithium iodide (LiJ) – **Ethanol** (C_2H_6O)

[377], $t=20$ °C, $\lambda=3,22$ cm

N	ε	ε'	ε''	λ_m	α
0,1	22,9	4,52	2,25	27,4	0,01
0,2	17,15	4,58	2,09	20,4	—
0,4	13,6	4,66	1,93	15,1	—
0,8	11,2	4,71	1,75	12,1	—

Magnesium chloride ($MgCl_2$) – **Ethanol** (C_2H_6O)

[377], $t=20$ °C, $\lambda=3,22$ cm

N	ε	ε'	ε''	λ_m	α
0,1	21,7	4,6	2,45	23,2	0,01
0,15	20,1	4,63	2,42	21	—
0,25	19,8	4,65	2,45	20,1	—
0,5	17,7	4,41	2,29	18,1	—
0,76	16,1	4,77	2,29	16,2	—

Magnesium perchlorate ($Mg(ClO_4)_2$) – **Ethanol** (C_2H_6O)
[381], $t = 20$ °C, $\nu = 300—3300$ MHz

N	ε	ε_∞	λ_m	N	ε	ε_∞	λ_m
0	24,3	4,4	31	0,24	20,2	4,9	33
0,045	23,4	4,6	32	0,40	16,1	5,1	30
0,12	22,8	4,55	34	0,81	11,0	5,0	23

Magnesium(II)nitrate ($Mn(NO_3)_2$) – **Ethanol** (C_2H_6O)
[378], $t = 28$ °C, $\lambda = 3,14$ cm

N	ε'	ε''	ε	λ_m
0,25	4,66	2,73	20,185	18,63
0,5	4,62	2,37	17,99	17,72
0,75	4,58	1,82	13,282	15,04
1,0	4,57	1,44	10,174	12,22

Sodium (Na) – **Ammonia** (NH_3)
[383], $t = 30$ °C, $\nu = 3,088$ GHz, fig. 21

N	ε'	ε''	N	ε'	ε''
$2,1 \cdot 10^{-5}$	19,9	1,03	$7,0 \cdot 10^{-4}$	21,3	1,4
$3,4 \cdot 10^{-5}$	19,2	1,3	$4,2 \cdot 10^{-3}$	21,4	1,65
$8,1 \cdot 10^{-5}$	18,4	1,2	$1,5 \cdot 10^{-2}$	22,03	2,52
$1,4 \cdot 10^{-4}$	21,2	1,0	$4,25 \cdot 10^{-2}$	26,15	—
$2,9 \cdot 10^{-4}$	20,8	1,2	$6,3 \cdot 10^{-2}$	29,2	—

Sodium bromide (NaBr) – **Methanol** (CH_4O)
[377, 376], $t = 20$ °C, $\lambda = 3,22$ cm, fig. 18

N	ε	ε'	ε''	λ_m	α
0,1	28,9	7,66	7,04	9,5	0,01
0,2	27,3	7,57	6,68	9,35	—
0,4	22,4	7,34	5,55	8,6	—
0,8	15,7	7,00	3,87	7,1	—
1,0	14,8	7,00	3,67	6,28	—

Sodium chloride (NaCl) – **Glycerine** ($C_3H_8O_3$)
[386], $t=-35,5$ °C, $\nu=100$ Hz—600 kHz

N	ε	ε_∞	β	$10^{-6}\tau_0$	$10^{-6}\tau_D$*)	N	ε	ε_∞	β	$10^{-6}\tau_0$	$10^{-6}\tau_D$*)
0	58,9	4,2	0,61	4,13	2,52	5	49,1	3,7	0,57	7,7	4,39
1	53,7	4,0	0,64	4,37	2,80	5	46,9	3,7	0,57	7,4	4,22
3	51,7	3,9	0,61	5,78	3,53	5	48,6	3,7	0,57	7,21	4,11
5	48,9	3,7	0,57	7,6	4,33	7	45,5	3,6	0,54	9,37	5,06

N	t °C	ε	ε_∞	β	$10^{-12}\tau$	τ_D
1	—27,5	52,5	4,0	0,64	$9,77\cdot10^{-7}$	$6,25\cdot10^{-7}$
	—35,6	53,7	4,0	0,64	$4,16\cdot10^{-6}$	$2,66\cdot10^{-6}$
	—43,4	58,6	4,0	0,62	$2,24\cdot10^{-5}$	$1,39\cdot10^{-5}$
	—49,1	58,7	4,0	0,62	$8,85\cdot10^{-4}$	$5,49\cdot10^{-4}$
5	—22,2	45,9	3,5	0,57	$5,61\cdot10^{-7}$	$3,20\cdot10^{-7}$
	—31,8	46,4	3,7	0,57	$2,63\cdot10^{-6}$	$1,5\cdot10^{-6}$
	—35,2	49,1	3,7	0,57	$7,24\cdot10^{-6}$	$4,13\cdot10^{-6}$
	—35,25	46,9	3,7	0,57	$6,75\cdot10^{-6}$	$3,85\cdot10^{-6}$
	—35,5	48,6	3,7	0,57	$7,21\cdot10^{-6}$	$4,11\cdot10^{-6}$
	—35,6	48,9	3,7	0,57	$7,96\cdot10^{-6}$	$4,54\cdot10^{-6}$
	—43,3	48,3	3,9	0,57	$4,03\cdot10^{-5}$	$2,30\cdot10^{-5}$
	—49,9	50,5	4,0	0,56	$2,27\cdot10^{-4}$	$1,27\cdot10^{-4}$

*) τ_D —mean value τ, obtained as product $\beta\tau_0$.

Sodium iodide (NaJ) – **Methanol** (CH_4O)
[377], $t=20$ °C, $\lambda=3,22$ cm, fig. 18

N	ε	ε'	ε''	λ_m	α	N	ε	ε'	ε''	λ_m	α
0,05	31,9	7,84	7,77	9,75	0,01	0,75	15,1	7,21	4	6,4	—
0,2	27,3	7,62	6,7	8,9	—	1,0	11,7	7,31	2,9	4,56	—
0,5	18,8	7,62	4,8	6,94	—						

Sodium iodide (NaJ) – **Ethanol** (C_2H_6O)
[377], $t=20$ °C, $\lambda=3,22$ cm

N	ε	ε'	ε''	λ_m	α
0,13	19,5	4,56	1,9	22,7	0,01
0,27	13,7	4,40	1,88	15,8	—

Nickel nitrate ($Ni(NO_3)_2$) – **Ethanol** (C_2H_6O)
[378], $t=28$ °C, $\lambda=3,14$ cm

N	ε'	ε''	λ_m	ε
0,25	4,66	2,51	17,22	18,49
0,50	4,63	2,24	16,39	16,35
0,75	4,61	1,93	14,78	13,70
1,00	4,59	1,71	13,77	12,09

Strontium nitrate ($Sr(NO_3)_2$) – **Ethanol** (C_2H_6O)
[378], $t=28$ °C, $\lambda=3,14$ cm

N	ε'	ε''	λ_m	ε
0,25	4,69	2,63	16,86	18,81
0,5	4,65	2,24	15,63	15,8
0,75	4,63	1,89	13,80	13,80
1,00	4,60	1,62	12,72	12,72

Supplementary information on the dielectric dispersion parameters of non-aqueous solutions of inorganic compounds

Barium bromide ($BaBr_2$) – Methanol (CH_4O), fig. 18.
Calcium nitrate ($Ca(NO_3)_2$) – Methanol (CH_4O), fig. 18.
Hydrogen chloride (HCl) – Tributyl phosphate ($(C_4H_9O)_3PO$) [379].
Nitric acid (HNO_3) – Tributyl phosphate ($(C_4H_9O)_3PO$) [379].
Sulphuric acid (H_2SO_4) – Tributyl phosphate ($(C_4H_9O)_3PO$) [379].
Potassium iodide (KI) – Ammonia (NH_3) [380], t = –75 to –35°C, ν = 0,5 – 70 GHz, N = 0,1 – 1.
Lithium bromide (LiBr) – Ammonia (NH_3) [380], t = –75 to –35°C, ν = 0,5 – 70 GHz, N = 0,1 – 1.
Lithium nitrate ($LiNO_3$) – Methylformamide (C_2H_5NO), fig. 19.
Lithium nitrate ($LiNO_3$) – Dimethylsulphoxide (C_2H_6OS), fig. 20.
Magnesium perchlorate ($Mg(ClO_4)_2$) – Methanol (CH_4O) [592].
Sodium bromide (NaBr) – Ammonia (NH_3) [380], t = –75 to –35°C, ν = 0,5 – 70 GHz, N = 0,1 – 1.
Strontium bromide ($SrBr_2$) – Methanol (CH_4O), fig. 18.

§ 4. Dielectric dispersion parameters of non-aqueous solutions of organic compounds

Trichlorofluoromethane (CCl_3F) – Carbon tetrachloride (CCl_4)

[388]

x_1	t °C	$\lambda=3,22$ cm		$\lambda=1,24$ cm		ε	α	λ_m
		ε′	ε″	ε′	ε″			
48	0	2,322	0,027	2,292	0,056	2,326	0	0,55
	20	2,272	0,021	2,254	0,047	2,273	0	0,43
72,4	0	2,343	0,035	2,312	0,081	2,352	0	0,49
	20	2,288	0,028	2,269	0,069	2,289	0	0,41

Trichlorofluoromethane (CCl_3F) – Dioxan ($C_4H_8O_2$)

[387], $t=20$ °C

x_1	$\lambda=3,22$ cm		$\lambda=1,25$ cm		ε
	ε′	ε″	ε′	ε″	
12,1	—	—	—	—	2,2514
18,5	2,2623	0,0145	2,2465	0,0230	2,2669
30,1	2,2824	0,0188	2,2613	0,0293	2,2877
35,1	2,2893	0,0215	2,2681	0,0318	2,2960

Carbon tetrachloride (CCl_4) – Nitromethane (CH_3NO_2)

[66], $\nu=9,23$ GHz

N	$t=10$ °C		$t=20$ °C		$t=30$ °C		$t=40$ °C	
	ε	10^{-3} tg δ	ε	10^{-3} tg δ	ε	10^{-3} tg δ	ε	10^{-3} tg δ
0	2,234	1,26	2,232	1,26	2,229	1,27	2,226	1,27
0,1	2,243	3,91	2,240	4,04	2,236	4,04	2,233	3,91
0,15	2,250	5,84	2,244	6,05	2,240	6,06	2,237	5,80
0,22	2,254	7,77	2,250	8,06	2,245	8,09	2,241	7,82
0,3	2,261	11,66	2,257	12,09	2,253	12,10	2,250	11,70

Carbon tetrachloride (CCl_4) – Acetonitrile (C_2H_3N)
[390], $x_1=25$

t °C	λ=3,23 cm		λ=1,95 cm		λ=1,25 cm		λ=0,4 cm		ε	$ε_\infty$	$10^{-12}τ$
	ε′	ε″	ε′	ε″	ε′	ε″	ε′	ε″			
24	5,55	1,10	5,05	1,40	4,70	1,30	3,52	0,72	5,97	3,3	7,6
30	5,33	0,99	5,02	1,12	4,68	1,24	—	—	5,91	—	7,2
40	5,46	0,92	5,07	1,10	4,60	1,30	3,60	0,92	5,77	3,2	6,3
50	5,35	0,79	5,08	0,92	4,61	1,26	—	—	5,55	—	5,3
60	5,22	0,68	5,05	0,90	4,62	1,07	4,08	1,14	5,34	3,0	4,5

Carbon tetrachloride (CCl_4) – Ethyl cyanide (C_3H_5N)
[390], $x_1=25$

t °C	λ=3,23 cm		λ=1,95 cm		λ=1,25 cm		λ=0,4 cm	
	ε′	ε″	ε′	ε″	ε′	ε″	ε′	ε″
24	5,27	1,57	4,81	1,80	4,10	1,67	3,06	1,21
30	5,36	1,43	4,95	1,70	4,14	1,52	—	—
40	5,42	1,30	5,00	1,50	4,25	1,43	3,25	1,05
50	5,46	1,18	5,12	1,33	4,47	1,32	—	—
60	5,28	1,10	5,00	1,31	4,35	1,35	3,49	0,88

t °C	$10^{-12}τ_1$	$10^{-12}τ_2$	C_1	ε	$ε_\infty$
24	11,0	2,1	0,70	6,48	2,3
40	10,0	1,6	0,70	6,14	2,3
60	7,0	1,6	0,70	5,72	2,3

Carbon tetrachloride (CCl_4) – Acetone (C_3H_6O) *)
[391—394], λ=3,21 cm

x_2	t=0 °C		t=10 °C		t=20 °C		t=30 °C		t=40 °C	
	ε′	ε″	ε′	ε″	ε′	ε″	ε′	ε″	ε′	ε″
100	22,46	4,69	21,30	3,99	20,92	3,54	19,76	2,94	18,81	2,65
84,21	17,01	3,60	16,4	3,10	16,04	2,67	15,36	2,28	14,88	2,03
66,31	12,79	2,64	12,40	2,26	11,97	1,90	11,56	1,64	11,10	1,46
46,58	8,92	1,68	8,64	1,45	8,30	1,23	8,18	1,03	7,95	0,94
23,53	5,43	0,72	5,25	0,67	5,10	0,55	4,95	0,47	4,84	0,39
0	2,25	—	2,24	—	2,22	—	2,20	—	2,18	—

*) For t=21 °C ν=300—510 MHz, see [396].

x_2	$t=0$ °C		$t=10$ °C		$t=20$ °C		$t=30$ °C		$t=40$ °C	
	α	$10^{-12}\tau$	α	$10^{-12}\tau$	α	$10^{-12}\tau$	α	$10^{-12}\tau$	α	$10^{-12}\tau$
100	0,04	3,73	0,036	3,22	0,028	3,01	0	2,80	0	2,262
84,21	0,055	3,91	0,044	3,44	0,033	3,07	0,022	2,82	0,011	2,63
66,31	0,074	3,94	0,069	3,42	0,050	3,08	0,040	2,77	0,028	2,62
46,58	0,112	3,85	0,10	3,26	0,090	2,96	0,076	2,64	0,066	2,44
23,53	0,1	3,37	0,089	2,90	0,078	2,67	0,067	2,34	0,055	2,16

[395, 392, 393], $\lambda=8,15$ mm

x_1	$t=0$ °C		$t=10$ °C		$t=20$ °C		$t=30$ °C		$t=40$ °C	
	ε'	ε''	ε'	ε''	ε'	ε''	ε'	ε''	ε'	ε''
0	2,29	—	2,27	—	2,24	—	2,22	—	2,20	—
18,96	4,0	1,52	4,1	1,43	4,1	1,35	4,1	1,26	4,1	1,17
39,86	5,8	3,1	6,0	3,0	6,2	2,9	6,2	2,8	6,2	2,65
59,94	7,9	5,1	8,2	4,9	8,4	4,7	8,4	4,5	8,4	4,3
80,22	10,3	7,4	10,7	7,1	11,1	6,8	11,2	6,4	11,1	6,0
100	13,5	10	14,1	9,6	14,5	9,2	14,5	8,6	14,1	7,7

x_1	$t=0$ °C		$t=10$ °C		$t=20$ °C		$t=30$ °C		$t=40$ °C	
	α	$10^{-12}\tau$	α	$10^{-12}\tau$	α	$10^{-12}\tau$	α	$10^{-12}\tau$	α	$10^{-12}\tau$
18,96	0,10	4,21	0,09	3,59	0,08	3,17	0,07	—	0,06	2,26
39,86	0,11	4,52	0,10	3,87	0,09	3,50	0,08	—	0,07	2,86
59,94	0,07	4,39	0,07	3,80	0,05	3,38	0,04	—	0,03	2,91
80,22	0,05	4,24	0,04	3,56	0,03	3,31	0,02	—	0,01	2,88
100	0,04	4,07	0,03	3,54	0,02	3,23	0	—	0	2,76

Carbon tetrachloride (CCl_4) – Tetrahydrofuran (C_4H_8O)
[398], $t=22$ °C, $\nu=23,8$ GHz

x_2	ε'	$10^{-3}\varepsilon''$	x_2	ε'	$10^{-3}\varepsilon''$	x_2	ε'	$10^{-3}\varepsilon''$
40,0	2,73	335	10,0	2,65	170	3,0	2,36	54,5
30,0	3,33	298	8,0	2,62	156	0,0	2,24	6,8
20,0	2,95	251	6,0	2,57	116	—	—	—
15,0	2,69	217	5,0	2,45	95,0	—	—	—

Carbon tetrachloride (CCl_4) – Dioxan ($C_4H_8O_2$)
[398], $t=22$ °C, $\nu=23,8$ GHz

x_2	100	65,0	40,0	30,0	15,0	10,00	5,00	0,0
ε'	2,39	2,36	2,32	2,30	2,28	2,27	2,26	2,24
$10^{-3}\varepsilon''$	21,4	20,9	20,4	19,6	14,8	12,6	10,6	6,8

Carbon tetrachloride (CCl_4) – 2–Chloroaniline (C_6H_5ClN)
[399]

x_2	t °C	λ=10,0 cm		λ=3,22 cm		λ=1,95 cm		λ=1,25 cm	
		ε′	ε″	ε′	ε″	ε′	ε″	ε′	ε″
100	24	6 72	1,60	5,06	1,76	4,40	1,35	4,06	0,96
	40	6,95	1,25	5,27	1,67	4,50	1,40	4,01	1,02
	60	6,39	0,97	5,57	1,42	4,62	1,39	4,16	1,15
47,9	24	—	—	3,78	0,70	3,47	0,67	3,28	0,50
	40	—	—	3,81	0,58	3,57	0,68	3,23	0,53
	60	—	—	3,80	0,58	3,48	0,61	3,16	0,53
20,7	24	—	—	2,94	0,29	2,84	0,28	2,71	0,18
	40	—	—	2,91	0,24	2,80	0,26	2,66	0,19
	60	—	—	2,86	0,19	2,75	0,21	2,65	0,18

x_2	t °C	$10^{-12}\tau_1$	$10^{-12}\tau_2$	C_1	ε	ε_∞
100	24	26,5	0,6	0,82	7,70	3,00
	40	21,7	0,5	0,78	7,32	2,90
	60	16,4	0,5	0,74	6,94	2,80
47,9	24	17,5	0,6	0,71	4,55	2,60
	40	14,8	0,6	0,74	4,38	2,55
	60	12,2	0,6	0,72	4,16	2,50
20,7	24	19,6	0,6	0,75	3,25	2,50
	40	16,4	0,6	0,70	3,16	2,45
	60	13,8	0,5	0,65	3,04	2,40

Carbon tetrachloride (CCl_4) – 3–Chloroaniline (C_6H_5ClN)
[399]

x_2	t °C	λ=10,0 cm		λ=3,22 cm		λ=1,95 cm		λ=1,25 cm	
		ε′	ε″	ε′	ε″	ε′	ε″	ε′	ε″
100	24	8,75	4,20	5,29	2,49	4,55	1,90	4,25	1,30
	40	9,37	3,80	5,57	2,80	4,80	2,06	4,35	1,40
	60	9,35	3,00	6,33	2,87	4,87	2,16	4,46	1,50
32,7	24	—	—	3,58	1,08	3,32	0,82	3,06	0,51
	40	—	—	3,66	1,06	3,31	0,85	2,93	0,53
	60	—	—	3,98	1,05	3,37	0,88	3,13	0,62
8,9	24	—	—	2,72	0,28	2,57	0,21	2,54	0,13
	40	—	—	2,69	0,25	2,56	0,20	2,50	0,14
	60	—	—	2,70	0,26	2,53	0,20	2,45	0,14

x_2	t °C	$10^{-12}\tau_1$	$10^{-12}\tau_2$	C_1	ε	ε_∞
100	24	45,6	0,6	0,88	12,57	3,00
	40	44,0	0,4	0,86	12,19	2,95
	60	3,07	0,3	0,84	11,31	2,90
32,7	24	31,3	0,3	0,82	5,56	2,45
	40	27,0	0,3	0,82	5,27	2,45
	60	21,2	0,3	0,80	5,08	2,45
8,9	24	25,4	0,3	0,87	3,07	2,40
	40	23,3	0,3	0,75	2,99	2,30
	60	21,7	0,3	0,77	2,90	2,25

Carbon tetrachloride (CCl_4) – Cyclohexanone ($C_6H_{10}O$)

[400], $t=21$ °C, $\lambda=14$—16 cm

x_2	100	60,19	39,77	20,93
ε	15,9	10,7	7,84	5,06
$10^{-11}\tau$	1,36	1,57	1,36	1,29

Carbon tetrachloride (CCl_4) – Nitrocyclohexane ($C_6H_{11}NO_2$)

[400], $t=21$ °C, $\lambda=14$—67 cm

x_2	100	59,67	40,85	20,76
ε	16,1	11,6	8,67	5,71
$10^{-11}\tau$	2,33	2,26	2,24	2,20

Carbon tetrachloride (CCl_4) – Bromocyclohexane ($C_6H_{11}Br$)

[400], $t=21$ °C, $\lambda=14$—67 cm

x_2	100	62,67	40,30	20,97
ε	8,010	6,182	4,950	3,717
$10^{-11}\tau$	2,37	2,00	1,86	1,53

Carbon tetrachloride (CCl_4) – Chlorocyclohexane ($C_6H_{11}Cl$)

[400], $t=21$ °C, $\lambda=14$—67 cm

x_2	100	60,70	40,35	20,26
ε	8,056	5,969	4,875	3,514
$10^{-11}\tau$	1,43	1,35	1,29	1,18

Carbon tetrachloride (CCl_4) – **Toluene** (C_7H_8)
[401], $t=25$ °C, $w_2=48,17$

ν, Гец	ε′	ε″	ε	$ε_∞$	$10^{-12}τ_1$	$10^{-12}τ_2$	α	C_1
9,313	2,319	0,026	2,3367	2,24	14,0	3,5	0,13	0,55
16,20	2,305	0,036	—	—	—	—	—	—
23,98	2,292	0,036	—	—	—	—	—	—
34,86	2,275	0,034	—	—	—	—	—	—

Carbon tetrachloride (CCl_4) – *o*–**Cresol** (C_7H_8O)
[402], $t=21$ °C, $ν=60$ MHz, fig. 25

x_2	50	40	30	20	10
ε	3,930	3,498	3,120	2,800	2,496
$ε_∞$	2,389	2,362	2,333	2,300	2,267
$10^{-11}τ$	5,80	4,97	4,05	3,10	2,10

Carbon tetrachloride (CCl_4) – **Dibutyl ether** ($C_8H_{18}O$)
[404], $t=20$ °C, $ν=9290$ MHz

$φ_2$	ε′	ε″	ε	$ε_∞$	$φ_2$	ε′	ε″	ε	$ε_∞$
100	2,814	0,421	3,099	1,98	50,0	2,532	0,209	2,675	2,11
87,5	2,745	0,370	2,990	2,01	37,5	2,463	0,155	2,565	2,15
75,0	2,677	0,317	2,890	2,03	25,0	2,390	0,103	2,456	2,18
62,5	2,606	0,263	2,784	2,08	12,5	2,315	0,053	2,349	2,21

Carbon tetrachloride (CCl_4) – **Eugenol** ($C_{10}H_{12}O$)
[402], $t=27$ °C, $ν=60$ MHz, fig. 26

x_2	70	50	25	10
ε	7,280	5,940	4,220	3,015
$ε_∞$	2,403	2,363	2,299	2,257
$10^{-11}τ$	14,53	11,22	6,95	4,30

Carbon tetrachloride (CCl_4) – Decqn–1–ol ($C_{10}H_{22}O$)
[406] *)

x_2	t °C	λ=215,6 cm		λ=52,0 cm		ε	$ε_∞$	$10^{-10}τ$
		ε′	ε″	ε′	ε″			
74,8	20	3,95	1,48	2,98	0,76	6,34	2,95	17,5
	30	4,63	1,33	3,21	0,96	5,78	3,05	9,8
	40	4,68	0,90	3,54	1,15	5,28	3,10	5,6
	50	4,78	0,46	3,92	1,20	4,86	3,15	3,2
	60	4,54	0,27	4,07	0,87	—	—	—
49,8	20	3,52	0,65	2,91	0,48	4,18	2,90	11,7
	30	3,71	0,46	3,07	0,55	4,00	3,00	7,4
	40	3,72	0,26	3,30	0,57	3,82	3,05	4,2
	50	3,64	0,14	3,38	0,46	3,68	3,10	2,9
	60	3,54	0,08	3,42	0,28	—	—	—

*) For x_2=64,8, $ΔH$=10,4 kcal/mole, $ΔS$=17,0; for x_2=49,8, $ΔH$=9,2 kcal/mole, $ΔS$=13,9.

Carbon tetrachloride (CCl_4) – Diphenyl ether ($C_{12}H_{10}O$)
[400], t=21 °C, λ=14—16 *cм*

x_2	50,69	29,85	20,12
ε	3,139	2,809	2,646
$10^{-11}τ$	0,67	0,61	0,54

Carbon tetrachloride (CCl_4) – Benzophenone ($C_{13}H_{10}O$)
[66], ν=9,23 GHz

N	t=10 °C		t=20 °C		t=30 °C	
	ε	10^{-3} tg δ	ε	10^{-3} tg δ	ε	10^{-3} tg δ
0	2,244	0,98	2,234	0,95	2,224	0,92
0,25	2,253	2,96	2,242	3,09	2,233	2,97
0,5	2,261	6,05	2,251	6,16	2,240	5,94
0,75	2,270	9,02	2,261	9,24	2,249	8,93
1	2,279	12,00	2,269	12,38	2,258	11,87

Carbon disulphide (CS_2) – Cyclohexane (C_6H_{12})
[398], t=22 °C, ν=23,8 GHz

x_1	0,0	5,0	10,0	20,0	30,0	40,0	50,0	80,0	100,0
ε′	2,04	2,06	2,10	2,15	2,21	2,28	2,34	9,50	2,64
10^{-3}ε″	0,7	4,8	9,3	11,0	12,2	12,5	13,1	13,5	13,8

Chloroform ($CHCl_3$) – **Acetone** (C_3H_6O)
[408, 409]

φ_2	t °C	λ=3,21 cm		λ=0,815 cm		ε	$10^{-12}\tau$ *)	$10^{-12}\tau$ **)	ΔF	ΔH	TΔS
		ε'	ε''	ε'	ε''						
100	10	21,3	4,11	14,1	9,6	22,3	3,70	3,70	1,7	0,8	0,9
	20	20,4	3,64	14,3	9,3	21,2	3,40	3,40	1,7	0,8	0,9
	30	19,6	3,11	14,4	8,3	20,2	3,20	3,20	1,8	0,8	1,0
	40	18,7	2,58	14,0	7,8	19,1	2,76	2,76	1,8	0,7	1,1
90	10	19,4	4,54	12,5	8,8	20,8	5,25	4,07	1,9	1,6	0,3
	20	18,7	3,89	13,1	8,3	19,8	4,60	3,55	1,9	1,5	0,4
	30	18,0	3,34	13,0	7,6	18,8	4,04	3,20	1,9	1,5	0,4
	40	17,2	2,87	13,0	7,0	17,8	3,60	2,90	1,9	1,5	0,4
70	10	15,1	5,37	7,90	5,9	18,2	9,90	7,60	2,3	2,0	0,3
	20	14,9	4,73	8,10	5,7	17,3	8,60	7,00	2,3	2,0	0,3
	30	14,5	4,31	8,30	5,4	16,4	7,40	6,50	2,3	2,0	0,3
	40	14,2	3,70	8,35	5,2	15,6	6,45	6,0	2,3	1,9	0,4
50	10	11,0	5,32	5,45	4,0	15,4	14,1	10,9	2,5	2,3	0,2
	20	11,2	4,85	5,46	3,9	14,6	11,9	10,1	2,5	2,3	0,2
	30	11,2	4,39	5,44	3,8	13,8	10,1	9,30	2,5	2,2	0,3
	40	11,3	3,93	5,6	3,8	13,2	8,6	8,65	2,5	2,2	0,3
30	10	7,84	4,40	4,30	3,8	12,1	16,6	12,5	2,6	2,4	0,2
	20	8,08	4,06	4,33	2,8	11,5	14,1	11,4	2,6	2,4	0,2
	30	8,16	3,74	4,36	2,8	10,9	11,9	10,4	2,5	2,4	0,2
	40	8,3	3,27	4,30	2,8	10,3	10,1	3,60	2,6	2,3	0,3
10	10	5,68	2,28	3,27	1,6	7,6	14,3	11,6	2,5	2,2	0,3
	20	5,72	2,07	3,28	1,6	7,23	12,1	10,8	2,5	2,2	0,3
	30	5,74	1,86	3,30	1,5	6,88	10,3	10,1	2,5	2,2	0,3
	40	5,74	1,65	3,31	1,5	6,56	8,80	9,40	2,5	2,2	0,3
0	10	4,70	0,91	3,22	1,1	5,06	7,10	6,90	2,1	1,8	0,3
	20	5,56	0,83	3,23	1,1	4,86	6,20	6,30	2,1	1,8	0,3
	30	4,44	0,71	3,22	1,1	4,67	5,40	5,80	2,1	1,8	0,3
	40	4,34	0,58	3,25	1,0	4,50	4,70	5,40	2,1	1,7	0,4

*) Value of τ obtained from measurement at λ = 3,21 cm.

**) From results for λ = 0,815 cm.

Chloroform ($CHCl_3$) – **Butan–2–one** (C_4H_8O)
[410]

φ_2	t °C	λ=3,2 cm		λ=1,8 cm		λ=0,408 cm		ε	ε_∞	$10^{-12}\tau$
		ε'	ε''	ε'	ε''	ε'	ε''			
100	10	17,7	5,10	14,7	7,47	5,07	5,56	19,4	2,90	5,9
	20	17,2	4,40	14,8	6,60	5,30	5,71	18,5	2,82	5,3
	30	16,6	3,82	14,9	5,73	5,61	5,88	17,6	2,74	4,6
	40	16,0	3,31	14,5	5,00	5,95	5,92	16,8	2,66	4,2

φ_2	t °C	λ=3,2 cm		λ=1,8 cm		λ=0,408 cm		ε	ε_∞	$10^{-12}\tau$
		ε'	ε''	ε'	ε''	ε'	ε''			
90	10	15,6	5,52	12,3	7,37	4,61	4,57	18,3	2,83	7,3
	20	15,3	4,95	12,6	6,78	4,68	4,82	17,4	2,76	6,6
	30	14,9	4,35	12,5	6,28	4,81	4,96	16,6	2,69	6,00
	40	14,5	3,76	12,3	5,70	5,05	5,02	15,8	2,61	5,3
70	10	11,5	5,85	8,50	6,30	3,85	2,92	15,8	2,72	11,1
	20	11,7	5,37	8,78	6,08	3,70	3,02	15,2	2,67	10,1
	30	11,8	4,81	9,12	5,84	3,99	3,18	14,5	2,58	8,11
	40	11,7	4,27	9,22	5,48	3,99	3,22	13,9	2,52	7,81
50	10	8,50	5,24	6,01	4,63	3,17	1,80	13,8	2,56	15,0
	20	8,85	4,94	6,25	4,65	3,27	1,93	13,2	2,54	13,2
	30	9,05	4,57	6,66	4,61	3,35	2,06	12,5	3,48	11,7
	40	9,14	4,10	7,06	4,51	3,30	2,09	11,9	2,43	10,3
30	10	6,45	3,99	4,70	3,42	2,82	1,26	11,0	2,45	18,9
	20	6,77	3,79	5,05	3,50	2,82	1,31	10,8	2,41	15,9
	30	6,95	3,50	5,19	3,46	2,85	1,38	9,84	2,38	12,9
	40	7,09	3,18	5,48	3,33	2,88	1,40	9,33	2,34	11,3
10	10	5,25	2,14	4,27	2,29	2,64	0,938	7,22	2,32	13,0
	20	5,27	1,98	4,44	2,12	2,65	0,984	6,85	2,30	11,4
	30	5,26	1,8	4,50	2,00	2,65	0,997	6,52	2,27	10,2
	40	5,23	1,58	4,59	1,92	2,63	1,02	6,24	2,25	8,9
0	10	4,58	0,961	4,12	1,25	2,56	0,794	4,99	2,25	—
	20	4,42	0,835	4,06	1,16	2,58	0,817	4,80	2,23	—
	30	4,39	0,722	3,98	1,06	2,62	0,838	4,64	2,22	—
	40	4,26	0,632	3,88	0,97	2,65	0,864	4,48	2,20	—

Chloroform ($CHCl_3$) – Dioxan ($C_4H_8O_2$)

[388], $t=20$ °C

x_1	λ=9,96 cm		λ=3,22 cm		λ=1,25 cm		ε	$10^{-12}\tau$	α
	ε'	ε''	ε'	ε''	ε'	ε''			
4,94	—	—	2,2934	0,0475	2,2623	0,0346	2,3409	16,2	0
9,9	—	—	—	—	—	—	2,4575	—	—
19,9	—	—	—	—	—	—	2,6834	—	—
30,0	—	—	—	—	—	—	2,9039	—	—
40,0	3,0854	0,155	2,8578	0,366	2,2566	0,348	3,1115	11,2	0
49,5	—	—	3,0457	0,433	2,6621	0,452	3,3099	9,9	0
59,1	3,4885	0,188	3,2394	0,487	2,7889	0,572	3,5110	8,3	0,02
70,0	3,7355	0,202	3,4774	0,549	2,9582	0,714	3,7652	7,1	0,04
80,5	—	—	3,7702	0,604	3,1484	0,878	3,0493	6,5	0,04
90,3	4,3491	0,241	4,0958	0,683	3,4184	1,023	4,3378	5,7	0,04
100,0	—	—	4,5720	0,775	3,6518	1,28	—	—	—

Chloroform ($CHCl_3$) – **Pyridine** (C_5H_5N)

[411], $\lambda = 3,21$ cm

φ_2	$t=10$ °C		$t=20$ °C		$t=30$ °C	
	ε′	ε″	ε′	ε″	ε′	ε″
100	12,0	4,5	11,8	3,8	11,8	3,3
70	8,2	4,3	8,3	3,8	8,7	3,7
50	6,5	3,9	6,6	3,5	7,2	3,2
30	5,8	3,0	6,2	3,1	6,3	2,7
0	4,6	0,96	4,5	0,83	4,4	0,72

Chloroform ($CHCl_3$) – **Cyclohexane** (C_6H_{12})

[388], $t = 20$ °C

x_1	$\lambda=3,22$ cm		$\lambda=1,25$ cm		ε	$10^{-12}\tau$	α
	ε′	ε″	ε′	ε″			
4,4	—	—	—	—	2,0829	—	—
7,7	2,1252	0,0198	2,1123	0,0275	2,1279	3,7	0
28,7	2,4573	0,088	2,3875	0,182	—	3,7	0
42,1	—	—	—	—	2,7477	—	—
46,2	2,8058	0,189	2,6357	0,349	2,8544	4,4	0
66,8	3,2776	0,340	2,9135	0,578	3,3747	5,2	0
87,4	3,8946	0,584	3,3403	0,919	—	5,6	0

Chloroform ($CHCl_3$(– **Diazabicyclo–[2, 2, 2]–octane** ($C_6H_{12}N_2$)

[388], $t = 26$ °C

x_1	λ, см	ε′	ε″	α	$10^{-12}\tau$
60,9	57500	3,6215	—	0	—
62,9	1,25	2,6948	0,306	0,22	22,8
	3,22	2,9551	0,416	—	—
	9,96	3,3455	0,338	—	—
90,0	1,25	3,3564	0,899	0,05	8,2
	3,22	4,0468	0,695	—	—
	9,96	4,3334	0,346	—	—
93,3	57500	4,5250	—	—	—

Nitromethane (CH_3NO_2) – Methanol (CH_4O)
[413]

x_2	t °C	λ=15,325 cm		λ=10,0 cm		λ=3,20 cm		λ=1,80 cm		λ=0,408 cm		ε
		ε′	ε″	ε′	ε″	ε′	ε″	ε′	ε″	ε′	ε″	
100	10	22,7	14,7	16,6	14,9	7,4	7,28	6,13	4,58	4,59	2,26	35,5
	20	25,3	12,9	18,9	14,3	8,0	8,08	6,40	5,29	4,51	2,34	33,2
	30	25,9	11,4	20,8	13,5	8,8	8,66	6,49	5,85	4,55	2,47	31,4
	40	26,1	9,7	22,2	12,2	9,6	9,36	6,56	6,25	4,72	2,52	29,2
	50	25,8	7,8	23,0	10,0	—	—	—	—	4.75	2,57	—
90	10	25,7	13,3	20,6	13,6	10,6	7,8	8,46	5,89	5,18	3,21	34,9
	20	26,4	11,3	22,4	12,9	11,1	8,4	8,62	6,21	5,12	3,36	33,5
	30	27,5	10,3	23,4	11,9	11,9	9,2	8,75	6,61	5,12	3,56	32,2
	40	27,2	8,1	23,9	10,5	12,2	9,7	9,02	7,17	5,13	3,72	30,5
	50	25,9	6,4	24,3	8,80	—	—	—	—	5,26	3,93	—
70	10	30,2	9,77	26,2	10,7	16,8	9,3	13,7	8,69	6,21	5,70	35,6
	20	30,1	8,23	27,1	9,8	17,2	9,5	13,9	8,70	6,32	5,72	33,8
	30	29,8	6,80	27,3	8,4	17,7	9,5	14,2	8,74	6,62	5,97	32,1
	40	29,3	5,37	27,1	7,2	18,4	9,4	14,5	8,76	7,12	6,26	30,6
	50	—	—	26,8	5,7	—	—	—	—	7,75	6,66	—
50	10	33,1	7,87	31,2	7,96	23,6	9,82	19,5	10,9	6,97	8,10	35,7
	20	32,5	5,18	30,8	6,52	23,8	9,28	19,6	10,4	7,43	8,55	33,9
	30	31,8	4,24	30,2	5,48	23,9	8,76	19,7	9,93	7,93	8,90	32,1
	40	—	—	29,4	4,35	23,9	8,23	20,1	9,50	8,50	9,14	30,6
	50	—	—	28,3	3,46	—	—	—	—	9,15	9,30	—
30	10	—	—	34,6	5,29	29,7	9,6	24,8	12,9	7,70	10,5	36,4
	20	—	—	33,6	4,44	29,3	8,6	25,1	11,7	8,50	10,9	34,6
	30	—	—	32,3	3,65	28,8	7,9	25,3	10,8	9,30	11,2	33,0
	40	—	—	30,9	2,60	28,2	7,1	25,3	9,9	10,1	11,6	31,4
	50	—	—	29,4	1,63	—	—	—	—	10,8	11,9	—
10	10	—	—	36,8	3,52	34,6	9,42	30,2	14,3	8,47	13,1	37,9
	20	—	—	35,4	2,95	33,7	8,16	30,3	13,0	9,36	14,4	36,0
	30	—	—	33,9	2,49	33,1	7,47	29,7	11,1	10,3	13,5	34,4
	40	—	—	32,15	2,12	32,2	6,83	29,3	10,1	11,1	13,4	32,9
	50	—	—	31,3	1,82	—	—	—	—	11,5	13,3	—
0	10	—	—	38,7	3,43	36,3	9,49	32,0	14,5	8,6	13,8	39,0
	20	—	—	37,1	2,92	35,2	8,37	31,6	13,2	9,4	13,9	36,4
	30	—	—	35,7	2,51	34,1	7,32	31,0	11,8	10,1	14,1	35,9
	40	—	—	33,9	2,14	32,8	6,44	30,4	10,3	10,8	14,1	34,2
	50	—	—	32,6	1,87	—	—	—	—	11,3	14,0	—

Nitromethane (CH_3NO_2) – **Benzene** (C_6H_6)
[66], $\nu = 3{,}465$ GHz

N	$t=10$ °C		$t=20$ °C		$t=30$ °C		$t=40$ °C	
	ε'	10^{-3} tg δ	ε'	10^{-3} tg δ	ε'	10^{-3} tg δ	ε'	10^{-3} tg δ
0	2,2895	1,44	2,287	1,55	2,284	1,61	2,282	1,65
0,1	2,298	4,33	2,295	4,22	2,292	3,92	2,287	3,36
0,15	2,302	6,48	2,299	6,37	2,295	5,78	2,291	5,08
0,22	2,306	8,76	2,303	8,58	2,300	7,95	2,296	6,85
0,3	2,313	13,2	2,310	12,82	2,306	11,73	2,302	10,25

Nitromethane (CH_3NO_2) – **Heptan-1-ol** ($C_7H_{16}O$)
[414], $\nu = 0{,}5$—$23{,}8$ GHz, $w_2 = 11{,}2$

t °C	ε	ε_∞	$\varepsilon_{\infty 2}$	$10^{-10}\tau_1$	$10^{-10}\tau_2$
39,8	10,90	4,63	1,65	2,05	7,00
35,3	11,21	4,75	1,90	2,41	8,50
29,8	11,64	5,00	2,50	3,07	11,4
26,3	11,90	5,20	2,80	3,90	19,0
23,0	12,20	5,40	3,20	4,90	41,2

Methanol (CH_4O) – **Acetone** (C_3H_6O)
[392—394, 416], $\lambda = 3{,}21$ cm, fig. 22

x_1	$t=0$ °C		$t=20$ °C		$t=40$ °C	
	ε'	ε''	ε'	ε''	ε'	ε''
0,0	22,46	4,69	20,92	3,54	18,61	2,65
31,21	20,14	7,05	19,44	6,22	18,73	4,90
54,50	17,04	8,43	16,81	7,97	17,22	6,93
73,22	13,68	8,69	14,00	9,02	15,26	8,45
88,16	10,62	8,14	11,14	9,27	12,70	9,53
100	7,41	6,52	8,37	8,25	10,15	9,66

x_1	t °C	α	$10^{-12}\tau$	ΔF	ΔH	$T\Delta S$
0,0	0	0,040	3,73	1,47	0,64	—0,83
	20	0,028	3,01	1,49	0,60	—0,89
	40	0	2,62	1,55	0,56	—0,99
31,21	0	0,122	6,99	1,82	0,56	—1,26
	20	0,073	6,07	1,87	0,52	—1,35
	40	0,039	5,22	2,01	0,48	—1,53

x_1	t °C	α	$10^{-12}\tau$	ΔF	ΔH	$T\Delta S$
54,50	0	0,221	14,11	2,20	0,86	—1,14
	20	0,133	11,47	2,25	0,82	—1,43
	40	0,094	9,39	2,38	0,78	—1,60
73,22	0	0,222	31,16	2,64	1,66	—0,98
	20	0,122	21,86	2,64	1,62	—1,02
	40	0,067	16,27	2,73	1,58	—1,15
88,16	0	0,183	52,35	2,93	2,61	—0,32
	20	0,089	35,10	2,91	2,57	—0,34
	40	0,035	24,23	2,98	2,53	—0,45
100	0	0,106	88,13	3,20	3,61	0,41
	20	0,051	51,90	3,14	3,57	0,43
	40	0	32,44	3,15	3,53	0,38

[395, 397, 393, 416], $\lambda=8{,}15$ mm

x_1	$t=0$ °C		$t=20$ °C		$t=40$ °C	
	ε'	ε''	ε'	ε''	ε'	ε''
0	13,5	10,0	14,5	9,2	14,1	7,7
31,21	11,3	8,8	11,8	8,8	11,8	7,8
54,50	9,3	7,0	9,6	7,6	10,0	7,1
73,22	7,6	5,3	7,8	5,6	8,0	5,7
88,16	6,5	3,9	6,5	4,2	6,6	4,5
100	5,4	2,72	5,4	3,10	5,5	3,50

[418], $t=21$ °C

ν, *Мгц*	$w_2=78$		$w_2=48{,}5$		$w_2=34$		$w_2=23{,}4$	
	ε'	ε''	ε'	ε''	ε'	ε''	ε'	ε''
900	13,01	9,95	21,23	9,46	23,39	9,45	25,40	8,95
1000	12,70	9,92	20,28	9,54	23,01	9,58	24,65	8,97
1100	—	—	19,61	9,31	20,01	5,51	23,42	8,80
1200	12,04	8,12	17,51	7,57	18,77	7,74	19,74	8,63
1300	11,15	6,97	16,51	6,27	18,12	6,31	19,23	5,90
1400	10,50	5,54	13,80	4,91	14,72	4,78	16,53	5,05
1500	9,74	4,63	12,29	4,09	13,21	3,94	16,25	7,44
1600	8,52	4,49	11,49	4,28	12,25	4,51	13,03	6,56
1700	7,92	4,43	10,37	4,40	11,00	4,62	—	—
1800	7,60	4,53	10,05	4,82	10,81	5,18	11,16	5,78
1900	7,14	4,96	9,63	5,97	—	—	9,40	7,20
2000	7,25	4,60	9,49	5,22	9,72	5,57	8,98	6,99
6900	3,92	2,35	6,17	2,90	—	—	—	—
9200	3,44	1,85	5,80	2,32	—	—	—	—

Methanol (CH_4O) – Benzene (C_6H_6)
[419], $t=25$ °C

φ_1	λ=1,25 cm		λ=3,20 cm		φ_1	λ=1,25 cm		λ=3,20 cm	
	ε′	ε″	ε′	ε″		ε′	ε″	ε′	ε″
10	2,73	0,266	2,90	0,402	70	4,69	2,42	5,74	4,84
25	3,20	0,69	3,51	1,16	85	5,25	3,16	7,06	6,42
40	3,63	1,18	4,20	2,16	100	6,04	4,13	8,18	8,00
55	4,17	1,76	4,96	3,39	—	—	—	—	—

1, 1, 1–Trichloroethane ($C_2H_3Cl_3$) – 1, 2–Dichlorobenzene ($C_6H_4Cl_2$) *)
[422], $t=20$ °C

λ, cm	$x_1=0$		$x_1=25$		$x_1=50$		$x_1=75$		$x_1=100$	
	ε′	ε″	ε′	ε″	ε′	ε″	ε′	ε″	ε′	ε″
14990	10,573	—	9,619	—	8,888	—	8,185	—	7,228	—
2,9058	5,066	3,428	5,375	3,163	6,208	2,992	6,727	2,382	6,581	1,766
2,6187	4,413	3,448	5,049	3,069	5,848	3,059	6,417	2,546	6,480	1,719
2,4385	4,532	3,176	5,066	3,083	5,549	3,031	6,046	2,677	6,391	2,048
1,9440	3,953	2,708	4,356	2,718	4,939	2,705	5,488	2,463	6,154	2,129
1,7000	3,735	2,495	4,203	2,609	4,717	2,641	5,263	2,620	5,820	2,412
1,4010	3,537	2,056	3,775	2,226	4,170	2,335	4,677	2,351	5,329	2,355
1,2602	3,264	1,910	3,692	2,040	4,044	2,212	4,594	2,292	5,183	2,378
0,2174	2,576	0,435	2,529	0,519	2,519	0,531	2,510	0,765	2,525	0,884
$5{,}893\cdot10^{-11}$	2,408	—	2,329	—	2,246	—	2,166	—	2,067	—

*) In [422] there are values of ϵ' and ϵ'' for $\lambda = 7{,}5$; 15 cm for $x_1 = 0{,}5$ and $x_1 = 0.75$.

x_1	$10^{-12}\tau_1$	$10^{-12}\tau_2$	C_1	$\frac{n_1\mu_1^2}{n_2\mu_2^2}$
25	20,0	2,6	0,74	0,5
50	14,1	2,2	0,84	1,5
75	11,1	2,3	0,84	4,4

1, 1, 1-Trichloroethane ($C_2H_3Cl_3$) – Benzene (C_6H_6) *)
[422], $t=20$ °C

λ, см	$x_1=25$		$x_1=50$		$x_1=75$		$x_1=100$	
	ε′	ε″	ε′	ε″	ε′	ε″	ε′	ε″
14990	3,288	—	4,391	—	5,670	—	7,288	—
2,9058	3,231	0,256	4,205	0,625	5,296	1,051	6,581	1,766
2,6187	—	—	4,200	0,697	5,219	1,117	6,480	1,719
2,4385	—	—	4,125	0,803	5,159	1,321	6,391	2,048
1,9440	3,129	0,366	4,000	0,797	5,137	1,352	6,154	2,129
1,7000	3,106	0,387	3,958	0,896	4,769	1,494	5,820	2,412
1,4010	3,028	0,425	3,760	0,943	4,532	1,534	5,329	2,355
1,2602	2,968	0,453	3,648	0,957	4,459	1,612	5,183	2,378
0,2174	2 339	0,247	2,346	0,414	2,396	0,578	2,525	0,884
$5,893\cdot10^{-11}$	2,195	—	2,145	—	2,100	—	2,067	—

*) For λ=7,5 cm, $x_1=25$, ε′=4,37, ε″=0,33.

Acetonitrile (C_2H_3N) – Benzene (C_6H_6) *)
[424], $t=30$ °C, ν=9950 MHz, fig. 33

x_1	13,52	30,51	39,67	46,34
ε′	2,3818	2,6523	2,7756	2,8991
ε″	0,03232	0,08314	0,10849	0,13098

*) See also [425], $t=24$—70 °C, λ=10, 3, 2; 1 cm.

[390], $x_1=25$

t °C	λ=3,23 cm		λ=1,95 cm		λ=1,25 cm		λ=0,4 cm		$10^{-12}\tau$	ε	ε_∞
	ε′	ε″	ε′	ε″	ε′	ε″	ε′	ε″			
24	6,59	0,98	5,95	1,43	5,54	1,50	4,36	0,84	6,7	7,04	4,1
30	6,41	0,94	5,88	1,35	5,39	1,44	—	—	6,5	6,77	—
40	6,28	0,84	5,82	1,20	5,68	1,44	4,61	1,51	4,5	6,41	3,3
50	6,18	0,66	5,82	1,06	5,55	1,40	—	—	3,4	6,21	—
60	5,94	0,59	5,70	0,91	5,55	1,18	4,90	1,46	3,1	6,00	2,8

Acetonitrile (C_2H_3N) – Benzonitrile (C_7H_5N)
[426, 425], $t=30$ °C

x_1	λ=9,84 cm		λ=3,22 cm		λ=1,96 cm		λ=1,15 cm		ε	ε_∞	$10^{-12}\tau_1$	$10^{-12}\tau_2$	C_1
	ε′	ε″	ε′	ε″	ε′	ε″	ε′	ε″					
30	22,0	8,5	12,7	9,7	9,7	8,7	7,6	5,2	27,0	3,0	5,3	32,5	0,27
40	22,6	8,2	14,6	9,4	11,4	8,9	9,1	7,8	27,6	2,8	6,4	33,5	0,41
50	23,5	8,6	17,5	10,1	13,9	9,3	10,9	8,3	28,5	2,6	5,3	33,0	0,49
60	24,9	8,4	19,5	9,4	18,1	9,2	12,8	10,0	29,3	2,4	5,3	34,5	0,61
82	31,8	2,5	27,2	8,9	24,0	10,5	18,8	13,7	32,3	2,4	4,8	14,3	0,83
100	34,3	1,6	33,6	6,9	32,1	8,8	28,5	12,5	—	—	3,3	—	—

Acetic acid ($C_2H_4O_2$) – Benzene (C_6H_6)
[427], $t = 20$ °C, $\lambda = 3,26$ cm

x_1	ε	ε′	ε″	$10^{-11}\tau$	x_1	ε	ε′	ε″	$10^{-11}\tau$
3,28	2,30	—	—	0,88	39,8	3,28	2,62	0,32	3,65
6,67	2,40	2,33	0,06	2,04	49,1	3,74	2,80	0,41	3,98
10,20	2,46	—	—	—	60,4	4,25	2,99	0,50	4,35
13,85	2,57	2,35	0,13	2,72	72,2	4,84	3,16	0,58	4,86
17,53	2,68	—	—	—	85,6	5,56	3,40	0,74	5,16
21,65	2,75	2,41	0,20	2,95	100	6,20	3,57	1,02	4,55
30,00	2,99	2,48	0,26	3,40	—	—	—	—	—

Bromoethane (C_2H_5Br) – Hexadecane ($C_{16}H_{34}$)
[428], $\lambda = 3,22$ cm

x_1	$t=20$ °C		$t=30$ °C		$t=40$ °C	
	ε′	ε″	ε′	ε″	ε′	ε″
0	2,051	0	2,039	0	2,026	0
1,98	2,077	0,006	2,065	0,005	2,052	0,004
4,50	2,11	0,0125	2,095	0,0104	2,080	0,0082
6,00	2,132	0,0166	2,116	0,0144	2,101	0,0122
8,16	2,163	0,0226	2,146	0,0197	2,128	0,0168

Ethanol (C_2H_6O) – Isopentyl bromide ($C_5H_{11}Br$)
[330, 429], $t = 133,4$ °K, $\nu = 50$ Hz – 5 MHz, fig. 34

x_1	ε_∞	ε	ν_{m1}, kHz	ν_{m2}, kHz	β
0	2,75	12,6	—	25	0,58
20	2,8	15,3	0,2	28	0,52
33	2,8	21,6	0,35	50	0,50
50	3	30,6	2	180	0,48
67	3	44,3	3	600	0,46
100	—	83 *)	40 *)	—	—

*) Extrapolated values.

Ethanol (C_2H_6O) – Benzene (C_6H_6)
[430], $t = 20$ °C, $\nu = 50$ MHz

x_1	10	20	30	40	50	60	70	80	90	100
ε′	4,00	6,20	7,30	10,53	12,85	15,40	18,47	20,50	24,06	25,70
ε″	0,04	0,135	0,240	0,455	0,590	0,693	0,818	0,887	1,028	1,100

Ethanol (C_2H_6O) – Cyclohexane (C_6H_{12})
[431], ν = 10 kHz—14 kHz

x_1	t °C	ε	ν_m*)	ε_∞	ΔH	x_1	t °C	ε	ν_m*)	ε_∞	ΔH
25	—5	3,69	289	2,50	6,7	75	25	15,05	669	3,6	—
	5	3,60	366	2,48	—		35	14,0	903	3,6	—
	15	3,48	600	2,42	—		50	12,63	1350	3,4	—
	25	3,34	857	2,40	—	90	—5	24,6	300	4,4	5,4
	35	3,21	1178	2,37	—		5	23,2	432	4,2	—
	50	3,05	1940	2,35	—		15	21,7	588	4,4	—
50	—5	9,55	183	3,2	6,8		25	20,3	847	4,2	—
	5	8,90	240	3,1	—		35	19,1	1094	3,9	—
	15	8,30	385	3,1	—		50	17,3	1621	4,6	—
	25	7,61	627	2,8	—	100	—5	29,5	366	5,0	4,95
	35	7,15	898	2,8	—		5	27,7	500	4,7	—
	50	6,41	1380	2,8	—		15	25,9	673	4,9	—
75	—5	18,3	238	3,9	5,55		25	24,35	935	4,5	—
	5	17,1	333	3,9	—		35	22,9	1180	4,7	—
	15	16,0	469	4,0	—		50	20,9	1695	5,0	—

*) ν_m — in MHz.

Ethanol (C_2H_6O) – Heptane (C_7H_{16}) *)
[435], ν = 9,3477 GHz

x_2	t = 2 °C				t = 20 °C			
	ε′	tg δ	ε	ε_∞	ε′	tg δ	ε	ε_∞
90	2,04	0,040	2,27	2,01	2,04	0,040	2,24	2,00
70	2,25	0,089	3,21	2,21	2,25	0,100	3,08	2,19
50	2,53	0,139	4,68	2,47	2,56	0,169	4,51	2,46
30	2,94	0,238	7,30	2,83	2,96	0,290	6,91	2,77
20	3,22	0,336	9,00	3,02	3,23	0,431	8,72	2,88
10	3,90	0,471	13,93	3,64	4,04	0,555	13,44	3,51

x_2	t = 30 °C				t = 40 °C			
	ε′	tg δ	ε	ε_∞	ε′	tg δ	ε	ε_∞
90	2,04	0,040	2,23	2,00	2,04	0,040	2,22	2,00
70	2,27	0,110	3,07	2,19	2,27	0,119	3,02	2,17
50	2,59	0,193	4,45	2,45	2,60	0,218	4,34	2,41
30	2,99	0,332	6,71	2,73	3,02	0,373	6,50	2,66
20	3,23	0,483	8,36	2,75	3,24	0,545	7,95	2,58
10	4,10	0,622	13,08	3,37	4,16	0,697	12,66	3,17

*) The dependence of relaxation time on the concentration of the solution is given by the expression
$\tau = A_0 + A_1 x + A_2 x^2 + A_3 x^3 + \ldots + A_n x^n$,
$A_0 = 0,772$; $A_1 = 1,223$; $A_2 = -0,776$; $A_3 = -1,171$; $n = 3$ при $t = 2$ °C.

Ethyl cyanide (C_3H_5N) – **Benzene** (C_6H_6)
[424], $t=30$ °C, $\nu=9{,}95$ GHz, fig. 33

x_1	20,45	30,68	40,91	50,54
ε'	2,5155	2,6391	2,8242	2,9611
ε''	0,07113	0,12880	0,16200	0,19690

[390], $x_1=25$

t° C	λ=3,23 cm		λ=1,95 cm		λ=1,25 cm		λ=0,4 cm	
	ε'	ε''	ε'	ε''	ε'	ε''	ε'	ε''
24	6,36	1,51	5,49	1,86	4,71	1,85	3,61	1,21
30	6,27	1,39	5,49	1,78	5,02	1,69	—	—
40	6,12	1,26	5,56	1,59	4,94	1,69	3,65	1,07
50	6,05	1,04	5,47	1,35	4,93	1,67	—	—
60	5,92	0,91	5,44	1,23	4,95	1,45	3,57	1,35

t° C	$10^{-12}\tau_1$	$10^{-12}\tau_2$	C_1	ε	ε_∞
24	10,0	2,1	0,75	6,98	2,8
40	8,5	1,6	0,75	6,69	2,8
60	7,0	1,6	0,70	6,39	2,8

Ethyl cyanide (C_3H_5N) – **Benzonitrile** (C_7H_5N) *)
[426], $t=30$ °C

x_1	λ=10 cm		λ=3,22 cm		λ=1,96 cm		λ=1,25 cm	
	ε'	ε''	ε'	ε''	ε'	ε''	ε'	ε''
30	23,7	7,9	12,4	9,0	8,6	7,5	7,3	5,4
50	23,2	6,00	17,1	9,4	11,5	9,10	8,9	7,3
70	26,3	4,4	21,2	9,0	16,0	9,7	11,1	8,4
100	—	—	26,0	6,8	23,8	9,0	18,2	11,5

x_1	ε	ε_∞	$10^{-12}\tau_1$	$10^{-12}\tau_2$	α	C_1
30	25,6	2,4	3,7	25,5	0,4	0,74
50	26,3	2,4	8,0	17,5	0,35	0,57
70	26,9	2,4	5,8	13,3	0,20	0,40

x_1	ε	ε_∞	$10^{-12}\tau_1$	$10^{-12}\tau_2$	$10^{-12}\tau_3$	C_1	C_2	C_3
30	25,6	2,4	1	14	25,5	0,16	0,10	0,74
50	26,3	2,4	1	14	17,5	0,21	0,22	0,57
70	26,9	2,4	1	13	13,3	0,24	0,36	0,40

*) See also [425], $t=24-70$ °C, λ=10; 3,2; 1 cm.

Acetone (C_3H_6O) – Nitrobenzene ($C_6H_5O_2$)
[392, 393, 416, 436]

x_2	t °C	$\lambda=3,21$ cm		$\lambda=0,815$ cm		$10^{-11}\tau$	α	$10^{-11}\tau_1$	$10^{-11}\tau_2$	C_2
		ε'	ε''	ε'	ε''					
100	10	6,49	8,45	4,18	2,52	6,23	0	—	—	—
	20	7,78	9,98	4,22	2,82	4,74	0	—	—	—
	30	9,29	10,98	4,27	3,15	3,77	0	4,52	—	1,0
	40	16,39	11,25	4,32	3,55	3,21	0	—	—	—
68,94	10	11,62	11,72	5,14	4,74	3,16	0,084	—	—	—
	20	12,74	11,83	5,30	5,14	2,68	0,077	—	—	—
	30	13,51	11,56	5,52	5,63	2,34	0,07	2,90	0,60	0,86
	40	14,01	11,11	5,69	6,13	2,08	0,064	—	—	—
41,69	10	16,33	11,03	6,38	6,56	1,73	0,139	—	—	—
	20	17,03	10,69	6,89	6,95	1,50	0,128	—	—	—
	30	17,41	10,11	7,52	7,46	1,31	0,111	2,10	0,5	0,68
	40	17,61	9,55	8,04	7,84	1,17	0,097	—	—	—
18,65	10	19,88	8,39	9,07	8,55	0,84	0,134	—	—	—
	20	19,72	7,49	9,95	8,63	0,76	0,125	—	—	—
	30	19,51	6,74	10,44	8,50	0,66	0,117	1,78	0,39	0,38
	40	19,22	6,19	10,73	7,34	0,59	0,110	—	—	—
0	10	21,38	3,99	14,12	9,61	0,32	0,036	—	—	—
	20	20,92	3,54	14,45	9,24	0,30	0,028	—	—	—
	30	19,76	2,95	14,45	8,59	0,28	0	—	0,31	0,0
	40	18,81	2,65	14,05	7,65	0,26	0	—	—	—

x_2	ΔF at t °C				ΔH at t °C			
	10	20	30	40	10	20	30	40
100	3,14	3,12	3,11	3,13	2,81	2,79	2,77	2,75
68,94	2,76	2,78	2,81	2,86	2,24	2,22	2,20	2,18
41,69	2,41	2,44	2,46	2,49	1,74	1,72	1,70	1,68
18,65	2,00	2,04	2,05	2,06	1,18	1,16	1,14	1,12
0	1,46	1,49	1,52	1,55	0,46	0,60	0,58	0,56

Acetone (C_3H_6O) – Benzene (C_6H_6)
[437, 438, 392, 393], $\lambda=3,21$ cm

x_1	$t=-10$ °C						$t=0$ °C					
	ε'	ε''	α	$10^{-12}\tau$	ΔF	ΔH	ε'	ε''	α	$10^{-12}\tau$	ΔF	ΔH
20	—	—	—	—	—	—	4,96	0,6	0,10	3,56	1,53	1,90
40	8,15	1,74	0,124	4,32	1,51	1,32	7,91	1,4	0,111	3,97	1,55	1,30
60	12,22	2,89	0,108	4,77	1,54	1,08	11,68	2,39	0,095	4,02	1,53	1,05
80	17,18	4,13	0,063	4,33	1,48	0,90	16,52	3,50	0,055	3,88	1,50	0,88
100	23,06	5,90	0,05	4,49	1,49	0,66	22,46	4,69	0,04	3,13	1,47	0,64

x_1	t=10 °C				t=20 °C					
	ε'	ε''	α	$10^{-12}\tau$	ε'	ε''	α	$10^{-12}\tau$	ΔF	ΔH
0	2,306	—	—	—	2,29	—	—	—	—	—
20	4,85	0,49	0,08	2,96	4,69	0,45	0,094	2,76	1,54	1,86
40	7,72	1,17	0,102	3,96	7,42	1,03	0,096	3,08	1,56	1,26
60	11,35	1,99	0,076	3,56	11,10	1,74	0,055	3,21	1,55	1,02
80	16,05	3,02	0,045	3,51	15,48	2,55	0,035	3,13	1,53	0,84
100	21,38	3,99	0,036	3,22	20,92	3,54	0,028	3,01	1,49	0,60

x_1	t=30 °C				t=40 °C					
	ε'	ε''	α	$10^{-12}\tau$	ε'	ε''	α	$10^{-12}\tau$	ΔF	ΔH
0	2,275	—	—	—	2,26	—	—	—	—	—
20	4,58	0,35	0,045	2,32	4,41	0,27	0,02	2,01	1,50	1,82
40	7,25	0,85	0,06	2,69	7,07	0,73	0,04	2,39	1,55	1,22
60	10,61	1,51	0,045	2,94	10,34	1,38	0,034	2,70	1,6	0,98
80	14,66	2,23	0,026	2,93	14,05	1,97	0,016	2,69	1,58	0,80
100	19,76	2,94	0	2,80	18,81	2,65	0	2,62	1,55	0,56

[437, 392, 393, 438], λ=8,15 mm

x_1	t=0 °C				t=20 °C				t=40 °C			
	ε'	ε''	α	$10^{-12}\tau$	ε'	ε''	α	$10^{-12}\tau$	ε'	ε''	α	$10^{-12}\tau$
0	2,30	—	—	—	2,28	—	—	—	2,26	—	—	—
20	3,90	1,18	0,10	3,94	3,95	1,06	0,09	2,88	3,95	0,89	0,02	2,25
40	5,40	2,42	0,11	4,45	5,8	2,35	0,10	3,15	5,9	2,18	0,04	2,50
60	7,40	4,35	0,10	4,46	7,8	4,0	0,06	3,48	7,8	3,65	0,03	2,87
80	10,1	7,2	0,06	4,21	10,5	6,1	0,04	3,46	10,8	5,6	0,02	2,97
100	13,5	10,0	0,04	4,07	14,5	9,2	0,02	3,23	14,1	7,7	0	2,76

Acetone (C_3H_6O) – Hexane (C_6H_{14}) *)

[439]

φ_1	t °C	λ=3,21 cm		λ=0,815 cm		ε	φ_1	t °C	λ=3,21 cm		λ=0,815 cm		ε
		ε'	ε''	ε'	ε''				ε'	ε''	ε'	ε''	
0,0	10	1,89	0	1,89	0	1,89	30,0	10	6,17	0,89	4,80	1,9	6,31
	20	1,87	0	1,87	0	1,87		20	5,96	0,74	4,82	1,7	6,07
	30	1,85	0	1,85	0	1,85		30	5,75	0,67	4,83	1,6	5,84
	40	1,83	0	1,83	0	1,83		40	5,54	0,58	4,81	1,5	5,61
15,0	10	3,86	0,36	3,25	0,7	3,15	50,0	10	9,89	1,78	7,20	3,9	10,2
	20	3,76	0,31	3,22	0,6	3,07		20	9,49	1,48	7,17	3,5	9,75
	30	3,65	0,26	3,19	0,5	2,98		30	9,17	1,31	7,15	3,2	9,35
	40	3,56	0,22	3,13	0,4	2,91		40	8,75	1,12	7,10	2,9	8,9

*) See also [396], ν=300—510 MHz.

φ_1	t °C	λ=3,21 cm		λ=0,815 cm		ε	φ_1	t °C	λ=3,21 cm		λ=0,815 cm		ε
		ε′	ε″	ε′	ε″				ε′	ε″	ε′	ε″	
70	10	14,4	2,63	9,69	6,8	14,9	89,9	30	17,4	2,60	12,8	7,2	17,8
	20	13,0	2,32	9,72	5,4	14,2		40	16,6	2,29	12,7	6,6	16,9
	30	13,3	1,92	9,68	5,0	13,6	100	10	21,3	4,11	14,1	9,6	22,3
	40	12,7	1,68	9,64	4,5	12,9		20	20,4	3,64	14,3	9,3	21,2
89,8	10	19,0	3,59	13,0	8,5	19,7		30	19,6	3,11	14,4	8,3	20,2
	20	18,1	3,09	13,0	7,9	18,6		40	18,7	2,58	14,1	7,8	19,1

Acetone (C_3H_6O) – Tetramethylammonium bromide ($C_{16}H_{10}BrN$)
[440], $t=25$ °C

$10^{-3}N$	ν=0,4 GHz		ν=0,6 GHz		ν=1,0 GHz		ν=1,5 GHz		ν=3,0 GHz	
	ε′	ε″	ε′	ε″	ε′	ε″	ε′	ε″	ε′	ε″
1,86	—	—	—	—	20,75	0,55	—	—	20,7	1,21
42,5	—	—	—	—	20,95	0,97	—	—	20,2	1,62
82,3	—	—	—	—	21,5	1,30	—	—	20,75	2,19
151	22,3	1,45	—	—	21,9	2,00	—	—	20,1	2,65
231	23,0	1,54	22,65	1,72	22,15	2,57	21,4	2,81	20,4	3,29
320	23,0	1,45	22,95	2,08	22,25	2,67	21,45	3,36	19,8	3,66
396	23,4	1,66	22,7	2,44	22,4	3,06	21,2	3,77	19,6	3,89

$10^{-3}N$	1,86	42,6	82,3	151	231	320	396
$10^{-12}\tau$	7	76	62	85	82	75	84
ε	—	21,2	21,8	22,7	23,2	23,3	23,7
ε_∞	—	19,8	19,8	19,1	18,7	17,95	17,65

Dimethylformamide (C_3H_7NO) – Benzene (C_6H_6)
[86], $t=34$ °C, $\nu=3,2$ cm

N	0,12	0,24	0,40	0,60	0,80	1,20	1,60	2,00
ε′	2,46	2,50	2,77	2,81	3,64	3,81	4,08	5,01
tg δ	0,027	0,046	0,083	0,137	0,141	0,205	0,233	0,270

***N*–Methylacetamide (C_3H_7NO) – Benzene (C_6H_6)**
[402], $t=25$ °C, $\nu=60$ MHz, $\tau=6,0\cdot10^{-11}$

x_1	1,001	1,989	3,504	5,009
ε	2,582	3,208	4,721	7,579
ε_∞	2,270	2,268	2,266	2,262

Propan-1-ol (C_3H_8O) – Glycerine ($C_3H_8O_3$)
[441]

x_1	t °C	ε	$ε_∞$	$10^{-8}τ$	β	x_1	t °C	ε	$ε_∞$	$10^{-8}τ$	β
0	—11,0	49,4	4,3	4,07	0,64	80	—49,9	37,2	4,3	1,75	0,82
20	—15,6	43,7	4,2	1,91	0,66	85	—55,0	38,0	4,2	1,97	0,87
40	—23,4	37,0	4,1	1,90	0,69	95	—70,0	41,0	4,4	2,04	0,95
70	—42,0	35,9	4,4	2,11	0,73	100	—75,0	43,5	4,35	3,07	1,00

[482, 442, 443], $t = -60$ °C, $ν = 50$ Hz—600 kHz

x_1	ε	$ε_∞$	β	$10^{-6}τ$
30	54,8	4,0	0,55	55
50	48,4	4,5	0,56	2,9
70	40,0	4,0	0,6	0,3

Propan-1-ol (C_3H_8O) – 1, 3-Butandiol ($C_4H_{10}O$)
[441]

x_1	t °C	ε	$ε_∞$	$10^{-8}τ$	β	x_1	t °C	ε	$ε_∞$	$10^{-8}τ$	β
0	—27,5	35,4	2,8	6,20	0,74	65	—53,4	38,0	3,8	3,75	0,84
10	—30,4	38,8	3,1	6,05	0,76	75	—60,0	34,5	3,6	3,12	0,87
25	—34,8	37,2	3,0	4,22	0,77	90	—71,2	36,8	3,6	3,20	0,95
45	—42,8	36,2	3,0	3,46	0,81	100	—75,0	43,5	4,8	3,07	1,00

Propan-1-ol (C_3H_8O) – Isopentyl bromide ($C_5H_{11}Br$)
[444], $ν = 50$ Hz—2 MHz $x_1 = 20$, fig. 39

T °K	120,3	122,2	126,0	131,0	133,8	136,2	138,6
β	0,57	0,57	0,60	0,60	0,60	0,60	0,61
$10^{-6}τ$	220,0	73,0	5,4	0,86	0,27	0,10	0,041

[330, 429], $ν = 50$ Hz—5 MHz

x_1	t °K	ε	$ε_∞$	$ν_{m1}$, kHz	$ν_{m2}$, kHz	β
0	139	12,2	2,75	—	250	0,58
	147	11,6	—	—	3200	0,59
20	139	15,9	2,75	0,35	260	0,51
33	139	22,9	2,75	0,35	350	0,50
	147	9,3	—	0,15	1300	0,50
100	139	63	—	1,5	—	—

Propan–1–ol (C_3H_8O) – Isopentyl chloride ($C_5H_{11}Cl$)

[330], $t=120$ °K, $\nu=50$ Hz—5 MHz

x_1	ε	ν_{m1}, kHz	ν_{m2}, kHz	β
0	19,3	—	2700	0,55
33	34,3	0,1	900	—

Propan-1-ol (C_3H_{16}) – Heptane (C_7H_{16})

[397], $\lambda=3.2$ cm

x_1	t °C	ε	$10^{-11}\tau$
30	51	8,68	1,32
50	65	9,50	1,28
70	72	10,97	1,22

[435], $\nu=9,3477$ GHz*)

x_2	$t=2$ °C				$t=20$ °C			
	ε'	tgδ	ε	ε_∞	ε'	tgδ	ε	ε_∞
90	2,04	0,040	2,25	2,00	2,04	0,040	2,23	2,00
80	2,23	0,051	3,09	2,21	2,24	0,062	3,06	2,21
50	2,46	0,091	4,2	2,43	2,49	0,116	4,22	2,43
30	2,82	0,169	6,26	2,75	2,85	0,209	6,02	2,74
20	3,03	0,197	7,67	2,95	3,08	0,250	7,43	2,94
10	3,17	0,293	9,21	3,03	3,20	0,351	8,70	2,97

x_2	$t=30$ °C				$t=40$ °C			
	ε'	tgδ	ε	ε_∞	ε'	tgδ	ε	ε_∞
90	2,04	0,040	2,21	2,00	2,04	0,040	2,20	2,00
70	2,26	0,070	3,04	2,21	2,27	0,080	2,99	2,22
50	2,51	0,134	4,16	2,43	2,56	0,152	4,13	2,46
30	2,89	0,240	5,93	2,73	2,99	0,280	5,90	2,75
20	3,12	0,290	7,26	2,91	3,19	0,345	7,15	2,78
10	3,22	0,393	8,44	2,92	3,22	0,451	8,09	2,82

*) The dependence of relaxation time on the concentration of the solution is given by the expression

$\tau=A_0+A_1x+A_2x^2+A_3x^3+\ldots+A_nx^n$,

$A_0=0,927$; $A_1=1,668$; $A_2=0,144$; $A_3=-2,764$; $n=3$ at $t=2$° C.

Propan–2–ol (C_3H_8O) – Heptane (C_7H_{16})

[397], $\lambda=3,2$ cm

x_1	t °C	ε	$10^{-11}\tau$
25	50	8,95	1,31
50	68	10,90	1,22
70	74	14,54	1,10

[435], $\nu = 9,3477$ GHz *)

x_2	$t=2$ °C				$t=20$ °C			
	ε'	tgδ	ε	ε_∞	ε'	tgδ	ε	ε_∞
90	2,06	0,038	2,28	2,03	2,06	0,038	2,26	2,03
70	2,23	0,050	3,11	2,22	2,24	0,065	3,03	2,21
50	2,43	0,093	4,20	2,41	2,47	0,125	4,15	2,41
30	2,65	0,143	5,88	2,61	2,73	0,195	5,73	2,63
20	2,84	0,193	7,14	2,77	2,91	0,250	6,85	2,77
10	3,01	0,229	8,61	2,93	3,08	0,304	8,25	2,91

x_2	$t=30$ °C				$t=40$ °C			
	ε'	tgδ	ε	ε_∞	ε'	tgδ	ε	ε_∞
90	2,06	0,038	2,24	2,03	2,06	0,038	2,23	2,02
70	2,23	0,075	3,02	2,21	2,27	0,080	3,01	2,22
50	2,48	0,139	4,09	2,40	2,52	0,153	4,05	2,40
30	2,79	0,222	5,71	2,65	2,90	0,259	5,70	2,65
20	2,95	0,279	6,76	2,77	2,99	0,328	6,62	2,72
10	3,11	0,344	8,09	2,88	3,20	0,392	8,02	2,87

*) The dependence of relaxation time on the concentration of the solution is obtained from the expression
$\tau = A_0 + A_1x + A_2x^2 + A_3x^3 + \ldots + A_nx^n$,
$A_0 = 0,948$; $A_1 = 2,331$; $A_2 = -1,904$; $A_3 = -1,383$; $n = 3$ at $t = 2$ °C.

1, 2-Propandiol ($C_3H_8O_2$) – **Dioxan** ($C_4H_8O_2$)
[447], $t = 29$ °C, $\beta = 0,537$, $\tau = 1,2 \cdot 10^{-10}$ sec, $x_2 = 66$

ν, *Гц*	0,001	1	1,2	1,5	1,7	2,0	9,36	13,8
ε'	22,00	17,40	19,95	14,61	12,00	10,00	5,20	4,30
ε"	—	7,31	6,80	7,84	7,00	6,17	2,60	2,10

Propyl cyanide (C_4H_7N) – **Benzene** (C_6H_6)
[424], $t = 30$ °C, $\nu = 9695$ MHz

x_1	10,12	20,27	30,40	40,55	50,69
ε'	2,3774	2,4370	2,6169	2,6833	2,8343
ε"	0,05193	0,08331	0,14570	0,17561	0,21673

Tetrahydrofuran (C_4H_8O) – **Benzene** (C_6H_6)
[398], $t = 22$ °C, $\nu = 23,8$ GHz

x_1	0,0	3,0	5,0	6,0	8,0	10,0	15,0	20,0	30,0	40,0	50,0
ε'	2,28	2,50	2,63	1,67	2,77	2,78	3,06	3,23	3,58	3,76	4,15
10^{-3}ε"	2,1	22,01	128	142	170	177	255	305	320	365	378

Tetrahydrofuran (C_4H_8O) – Cyclohexane (C_6H_{12})
[398,] $t=22$ °C, $\nu=23{,}8$ GHz

x_1	0,0	3,0	5,0	6,0	10,0	15,0	20,0	30,0	40,0	50,0
ε'	2,04	2,18	2,29	2,30	2,40	2,59	2,74	3,21	3,66	4,21
$10^{-3}\varepsilon''$	0,7	20,2	37,0	67,0	80,0	135	162	242	277	305

Tetrahydrofuran (C_4H_8O) – Benzophenone ($C_{13}H_{10}O$)
[449]

x_2	t °C	$\lambda=10{,}09$ cm		$\lambda=3{,}22$ cm		$\lambda=1{,}22$ cm		$\lambda=0{,}217$ cm	
		ε'	ε''	ε'	ε''	ε'	ε''	ε'	ε''
49,2	20	7,52	3,32	4,755	1,91	4,02	1,27	2,98	0,69
	40	7,94	3,12	5,025	2,24	4,055	1,42	2,97	0,70
23,5	20	8,69	2,21	6,25	2,19	4,97	1,81	2,94	1,05
	40	8,27	1,74	6,30	2,17	5,04	1,82	3,03	1,17
10,8	20	8,37	1,14	7,05	1,76	5,83	2,11	2,715	1,49
	40	7,93	0,88	6,87	1,54	5,70	1,83	2,965	1,75

x_2	t °C	$10^{-12}\tau_1$	$10^{-12}\tau_2$	C_1	ε_∞
49,2	20	67,9	3,13	0,80	2,71
	40	45,8	2,81	0,80	2,62
23,5	20	35,0	2,97	0,615	2,60
	40	25,8	2,53	0,595	2,48
10,8	20	25,5	2,96	0,36	2,22
	40	17,5	2,13	0,365	2,10

Dioxan ($C_4H_8O_2$) – Benzene (C_6H_6)
[338], $t=22$ °C, $\nu=23{,}8$ GHz

x_1	0,0	5,0	10,0	15,0	20,0	30,0	50,0	65,0	80,0	100
ε'	2,28	2,29	2,30	2,30	2,30	2,32	2,34	2,35	2,37	2,39
$10^{-3}\varepsilon''$	2,1	9,4	12,8	15,6	17,1	18,4	20,1	20,8	21,2	21,4

Isobutyl chloride (C_4H_9Cl) – Isopentyl bromide (C_4H_9Br)
[444], $\nu=50$ Hz–2 MHz $x_1=50$

t °K	ε	t °K	ε	ε_∞	β	t °K	ε	t °K	ε	ε_∞	β
118,9	18,22	105,6	—	2,88	0,50	130,8	16,74	111,4	18,8	2,83	0,55
121,5	17,78	108,2	19,1	2,85	0,51	133,7	16,39	113,6	18,7	2,81	0,56
124,0	17,59	109,6	18,9	2,84	0,53	142,3	15,35	115,9	18,57	2,78	0,57

Pyrrolidine (C_4H_9N) – Cyclohexane (C_6H_{12})
[450], $t=25$ °C

x_1	$\nu=9,313$ Гец		$\nu=16,20$ Гец		$\nu=23,98$ Гец		$\nu=34,86$ Гец		$\nu=70,01$ Гец	
	ε'	ε''	ε'	ε''	ε'	ε''	ε'	ε''	ε'	ε''
4,546	2,104	0,0135	2,092	0,0193	2,090	0,0245	2,085	0,0306	2,054	0,0310
6,509	2,147	0,0222	2,133	0,0311	2,124	0,0354	2,119	0,0442	2,062	0,0466
11,22	2,251	0,0458	2,230	0,0673	2,208	0,0765	2,183	0,0820	2,133	0,0791
22,89	2,523	0,115	2,480	0,151	2,414	0,184	2,337	0,184	2,250	0,150

x_1	ε	ε_∞	α	$10^{-12}\tau$	$10^{-12}\tau_1$	$10^{-12}\tau_2$	C_1
4,546	2,110	2,026	0,17	2,5	6,2	1,2	0,35
6,509	2,158	2,028	0,21	2,5	7,0	1,2	0,36
11,2	2,275	2,060	0,10	3,8	8,9	2,0	0,39
22,89	2,572	2,140	0,10	5,0	10,2	3,0	0,39

Butan–1–ol ($C_4H_{10}O$) – Isopentyl bromide ($C_5H_{11}Br$)
[330], $t=147$ °K, $\nu=50$ Hz—5 MHz

x_1	ε	ν_{m1}, kHz	ν_{m2}, kHz	β
33	21	0,85	2900	0,53
100	53	0,8	220	—

Butan–1–ol ($C_4H_{10}O$) – Chlorobenzene (C_6H_5Cl)
[451], $\nu=25$—1400 MHz, fig. 43

x_1	t °C	ε	$\varepsilon_{\infty1}$	$\varepsilon_{\infty2}$	$\frac{\tau_1}{10^{12}}$	$\frac{\tau_2}{10^{12}}$	x_1 *)	t °C	ε	ε_∞	β	$\frac{\tau}{10^9}$
0	20	—	5,72	2,58	—	12,5						
	0	—	6,08	2,62	—	16,3						
	–20	—	6,44	2,68	—	22,0						
	–40	—	6,80	2,74	—	30,2						
27	20	6,80	5,50	2,60	1,64	12,0	27	–70	4,90	2,6	0,42	0,16
	0	7,52	5,70	2,65	330	15,0		–75	4,86	2,6	0,40	0,27
	–20	8,15	5,92	2,70	790	21,0		–80	4,76	2,6	0,38	0,50
	–40	8,95	6,18	2,77	2100	29,4		–85	4,76	2,6	0,38	1,20
52,6	20	9,50	5,20	2,56	360	12,5	52,1	–70	8,78	2,6	0,42	0,10
	0	11,5	5,30	2,62	750	15,0		–75	8,42	2,5	0,41	0,17
	–20	13,4	5,44	2,72	2160	21,0		–80	8,30	2,5	0,40	0,33
	–40	15,7	5,62	2,75	6340	30,0		–85	8,10	2,6	0,39	0,66

*) At – 70 °C, $\Delta F=10,0$; 9,9; 9,8 for corresponding concentrations.

x_1	t °C	ε	$ε_{∞1}$	$ε_{∞2}$	$\frac{τ_1}{10^{12}}$	$\frac{τ_2}{10^{12}}$	x_1 *)	t °C	ε	$ε_∞$	β	$\frac{τ}{10^6}$
76,9	20	13,8	4,50	2,55	515	12,5	76,9	—70	23,7	3,7	0,21	0,06
	0	16,4	4,60	2,60	1200	15,6		—75	23,7	3,5	0,27	0,10
	—20	19,0	4,68	2,65	3160	21,0		—80	22,8	3,0	0,34	0,19
	—40	22,1	4,78	2,70	9600	30,0		—85	22,7	2,9	0,38	0,40
100	20	18,1	3,38	—	630	—						
	0	21,2	3,44	—	1400	—						
	—20	24,3	3,53	—	3700	—						
	—40	27,9	3,57	—	11000	—						

*) At —70 °C, $ΔF$=10,0; 9,9; 9,8 for corresponding concentrations.

Butan–1–ol ($C_4H_{10}O$) – Heptane (C_7H_{16})

[397], λ=3,2 cm

x_1	t °C	ε	$10^{-11}τ$
30	55	8,20	1,34
50	70	8,85	1,31
70	76	9,50	1,28

[435], ν=9,3477 GHz *)

x_2	t=2 °C				t=20 °C			
	ε′	tg δ	ε	$ε_∞$	ε′	tg δ	ε	$ε_∞$
90	2,04	0,037	2,24	2,01	2,04	0,037	2,22	2,01
70	2,22	0,067	3,07	2,21	2,23	0,075	3,00	2,19
50	2,43	0,094	4,14	2,40	2,45	0,119	3,99	2,40
30	2,69	0,139	5,60	2,64	2,72	0,172	5,44	2,63
20	2,82	0,185	6,52	2,75	2,86	0,217	6,31	2,74
10	2,92	0,213	7,54	2,84	2,98	0,256	7,29	2,84

x_2	t=30 °C				t=40 °C			
	ε′	tg δ	ε	$ε_∞$	ε′	tg δ	ε	$ε_∞$
90	2,04	0,037	2,21	2,01	2,04	0,037	2,20	2,20
70	2,25	0,082	2,95	2,20	2,26	0,090	2,93	2,20
50	2,48	0,132	3,98	2,40	2,49	0,149	3,89	2,39
30	2,75	0,198	5,32	2,63	2,76	0,226	5,19	2,60
20	2,89	0,243	6,19	2,74	2,95	0,273	6,12	2,73
10	3,04	0,287	7,21	2,83	3,05	0,327	6,97	2,80

*) The dependence of the relaxation time on the concentration of the solution is given by the expression

$$τ=A_0+A_1x+A_2x^2+A_3x^3+\ldots+A_nx^n,$$

A_0=1,126; A_1=1,570; A_2=−2,622; A_3=0,051; n=3 at t=2 °C.

Isobutanol ($C_4H_{10}O$) – Benzene (C_6H_6)
[453, 420], $t=20$ °C, $\lambda=9$ cm—300 m

w_1	ε_∞	ε	$10^{-10}\tau$	w_1	ε_∞	ε	$10^{-10}\tau$
2,52	2,40	2,244	0,33	49,88	8,2	2,068	1,10
5,00	2,51	2,234	0,25	79,61	14,0	1,992	1,25
10,02	2,76	2,215	0,34	90,02	16,2	1,967	1,23
19,90	3,49	2,178	0,56	94,88	17,0	1,955	1,40
—	—	—	—	100	18,1	1,948	1,52

Isobutanol ($C_4H_{10}O$) – Heptane (C_7H_{16})
[397], $\lambda=3,2$ cm

x_1	t °C	ε	$10^{-11}\tau$
25	55	8,25	1,34
50	72	9,65	1,28
70	78	10,57	1,24

[435], $\nu=9,3477$ GHz *)

x_2	$t=2$ °C				$t=20$ °C			
	ε'	tg δ	ε	ε_∞	ε'	tg δ	ε	ε_∞
90	2,04	0,035	2,25	2,02	2,04	0,035	2,23	2,01
70	2,21	0,065	3,01	2,21	2,24	0,074	2,96	2,27
50	2,40	0,092	4,05	2,37	2,45	0,117	3,95	2,39
30	2,56	0,129	5,31	2,52	2,66	0,172	5,26	2,57
20	2,65	0,158	5,94	2,60	2,74	0,202	5,93	2,65
10	2,77	0,164	7,04	2,72	2,89	0,239	6,98	2,77

x_2	$t=30$ °C				$t=40$ °C			
	ε'	tg δ	ε	ε_∞	ε'	tg δ	ε	ε_∞
90	2,04	0,035	2,22	2,01	2,04	0,035	2,21	2,01
70	2,27	0,082	2,95	2,22	2,28	0,090	2,94	2,20
50	2,48	0,132	3,94	2,40	2,51	0,147	3,90	2,41
30	2,70	0,207	5,14	2,57	2,76	0,242	5,13	2,57
20	2,84	0,238	5,92	2,67	2,89	0,279	5,91	2,67
10	2,97	0,278	6,94	2,80	3,08	0,325	6,92	2,80

*) The dependence of relaxation time on the concentration of the solution is given by the expression

$$\tau=A_0+A_1x+A_2x^2+A_3x^3+\ldots+A_nx^n,$$

$A_0=1,127$; $A_1=1,765$; $A_2=2,827$; $A_3=0,061$; $n=3$ at $t=2$ °C.

Pyridine (C_5H_5N) – Benzene (C_6H_6)

[454]

φ	t °C	ε	λ=3,21 cm		λ=0,815 cm		$10^{-12}\tau$
			ε'	ε''	ε'	ε''	
0	10	2,33	—	—	—	—	—
	20	2,28	—	—	—	—	—
	30	2,25	—	—	—	—	—
	40	2,23	—	—	—	—	—
30,0	10	5,13	4,80	0,84	3,49	1,17	5,1
	20	4,95	4,70	0,71	3,56	1,14	4,6
	30	4,77	4,60	0,62	3,53	1,10	4,1
	40	4,63	4,50	0,54	3,55	1,08	3,7
50,1	10	7,28	6,57	1,70	3,91	2,09	5,8
	20	7,00	6,52	1,45	4,24	2,13	5,1
	30	6,77	6,38	1,26	4,36	2,02	4,6
	40	6,52	6,30	1,13	4,44	2,06	4,2
71,0	10	9,68	8,55	2,61	4,68	3,08	6,3
	20	9,25	8,44	2,24	4,89	2,98	5,6
	30	8,88	8,24	1,98	5,01	2,97	5,0
	40	8,53	8,10	1,74	5,35	3,06	4,5
90,0	10	12,4	—	—	5,64	4,80	6,8
	20	11,8	—	—	6,00	4,72	5,9
	30	11,3	—	—	6,48	4,61	5,3
	40	10,9	—	—	6,68	4,56	4,8
100	10	13,8	11,7	4,31	—	—	7,5
	20	13,2	11,6	3,74	—	—	6,6
	30	12,6	11,5	3,19	—	—	5,9
	40	12,0	11,2	2,86	—	—	5,3

Pyridine (C_5H_5N) – 2–Chloronaphthalene ($C_{10}H_7Cl$) *)

[449], $x_2=24,3$

t °C	λ=10,1 cm		λ=3,22 cm		λ=0,217 cm		λ=1,22 cm		$10^{-12}\tau_1$	$10^{-12}\tau_2$	ε
	ε'	ε''	ε'	ε''	ε'	ε''	ε'	ε''			
20	9,48	1,56	7,865	2,71	2,95	1,135	5,195	3,13	39,8	7,11	10,23
40	9,16	1,22	7,795	2,92	2,98	1,185	5,57	2,92	30,3	5,52	9,58

*) $C_1=0,225$; 0,235 and $\varepsilon_\infty=2,69$; 2,69 at $t=20$ and 40 °C respectively.

Pyridine (C_5H_5N) – Benzophenone ($C_{13}H_{10}O$) *)
[449], $x_2 = 21,1$

t °C	λ=10,9 см		λ=3,22 см		λ=1,22 см		λ=0,216 см		$\frac{\tau_1}{10^{12}}$	$\frac{\tau_2}{10^{12}}$	ε	C_1	ε_∞
	ε′	ε″	ε′	ε″	ε′	ε″	ε′	ε″					
20	11,60	3,07	8,05	3,67	5,32	3,32	2,815	1,075	38,2	6,00	13,40	0,55	2,70
40	11,27	2,50	8,34	3,34	5,66	3,37	3,065	1,34	27,9	4,51	12,55	0,535	2,63

*) $C_1 = 0,55$; 0,535 and $\varepsilon_\infty = 2,70$; 2,63 at $t = 20$ and 40 °C respectively.

Butyl cyanide (C_5H_9N) – Benzene (C_6H_6)
[424], $t = 30$ °C, $\nu = 9950$ MHz

x_1	12,53	21,50	28,50	40,03	47,50
ε′	2,3424	2,4752	2,5149	2,6602	2,7487
ε″	0,05113	0,10850	0,14000	0,19840	0,22660

Piperidine ($C_5H_{11}N$) – Cyclohexane (C_6H_{12})
[450], $t = 25$ °C

x_1	ν=9,313 GHz		ν=16,20 GHz		ν=23,98 GHz		ν=34,86 GHz		ν=70,01 GHz	
	ε′	ε″	ε′	ε″	ε′	ε″	ε′	ε″	ε′	ε″
9,163	2,138	0,0254	2,124	0,0351	2,108	0,0436	2,091	0,0468	2,064	0,0398
15,15	2,21	0,0412	2,19	0,0633	2,17	0,0731	2,13	0,0768	2,09	0,0719

x_1	ε	ε_∞	α	$10^{-12}\tau$	$10^{-12}\tau_1$	$10^{-12}\tau_2$	C_1
9,163	2,146	2,040	0,09	4,3	7,2	2,7	0,40
15,15	2,228	2,051	0,09	4,3	7,2	2,7	0,43

Pentan-1-ol ($C_5H_{12}O$) – Heptane (C_7H_{14}) *)
[435], $\nu = 9,3477$ GHz

x_2	t=2 °C				t=20 °C			
	ε′	tg δ	ε	ε_∞	ε′	tg δ	ε	ε_∞
90	2,06	0,037	2,25	2,03	2,06	0,037	2,23	2,03
70	2,21	0,053	3,03	2,19	2,23	0,060	2,94	2,20
50	2,40	0,089	3,92	2,37	2,42	0,104	3,83	2,37

x_2	$t=2$ °C				$t=20$ °C			
	ε'	tg δ	ε	$ε_∞$	ε'	tg δ	ε	$ε_∞$
30	2,58	0,127	5,10	2,54	2,63	0,154	4,97	2,56
20	2,66	0,156	5,67	2,60	2,71	0,184	5,55	2,62
10	2,77	0,165	6,52	2,71	2,85	0,205	6,33	2,75

*) The dependence of relaxation time on the concentration of the solution is given by the expression

$$τ=A_0+A_1x+A_2x^2+A_3x^3+...+A_nx^n,$$

$A_0=1,275$; $A_1=-0,336$; $A_2=2,524$; $A_3=-3,438$; $n=3$ at $t=2$ °C.

x_2	$t=30$ °C				$t=40$ °C			
	ε'	tg δ	ε	$ε_∞$	ε'	tg δ	ε	$ε_∞$
90	2,06	0,037	2,22	2,03	2,06	0,037	2,21	2,03
70	2,24	0,073	2,91	2,20	2,26	0,080	2,90	2,21
50	2,43	0,116	3,77	2,37	2,46	0,128	3,73	2,38
30	2,68	0,167	4,94	2,59	2,69	0,190	4,81	2,56
20	2,75	0,205	5,46	2,63	2,78	0,229	5,39	2,62
10	2,91	0,233	6,32	2,77	2,98	0,264	6,29	2,79

Pentan–2–ol ($C_5H_{12}O$) – Heptane (C_7H_{16}) *)

[435], $ν=9,3477$ GHz

x_2	$t=2$ °C				$t=20$ °C			
	ε'	tg δ	ε	$ε_∞$	ε'	tg δ	ε	$ε_∞$
90	2,04	0,035	2,25	2,02	2,04	0,035	2,23	2,01
70	2,20	0,045	3,00	2,19	2,23	0,055	2,97	2,20
50	2,35	0,072	3,80	2,33	2,40	0,096	3,79	2,36
30	2,50	0,104	4,92	2,47	2,57	0,138	4,83	2,51
20	2,57	0,117	5,51	2,54	2,64	0,164	5,40	2,57
10	2,64	0,152	6,15	2,59	2,75	0,208	6,08	2,65

2	$t=30$ °C				$t=40$ °C			
	ε'	tg δ	ε	$ε_∞$	ε'	tg δ	ε	$ε_∞$
90	2,04	0,035	2,21	2,01	2,04	0,035	2,20	2,01
70	2,24	0,060	2,96	2,21	2,27	0,070	2,94	2,23
50	2,45	0,106	3,79	2,40	2,49	0,111	3,78	2,43
30	2,62	0,156	4,82	2,57	2,67	0,176	4,79	2,56
20	2,71	0,185	5,39	2,61	2,79	0,213	5,38	2,65
10	2,81	0,234	6,06	2,67	2,87	0,260	6,02	2,69

*) The dependence of relaxation time on the concentration of the solution is given by the expression

$$τ=A_0+A_1x+A_2x^2+A_3x^3+...+A_nx^n,$$

$A_0=1,337$; $A_1=1,337$; $A_2=-0,453$; $A_3=-2,228$; $n=3$ at $t=2$ °C.

2, 4, 6-Tribromophenol ($C_6H_3Br_3O$) – Benzene (C_6H_6)
[457], $\nu=0,2-8$ GHz

x_2	t °C	ε	β	$10^{-12}\tau$
32,3	18,5	2,4	0,91	24

Picric acid ($C_6H_3N_3O_7$) – Benzene (C_6H_6)
[457], $\nu=0,2-8$ GHz

x_1	t °C	ε	β	$10^{-12}\tau$
6,2	6,0	2,30	0,80	54
20,3	18,3	2,36	0,89	49
20,3	35,0	2,36	1,00	40
32,0	18,0	2,39	0,84	50

1, 2-Dichlorobenzene ($C_6H_4Cl_2$) – Benzene (C_6H_6)
[422], $t=20$ °C

λ, cm	$x_1=25$		$x_1=50$		$x_1=75$		$x_2=100$	
	ε′	ε″	ε′	ε″	ε′	ε″	ε′	ε″
14990	4,245	—	6,364	—	8,205	—	10,573	—
7,5	4,27	0,66	5,88	0,93	7,20	1,150	—	—
2,9058	3,451	0,884	4,240	1,742	4,673	2,574	5,066	3,428
2,6187	3,327	0,900	4,125	1,785	4,581	2,704	4,413	3,448
2,4385	3,343	0,978	4,186	1,793	4,560	2,418	4,532	3,176
1,9440	3,097	0,882	3,590	1,630	3,800	2,204	3,953	2,708
1,7000	2,985	0,808	3,472	1,522	3,698	2,073	3,735	2,495
1,4010	2,852	0,770	3,178	1,345	3,334	1,806	3,537	2,056
1,2602	2,777	0,727	3,078	1,285	3,262	1,674	3,264	1,910
0,2174	2,274	0,191	2,425	0,294	2,533	0,359	2,576	0,435
$5,893\cdot10^{-11}$	2,290	—	2,332	—	2,367	—	2,408	—

2, 6-Dinitrophenol ($C_6H_4N_2O_5$) – Benzene (C_6H_6)
[457], $\nu=0,2-8$ GHz

x_2	t °C	ε	β	$10^{-12}\tau$
29,1	19,0	2,82	1,00	25

Bromobenzene (C_6H_5Br) – Chlorobenzene (C_6H_5Cl)

[458, 459, 460], $t = 25$ °C, fig. 44

x_2	λ, cm	ε′	ε″	ε	$10^{-12}\tau_1$	$10^{-12}\tau_2$	C_1
60,8	18,00	5,42	0,56				
	3,15	4,27	1,58				
	0,60	2,48	0,78	5,52	11	18,5	0,60
50,7	18,00	5,44	0,33				
	3,22	4,27	1,43				
	1,92	3,57	1,43				
	1,27	3,01	1,18	5,51	9,5	18	0,40
40,7	3,00	4,20	1,36				
	1,92	3,54	1,37				
	1,36	3,09	1,15	5,50	9,5	18	0,40
30,6	3,22	4,06	1,5				
	1,95	3,50	1,35				
	1,27	3,00	1,30				
	0,6	2,50	0,69	5,47	10	17	0,30

Bromobenzene (C_6H_5Br) – Benzene (C_6H_6)

[461], λ = 12,8 cm

t °C	$\Delta H=1,43$ $x_1=45,9$			$\Delta H=1,49$ $x_1=56,0$			$\Delta H=1,56$ $x_1=71,8$			$\Delta H=1,65$ $x_1=100$		
	ε′	ε″	ΔF	ε′	ε″	ΔF	ε′	ε″	ΔF	ε′	ε″	ΔF
20	3,98	0,33	2,61	4,27	0,40	2,64	4,66	0,52	2,70	5,25	0,73	2,76
10	4,02	0,37	—	4,32	0,45	—	4,74	0,58	—	5,35	0,83	—
0	4,06	0,43	2,52	4,39	0,52	2,53	4,82	0,69	2,60	5,39	0,97	2,70
–10	4,10	0,51	—	4,45	0,63	—	4,90	0,82	—	5,43	1,12	—
–20	4,13	0,60	2,44	4,50	0,75	2,47	4,94	0,98	2,52	5,38	1,31	2,61
–30	4,15	0,70	—	4,52	0,88	—	4,93	1,16	—	5,29	1,49	—
–40	—	—	—	4,54	1,04	2,38	4,90	1,32	2,44	—	—	—

[462, 463]

x_1	ε	λ = 12,8 cm		λ = 3,26 cm		ε_∞	$10^{-11}\tau$
		ε′	ε″	ε′	ε″		
0,00	2,28	2,28	—	2,28	—	2,28	1,00
17,5	3,00	2,98	0,12	2,78	0,31	3,34	1,21
36,1	3,70	3,65	0,25	3,19	0,63	2,41	1,38
56,0	4,36	4,27	0,40	3,52	0,94	2,47	1,54
77,2	4,94	4,80	0,57	3,75	1,21	2,54	1,70
100	5,44	5,25	0,73	3,96	1,43	2,60	1,81

Bromobenzene (C_6H_5Br) – Hexane (C_6H_{14})
[463, 462], $t=25$ °C

x_1	λ=12,8 *см*		λ=3,26 *см*		ε	$ε_\infty$	$10^{-11}τ$
	ε′	ε″	ε′	ε″			
0,00	1,89	—	1,89	—	1,89	1,89	—
23,8	2,59	0,07	2,48	0,24	2,60	2,03	0,90
45,5	3,30	0,18	3,02	0,52	3,33	2,17	1,08
65,2	4,00	0,31	3,42	0,81	4,05	2,32	1,24
83,4	4,64	0,48	3,74	1,12	4,75	2,46	1,54
100	5,25	0,73	3,96	1,43	5,44	2,60	1,81

λ=12,8 cm

t °C	ΔH=1,06 x_1=29,4			ΔH=1,25 x_1=55,6			ΔH=1,42 x_1=74,5			ΔH=1,65 x_1=100		
	ε′	ε″	ΔF	ε′	ε″	ΔF	ε′	ε″	ΔF	ε′	ε″	ΔF
20	2,80	0,10	2,36	3,60	0,23	2,49	4,34	0,40	2,60	5,25	0,75	2,76
10	2,84	0,11	—	3,65	0,25	—	4,42	0,45	—	5,35	0,83	—
0	2,88	0,13	2,26	3,69	0,29	2,38	4,49	0,53	2,50	5,39	0,97	2,70
—10	2,92	0,15	—	3,72	0,34	—	4,53	0,62	—	5,43	1,12	—
—20	2,95	0,17	2,16	3,75	0,39	2,31	4,52	0,71	2,43	5,38	1,31	2,61
—30	2,99	0,20	—	3,75	0,44	—	4,49	0,81	—	5,29	1,49	—
—34	3,00	0,21	—	3,75	0,48	—	4,46	0,86	—	—	—	—
—40	3,02	0,24	2,08	3,75	0,51	2,23	—	—	—	—	—	—

Chlorobenzene (C_6H_5Cl) – Nitrobenzene ($C_6H_5NO_2$)
[456, 214], $t=20$ °C

x_1=0				x_1=27			
ν, GHz	ε′	ε″	$ε_\infty$	ν, GHz	ε′	ε″	$ε_\infty$
0,1	35,4	2,05	4,07	0,001	25,5	—	3,35
0,4	34,1	5,13	—	0,2	24,8	1,51	—
2,02	25,79	14,22	—	0,5	25,4	2,27	—
2,4	23,22	15,0	—	2,02	21,06	8,32	—
3,4	21,65	15,0	—	4,02	14,52	10,10	—
4,02	16,2	15,0	—	9,75	7,22	5,71	—

x_1=52				x_1=77				x_1=100			
ν, GHz	ε′	ε″	$ε_\infty$	ν, GHz	ε′	ε″	$ε_\infty$	ν, GHz	ε′	ε″	$ε_\infty$
0,01	17,3	—	3,16	0,001	10,70	—	2,84	0,001	6,69	—	2,56
0,2	17,3	0,68	—	0,2	10,71	0,225	—	0,5	5,60	0,16	—
0,4	17,3	1,03	—	0,5	10,68	0,372	—	2,4	5,57	0,57	—
2,02	15,35	4,73	—	2,78	9,35	2,8	—	4,03	5,31	0,95	—
2,78	14,04	5,3	—	4,02	8,30	3,4	—	9,7	4,34	1,6	
4,08	12,38	6,5	—	9,7	5,2	3,43	—	34,4	3,13	1,21	

Chlorobenzene (C_6H_5Cl) – Benzene (C_6H_6)

[461], $\lambda = 12,8$ cm

t °C	$\Delta H=1,24$ $x_1=17,9$			$\Delta H=1,34$ $x_1=36,8$			$\Delta H=1,42$ $x_1=56,7$			$\Delta H=1,46$ $x_1=77,8$			$\Delta H=1,49$ $x_1=100$		
	ε'	ε''	ΔF	ε'	ε''	ΔF	ε'	ε''	ΔF	ε'	ε''	ΔF	ε'	ε''	ΔF
20	3,05	0,10	2,37	3,81	0,22	2,44	4,54	0,35	2,48	5,12	0,45	2,50	5,60	0,54	2,50
10	3,08	0,11	—	3,86	0,25	—	4,60	0,40	—	5,22	0,52	—	5,72	0,65	—
0	3,12	0,13	2,28	3,91	0,29	2,35	4,67	0,47	2,40	5,32	0,61	2,42	5,83	0,78	2,47
—10	3,15	0,15	—	3,94	0,33	—	4,73	0,54	—	5,41	0,72	—	5,95	0,90	—
—20	—	—	—	3,98	0,39	2,27	4,79	0,63	2,32	5,50	0,86	2,35	6,07	1,09	2,39
—26	—	—	—	4,02	0,42	—	4,82	0,70	—	5,54	0,95	—	6,13	1,18	—
—30	—	—	—	—	—	—	4,84	0,75	2,24	5,57	1,02	2,27	6,15	1,31	2,31
—40	—	—	—	—	—	—	4,87	0,89	—	5,61	1,21	—	6,16	1,51	—
—45	—	—	—	—	—	—	4,87	0,96	—	5,62	1,32	—	6,12	1,70	—

[463, 462], $t = 20$ °C

x_1	$\lambda=12,8$ cm		$\lambda=3,26$ cm		ε	ε_∞	$10^{-11}\tau$
	ε'	ε''	ε'	ε''			
0,0	2,28	—	2,28	—	2,28	2,28	—
17,9	3,06	0,10	2,90	0,31	3,07	2,34	0,95
36,8	3,81	0,22	3,45	0,64	3,84	2,40	1,05
56,7	4,54	0,35	3,97	0,98	4,60	2,46	1,14
77,8	5,12	0,45	4,34	1,25	5,20	2,52	1,19
100	5,60	0,54	4,63	1,47	5,70	2,58	1,25

[466], $\lambda = 8,6$ mm

x_1	20	30	40	50	60	70	80	90	100
ε'	2,43	2,49	2,55	2,61	2,66	2,72	2,79	2,85	2,92
ε''	0,27	0,40	0,52	0,61	0,68	0,76	0,85	0,92	1,05

Chlorobenzene (C_6H_5Cl) – Hexane (C_6H_{14})

[463, 462], $t = 20$ °C

x_1	$\lambda=12,8$ cm		$\lambda=3,26$ cm		ε	ε_∞	$10^{-11}\tau$
	ε'	ε''	ε'	ε''			
0,0	1,89	—	1,89	—	1,89	1,89	—
24,3	2,62	0,06	2,57	0,21	2,63	2,03	0,68
46,1	3,38	0,14	3,20	0,46	3,40	2,16	0,77
65,8	4,12	0,24	3,76	0,77	4,15	2,31	0,88
83,7	4,86	0,38	4,24	1,11	4,92	2,44	1,07
100	5,60	0,54	4,63	1,47	5,70	2,58	1,25

[461], $\lambda=12,8$ cm

t °C	$x_1=30,0$			$x_1=56,2$			$x_1=79,4$			$x_1=100$		
	ε'	ε''	ΔF	ε'	ε''	ΔF	ε'	ε''	ΔF	ε'	ε''	ΔF
20	2,84	0,08	2,20	3,74	0,18	2,28	4,75	0,34	2,40	5,60	0,54	2,50
10	2,88	0,09	—	3,80	0,20	—	4,83	0,38	—	5,72	0,65	—
0	2,92	0,10	2,09	3,86	0,23	2,19	4,91	0,45	2,31	5,83	0,78	2,47
—10	2,96	0,12	—	3,92	0,27	—	4,99	0,53	—	5,95	0,90	—
—20	3,00	0,14	2,03	3,98	0,32	2,11	5,07	0,63	2,25	6,07	1,09	2,39
—30	3,04	0,16	—	4,03	0,37	—	5,14	0,73	—	6,15	1,31	—
—40	3,07	0,19	1,95	4,08	0,44	2,04	5,20	0,88	2,17	6,16	1,51	2,31
—45	3,09	0,20	—	4,10	0,48	—	5,22	0,96	—	6,12	1,70	—
—52	3,11	0,23	—	4,12	0,53	—	5,24	1,07	—	—	—	—
—59	3,12	0,26	—	4,12	0,61	—	—	—	—	—	—	—

Chlorobenzene (C_6H_5Cl) – *p*–Xylene (C_8H_{10})

[467], $t=20$ °C, $\lambda=0,337$ mm

φ_1	0	5	25	50	75	100
ε'	2,241	2,247	2,253	2,304	2,323	2,365
10^{-3}ε''	15,4	21,7	43,8	75,8	101,3	125,6

Chlorobenzene (C_6H_5Cl) – 1–Bromonaphthalene ($C_{10}H_7Br$)

[458], $t=25$ °C, $x_1=19,8$

λ, cm	ε'	ε''	ε	ε_∞	$10^{-11}\tau_1$	$10^{-11}\tau_2$	C_1
10,0	5,28	0,81	5,458	2,58	35	12	0,2
3,22	4,25	1,30	—	—	—	—	—
1,25	3,23	1,06	—	—	—	—	—

Chlorobenzene (C_6H_5Cl) – 1–Chloronaphthalene ($C_{10}H_7Cl$)

[458], $t=25$ °C

x_2	λ=1,25 cm		λ=3,22 cm		λ=10,0 cm		ε	ε_∞	$10^{-11}\tau_1$	$10^{-11}\tau_2$	C_1
	ε'	ε''	ε'	ε''	ε'	ε''					
50	3,05	0,82	3,74	1,17	4,97	1,10	5,341	2,61	35	13	0,5
19,6	3,24	1,19	4,26	1,36	5,34	0,79	5,497	2,50	32	11	0,2

Iodobenzene (C_6H_5I) – Benzene (C_6H_6)
[462], $t=20$ °C

x_1	ε	λ=12,8 cm		λ=3,26 cm		$ε_\infty$	10^{-11} τ
		ε′	ε″	ε′	ε″		
0,0	2,28	2,28	—	2,28	—	2,28	1,12
16,7	2,77	2,75	0,09	2,60	0,21	2,34	1,46
34,7	3,31	3,25	0,21	2,87	0,43	2,45	1,82
54,5	3,78	3,67	0,35	3,04	0,60	2,55	2,13
76,2	4,23	4,05	0,50	3,19	0,74	2,64	2,43
100	4,64	4,38	0,67	3,30	0,86	2,76	2,71

Iodobenzene (C_6H_5I) – Hexane (C_6H_{14})
[462], $t=20$ °C

x_1	ε	λ=12,8 cm		λ=3,26 cm		$ε_\infty$	10^{-11} τ
		ε′	ε″	ε′	ε″		
0,0	1,89	1,89	—	1,89	—	1,89	0,90
22,7	2,40	2,39	0,06	2,29	0,16	2,05	1,17
43,9	2,95	2,92	0,15	2,67	0,35	2,23	1,39
63,7	3,49	3,41	0,28	2,98	0,54	2,40	1,85
82,4	4,04	3,90	0,43	3,13	0,71	2,58	2,24

Nitrobenzene ($C_6H_5NO_2$) – Benzene (C_6H_6)
[419], $t=20$ °C

λ=3,99 cm			λ=3,2 cm			λ=1,25 cm		
$φ_1$	ε′	ε″	$φ_1$	ε′	ε″	$φ_1$	ε′	ε″
0	2,28	—	0	2,28	—	0	2,28	—
10,0	3,72	0,912	4,98	2,87	0,50	15,0	2,97	1,01
19,8	4,82	1,88	14,97	3,97	1,49	30,0	3,42	1,88
33,1	6,40	3,51	24,9	4,87	2,58	44,9	3,79	2,64
49,7	7,78	5,64	34,6	5,59	3,71	59,6	4,12	3,42
65,0	8,69	7,70	49,6	6,50	5,40	79,8	4,48	4,13
76,8	9,23	9,30	65,0	7,07	7,18	100	4,73	4,58
100	10,15	12,36	80	7,40	8,73	—	—	—
—	—	—	100	7,45	9,51	—	—	—

$φ_1$	10	20	30	40	50	60	70	80	90	100
10^{-12}τ	11,4	10,1	8,9	8,4	8,2	7,8	7,6	7,4	7,4	7,3

[430], $t=16$ °C, $\nu=50$ MHz

x_1	20	40	60	80	100
ε'	6,71	11,70	18,17	24,06	32,88
ε''	0,135	0,270	0,436	0,607	0,759

Nitrobenzene ($C_6H_5NO_2$) – **Hexane** (C_6H_{14})
[468, 469], $\lambda=8,15$ mm

x_1	$t=20$ °C			$t=30$ °C			$t=40$ °C		
	ε'	ε''	$10^{-11}\tau$	ε'	ε''	$10^{-11}\tau$	ε'	ε''	$10^{-11}\tau$
0,00	1,896	—	—	1,878	—	—	1,871	—	—
20,27	2,38	0,53	1,21	2,37	0,55	1,07	2,36	0,58	0,97
45,86	2,92	1,27	1,85	2,94	1,36	1,58	2,96	1,25	1,38
57,50	—	—	—	3,24	1,72	2,00	3,28	1,83	1,85
70,96	3,54	2,14	2,65	3,56	2,29	2,29	3,59	2,44	2,00
75,03	3,62	3,32	2,78	3,65	2,51	2,39	3,68	2,70	2,07
85,27	3,88	2,68	3,25	3,90	2,94	2,81	3,92	3,23	2,41
100	4,22	2,82	4,90	4,27	3,15	4,15	4,32	3,55	3,47

Benzene (C_6H_6) – **2-Chloroaniline** (C_6H_6ClN)
[32], $t=24$ °C, $\nu=10$ GHz

x_2	ε'	ε''	x_2	ε'	ε''
100	2,614	0,166	20	2,369	0,315
60	2,478	0,105	0	2,305	—

Benzene (C_6H_6) – **3-Chloroaniline** (C_6H_6ClN)
[32], $t=24$ °C, $\nu=10$ GHz

x_2	ε'	ε''	$10^{-12}\tau$
60	2,648	0,105	5,02
30	2,456	0,097	—
0	2,305	—	—

Benzene (C_6H_6) – **4-Chloroaniline** (C_6H_6ClN)
[32], $t=24$ °C, $\nu=10$ GHz

x_2	ε'	ε''	$10^{-12}\tau$
60	2,705	0,312	4,95
30	2,501	0,155	—
0	2,305	—	—

Benzene (C_6H_6) – Benzyl chloride (C_7H_7Cl)
[100, 101]

x_1	λ, cm	t=20 °C		t=30 °C		t=40 °C	
		ε′	ε″	ε′	ε″	ε′	ε″
25	14990	3 ,577	—	3,500	—	3,448	—
	15,0	3,477	0,208	3,461	0,195	3,394	0,127
	7,5	3,421	0,406	3,498	0,384	3,294	0,267
	2,905	3,057	0,445	3,075	0,431	3,049	0,397
	2,618	3,022	0,469	2,999	0,434	3,003	0,422
	1,948	2,924	0,446	2,938	0,444	2,944	0,438
	1,700	2,921	0,473	2,920	0,477	2,884	0,447
	1,401	2,793	0,416	2,818	0,424	2,806	0,410
	1,260	2,753	0,392	2,771	0,392	2,762	0,387
	0,219	2,446	0,171	2,158	0,197	2,505	0,249
50	14990	4,787	—	4,664	—	4,585	—
	15,0	4,470	0,530	4,514	0,494	4,361	0,373
	7,50	4,397	0,662	4,375	0,620	4,387	0,574
	2,905	3,676	0,875	3,678	0,842	3,719	0,839
	2,618	3,565	0,870	3,574	0,844	3,615	0,840
	1,948	3,389	0,854	3,415	0,821	3,434	0,811
	1,700	3,328	0,769	3,363	0,831	3,394	0,837
	1,401	3,217	0,779	3,166	0,753	3,264	0,787
	1,260	3,127	0,732	3,164	0,736	3,179	0,739
	0,219	2,600	0,318	2,630	0,312	2,581	0,334
75	14990	5848	—	5,659	—	5,501	—
	2,905	4,173	1,231	4,216	1,218	4,217	1,711
	2,618	4,113	1,221	4,167	1,194	4,183	1,164
	1,948	3,779	1,167	3,833	1,171	3,855	1,159
	1,700	3,669	1,135	3,711	1,140	3,734	1,137
	1,401	3,474	1,059	3,506	1,059	3,554	1,066
	1,260	3,362	0,997	3,415	1,022	3,420	1,027
	0,219	2,785	0,419	2,706	0,370	2,759	0,431
100	14990	7,095	—	6,822	—	6,618	—
	2,905	4,564	1,559	4,593	1,553	4,631	1,522
	2,618	4,362	1,534	4,445	1,535	4,457	1,500
	1,948	4,076	1,497	4,104	1,639	4,192	1,431
	1,700	3,879	1,380	3,939	1,374	4,015	1,389
	1,401	3,689	1,264	3,752	1,294	3,830	1,310
	1,260	3,610	1,202	3,655	1,298	3,736	1,270
	0,219	2,817	0,507	2,826	0,477	2,815	0,470

x_1	t °C	$10^{-12}\tau_1$	$10^{-12}\tau_2$	C_1 *)	$10^{-12}\tau_1$	$10^{-12}\tau_2$	C_1 **)	ε_∞
25	20	20,9	4,3	0,57	20,3	4,3	0,57	2,40
	30	13,2	1,0	0,73	12,8	1,0	0,73	2,34
	40	12,9	0,9	0,69	12,6	0,9	0,69	2,30

x_1	t °C	$10^{-12}\tau_1$	$10^{-12}\tau_2$	C_1 *)	$10^{-12}\tau_1$	$10^{-12}\tau_2$	C_1 **)	ε_∞
50	20	23,6	4,2	0,60	22,1	4,2	0,60	2,52
	30	20,1	4,0	0,62	18,7	4,0	0,62	2,54
	40	17,4	3,2	0,62	16,3	3,2	0,62	2,47
75	20	26,4	5,6	0,59	24,8	6,0	0,55	2,72
	30	21,7	4,7	0,60	20,1	4,9	0,57	2,65
	40	17,1	3,3	0,67	15,1	3,0	0,70	2,63
100	20	29,7	4,6	0,62	26,6	4,8	0,64	2,74
	30	26,1	4,9	0,62	23,7	5,2	0,59	2,74
	40	23,7	4,3	0,66	21,5	4,3	0,60	2,71

*) Calculation made by the formula

$$\mathscr{L}_{i\omega}(-\dot{\gamma})=\varepsilon(\varepsilon^*-\varepsilon_\infty)(2\varepsilon^*+\varepsilon_\infty)/\varepsilon^*(\varepsilon-\varepsilon_\infty)(2\varepsilon+\varepsilon_\infty), \quad (1)$$

**) Calculation made by the formula

$$\mathscr{L}_{i\omega}(-\dot{\gamma})=3\varepsilon(\varepsilon^*-\varepsilon_\infty)/(\varepsilon-\varepsilon_\infty)(\varepsilon^*+2\varepsilon), \quad (2)$$

where $\mathscr{L}_{i\omega}(-\dot{\gamma})$ — is the autocorrelation function.

Benzene (C_6H_6) – **Anisole** (C_7H_8O)

[473]

x_2	λ, cm	t=10 °C		t=20 °C		t=40 °C	
		ε′	ε″	ε′	ε″	ε′	ε″
100	14990	4,488	—	4,410	—	4,259	—
	2,905	3,636	0,837	3,688	0,800	3,726	0,699
	2,619	3,528	0,822	3,563	0,784	3,655	0,709
	2,438	3,458	0,813	3,559	0,798	3,586	0,680
	1,950	3,321	0,822	3,384	0,813	3,466	0,762
	1,700	3,224	0,796	3,273	0,801	3,371	0,779
	1,401	3,079	0,719	3,145	0,740	3,232	0,745
	2,260	3,002	0,687	3,035	0,701	3,132	0,739
	0,216	2,696	0,234	2,643	0,294	2,621	0,325
84,6	14990	4,201	—	4,106	—	3,956	—
	2,905	3,499	0,719	3,527	0,681	3,556	0,593
	2,619	3,386	0,712	3,424	0,678	3,460	0,607
	2,438	3,352	0,728	3,416	0,695	3,493	0,612
	1,950	3,225	0,716	3,255	0,696	3,343	0,654
	1,700	3,135	0,701	3,179	0,705	3,289	0,688
	1,401	3,001	0,641	3,061	0,658	3,144	0,653
	0,260	2,955	0,617	2,977	0,627	3,063	0,641
	0,216	2,625	0,235	2,622	0,271	2,595	0,291
72,1	14990	3,956	—	3,881	—	3,737	—
	2,905	3,356	0,621	3,369	0,581	3,387	0,502
	2,619	3,242	0,642	3,267	0,610	3,301	0,551
	2,438	3,268	0,626	3,280	0,596	3,336	0,519
	1,950	3,137	0,632	3,175	0,620	3,228	0,570
	1,700	3,048	0,613	3,092	0,612	3,153	0,580
	1,401	2,944	0,576	2,989	0,586	3,062	0,578
	1,260	2,891	0,551	2,922	0,565	2,989	0,554
	0,216	2,600	0,213	2,533	0,234	2,506	0,250

x_2	λ, cm	t=10 °C		t=20 °C		t=40 °C	
		ε′	ε″	ε′	ε″	ε′	ε″
54,9	14990	3,585	—	3,536	—	3,406	—
	2,905	3,149	0,489	3,170	0,450	3,163	0,373
	2,619	3,098	0,500	3,114	0,468	3,111	0,397
	2,438	3,046	0,487	3,087	0,454	3,089	0,406
	1,905	2,994	0,498	3,008	0,482	3,032	0,432
	1,700	2,921	0,482	2,939	0,475	2,990	0,436
	1,401	2,823	0,461	2,850	0,462	2,896	0,447
	1,260	2,786	0,449	2,813	0,454	2,863	0,449
	0,216	2,507	0,176	2,478	0,172	2,471	0,219
38,7	14990	3,227	—	3,166	—	3,082	—
	2,905	2,938	0,342	2,944	0,313	2,894	0,262
	2,619	2,942	0,342	2,901	0,324	2,882	0,266
	2,438	2,871	0,361	2,875	0,335	2,877	0,295
	1,950	2,806	0,358	2,823	0,344	2,829	0,300
	1,700	2,772	0,349	2,798	0,341	2,808	0,309
	1,401	2,704	0,346	2,725	0,345	2,741	0,318
	1,260	2,677	0,329	2,669	0,332	2,700	0,313
	0,216	2,452	0,147	—	—	2,341	0,177
32,9	14990	3,096	—	3,048	—	2,966	—
	2,905	2,862	0,293	2,861	0,266	2,839	0,216
	2,619	2,825	0,301	2,827	0,271	2,797	0,201
	2,438	2,809	0,311	2,805	0,288	2,783	0,261
	1,950	2,734	0,307	2,742	0,292	2,749	0,254
	1,700	2,704	0,297	2,718	0,289	2,722	0,260
	1,401	2,655	0,300	2,664	0,294	2,676	0,280
	1,260	2,632	0,286	2,639	0,293	2,654	0,271
	0,216	2,442	0,136	2,390	0,133	2,371	0,148

x_2	t °C	$10^{-12}T_1$	$10^{-12}T_2$	z_1	ε_∞	$10^{-12}\tau_1$	$10^{-12}\tau_2$	C_1
100	10	15,4	1,22	0,89	2,58	13,5	1,17	0,87
	20	13,9	1,17	0,83	2,48	12,2	1,17	0,81
	40	11,6	1,43	0,79	2,45	10,3	1,43	0,76
84,6	10	14,9	1,17	0,86	2,49	13,2	1,17	0,83
	20	12,9	0,80	0,81	2,39	11,5	0,80	0,79
	40	10,0	0,85	0,80	2,38	9,0	0,85	0,78
72,1	10	14,3	0,96	0,85	2,45	12,7	0,96	0,83
	20	13,7	2,39	0,78	2,44	12,4	2,28	0,75
	40	10,9	2,44	0,73	2,41	10,0	2,34	0,69
54,9	10	13,6	1,59	0,83	2,42	12,4	1,54	0,81
	20	14,4	4,25	0,67	2,43	13,5	4,09	0,62
	40	9,9	1,49	0,75	2,35	9,1	1,49	0,72
32,9	10	12,3	0,85	0,78	2,32	11,5	0,80	0,76
	20	11,7	2,23	0,72	2,34	11,1	2,18	0,70
	40	10,7	2,44	0,60	2,31	10,2	2,34	0,57

Benzene (C_6H_6) – Bromonaphthalene ($C_{10}H_7Br$)

[458], $t = 25$ °C, $x_1 = 19{,}9$

λ, cm	ε′	ε″	ε	ε_∞	$10^{-12}\tau$
10,0	2,93	0,20	3,026	2,47	23,1
3,22	2,66	0,28	—	—	—
1,25	2,52	0,17	—	—	—

Benzene (C_6H_6) – 1-Chloronaphthalene ($C_{10}H_7Cl$)

[458], $t = 25$ °C, $x_1 = 19{,}9$

λ, cm	ε′	ε″	ε	ε_∞	$10^{-12}\tau$
10	2,95	0,20	3,02	2,44	18,2
3,22	2,69	0,29	—	—	—
1,25	2,52	0,20	—	—	—

Benzene (C_6H_6) – Benzophenone ($C_{13}H_{10}O$)

[66], ν = 9,23 GHz

t °C	N=0		N=0,25		N=0,5		N=0,75		N=1,0	
	ε′	10^{-3}tgδ	ε′	10^{-3}tgδ	ε′	10^{-3}tgδ	ε′	10^{-3}tgδ	ε′	10^{-3} tgδ
10	2,298	1,33	2,308	3,41	2,318	6,885	2,327	10,32	2,336	13,70
15	2,293	1,32	2,304	3,415	2,315	6,84	2,322	10,35	2,332	13,67
20	2,289	1,31	2,299	3,335	2,308	6,67	2,319	10,02	2,327	13,36
25	2,284	1,29	2,295	3,22	2,304	6,44	2,313	9,71	2,322	12,89
30	2,278	1,28	2,289	3,11	2,299	6,20	2,307	9,37	2,317	12,37

Phenol (C_6H_6O) – *p*-Xylene (C_8H_{10})

[128], $t = 25$ °C

x_1	ν=9,313 GHz		ν=16,20 GHz		ν=23,98 GHz		ν=34,86 GHz	
	ε′	ε″	ε′	ε″	ε′	ε″	ε′	ε″
2,21	2,30	0,020	2,30	0,028	2,27	0,020	2,26	0,015
5,10	2,35	0,051	2,33	0,053	2,30	0,043	2,30	0,037
8,0	2,40	0,092	2,37	0,085	2,32	0,065	2,30	0,049
16,74	2,51	0,194	2,46	0,156	2,38	0,122	2,35	0,094
27,51	2,63	0,334	2,55	0,264	2,47	0,199	2,42	0,143

x_1	$10^{-12}\tau_0$	$10^{-12}\tau'$ *)	α	ε	ε_∞
2,21	11,7	10,6	0	2,319	2,17
5,10	13,8	12,1	0,11	2,408	2,28
8,00	16,7	14,1	0,02	2,501	2,30
16,74	28,5	21,1	0,15	2,866	2,33
27,51	41,4	26,5	0,17	3,468	2,38

*) $\tau' = \tau_0 \frac{\eta_1}{\eta}$, where η_1 is the viscosity of the dissolved substance, η is the viscosity of the solution.

Cyclohexane (C_6H_{12}) – Cyclohexanone ($C_6H_{10}O$)
[476]

φ_2	t °C	λ=10,67 cm		λ=3,2 cm		λ=0,822 cm		λ=0,42 cm		ε
		ε′	ε″	ε′	ε″	ε′	ε″	ε′	ε″	
100	10	15,74	3,29	11,30	6,40	4,75	4,02	3,37	2,53	—
	20	15,52	2,84	11,67	5,84	4,95	4,15	3,50	2,67	16,13
	30	15,00	2,50	11,96	5,44	5,21	4,27	3,61	2,73	15,37
	40	14,50	2,19	12,06	4,76	5,46	4,46	3,74	2,95	14,79
80	10	12,55	2,38	9,50	5,0	4,38	3,18	3,28	2,14	—
	20	12,36	2,02	9,35	4,46	4,53	3,24	3,34	2,24	12,86
	30	12,19	1,84	9,60	4,16	4,75	3,38	3,40	2,33	12,24
	40	11,82	1,64	9,90	3,75	4,91	3,50	3,46	2,42	11,76
60	10	9,55	1,56	7,55	3,35	3,76	3,23	3,05	1,63	—
	20	9,44	1,38	7,63	3,10	3,92	2,31	3,08	1,71	9,60
	30	9,21	1,12	7,70	2,81	4,04	2,40	3,11	1,80	9,21
	40	8,97	1,02	7,76	2,49	4,17	2,49	3,14	1,87	8,92
40	10	7,35	0,90	5,59	1,89	3,30	1,47	2,74	1,07	—
	20	6,67	0,76	5,63	1,74	3,40	1,53	2,75	1,12	6,76
	30	6,48	0,66	5,66	1,59	3,46	1,58	2,76	1,16	6,56
	40	6,32	0,57	5,69	1,45	3,53	1,62	2,78	1,22	6,36
20	10	4,24	0,34	3,76	0,84	2,71	0,74	2,39	0,52	—
	20	4,17	0,30	3,78	0,76	2,77	0,76	2,40	0,54	4,19
	30	4,09	0,26	3,79	0,68	2,80	0,77	2,42	0,57	4,10
	40	4,00	0,23	3,80	0,60	2,85	0,79	2,45	0,60	3,99

Cyclohexane (C_6H_{12}) – Cyclohexanol ($C_6H_{12}O$)
[477], t=25 °C

w_2	ε	ε_∞	ν_m, MHz	w_2	ε	ε_∞	ν_m, MHz
100	16,8	4,3	65,5	45	4,75	2,8	161
86	15,3	3,1	75	26	3,0	2,3	200
70	9,75	2,95	110	10	2,2	2,0	—
54	6,1	2,9	141	0	2,04	—	—

Cyclohexane (C_6H_{12}) – Toluene (C_7H_8)
[478], рис. 47

w_2	t °C	ν=9,313 GHz		ν=16,20 GHz		ν=23,98 GHz		ν=34,86 GHz		ν=70,00 GHz	
		ε′	ε″	ε′	ε″	ε′	ε″	ε′	ε″	ε′	ε″
100	25	2,364	0,039	2,339	0,055	2,310	0,057	2,293	0,053	—	—
90,15	25	2,321	0,034	2,30	0,047	2,273	0,049	2,252	0,047	—	—
78,38	25	2,274	0,029	2,251	0,039	2,235	0,041	2,211	0,039	—	—

w_2	t °C	ν=9,313 GHz		ν=16,20 GHz		ν=23,98 GHz		ν=34,86 GHz		ν=70,00 GHz	
		ε′	ε″	ε′	ε″	ε′	ε″	ε′	ε″	ε′	ε″
69,32	15	2,260	0,028	2,238	0,036	2,223	0,038	2,206	0,035	2,198	0,025
	25	2,235	0,025	2,216	0,034	2,205	0,036	2,181	0,035	2,172	0,026
	40	2,216	0,021	2,196	0,029	2,185	0,035	2,172	0,034	2,155	0,029
	60	2,171	0,016	2,161	0,025	2,148	0,029	2,139	0,031	2,118	0,026
59,46	15	2,222	0,023	2,199	0,030	2,186	0,031	2,172	0,029	2,170	0,022
	25	2,199	0,021	2,179	0,027	2,171	0,030	2,155	0,029	2,153	0,022
	40	2,181	0,017	2,161	0,025	2,145	0,028	2,135	0,028	2,125	0,026
	60	2,134	0,014	2,122	0,021	2,121	0,026	2,108	0,026	2,089	0,022
49,67	25	2,163	0,017	2,148	0,024	2,139	0,025	2,124	0,023	—	—
40,78	25	2,132	0,014	2,112	0,019	2,108	0,021	2,105	0,019	—	—
29,96	25	2,099	0,010	2,082	0,014	2,078	0,015	2,076	0,014	—	—
21,41	25	2,074	0,007	2,065	0,010	2,059	0,010	2,059	0,010	—	—

w_2	t °C	ε	ε_∞	$10^{-12}\tau_1$	$10^{-12}\tau_2$	α *)
100	25	2,3795	2,26	9,5	4,3	0,03
90,15	25	2,3324	2,23	9,5	3,7	0,04
78,38	25	2,2843	2,18	9,2	2,5	0,13
69,32	15	2,2713	2,17	9,9	1,8	0,14
	25	2,2440	2,15	9,0	1,6	0,13
	40	2,2260	2,127	8,0	1,1	0,23
	60	2,1776	2,093	6,6	1,1	0,22
59,46	15	2,2312	2,15	10,2	1,6	0,14
	25	2,2070	2,13	9,0	1,6	0,16
	40	2,1870	2,103	8,0	1,1	0,24
	60	2,1420	2,070	6,6	1,1	0,25
49,67	25	2,1780	2,11	9,0	1,6	0,19
40,78	25	2,1390	2,08	9,0	1,6	0,18
29,96	25	2,1012	2,06	9,0	1,6	0,19
21,41	25	2,0767	2,05	9,0	1,6	0,17

*) C_1 of toluene is 0,55.

Cyclohexane (C_6H_{12}) – Ethylbenzene (C_8H_{10})
[478], $t = 25$ °C

ν, GHz	w_2=100		w_2=70,03		w_2=60,27	
	ε′	ε″	ε′	ε″	ε′	ε″
9,313	2,349	0,063	2,235	0,041	2,198	0,034
16,20	2,309	0,067	2,200	0,048	2,165	0,036
23,98	2,286	0,058	2,182	0,037	2,156	0,032
34,86	2,271	0,047	2,174	0,030	2,150	0,026
70,00	2,254	0,032	—	—	2,140	0,017

w_2	ε	$ε_∞$	$10^{-12}τ_1$	$10^{-12}τ_2$	α	C_1
100	2,4011	2,246	18,8	6,2	0,09	0,595
70,03	2,2584	2,160	16,4	7,6	0,02	—
60,27	2,2201	2,138	16,2	8,0	0,03	—

Cyclohexane (C_6H_{12}) – Isopropylbenzene (C_9H_{12})
[478], $t = 25$ °C

ν, GHz	$w_2 = 100$		$w_2 = 70,03$		$w_2 = 60,27$	
	ε′	ε″	ε′	ε″	ε′	ε″
9,313	2,320	0,066	2,207	0,042	2,176	0,035
16,20	2,280	0,062	2,182	0,037	2,156	0,032
23,98	2,252	0,047	2,166	0,032	2,145	0,026
34,86	2,242	0,039	2,152	0,025	2,140	0,022
70,00	2,243	0,025	—	—	2,137	0,016

w_2	ε	$ε_∞$	$10^{-12}τ_1$	$10^{-12}τ_2$	α	C_1
100	2,3761	2,229	21,5	8,4	0,06	0,667
70,03	2,2451	2,147	24,2	6,7	0,11	—
60,27	2,2058	2,130	20,4	8,1	0,07	—

Cyclohexane (C_6H_{12}) – *tert*–Butylbenzene ($C_{10}H_{14}$)
[478], $t = 25$ °C

ν, GHz	$w_2 = 100$		$w_2 = 69,31$		$w_2 = 59,27$	
	ε′	ε″	ε′	ε″	ε′	ε″
9,313	2,287	0,054	2,192	0,036	2,165	0,029
16,20	2,265	0,042	2,172	0,029	2,143	0,024
23,98	2,256	0,034	2,164	0,023	2,139	0,019
34,86	2,241	0,027	2,156	0,016	2,133	0,014
70,00	2,238	0,018	—	—	2,128	0,010

w_2	ε	$ε_∞$	$10^{-12}τ_1$	$10^{-12}τ_2$	α	C_1
100,0	2,3588	2,239	27,4	9,2	0,04	0,75
69,31	2,2306	2,154	26,0	10,4	0,06	—
59,27	2,1940	2,126	30,0	8,5	0,09	—

[480]

w_1	t °C	ν=6,700GHz ε′	ν=6,700GHz ε″	ν=9,313 GHz ε′	ν=9,313 GHz ε″	ν=16,20 GHz ε′	ν=16,20 GHz ε″	ν=23,98 GHz ε′	ν=23,98 GHz ε″	ν=35,09 GHz ε′	ν=35,09 GHz ε″
67,57	15	—	—	2,20	0,035	2,18	0,026	2,17	0,023	2,17	0,017
	50	2,17	0,031	2,16	0,030	2,15	0,030	2,13	0,025	2,13	0,021
68,00	45	2,18	0,31	2,17	0,034	2,16	0,031	2,14	0,025	2,14	0,021
68,40	37,5	2,20	0,033	2,18	0,034	2,17	0,030	2,15	0,024	2,14	0,019

w_1	t °C	$10^{-12}\tau$	$10^{-12}\tau_1$	$10^{-12}\tau_2$	C_1	α	ε	ε_∞
67,57	15	24,0	31,7	8,1	0,8	0,12	2,255	2,16
	50	12,3	17,1	7,0	0,7	0,04	2,184	2,11
68,00	45	14,7	16,6	7,5	0,8	0,02	2,201	2,13
68,40	37,5	16,6	20,9	8,1	0,8	0,05	2,215	2,14

Cyclohexane (C_6H_{12}) – Dicyclohexylamine ($C_{12}H_{23}N$)

[450], $t=25$ °C

ν, GHz	x_2=32,22 ε′	x_2=32,22 ε″	x_2=17,82 ε′	x_2=17,82 ε″	ν, GHz	x_2=32,22 ε′	x_2=32,22 ε″	x_2=17,82 ε′	x_2=17,82 ε″
9,313	2,297	0,1190	2,185	0,0662	34,86	2,177	0,0769	2,113	0,0475
16,20	2,235	0,1147	2,143	0,0653	70,01	2,150	0,0503	2,099	0,0350
23,98	2,196	0,0941	2,123	0,0590					

Hexan-1-ol ($C_6H_{14}O$) – Heptane (C_7H_{16})

[435], ν=9,3477 GHz *)

x_2	t=2 °C ε′	t=2 °C tg δ	t=2 °C ε	t=2 °C ε_∞	t=20 °C ε′	t=20 °C tg δ	t=20 °C ε	t=20 °C ε_∞
90	2,04	0,035	2,24	2,01	2,04	0,035	2,22	2,01
70	2,17	0,060	2,91	2,15	2,20	0,070	2,86	2,16
50	2,35	0,084	3,78	2,32	2,40	0,098	3,68	2,35
30	2,48	0,112	4,65	2,44	2,53	0,135	4,53	2,47
20	2,55	0,119	5,11	2,51	2,58	0,149	4,94	2,51
10	2,63	1,135	5,66	2,59	2,69	0,162	5,53	2,62

*) The dependence of relaxation time on the concentration of the solution is given by the expression

$$\tau=A_0+A_1x+A_2x^2+A_3x^3+\ldots+A_nx^n,$$

$A_0=1,484$; $A_1=-0,291$; $A_2=0,490$; $A_3=-1,164$; $n=3$ at $t=2$ °C.

x_2	t=30° C				t=40 °C			
	ε′	tg δ	ε	$ε_∞$	ε′	tg δ	ε	$ε_∞$
90	2,04	0,035	2,21	2,01	2,04	0,035	2,20	2,01
70	2,21	0,075	2,83	2,16	2,23	0,081	2,82	2,17
50	2,42	0,105	3,65	2,37	2,43	0,113	3,60	2,36
30	2,59	0,145	4,51	2,51	2,62	0,156	4,50	2,53
20	2,63	0,163	4,93	2,55	2,69	0,187	4,92	2,57
10	2,75	0,178	5,52	2,66	2,81	0,200	5,51	2,69

Di–isopropyl ether ($C_6H_{14}O$) – Dibutyl ether ($C_8H_{18}O$)

[404], t=20 °C

$φ_2$	ε′	ε″	ε	$ε_∞$	$φ_2$	ε′	ε″	ε	$ε_∞$
100	3,489	0,744	3,960	2,28	36	3,038	0,534	3,401	2,09
88	3,405	0,700	3,850	2,24	24	2,958	0,505	3,308	2,06
76	3,320	0,652	3,745	2,21	12	2,889	0,467	3,213	2,03
64	3,230	0,619	3,638	2,17	0	2,833	0,443	3,121	2,00
50	3,131	0,585	3,518	2,13	—	—	—	—	—

Heptane (C_7H_{16}) – Heptan–1–ol ($C_7H_{16}O$)

[397], λ=3,18 cm

x_2	t °C	ε	$10^{-11}τ$
75	87	4,45	1,56
50	80	3,82	1,61
30	75	3,42	1,63

Heptane (C_7H_{16}) – 2–Methylheptan–3–ol ($C_8H_{18}O$)

[483, 482], t=25 °C, fig. 49

λ, cm	x_1=3,398		x_1=10,30		x_1=20,60		x_1=34,66	
	ε′	ε″	ε′	ε″	ε′	ε″	ε′	ε″
49,88	—	—	2,058	0,0039	2,168	0,0111	2,325	0,0259
24,77	—	—	2,057	0,0063	2,166	0,0174	2,322	0,0444
9,990	1,970	0,0040	2,052	0,0137	2,150	0,0309	2,269	0,0583
3,226	1,967	0,088	2,044	0,0244	2,115	0,0412	2,218	0,0710
1,233	1,960	0,0156	2,022	0,0361	2,095	0,0565	2,169	0,0809
0,204	1,943	0,0161	1,980	0,0265	2,023	0,0395	2,077	0,0560

Heptane (C_7H_{16}) – 3-Methylheptan-3-ol ($C_8H_{18}O$)

[483], $t = 25$ °C

λ, cm	$x_1=3,497$		$x_1=10,59$		$x_1=20,74$		$x_1=34,69$	
	ε′	ε″	ε′	ε″	ε′	ε″	ε′	ε″
49,88	—	—	2,068	0,0039	2,178	0,0091	2,328	0,0156
24,77	—	—	2,069	0,0061	2,178	0,0125	2,331	0,0290
9,990	1,974	0,0043	2,065	0,0131	2,166	0,0272	2,299	0,0489
3,226	1,968	0,0103	2,045	0,0263	2,130	0,0458	2,249	0,0766
1,233	1,965	0,0137	2,028	0,0393	2,103	0,0674	2,188	0,0928
0,204	1,944	0,0147	1,980	0,0296	2,022	0,0452	2,081	0,0646

x_1	ε	ε_∞	$10^{-12}\tau_2$	$10^{-12}\tau_3$	C_1	C_2	C_3
3,497	1,9740	1,934	14	2,0	0,0	0,31	0,69
10,59	2,069	1,966	31	3,2	0,0	0,21	0,79
20,74	2,180	2,004	41	3,6	0,0	0,21	0,79
34,69	2,334	2,057	41	3,3	0,0	0,30	0,70

Heptane (C_7H_{16}) – 4-Methylheptan-3-ol ($C_8H_{18}O$)

[483], $t = 25$ °C

λ, cm	$x_1=3,621$		$x_1=10,52$		$x_1=21,19$		$x_1=35,06$	
	ε′	ε″	ε′	ε″	ε′	ε″	ε′	ε″
49,88	—	—	2,061	0,0040	2,183	0,0130	2,347	0,0236
24,77	—	—	2,066	0,0072	2,182	0,0188	2,348	0,0454
9,990	1,975	0,0046	2,059	0,0147	2,167	0,0338	2,289	0,0572
3,226	1,970	0,0091	2,045	0,0262	2,123	0,0472	2,233	0,0766
1,233	1,964	0,0146	2,025	0,0375	2,104	0,0616	1,180	0,0849
0,204	1,944	0,0137	1,981	0,0264	2,026	0,0410	2,085	0,0570

x_1	ε	ε_∞	$10^{-12}\tau_2$	$10^{-12}\tau_3$	C_1	C_2	C_3
3,621	1,976	1,936	19	2,5	0,0	0,23	0,77
10,52	2,065	1,969	30	3,2	0,0	0,27	0,73
21,19	2,185	2,011	44	3,4	0,0	0,32	0,68
35,06	2,352	2,065	59	3,6	0,0	0,35	0,65

Heptane (C_7H_{16}) – 5–Methylheptan–3–ol ($C_8H_{18}O$)
[483], $t=25$ °C

λ, cm	$x_1=9,719$		$x_1=20,31$		$x_1=32,95$	
	ε′	ε″	ε′	ε″	ε′	ε″
49,88	2,052	0,0054	2,164	0,0127	2,309	0,0343
24,77	2,048	0,0072	2,161	0,0232	—	—
9,990	2,040	0,0133	2,138	0,0337	2,242	0,0554
3,226	2,024	0,0225	2,099	0,0441	2,183	0,0706
1,254	2,008	0,0296	2,075	0,0511	2,148	0,0676
0,204	1,972	0,0266	2,020	0,0379	2,066	0,0526

$x_1=32,95$, $t=25$ °C

λ, cm	ε′	ε″	λ, cm	ε′	ε″	λ, cm	ε′	ε″
599,8	2,319	0,0024	171,3	2,311	0,0135	132,2	2,297	0,0142
299,8	2,312	0,0074	149,9	2,306	0,0143	119,9	2,311	0,0193

x_1	ε	ε_∞	$10^{-12}\tau_1$	$10^{-12}\tau_2$	$10^{-12}\tau_3$	C_1	C_2	C_3
9,719	2,048	1,956	—	41	2,9	0,0	0,24	0,76
20,31	2,168	2,004	—	54	3,3	0,0	0,34	0,66
32,95	2,315	2,040	150	26	3,5	0,23	0,28	0,49

Heptane (C_7H_{16}) – Octan–1–ol ($C_8H_{18}O$)
[397], λ=3,18 cm

x_2	t °C	ε	$10^{-11}\tau$
75	95	3,95	1,60
50	90	3,45	1,63
30	84	3,12	1,65

Heptane (C_7H_{16}) – Octan–2–ol ($C_8H_{18}O$)
[483], $t=25$ °C

λ, cm	$x_1=9,784$		$x_1=20,23$		$x_1=32,45$	
	ε′	ε″	ε′	ε″	ε′	ε″
49,88	2,050	0,0063	2,180	0,0255	—	—
24,77	2,044	0,0098	2,170	0,0354	—	—
9,990	2,032	0,0156	2,133	0,0419	2,243	0,0791
3,226	2,015	0,0227	2,089	0,0467	2,169	0,0786
1,244	2,001	0,0245	2,068	0,0513	2,141	0,0680
0,204	1,970	0,0248	2,013	0,0380	2,058	0,0484

$x_1=32,45$, $t=25$ °C

λ, cm	ε′	ε″	λ, cm	ε′	ε″	λ, cm	ε′	ε″
599,6	2,409	0,0072	171,3	2,396	0,0433	133,2	2,383	0,0477
199,9	2,387	0,0320	149,9	2,395	0,0500	119,9	2,381	0,0542

x_1	ε	ε_∞	$10^{-12}\tau_1$	$10^{-12}\tau_2$	$10^{-12}\tau_3$	C_1	C_2	C_3
9,784	2,049	1,954	—	41	2,4	0,0	0,33	0,67
20,23	2,195	1,996	—	78	3,3	0,0	0,40	0,60
32,45	2,410	2,036	210	39	3,5	0,45	0,22	0,23

Heptane (C_7H_{16}) – Octan–4–ol ($C_8H_{18}O$)
[483], $t=25$ °C

λ, cm	$x_1=9,501$		$x_1=19,93$		$x_1=32,39$	
	ε′	ε″	ε′	ε″	ε′	ε″
49,88	2,042	0,0038	2,141	0,0136	—	—
24,77	2,038	0,0068	2,138	0,0206	—	—
9,990	2,029	0,0129	2,117	0,0302	2,212	0,0549
3,226	2,011	0,0198	2,084	0,0397	2,159	0,0616
1,244	2,00	0,0265	2,067	0,0449	2,140	0,0659
0,204	1,968	0,0260	2,009	0,0356	2,050	0,0484

$x_1 = 32{,}39$, $t = 25$ °C

λ, cm	ε′	ε″	λ, cm	ε′	ε″	λ, cm	ε′	ε″
599,6	2,282	0,0024	171,3	2,272	0,0163	133,2	2,272	0,0178
299,8	2,278	0,0078	149,9	2,282	0,0189	119,9	2,270	0,0189

x_1	ε	ε_∞	$10^{-12}\tau_1$	$10^{-12}\tau_2$	$10^{-12}\tau_3$	C_1	C_2	C_3
9,501	2,040	1,948	—	36	2,5	0,0	0,24	0,76
19,93	2,148	1,997	—	58	3,2	0,0	0,33	0,67
32,39	2,282	2,031	160	26	3,5	0,20	0,28	0,52

Supplementary information on the dielectric dispersion parameters of non-aqueous solutions of organic compounds

Carbon tetrachloride (CCl_4) – Methanol (CH_4O), fig. 22.
Carbon tetrachloride (CCl_4) – Ethanol (C_2H_6O), fig. 23.
Carbon tetrachloride (CCl_4) – Propan–1–ol (C_3H_8O), [397].
Carbon tetrachloride (CCl_4) – Propan–2–ol (C_3H_8O), [397].
Carbon tetrachloride (CCl_4) – Butan–2–one (C_4H_8O), [396], ν = 250 – 510 MHz.
Carbon tetrachloride (CCl_4) – Butan–1–ol ($C_4H_{10}O$) [389, 397], fig. 23.
Carbon tetrachloride (CCl_4) – Isobutanol ($C_4H_{10}O$) [397].
Carbon tetrachloride (CCl_4) – Isopentanol ($C_5H_{12}O$), fig. 23.
Carbon tetrachloride (CCl_4) – Hexanol ($C_6H_{14}O$), fig. 24.
Carbon tetrachloride (CCl_4) – Heptanol ($C_7H_{16}O$) [397].
Carbon tetrachloride (CCl_4) – 2, 6–Dimethoxyphenol ($C_8H_{10}O_3$) [403], t = 18°C, ν = 0.1 – 8.5 GHz.
Carbon tetrachloride (CCl_4) – Octanol ($C_8H_{18}O$) [397].
Carbon tetrachloride (CCl_4) – Camphor ($C_{10}H_{16}O$) [405], ν = 1 – 100 kHz.
Carbon tetrachloride – 2, 2–Bipyridine ($C_{10}H_8N_2$) [407], t = 10, 20, 40°C, λ = 0.216; 1.26; 1.7; 1.95; 2.44; 2.62; 2.905; 14.99 cm.
Chloroform ($CHCl_3$) – Methanol (CH_4O), fig. 22.
Chloroform ($CHCl_3$) – Butan–1–ol ($C_4H_{10}O$), fig. 23.
Chloroform ($CHCl_3$) – Diethyl ether ($C_4H_{10}O$) [409].
Chloroform ($CHCl_3$) – Ketene ($C_6H_8N_2$), fig. 27, 28.
Chloroform ($CHCl_3$) – Heptane (C_7H_{16}) [412], λ = 1.3; 3.2; 10 cm.
Chloroform ($CHCl_3$) – *p*–Xylene (C_8H_{10}), fig. 29.

Urea (CH_4NO_2) – Acetic acid ($C_2H_4O_2$) [415], t = 10 –40°C, ν = 9.3 GHz
Methanol (CH_4O) – Dioxan ($C_4H_8O_2$), fig. 22, 30.
Methanol (CH_4O) – Pyridine (C_5H_5N) [598].
Methanol (CH_4O) – Benzene (C_6H_6) [420], t = 20 –60°C, λ = 30 cm, x_1 = 14.7; [389], t = 20°C, ν = 137 MHz; [421], ν = 5 GHz, fig.22.
Methanol (CH_4O) – 2–Ethylhexan–1–ol ($C_8H_{18}O$), fig. 22.
Trichloroethylene (C_2HCl_3) – Tributylammonium iodide ($C_{12}H_{28}NI$), fig. 32.
Trichloroethylene (C_2HCl_3) – Tributylammonium picrate ($C_{18}H_{30}N_4O_7$), fig. 32.
Pentachloroethane (C_2HCl_5) – Decahydronaphthalene ($C_{10}H_{18}$), [118].
Dichloroethylene ($C_2H_2Cl_2$) – Tributylammonium picrate ($C_{18}H_{30}N_4O_7$), fig. 32.
1, 1, 1–Trichloroethane ($C_2H_3Cl_3$) – Hexane (C_6H_{14}) [423].
1, 2–Dichloroethane ($C_2H_4Cl_2$) – Tributylammonium iodide ($C_{12}H_{28}NI$), fig. 32.
1,2–Dichloroethane ($C_2H_4Cl_2$) – Tributylammonium picrate ($C_{18}H_{30}N_4O_7$), fig. 32.
Ethanol (C_2H_6O) – Cyclohexanol ($C_6H_{12}O$) [432]. fig. 35–37, ν = 100 kHz.
Ethanol (C_2H_6O) – Cresol (C_7H_8O) [433, 434], ν = 900–2000 MHz.
Ethanol (C_2H_6O) – 1–Bromonaphthalene ($C_{10}H_7Br$) [434], ν = 900 – 2000 MHz.
Ethanol (C_2H_6O) – Tolyl phosphate ($C_{21}H_{21}PO_4$), fig. 38.
Acetone (C_3H_6O) – Butan–1–ol ($C_4H_{10}O$), fig. 23.
Acetone (C_3H_6O) – Cyclohexane (C_6H_{12}) [396] ν = 300–510 MHz.
Acetone (C_3H_6O) – Heptane (C_7H_{16}) [396], ν = 300–510 MHz.
Acetone (C_3H_6O) – Tributylammonium iodide ($C_{12}H_{28}NI$), fig. 32.
Propyl bromide (C_3H_7Br) – Isobutanol ($C_4H_{10}O$) [421], ν = 5 GHz.
Propane–1–ol (C_3H_8O) – Cyclohexanol ($C_6H_{12}O$) [423].
Propane–1–ol (C_3H_8O) – 2–Methypentane (C_6H_{14}), fig 40.
Propane–2–ol (C_3H_8O) – Glycerine ($C_3H_8O_3$) [445, 446], t = –19 to 50° C, ν = 0.1 – 10 MHz.
Propane–2–ol (C_3H_8O) – Dioxan ($C_4H_8O_2$), fig. 30.
Propane–2–ol (C_3H_8O) – Pyridine (C_5H_5N) [598].
Propane–2–ol (C_3H_8O) – Benzene (C_6H_6) [421], ν = 5 GHz.
Glycerin ($C_3H_8O_3$) – Butan–1–ol ($C_4H_{10}O$) [448]
Glycerin ($C_3H_8O_3$) – Isobutanol ($C_4H_{10}O$) [445, 446], t = –19 to 50°C, ν = 0.1 – 10 MHz.
Allyl isothiocyanate (C_4H_5NS) – Ethylaniline ($C_8H_{11}N$) [212].
Allyl isothiocyanate (C_4H_5NS) – Diethylaniline ($C_{10}H_{15}N$) [212].

Butan–2–one (C_4H_8O) – Benzene (C_6H_6) [369], ν = 250 – 510 MHz.

Butan–2–one (C_4H_8O) – Heptane (C_7H_{16}) [369], ν = 250 – 510 MHz.

Tetrahydrofuran (C_4H_8O) – Tributylammonium picrate ($C_{18}H_{30}N_4O_7$), fig. 32.

Dioxan ($C_4H_8O_2$) – Butan–1–ol ($C_4H_{10}O$) [114], fig. 30.

Dioxan ($C_4H_8O_2$) – Hexan–1–ol ($C_6H_{14}O$), fig. 24.

Isobutyl chloride (C_4H_9Cl) – tri(o–tolyl) phosphate ($C_{21}H_{21}O_4P$), fig. 41, 42.

Butan–1–ol ($C_4H_{10}O$) – Pyridine (C_5H_5N) [589].

Butan–1–ol ($C_4H_{10}O$) – Benzene (C_6H_6) [420, 453], t = 20°C, λ = 9 cm – 300 m.

Butan–1–ol $C_4H_{10}O$) – Octan–2–ol ($C_8H_{18}O$) [452], t = 10 to –30°C, ν = 0.1 – 10 MHz.

Pentan–3–one ($C_5H_{10}O$) – Benzene (C_6H_6) [386], ν = 250 – 490 MHz.

Isopentyl bromide ($C_5H_{11}Br$) – 2–Methylpentane (C_6H_{14}), [444, 455], ν = 50 Hz – 300 kHz.

Trichlorobenzene ($C_6H_3Cl_3$) – Decahydronaphthalene ($C_{10}H_{18}$) [118].

Chlorobenzene (C_6H_5Cl) – Iodobenzene (C_6H_5I), fig. 45.

Nitrobenzene ($C_6H_5NO_2$) – Heptane (C_7H_{16}) [470], t = 16 – 32°C, ν = 10 kHz.

Nitrobenzene ($C_6H_5NO_2$) – Octane (C_8H_{18}) [472, 602], t = 20 – 35°C, ν = 5, 20 kHz, 2 MHz.

Nitrobenzene ($C_6H_5NO_2$) – 2–Methylheptane (C_8H_{18}) [471, 603, 604].

Benzene (C_6H_6) – Hexan–1–ol ($C_6H_{14}O$), fig. 24.

Benzene (C_6H_6) – Cresol (C_7H_8O) [474].

Benzene (C_6H_6) – Heptan–1–one ($C_7H_{14}O$) [396], ν = 250 – 400 MHz.

Benzene (C_6H_6) – 1, 4–Dimethoxybenzene ($C_8H_{10}O_2$) [403], t = 21.5°C, ν = 0.1 – 8.5 MHz.

Benzene (C_6H_6) – Diphenyl ether ($C_{12}H_{10}O$), [400].

Chloroaniline (C_6H_6NCl) – Diphenyl ether ($C_{12}H_{10}O$), fig. 46.

Hexachlorocyclohexane ($C_6H_6Cl_6$) – Decahydronaphthalene ($C_{10}H_{18}$) [118].

Aniline (C_6H_7N) – Cyclohexane (C_6H_{12}) [470].

Aniline (C_6H_7N) – Diphenyl ether ($C_{12}H_{10}O$), fig. 46.

Cyclohexane (C_6H_{12}) – *o*–Xylene (C_8H_{10}) [479].

Cyclohexane (C_6H_{12}) – *m*–Xylene (C_8H_{10}) [479].

Cyclohexane (C_6H_{12}) – 1, 2, 3, 4–Tetrahydronaphthalene ($C_{10}H_{18}$) [479].

Cyclohexane (C_6H_{12}) – Methylnaphthalene ($C_{11}H_{10}$) [479].

Cyclohexane (C_6H_{12}) – 2–Methylnaphthalene ($C_{11}H_{10}$) [479].

Cyclohexanol ($C_6H_{12}O$) – Heptane (C_7H_{16}) [299].

Hexanoic acid ($C_6H_{12}O_2$) – Xylene (C_8H_{10}), [481].
Hexane (C_6H_{14}) – Hexan–1–ol ($C_6H_{14}O$), fig. 24.
Hexan–1–ol ($C_6H_{14}O$) – Heptane (C_7H_{16}), fig. 48.
Toluidine (C_7H_9N) – Diphenyl ether ($C_{12}H_{10}O$), fig. 46.
Heptane (C_7H_{16}) – Benzophenone ($C_{13}H_{10}O$) [484].
1, 4–Dimethoxybenzene ($C_8H_{10}O_2$) – Decahydronaphthalene ($C_{10}H_{18}$), [403], t = 23°C, ν = 0.1 – 8.5 GHz.
3–Methyl–2, 2–heptandiol ($C_8H_{18}O_2$) – Dibutyl phthalate ($C_{16}H_{22}O_4$), fig. 50.
2–Ethylhexan–1–ol ($C_8H_{18}O$) – tri(o–tolyl) phosphate ($C_{21}H_{21}O_4P$), fig. 51.
1, 3, 5–Trimethylbenzene (C_9H_{12}) – Durene ($C_{10}H_{14}$) [485], t = 60°C, ν = 9 – 24 GHz.
1, 3, 5–Trimethylbenzene (C_9H_{12}) – Hexamethylbenzene ($C_{12}H_{18}$) [485], t = 60°C, ν = 9 – 24 GHz.
Decahydronaphthalene ($C_{10}H_{18}$) – Diphenyl ether ($C_{12}H_{10}O$) [400].

CHAPTER III

DIELECTRIC DATA FOR BINARY SYSTEMS – AQUEOUS SOLUTIONS

§ 1. Permittivity of aqueous solutions of inorganic compounds

Water – Aluminium chloride ($AlCl_3$)
[486]

$10^{-3}w_2$	0,00	0,10	0,15	0,25	0,30	0,35	0,40	0,70	1,00
$10^{-3}N$	0,00	2,25	3,37	5,62	6,75	7,87	9,00	15,75	22,50
t °C	21,4	21,6	21,4	21,4	21,5	21,3	21,6	22,70	22,6
ε	80,0	69,8	66,2	61,3	59,1	62,4	66,7	75,2	75,3

Water – Barium chloride ($BaCl_2$)
[486]

$10^{-3}w_2$	0,00	0,10	0,20	0,25	0,30	0,35	0,40	0,70	1,00
$10^{-3}N$	0,00	0,96	1,92	2,40	2,88	3,36	3,84	6,72	9,60
t °C	20,8	20,6	20,5	21,4	21,6	24,4	21,6	21,1	21,0
ε	80,0	76,1	73,1	69,0	66,2	68,0	73,0	76,0	76,6

[487, 488], $t = 18$ °C

$10^{-3}N$	0,75	0,50	0,25
ε	79,80	80,25	80,59

[489, 490]

$10^{-3}N$	0,25	0,50	1,0
t °C	3	10	25
ε	79,1	77,8	63,4

Water – Barium nitrate ($Ba(NO_3)_2$)
[486]

$10^{-3}w_2$	0,00	0,10	0,15	0,20	0,25	0,30	0,40	0,70	1,00
$10^{-3}N$	0,00	0,76	1,15	1,53	1,91	2,29	3,06	5,36	7,65
t °C	22,5	21,8	22,4	21,9	22,4	22,0	22,5	22,8	23,4
ε	80,0	74,9	70,1	68,5	70,1	72,7	73,4	77,6	78,4

[489], $t = 25$ °C

$10^{-3}N$	1	2	4
ε	57,5	54,0	49,0

Water – Calcium Hydroxide ($Ca(OH)_2$)
[486]

$10^{-3}w_2$	0,00	0,10	0,15	0,20	0,25	0,30	0,40	0,50
$10^{-3}N$	0,00	2,70	4,05	5,40	6,75	8,10	10,80	13,50
t °C	21,6	22,0	21,8	21,8	21,0	22,3	22,9	23,0
ε	80,0	76,1	72,1	75,3	76,6	79,0	83,3	84,1

Water – Copper(I)chloride (CuCl)
[486]

$10^{-3}w_2$	0,00	0,15	0,35	0,45	0,70	1,25	1,50	1,75
$10^{-3}N$	0,00	2,33	5,20	6,69	10,41	18,59	22,31	26,02
t °C	21,7	21,5	21,7	21,8	21,6	21,8	22,0	21,9
ε	80,0	72,0	68,0	63,9	71,0	74,7	74,5	75,4

Water – Copper(II)sulphate ($CuSO_4$)
[487, 488, 490—492], $t = 18$ °C

$10^{-3}N$	0,25	0,32	0,5	0,64	0,75
ε	80,46	80,28	80,13	79,82	79,76

Water – Iron(II)chloride ($FeCl_2$)
[486]

$10^{-3}w_2$	0,00	0,10	0,20	0,25	0,35	0,40	0,45	0,70	1,0
$10^{-3}N$	0,00	1,58	3,16	3,94	5,52	6,31	7,10	11,04	15,78
t °C	20,3	20,5	20,6	21,1	20,8	21,0	21,1	21,1	21,4
ε	80,0	74,4	71,3	70,1	68,2	72,3	73,0	75,2	77,0

Water – Iron(III)chloride ($FeCl_3$)
[486]

$10^{-3}w_2$	0,00	0,15	0,20	0,25	0,35	0,40	0,45	0,70	1,0
$10^{-3}N$	0,00	2,77	3,70	4,62	6,47	7,40	8,32	12,95	18,49
t °C	22,2	22,2	22,2	22,2	22,2	22,2	22,2	22,2	22,2
ε	80,0	69,1	63,9	61,7	60,1	63,1	64,6	65,7	67,4

Water – Hydrogen chloride (HCl)
[486]

$10^{-3}w_2$	0,00	0,10	0,15	0,20	0,25	0,30	0,40	0,50
$10^{-3}N$	0,00	2,74	4,11	5,48	6,86	8,23	10,97	13,71
t °C	21,6	19,9	21,3	23,0	23,3	24,0	23,1	23,6
ε	80,0	77,1	74,4	69,6	71,4	80,6	82,2	82,9

Water – Nitric acid (HNO_3)
[486]

$10^{-3}w_2$	0,00	0,10	0,15	0,20	0,25	0,30	0,40	0,50
$10^{-3}N$	0,00	1,59	2,38	3,17	3,97	4,76	6,35	7,93
t °C	20,2	22,1	20,6	21,9	22,9	22,5	22,6	23,3
ε	80,0	76,3	69,7	74,8	80,2	80,8	81,3	81,4

Water – Hydrogen peroxide (H_2O_2)
[493], fig. 53

w_2	ε at t °C							
	−40	−30	−20	−10	0	10	20	30
99,2	—	—	—	91,4	84,9	79,0	73,6	68,6
90,0	—	106,3	98,7	91,3	84,7	78,7	73,2	69,5
96,6	—	—	—	92,4	85,9	80,1	74,7	69,9
85,2	—	—	—	—	90,4	84,3	78,8	73,6
75,4	125,3	115,9	107,7	100,1	93,5	87,3	81,3	75,9
64,9	126,5	117,0	109,0	101,7	95,0	89,5	—	—
53,5	125,0	116,9	109,6	102,6	95,9	89,9	83,9	78,6
51,8	125,0	117,0	109,9	102,3	95,3	88,9	83,3	77,7
46,5	127,7	115,4	108,0	101,2	94,7	88,6	82,6	76,8
35,3	—	—	113,3	106,7	94,6	88,8	83,1	77,5
26,5	—	—	—	98,4	93,1	87,7	82,4	—
26,2	—	—	103,5	98,2	92,8	87,3	81,9	77,6
11,4	—	—	—	95,3	90,4	85,7	81,3	76,9
7,4	—	—	—	93,7	89,4	85,0	80,7	76,3
0,00	—	—	—	—	88,0	84,1	80,4	76,7

[494], $t=0$ °C

w_2	ε	w_2	ε	w_2	ε	w_2	ε
99,45	89,2	50,23	115,0	32,0	119,0	14,0	108,5
81,27	101,6	43,25	116,2	25,8	116,0	6,9	94,0
63,80	108,8	36,3	121,1	20,8	113,5	0,0	84,4

Water – Sulphuric acid (H_2SO_4)
[486]

$10^{-3}w_2$	0,00	0,10	0,15	0,20	0,25	0,30	0,40	0,70
$10^{-3}N$	0,00	2,04	3,06	4,08	5,10	6,12	8,16	14,27
t °C	19,6	19,4	21,4	21,8	21,6	23,1	24,4	23,6
ε	80,0	77,5	71,5	68,8	75,3	81,4	84,5	84,5

Water – Orthoboric acid (H_3BO_3)
[486]

$10^{-3}w_2$	0,00	0,10	0,15	0,20	0,25	0,30	0,40	0,50
$10^{-3}N$	0,00	4,85	7,28	9,70	12,13	14,55	19,40	24,25
t °C	21,5	21,8	21,8	21,6	21,6	21,4	21,3	21,4
ε	80,0	77,0	76,4	74,6	68,6	65,3	78,8	79,9

Water – Phosphoric acid (H_3BO_4)
[486]

$10^{-3}w_2$	0,00	0,10	0,15	0,20	0,25	0,30	0,40	0,50
$10^{-3}N$	0,00	3,06	4,59	6,12	7,65	9,18	12,24	15,30
t °C	20,4	21,0	21,0	22,5	22,3	20,3	20,2	20,6
ε	80,0	77,2	75,2	68,8	74,3	78,9	80,8	81,2

Water – Potassium chloride (KCl)
[495, 487—490, 499, 496], $t=25$ °C, $\nu=80$ MHz

$10^{-3}N$	0	0,2	0,5	1,0	2,5	4,0	5,0
ε	79,00	78,19	78,07	78,28	79,01	81,57	82,57

Water – Potassium hydroxide (KOH)
[486]

$10^{-3}w_2$	0,00	0,10	0,20	0,25	0,30	0,40	0,50	0,70
$10^{-3}N$	0,00	1,78	3,56	4,45	5,35	7,13	8,91	12,47
t °C	19,2	20,9	20,3	20,6	21,1	21,6	22,2	22,4
ε	80,0	72,6	69,2	67,5	65,5	64,5	74,8	79,6

Water – Potassium sulphate (K_2SO_4)
[495], $t=25$ °C, $\nu=80$ MHz

$10^{-3}N$	0,1	0,2	0,5	1	2,5	4	5
ε	79,00	78,48	78,47	79,28	80,07	81,11	81,43

Water – Lanthanum chloride
($LaCl_3$)
[487], $t=18$ °C

$10^{-3}N$	0,50	0,33	0,17
ε	79,81	80,04	80,47

Water – Lanthanum nitrate
($La(NO_3)_3$)
[487], $t=18$ °C

$10^{-3}N$	0,50	0,33	0,17
ε	79,92	80,17	80,42

Water – Lithium chloride
(LiCl)
[487], $t=18$ °C

$10^{-3}N$	1,32	0,99	0,66
ε	80,05	80,17	80,59

Water – Magnesium chloride
($MgCl_2$)
[489], $t=25$ °C

$10^{-3}N$	1	3	4
ε	57,1	44,3	42,0

Water – Magnesium sulphate ($MgSO_4$)
[321], $t=25$ °C

$10^{-3}N$	0	0,2	0,5	1,0	2,5	4,0	5,0
ε	79,00	78,11	78,29	78,18	79,03	80,26	80,67

[487], $t=18$ °C

$10^{-3}N$	0,75	0,5	0,25
ε	79,73	80,23	80,57

Water – Sodium bromide (NaBr)
[486, 489]

$10^{-3}w_2$	0,00	0,10	0,25	0,30	0,35	0,40	0,45	0,70	1,00
$10^{-3}N$	0,00	0,97	2,43	2,91	3,40	3,89	4,37	6,80	9,72
t °C	21,0	20,8	21,5	20,8	21,5	21,1	23,0	21,5	21,2
ε	80,0	76,1	68,5	62,4	60,5	70,2	75,7	77,5	78,0

Water – Sodium chloride (NaCl)
[486, 487, 489—491]

$10^{-3}w$	$10^{-3}N$	t °C	ε	$10^{-3}w$	$10^{-3}N$	t °C	ε
0,00	0,00	20,6	80,0	1,50	25,66	21,6	83,2
0,15	2,57	21,5	78,5	1,75	29,94	21,7	83,8
0,25	4,28	21,2	76,0	2,00	34,21	22,0	84,3
0,35	5,99	21,3	73,0	2,25	38,49	22,0	83,6
0,45	7,70	21,0	72,5	2,50	42,77	22,0	84,1
0,70	11,97	21,7	80,2	2,75	47,04	21,7	84,1
1,25	21,38	22,0	83,0	3,25	55,60	22,1	84,3

Water – Sodium iodide (NaI)
[486]

$10^{-3}w_2$	0,00	0,10	0,20	0,25	0,30	0,35	0,40	0,70	1,00
$10^{-3}N$	0,00	0,67	1,33	1,67	2,00	2,33	2,67	4,67	6,67
t °C	18,3	18,3	18,7	18,6	19,8	20,0	18,5	21,4	20,0
ε	80,0	68,9	64,7	67,8	76,0	71,9	70,2	68,3	68,7

Water – Sodium nitrate ($NaNO_3$)
[486], fig. 52

$10^{-3}w_2$	0,00	0,10	0,20	0,25	0,40	0,50	0,70	1,00
$10^{-3}N$	0,00	1,18	2,35	2,94	4,71	5,88	8,23	11,76
t °C	20,8	20,8	20,9	21,2	22,0	21,6	22,4	22,4
ε	80,0	77,4	76,9	76,1	72,3	72,6	81,0	84,5

Water – Sodium hydroxide (NaOH)
[486, 489]

$10^{-3}w_2$	0,00	0,10	0,15	0,20	0,25	0,30	0,40	0,50
$10^{-3}N$	0,00	2,50	3,75	5,00	6,25	7,50	10,0	12,50
t °C	19,1	21,0	20,0	22,0	22,0	22,0	22,4	22,0
ε	80,0	76,9	75,3	80,4	81,3	82,6	84,6	86,8

Water – Lead(II)chloride ($PbCl_2$)
[486]

$10^{-3}w_2$	$10^{-3}N$	t °C	ε	$10^{-3}w_2$	$10^{-3}N$	t °C	ε
0,00	0,00	20,9	80,0	0,40	2,88	21,5	61,8
0,10	0,72	20,8	67,0	0,45	3,24	21,5	66,6
0,15	1,08	21,5	62,4	0,70	5,03	21,4	73,0
0,25	1,80	21,7	60,3	1,00	7,19	21,3	75,5
0,30	2,16	21,5	59,4	1,25	8,99	21,4	76,2
0,35	2,52	21,4	60,4	1,50	10,79	21,6	76,3

Supplementary information on the dielectric permittivity of aqueous solutions of inorganic compounds

Water – Caesium nitrate ($CsNO_3$), fig. 52.
Water – Potassium nitrate (KNO_3), fig. 52.
Water – Lithium nitrate ($LiNO_3$), fig. 2, 52.
Water – Rubidium chloride (RbCl), [489].
Water – Rubidium nitrate ($RbNO_3$), fig. 32.

§ 2. Permittivity of aqueous solutions of organic compounds

Water – Formic acid (CH_2O_2)
[157], $t=25$ °C, $\nu=15$ MHz

φ_1	0	20	40	60	80	100
ε	56,2	69,0	77	82	80	78,54

Water – Methanol (CH_4O)

$t=25$ °C [281]		$t=35$ °C [260]		$t=39,9$ °C [281] *)			
x_2	ε	x_2	ε	x_2	ε	x_2	ε
81,37	36,8	88,16	35,9	90	31,33	10	65,95
65,79	41,6	70,82	41,2	80	33,75	5	69,53
39,79	53,1	57,12	45,8	70	36,57	0	73,15
20,20	63,9	44,63	50,6	60	40,00	—	—
11,50	69,9	31,20	56,0	50	44,13	—	—
7,90	72,5	19,56	62,0	40	48,80	—	—
6,20	73,8	13,48	66,6	30	53,70	—	—
2,00	77,0	5,36	71,2	20	59,35	—	—
0,00	78,5	—	—	—	—	—	—

*) $\nu=180$ kHz

t=17 °C [497], λ=73 cm						t=18 °C [498]		t=30 °C [345]*)	
w_2	ε	w_2	ε	w_2	ε	w_2	ε	w_2	ε
100,0	33,2	65,2	50,1	31,4	66,6	100	33,78	100	31,72
94,6	36,1	60,5	52,0	26,0	69,4	84,0	40,96	80	43,72
89,9	38,2	56,0	55,1	19,5	72,3	75,9	45,31	60	53,02
84,8	40,6	51,1	57,7	15,0	74,5	48,0	57,44	40	60,48
80,1	42,8	46,8	59,1	11,0	76,4	33,1	64,47	20	69,01
78,4	44,0	45,2	60,3	5,7	79,1	17,4	71,11	0	76,65
74,6	46,0	39,7	62,5	0	81,7	0,0	78,73	—	—
69,8	47,6	35,0	65,0	—	—	—	—	—	—

w_2	ε at t °C [500] **)					ln a	b
	20	30	40	50	60		
100	32,35	30,68	29,03	27,44	25,97	1,5099	0,00234
90	36,80	34,62	32,56	30,67	28,91	1,5648	0,00242
80	41,46	38,98	36,66	34,62	32,74	1,6160	0,00248
70	46,46	43,63	41,04	38,81	36,68	1,6658	0,00252
60	51,53	48,58	45,64	43,22	41,22	1,7120	0,00244
50	56,53	53,47	50,40	47,82	45,28	1,7513	0,00234
40	61,24	58,06	54,82	52,17	49,52	1,7865	0,00225
30	66,01	62,71	59,53	56,59	53,94	1,8190	0,00218
20	71,02	67,48	64,13	61,06	58,24	1,8505	0,00212
10	75,84	72,37	68,90	65,66	62,77	1,8799	0,00208
0	80,37	76,73	73,12	69,85	66,62	1,9051	0,00205

*) In [345] there are values of ε depending on the pressure.

**) The temperature dependence is

[501], ν=0,1−3 MHz

w_2	ε at t °C					
	5	15	25	35	45	55
100	36,88	34,70	32,66	30,74	28,92	27,21
95	39,38	37,61	35,38	33,28	31,29	29,43
90	42,90	40,33	37,91	35,65	33,53	31,53
80	48,01	45,24	52,60	40,08	37,70	35,46
70	52,96	49,97	47,11	44,42	41,83	39,38
60	57,92	54,71	51,67	48,76	46,02	43,42
50	62,96	59,54	56,28	53,21	50,29	47,53
40	67,91	64,31	60,94	57,72	54,62	51,69
30	72,80	69,05	65,55	62,20	58,97	55,92
20	77,38	73,59	69,99	66,52	63,24	60,06
10	81,68	77,83	74,18	70,68	67,32	64,08
0	86,10	82,19	78,48	74,94	71,50	68,13

w_2	ε at t °C [499] *)				t=25 °C [249]			
	20	20	25	25	w_2	ε	w_2	ε
100,00	33,580	—	32,610	—	99,7	34,05	45	59,41
94,89	36,484	36,492	35,404	35,419	95	34,05	40	60,22
90,71	30,649	38,677	37,493	37,520	90	36,30	37	61,50
79,61	44,049	44,056	42,745	42,760	85	38,05	35	64,14
69,82	48,584	48,586	47,200	47,211	80	40,98	30,8	67,94
59,83	53,247	53,270	51,760	51,815	78	44,62	25	67,94
49,96	57,861	57,856	56,283	56,287	75	44,11	20	69,20
39,86	62,681	62,721	61,025	61,071	70	45,16	15	70,73
29,96	67,300	67,332	65,581	63,618	65	46,76	10	72,30
19,87	71,783	71,880	69,994	70,091	60	50,00	5	73,24
9,99	76,009	76,142	74,230	74,329	55	52,51	0	75,50
4,95	78,146	78,215	76,313	76,401	50	57,14	—	—
0	80,290	80,320	78,480	78,500	47,5	60,22	—	—

*) Two sets of measurements for the same temperature.

[503], t = 25 °C

w_2	ε	w_2	ε	w_2	ε	w_2	ε
99,166	33,55	75,045	44,06	50,501	57,29	30,018	68,05
95,097	35,04	74,906	44,56	50,061	56,86	25,018	69,49
93,489	36,63	70,020	45,93	49,986	56,85	20,011	72,14
89,907	39,10	70,000	45,78	45,149	60,11	15,016	73,81
84,991	39,29	64,966	48,79	44,967	60,58	10,014	76,63
80,013	41,45	60,041	51,23	40,037	63,30	4,999	78,53
79,734	42,01	54,954	54,03	33,893	66,53	0,000	81,12

φ_2	ε at t °C [502]							
	20	10	0	−10	−20	−30	−40	−50
100	33,6	35,4	37,9	40,6	42,7	45,4	48,3	51,3
80	43,7	46,4	49,5	52,3	55,4	58,6	61,9	65,7
70	46,3	49,4	53,0	56,6	60,1	63,5	66,8	70,9
60	55,1	58,7	62,5	66,0	70,5	73,8	78,2	82,2
50	60,3	64,0	67,8	71,2	75,5	79,2	83,9	87,9
40	63,8	67,7	71,9	75,6	79,5	83,5	87,5	—
0	80,4	84,2	88,1	—	—	—	—	—

φ_2	ε at t °C					a	$10^{-3}b$ *)
	−60	−70	−80	−90	−100		
100	54,2	58,0	62,0	66,5	—	1,580	2,65
80	69,3	73,5	77,8	82,8	88,4	1,695	2,50
70	74,9	79,2	83,5	88,7	94,0	1,730	2,45
60	86,7	92,1	97,5	103,8	—	1,790	2,50
50	92,5	—	—	—	—	1,830	2,30
40	—	—	—	—	—	1,855	2,20
0	—	—	—	—	—	1,945	2,00

*) Permittivity calculated by the formula $\lg \varepsilon = a - bt°$.

[175], $t = 20$ °C, $\nu = 1,8$ MHz

φ_1	0	20	40	60	80	100
ε	33,60	44,90	55,20	64,70	73,00	80,37

Water – Urea (CH_4N_2O)

[498, 504], $t = 18$ °C, $\nu = 1$ MHz

x_2	3,0	2,5	2,0	1,5	1,0	0,5
ε	86,17	85,16	83,98	82,81	81,51	80,22

[505, 490], $t = 20$ °C

x_2	37,5	33,3	23,1	16,6	9,1	2,91	0,0
ε	90,9	90,5	88,6	87,5	85,7	82,2	80,5

[506], $t = 25°$ C, $\lambda = 3,6—20$ m*)

w_2	N	ε	b *)	$10^{-5}c$	w_2	N	ε	b	$10^{-5}c$
0	0,	78,54	0,362	70	29,64	5,325	91,76	0,391	85
11,52	1,973	83,90	0,376	75	36,83	6,743	94,43	0,394	89
20,31	3,559	87,95	0,385	81	42,47	7,892	96,58	0,397	92

*) Temperature dependence is given by $\varepsilon_t = \varepsilon - b\,(t - 25°) + c\,(t - 25°)^2$.

Water – Acetonitrile (C_2H_3N)

t=25 °C [247]				t=25 °C [507]					
x_1	ε	x_1	ε	φ_2	x_2	ε	φ_2	x_2	ε
0,00	35,95	75,69	59,25	100	100	36,30	30	15,8	65,5
0,75	36,03	82,43	64,01	90	79,8	39,4	20	9,9	70,7
4,62	36,62	91,66	71,85	80	63,8	42,9	10	4,6	75,1
13,31	38,10	96,57	75,87	70	50,6	46,8	—	—	—
23,50	40,05	97,97	76,91	60	39,7	51,0	—	—	—
37,09	43,18	99,76	78,15	50	30,5	55,5	—	—	—
57,67	49,89	100,0	78,35	40	22,6	60,3	—	—	—

[175], $t=20\,°C$, $\nu=1{,}8$ MHz

φ_1	0	20	40	60	80	100
ε	36,80	45,55	54,60	64,75	73,20	80,37

Water – Acetic acid ($C_2H_4O_2$)

[249], $t=15\,°C$, fig. 55

w_2	ε	w_2	ε	w_2	ε	w_2	ε
100	10,30	74	36,62	60	60,80	30	64,20
80	22,90	70	39,48	50	60,80	0	75,50
77	36,87	62,5	61,25	40	61,79	—	—

[157], $t=25\,°C$, $\nu=15$ MHz

φ_1	0	20	40	60	80	90,0	100,0
ε	6,20	33,1	47,8	59,3	68,3	71,5	78,54

Water – Glycine ($C_2H_5NO_2$)

[508], $t=25\,°C$, $\lambda=3$ m

N	0,0	0,8750	1,750	2,000	2,500
ε	78,54	98,03	117,5	123,5	134,5

[509], $t=18\,°C$

N	0	0,25	0,50	1	1,5	2	2,5
ε	81,0	86,84	92,39	104,1	115,1	126,5	138,3

Water – Ethanol (C_2H_6O)

x_2	ε at t °C [281, 505, 511, 513], ν=180 kHz					t=25 °C [194] *)	
	20	25	40	55	75	x_2	ε
90	26,66	25,53	23,60	21,41	19,20	100	24,69
80	28,60	27,70	25,25	23,00	20,61	88,10	26,93
70	31,13	30,23	27,45	24,99	22,28	78,00	28,84
60	33,23	33,35	30,20	27,46	24,53	69,34	30,80
50	38,20	37,15	34,00	30,90	27,50	55,21	34,94
40	43,10	41,75	38,60	35,00	31,25	48,06	37,75
30	49,30	47,85	44,33	40,21	36,10	31,63	46,55
20	57,30	55,70	51,61	47,50	42,90	20,16	56,25
10	67,35	65,50	61,27	56,64	51,30	9,33	67,55
0	80,37	78,48	73,12	68,13	61,18	5,16	72,35
						0,0	79,45

x_2	ε at t °C [514]			x_2	ε at t °C [514]		
	25	35	45		25	35	45
100	24,44	22,26	18,73	21,1	54,85	52,16	48,66
82,6	28,05	25,63	22,14	19,2	56,53	53,75	49,91
70,0	30,52	28,45	25,33	17,5	58,22	55,33	51,06
59,8	33,11	31,24	28,09	16,3	59,55	56,61	52,37
39,3	41,96	39,41	36,39	7,0	69,34	66,00	61,63
29,8	47,83	44,76	41,53	0,0	78,54	75,04	71,70
22,4	53,83	50,70	47,51				

w_2	ε at t °C [500] **)					ln a	b
	20	40	50	60	80		
100	25,00	22,20	20,87	19,55	—	1,3979	0,00264
90	29,06	25,64	24,08	22,51	19,18	1,4625	0,00268
80	33,89	29,83	28,10	26,31	23,20	1,5300	0,00272
70	39,14	34,88	32,86	30,87	27,30	1,5926	0,00262
60	44,67	40,02	37,72	35,66	31,82	1,6500	0,00250
50	50,38	45,30	42,92	40,66	36,51	1,7024	0,00240
40	56,49	51,08	48,36	45,80	40,93	1,7520	0,00230
30	62,63	56,73	53,79	51,04	45,88	1,7968	0,00221
20	68,66	62,41	59,22	56,40	50,81	1,8367	0,00214
10	74,60	67,86	64,53	61,49	55,70	1,8727	0,00209
0	80,37	73,12	69,85	66,62	60,58	1,9051	0,00205

*) λ=450 m

**) Temperature dependence $\varepsilon = ae^{-b/t}$.

[511], $t = 25$ °C

w_2	41,076	34,864	25,299	14,497	8,378	0
ε	53,69	57,85	64,58	70,66	75,56	81,12

[510], $t = 20$ °C

w_2	99,8	98,7	93,0	81,1	57,1
ε	25,6	26,6	28,4	34,3	44,2

[515], ν=215—400 MHz *)

φ_2	w_2	ε at t°C						
		−5	0	10	20	25	30	40
100	100,0	29,17	28,32	26,68	25,07	24,28	23,50	21,97
90	85,7	36,26	34,93	32,86	30,89	29,95	29,01	27,24
80	73,5	42,92	41,62	39,13	36,81	35,72	34,64	32,62
70	62,0	50,00	48,49	45,64	42,99	41,76	40,53	38,27
60	52,1	57,01	55,30	52,03	49,06	47,70	46,33	43,87
50	42,4	63,83	61,92	58,30	54,97	53,44	51,91	49,13
40	33,3	70,38	68,33	64,47	60,90	59,16	57,62	54,64
30	24,6	76,05	73,93	69,91	66,17	64,45	62,73	59,57
20	16,2	81,22	79,08	75,02	71,23	69,47	67,72	64,42
10	8,0	85,56	83,46	79,45	75,67	73,89	72,15	68,82
0	0	90,04	88,03	84,13	80,37	78,54	76,75	73,27

*) Temperature dependence for water, ethanol and water-ethanol solutions is given by the formula

$$\varepsilon = a - b\,(t-20) + c\,(t-20)^2.$$

φ_2	w_2	a	b	10^4c	φ_2	w_2	a	b	10^4c
100	100	25,07	0,159	2	39,9	33,2	60,96	0,342	15
89,6	85,3	31,13	0,192	5	29,7	24,4	66,34	0,359	15
79,7	73,2	36,98	0,225	8	20,0	16,2	71,23	0,365	14
70,6	62,5	42,65	0,256	10	10,0	8,0	75,67	0,366	12
60,2	52,3	48,93	0,286	13	0	0	80,37	0,369	7
50,0	42,4	54,97	0,320	14					

[249] $t=15$ °C

w_2	ε	w_2	ε	w_2	ε	w_2	ε	w_2	ε	w_2	ε
99,8	25,02	80	28,15	65	34,60	46	48,40	30	59,55	10	67,95
95	25,27	75	30,30	60	36,31	40	48,40	25	60,21	5	72,30
90	25,71	72	33,86	55	39,93	35	50,52	20	61,79	0	75,50
85	26,58	70	33,66	50	44,11	32	55,20	15	65,36	—	—

$t=20$ °C

[175], ν=1,8 MHz				[512]			
φ_1	ε	φ_1	ε	φ_1	ε	φ_1	ε
0	25,07	60	60,00	0	26,0	80	68,6
20	35,80	80	70,90	20	36,2	90	74,2
40	47,30	100	80,37	40	46,4	100	80,5
—	—	—	—	60	58,0	—	—

Water – Dimethyl sulphoxide (C_2H_6OS)

$t=25$ °C

[516], ν=2,5 MHz						[507]			
x_2	w_2	ε	x_2	w_2	ε	φ_2	ε	φ_2	ε
100,0	100,0	46,4	47,60	79,75	64,8	100	46,0	40	76,0
91,18	97,82	49,4	31,77	66,88	70,9	90	57,4	30	76,8
81,54	95,04	52,8	26,49	60,97	73,0	80	64,7	20	77,4
77,57	93,75	54,2	16,23	45,65	75,8	70	69,7	10	78,0
71,77	91,68	56,1	5,43	19,93	77,9	60	72,9	—	—
62,12	87,67	59,6	0,00	0,00	78,54	50	74,9	—	—

Water – Glycol ($C_2H_6O_2$)

[500]

w_2	ε at t° C **)						ln a	b
	20	25 *)	40	60	80	100		
100	38,66	37,7	34,94	31,58	28,45	25,61	1,5872	0,00224
90	44,91	43,7	40,43	36,35	32,58	29,27	1,6523	0,00231
80	50,64	49,3	45,45	40,72	36,36	32,52	1,7045	0,00238

*) Values from paper [517].

**) Temperature dependence $\varepsilon=ae^{-b/t}$.

w_2	ε at t °C **) 20	25 *)	40	60	80	100	ln *a*	*b*
70	56,30	54,7	50,17	44,98	40,19	35,94	1,7505	0,00242
60	61,08	59,4	54,53	48,75	43,68	39,13	1,7859	0,00240
50	64,92	63,2	58,25	52,30	46,75	—	1,8124	0,00237
40	68,40	66,6	61,56	55,48	49,81	—	1,8351	0,00230
30	71,59	69,8	64,51	58,37	52,59	—	1,8548	0,00222
20	74,60	72,8	67,52	61,20	55,36	—	1,8727	0,00215
10	77,49	75,6	70,29	63,92	58,02	—	1,8893	0,00210
0	80,37	—	73,12	66,62	60,58	55,10	1,9051	0,00205

φ_2	ε at t °C [518] 15	25	35	φ_2	ε at t °C [518] 15	25	35
100	43,1	40,8	38,8	40	70,7	67,3	64,0
90	49,2	46,6	44,2	30	74,0	70,3	67,0
80	54,7	52,0	49,2	20	76,7	73,2	69,7
70	59,3	56,3	53,4	10	79,4	75,7	72,2
60	63,5	60,2	57,3	0	81,9	78,3	74,8
50	67,3	63,9	61,1	—	—	—	—

[502], $\nu = 300$ MHz $\varphi_2 = 50$

t °C	—60	—50	—40	—30	—20	—10	0	10	20
ε	98,8	94,0	89,3	85,0	80,7	76,5	72,4	68,4	64,5

Water – Acetone (C_3H_6O)

t=25 °C [521] *) w_2	ε	w_2	ε	t=25° C [288, 226] w_2	ε	t=20 °C [175] φ_2	ε	t=15 °C [520] φ_2	ε
100,0	20,74	19,96	67,62	100	20,5	100	21,07	100	19,8
91,23	25,41	10,01	73,08	94,9	23,5	80	34,55	60	33,4
80,03	31,50	0,00	78,48	89,9	26,2	60	47,60	30	54,1
70,07	37,33	—	—	80,2	31,3	40	59,90	15	69,5
60,06	43,39	—	—	66,9	38,8	20	70,40	0	81,1
50,24	49,34	—	—	50,0	50,6	0	80,37	—	—
39,93	55,74	—	—	25,0	67,0	—	—	—	—
29,98	61,90	—	—	0	80,9	—	—	—	—

*) Measurements made in [521] for ν = 0,57 MHz, in [288] for λ = 73 cm, in [175] for ν = 1,8 MHz and in ν = 1 MHz.

[500, 194, 517, 519], ν=2 MHz

w_2	ε at t °C				
	20	25	30	40	50
100	19,56	19,10	18,67	17,80	16,98
90	24,61	23,96	23,38	22,32	21,16
80	30,33	29,62	28,74	27,50	26,20
70	36,51	35,70	34,63	33,03	31,44
60	42,93	41,80	40,75	38,86	37,04
50	49,52	48,22	49,99	44,81	42,81
40	56,00	54,60	53,23	50,82	48,52
30	62,48	61,04	59,47	56,77	54,17
20	68,58	66,98	65,34	62,28	59,45
10	74,84	73,02	71,37	68,07	65,01
0	80,37	78,34	76,73	73,12	69,85

Water – Propionic acid ($C_3H_6O_2$)

[226], t=17 °C, λ=73 cm, fig. 55

w_2	ε	w_2	ε	w_2	ε	w_2	ε
100,0	3,15	92,9	8,55	74,4	24,0	50,3	46,3
99,0	3,80	89,9	10,92	68,5	29,6	0	81,7
97,5	5,01	86,0	14,22	62,5	35,2	—	—
96,1	6,13	80,0	18,9	56,3	41,5	—	—

[157], t=25 ° C

φ_2	100	79,01	70	59,6	39,12	20	0
ε	3,20	21,4	29,5	38,2	55,3	68,2	78,54

Water – Dimethylformamide (C_3H_7NO)

t=25 °C [522]				t=25 °C [507]					
w_2	ε	w_2	ε	φ_2	x_2	ε	φ_2	x_2	ε
95,00	40,69	39,84	68,92	100	100	37,0	50	19,8	65,2
90,14	43,75	29,90	72,00	90	68,9	43,6	40	14,1	69,0
77,42	51,04	19,98	74,65	80	49,6	49,7	30	9,5	72,2
67,14	56,73	9,91	76,78	70	36,5	55,5	20	5,8	74,9
60,00	60,48	0	78,54	60	27,0	60,6	10	2,7	76,9
50,01	64,96	—	—	—	—	—	—	—	—

Water – Propan-1-ol (C_3H_8O)

t=20 °C [175]		t=25 °C [281]				t=35 °C [298] *)			
φ_1	ε	x_1	ε	x_1	ε	x_1	ε	x_1	ε
0	20,65	5	20,4	90	59,7	0	18,90	79,28	43,30
20	28,80	10	20,8	95	68,28	17,50	20,95	86,08	51,38
40	40,60	20	21,8	100	78,48	41,61	25,00	89,56	58,50
60	54,50	40	25,0	—	—	57,27	29,71	94,00	64,37
80	68,70	60	31,15	—	—	66,17	34,00	97,57	69,75
100	80,37	80	46,2	—	—	73,71	38,75	100	74,94

w_2	ε at t °C [500, 281] **)					ln a	b
	20	40	50	60	80		
100	20,81	18,25	17,11	15,88	13,86	1,3183	0,00293
90	23,34	20,67	19,37	18,07	15,81	1,3710	0,00282
80	26,83	23,89	22,39	20,95	18,28	1,4300	0,00271
70	31,56	28,20	26,42	24,92	21,84	1,5000	0,00261
60	37,51	33,54	31,49	29,71	26,22	1,5740	0,00253
50	44,29	39,70	36,38	35,39	31,42	1,6455	0,00243
40	51,68	46,55	44,08	41,76	37,53	1,7133	0,00234
30	59,21	53,46	50,72	48,19	43,00	1,7724	0,00225
20	66,54	60,24	57,23	54,49	49,01	1,8231	0,00216
10	73,52	66,81	63,66	60,65	54,77	1,8664	0,00210
0	80,37	73,12	69,85	66,62	60,58	1,9051	0,00205

*) Measurements at t = 20 and 35 °C were made for ν = 1,8 MHz.

**) Temperature dependence $\varepsilon = ae^{-b/t}$.

[503], t=25 °C

w_2	ε	w_2	ε	w_2	ε
99,870	18,93	64,989	28,51	40,010	40,80
96,862	19,01	62,507	29,35	35,744	44,19
94,989	19,43	62,499	29,27	32,269	47,13
89,959	20,03	60,002	30,81	30,018	50,67
86,964	20,60	57,494	32,89	25,015	53,05
84,774	21,29	55,020	33,08	20,027	59,25
80,000	22,20	54,992	33,68	15,010	64,81
76,904	22,92	52,664	34,90	15,009	69,48
75,011	24,09	50,001	34,86	10,005	73,67
70,011	26,21	49,904	35,86	9,999	73,24
69,985	25,72	45,452	36,72	5,014	77,64
69,966	26,10	45,029	39,69	0,000	81,12

[249], $t = 15$ °C

w_2	ε	w_2	ε	w_2	ε	w_2	ε
99,8	20,45	77	36,00	55	42,00	20	62,80
95	20,73	70	36,87	50	44,63	10	69,50
90	21,60	65	37,62	35	53,35	0	75,50
80	28,92	60	38,75	25	60,08	—	—

[523], $\nu = 1$ MHz

w_2	ε at t °C			w_2	ε at t °C		
	15	25	35		15	25	35
100	21,92	20,40	19,22	40	52,45	49,70	47,02
90	24,50	22,86	21,34	20	68,39	64,89	62,14
80	28,19	26,41	24,70	0	81,95	78,30	74,62
60	38,60	36,61	34,49	—	—	—	—

Water – Propan–2–ol (C_3H_8O)

w_2	ε at t °C [500] *)					ln a	b
	20	40	50	60	80		
100	18,62	16,23	15,06	14,03	11,91	1,2700	—0,00311
90	20,95	18,48	17,11	16,02	13,83	1,3225	—0,00294
80	24,44	21,63	20,26	19,03	16,70	1,3894	—0,00277
70	29,57	26,30	24,85	23,34	20,67	1,4709	—0,00262
60	36,28	32,45	30,67	28,90	25,67	1,5597	—0,00249
50	43,68	39,16	37,03	35,05	31,49	1,6403	—0,00238
40	51,07	45,86	43,54	41,35	37,31	1,7082	—0,00228
30	58,40	52,71	50,18	47,58	43,13	1,7664	—0,00220
20	65,72	59,56	56,61	53,87	49,01	1,8177	—0,00213
10	73,11	66,33	63,12	60,24	54,83	1,8640	—0,00208
0	80,37	73,12	69,85	66,62	60,58	1,9051	—0,00205

*) Temperature dependence $\varepsilon = ae^{-b/t}$.

[197], $t = 35$ °C, $\nu = 1,8$ MHz

x_2	ε	x_2	ε	x_2	ε	x_2	ε
100	72,94	90,44	58,22	73,81	38,62	32,03	20,70
97,7	71,00	84,22	49,72	64,74	32,47	11,83	18,35
94,86	66,25	80,86	45,98	51,71	26,51	0	17,50

Water – Glycerine ($C_3H_8O_3$)

w_2	ε at *t* °C [500] *)					ln *a*	*b*
	20	40	60	80	100		
100	41,14	37,30	33,82	40,63	27,88	1,6142	0,00212
90	46,98	42,26	38,19	34,37	31,34	1,6688	0,00214
80	52,27	46,92	42,32	38,30	34,70	1,7154	0,00217
70	57,06	51,41	46,33	41,90	38,07	1,7563	0,00220
60	62,03	55,48	50,17	45,39	41,08	1,7890	0,00220
50	65,63	59,55	53,36	48,52	—	1,8171	0,00217
40	68,76	62,03	56,24	51,17	—	1,8373	0,00214
30	71,77	64,87	58,97	53,65	—	1,8560	0,00211
20	74,72	67,70	61,56	56,01	—	1,8634	0,00209
10	77,55	70,41	63,98	58,31	—	1,8896	0,00207
0	80,37	73,12	66,62	60,58	55,10	1,9051	0,00205

t=15 °C [249]				*t*=25 °C [521], ν=0,57 MHz			
w_2	ε	w_2	ε	w_2	ε	w_2	ε
94	56,54	48,5	71,46	100,0	42,48	30,19	71,44
89	57,80	40	71,36	90,42	48,66	20,33	73,86
80	59,40	30	71,36	79,86	54,08	9,88	75,98
73	61,79	20	72,30	70,00	58,52	0,00	78,48
65	64,14	10	73,24	60,15	62,38	—	—
58	69,20	0	75,50	50,23	65,72	—	—
56	72,30	—	—	39,67	68,93	—	—

*) Temperature dependence $\varepsilon = ae^{-b/t}$.

Water – 2-Pyrrolidone ($C_4H_5NO_2$)

[524], ν = 1—10 MHz

w_1	x_1	ε at *t* °C [524], ν=1—10 MHz				
		31	35	40	45	50
0,000	0,000	27,37	27,07	26,64	26,65	25,87
2,706	11,616	30,19	29,86	29,43	28,99	28,48
4,686	18,852	32,24	31,79	31,34	30,82	30,27
8,182	29,624	35,73	35,25	34,64	34,01	33,44
12,685	40,697	40,06	39,40	38,71	37,98	37,27
17,824	50,607	44,58	43,89	43,04	42,17	41,33
23,912	59,751	49,38	48,56	47,65	46,57	45,61
31,411	68,309	54,55	53,61	52,44	51,23	50,09
46,883	80,656	62,60	61,43	60,05	58,64	57,32
66,062	90,193	69,27	67,94	66,32	64,78	63,31
100,000	100,000	76,20	74,80	73,12	71,43	69,78

[524], ν=1—10 MHz *)

t °C	31	35	40	45	50
A	27,41	27,12	26,72	26,36	25,96
B	19,37	18,78	18,38	17,60	17,06
D	29,61	29,05	28,12	27,55	26,82

*) Temperature dependence $\varepsilon_t = A + Bx + Dx^2$.

Water – Tetrahydrofuran (C_4H_8O)

w_2	ε at t °C [525]				w_2	ε at t °C [525]			
	20	25	30	35		20	25	30	35
100	7,58	7,39	7,25	7,16	40	49,77	48,22	46,91	45,65
80	18,75	18,25	17,77	17,38	20	66,46	64,60	63,02	61,64
60	33,04	31,97	31,04	30,24	0	80,37	78,48	76,75	74,95

Water – Dioxan ($C_4H_8O_2$)

w_2	ε at t °C [525]				w_2	ε at t °C [525]			
	20	25	30	35		20	25	30	35
100	2,24	2,21	2,20	2,19	50	36,89	35,85	34,81	33,88
95	3,99	3,89	3,83	3,76	40	45,96	44,54	43,33	42,24
90	6,23	6,07	5,96	5,85	30	54,81	53,28	51,91	50,60
80	12,19	11,86	11,58	11,26	20	63,50	61,86	60,38	58,96
70	19,73	19,07	18,58	18,07	10	72,02	70,33	68,74	67,10
60	28,09	27,21	26,45	25,74	0	80,38	78,48	76,72	74,97

t=20 °C [530] *)				t=20 °C [175]		t=25 °C [527]		t=25 °C [526]	
x_1	ε	x_1	ε	φ_1	ε	w_1	ε	w_1	ε
0	2,28	83,35	41,80	0	2,22	15,04	9,00	20,1	12,12
35,22	6,65	91,31	52,00	20	12,40	20,05	11,99	22,3	13,51
45,06	9,15	97,88	67,80	40	26,80	30,06	19,10	26,5	16,67
52,26	11,60	100,0	81,00	60	43,90	40,06	27,35	29,4	18,74
67,92	22,15	—	—	80	61,80	55,02	40,25	39,5	26,85
76,60	30,00	—	—	100	80,37	74,98	55,50	46,3	32,77
—	—	—	—	—	—	100,0	78,54	65,8	49,54
—	—	—	—	—	—	—	—	100	78,54

*) Measurements made at frequency ν = 150 MHz.

w_2	ε at *t* °C [529]					
	0	5	10	15	20	25
70	20,37	19,81	19,25	18,72	18,20	17,69
45	44,28	43,05	41,86	40,70	39,57	38,48
20	69,16	67,39	65,68	64,01	62,38	60,79

w_2	ε at *t* °C				
	30	35	40	45	50
70	17,20	16,72	16,26	15,80	15,37
45	37,41	36,37	35,37	34,39	33,43
20	59,94	57,73	56,26	54,83	53,43

w_2	ε at *t* °C [289]*)								
	0	10	20	30	40	50	60	70	80
100	2,109	2,104	2,102	2,100	2,098	2,096	2,094	2,092	2,090
98	2,73	2,70	2,68	2,65	2,62	2,60	2,57	2,55	2,52
95	3,91	3,82	3,74	3,65	3,57	3,49	3,41	3,33	3,25
90	6,16	5,93	5,71	5,50	5,30	5,10	4,91	4,73	4,56
80	12,19	11,58	10,99	10,44	9,91	9,41	8,93	8,43	8,05
70	20,37	19,25	18,20	17,20	16,26	15,37	14,52	13,73	12,97
60	29,84	28,17	26,60	25,12	23,72	22,40	21,15	19,97	18,86
50	39,50	37,31	35,25	33,30	31,46	29,72	28,08	26,53	25,05
40	49,37	46,71	44,19	41,80	31,46	37,41	35,39	33,48	31,67
30	59,34	56,24	53,30	50,52	39,54	45,38	43,01	40,76	38,63
20	69,16	65,68	62,38	59,24	47,88	53,43	50,75	48,20	45,77
10	78,86	75,05	71,43	67,98	56,26	61,57	68,60	55,77	53,07
0	88,33	84,25	80,37	76,73	64,70	69,85	66,92	63,50	60,58

*) Temperature dependence ε is given by the formula

$$\lg \varepsilon = \lg a - bt.$$

w_2	ln *a*	*b*	w_2	ln *a*	*b*
100	0,3234	0,00004	50	1,5965	0,00247
95	0,5923	0,00100	40	1,6935	0,00241
90	0,7896	0,00164	30	1,7734	0,00233
80	1,0870	0,00225	20	1,8398	0,00224
70	1,3090	0,00245	10	1,8969	0,00215
60	1,4747	0,00249	0	1,9461	0,00205

[528], $t=18$ °C, $\lambda=73$ cm

φ_1	10	20	30	40	50	60	70	80	90
ε	76,5	65,0	49,0	41,0	32,0	25,0	17,0	10,3	6

Water – Butyric acid ($C_4H_8O_2$)

[157], $t=25$ °C, $\nu=15$ MHz

φ_2	100	77,99	59,29	39,68	20,12	0
ε	2,88	18,0	31,9	47,7	62,6	78,54

Water – Tetramethylene sulphone (C_4H_8S)

[529], $t=25$ °C, $\nu=1,8$ MHz

w_2	100	90,00	80,19	61,07	50,05	38,50	26,48	13,63	0
ε	42,0	49,83	54,30	61,07	65,42	69,00	72,35	75,54	78,54

Water – Aminobutyric acid ($C_4H_9NO_2$)

[508], $t=25$ °C, $\lambda=3$ m

N	ε	*N*	ε	*N*	ε
0	78,54	0,5815	92,42	1,167	106,08
0,2227	83,61	0,6844	94,46	1,595	115,66
0,2866	85,67	0,7371	95,65	1,823	121,03
0,2874	85,25	0,7825	97,10	2,042	126,68

Water – *tert*–Butyl alcohol ($C_4H_{10}O$) *)

[500], $\nu=2$ MHz

w_2	ε at *t* °C					ln *a*	*b*
	20	40	50	60	80		
100	—	8,44	7,67	6,96	5,90	0,9870	—0,00400
90	12,97	10,93	10,00	9,36	7,66	1,1120	—0,00380
80	17,23	14,60	13,44	12,51	10,45	1,2364	—0,00360
70	22,30	19,00	17,41	16,29	13,94	1,3485	—0,00340
60	28,91	25,02	23,01	21,30	18,41	1,4610	—0,00322
50	36,59	32,16	29,80	27,80	24,24	1,5634	—0,00298
40	45,38	40,01	37,35	34,93	30,59	1,6565	—0,00274
30	54,17	48,33	45,44	42,96	38,47	1,7338	—0,00250
20	62,93	56,77	53,82	50,87	45,14	1,7989	—0,00230
10	71,75	64,91	61,84	58,83	52,87	1,8558	—0,00216
0	80,37	73,12	69,85	66,62	60,58	1,9051	—0,00205

*) Temperature dependence $\varepsilon=ae^{-b/t}$.

Water – Diethyleneglycol ($C_4H_{10}O_3$)

[531]

φ_2	x_2	ε at t °C			φ_2	x_2	ε at t °C		
		15	25	35			15	25	35
100	100	32,6	30,7	29,2	50	14,5	62,9	59,6	56,6
95	76,2	36,5	34,3	32,5	40	10,2	67,4	63,7	60,7
90	60,4	40,0	37,7	35,8	30	6,8	71,6	67,7	64,6
80	40,4	46,4	44,0	41,8	20	4,1	75,3	71,5	68,3
70	28,4	52,5	49,7	47,3	10	1,8	78,8	74,9	71,8
60	20,3	57,8	54,8	52,1	0	0	81,9	78,3	74,8

Water – Pyridine (C_5H_5N)

[532, 533]

w_2	ε at t °C		w_2	ε at t °C		w_2	ε at t °C	
	0	18		0	18		0	18
96,0	13,0	12,5	67,0	27,7	25,1	20,0	68,3	64,1
92,5	15,3	16,0	60,0	40,0	37,0	—	—	—
80,0	22,9	21,0	40,0	56,4	52,2	—	—	—

[158], $t=25$ °C, $\nu=15$ MHz

φ_2	100	79,98	60,10	39,85	19,70	0
ε	12,91	22,2	43,3	56,2	66,1	78,54

Water – *N*–Methylpyrrolidine ($C_5H_{11}N$)

[522], $t=25$ °C, $\nu=1,8$ MHz

w_2	ε	w_2	ε	w_2	ε	w_2	ε
100	32,2	80,00	47,7	49,71	64,1	20,46	73,8
94,97	36,6	64,94	56,6	40,18	67,8	10,34	76,3
89,82	40,7	59,86	59,3	30,27	71,0	0	78,54

Water – Benzene (saturated solution of water in benzene) (C_6H_6)

[534]

t °C	ε	t °C	ε	t °C	ε	t °C	ε	t °C	ε
6,2	2,3186	15,3	2,3040	25,0	2,2858	35,0	2,2695	40,5	2,2506
10,2	2,3114	20,0	2,2942	30,0	2,2783	40,3	2,2576	50,1	2,2416

Water – Phenol (C_6H_6O)

[369], $t=70$ °C, $\nu=1$ MHz

w_2	x_2	ε	w_2	x_2	ε	w_2	x_2	ε
100	100	9,03	75,00	36,49	18,23	50,01	16,08	25,13
94,98	78,37	10,94	69,99	30,87	19,94	44,91	13,50	26,72
89,62	62,31	12,81	65,00	26,23	21,37	40,00	11,32	27,67
85,00	52,04	14,41	59,99	22,31	22,64	—	—	—
81,01	43,39	16,36	55,00	18,97	23,67	—	—	—

Water – α–Alanine ($C_3H_7NO_2$)

[509], $t=18$ °C, $\nu=0,9$ MHz

x_2	100	75	50	25	0
ε	103,8	98,08	92,64	86,51	81,0

Water – β–Alanine ($C_3H_7NO_2$)

[509], $t=18$ °C, $\nu=0,9$ MHz

x_2	50	20	10	0
ε	102,1	88,8	85,1	81,0

Water – Galactose ($C_6H_{12}O_2$)

[535]

w_2	ε at t °C					
	25	30	40	50	60	70
60	68,9	66,16	62,80	59,61	56,20	54,20
50	68,81	67,36	64,10	60,41	—	—
40	71,87	70,07	66,28	62,86	—	—
30	75,30	73,67	69,71	65,74	—	—
20	77,28	75,48	71,33	67,72	—	—
10	78,18	76,20	72,41	69,17	—	—
0	78,60	76,73	73,52	70,18	67,41	64,45

Water – Glucose ($C_6H_{12}O_6$)

[536, 505],

w_2	ε at t °C			w_2	ε at t °C		
	20	25	30		20	25	30
50	64,90	63,39	61,91	15	76,56	74,80	73,11
40	68,73	67,11	65,56	10	—	76,14	—
30	72,13	70,46	68,82	5	79,17	77,37	75,64
20	—	73,43	—	0	80,38	78,54	76,76

w_2	ε at t °C [537]								
	20	30	40	50	60	70	80	90	95
95	12,5	16,4	21,2	26,55	30,66	31,8	31,45	31,0	30,6
90	31,61	37,26	39,7	39,7	38,7	37,5	36,23	35,25	34,34
80	49,55	48,4	46,73	45,0	44,5	42,35	41,58	40,7	40,48
60	63,34	61,7	60,03	58,02	55,4	52,45	49,8	46,6	—
40	72,2	68,9	66,7	63,3	60,0	56,7	53,3	50,5	—
20	75,5	71,8	69,6	66,0	63,2	59,55	56,6	53,5	—
10	77,7	74,4	71,5	68,1	65,75	63,14	60,42	57,54	55,9

[505], $t = 20$ °C

w_2	44,4	33,3	28,6	23,1	16,6	9,1	0
ε	26,0	38,0	44,0	49,5	59,5	70,0	80,5

Water – Fructose ($C_6H_{12}O_6$)

[505], $t = 30$ °C

w_2	37,5	33,3	28,6	23,1	16,6	9,1	0
ε	34,5	41	45	54	64,5	71,5	80,5

Water – *N*–Butylacetamide ($C_6H_{13}NO$)

[522], $t = 25$ °C, ν = 1,8 MHz

w_2	100	94,80	84,40	76,86	56,37	36,88	18,21	10,10	0
ε	100,3	79,76	68,70	67,76	69,48	72,22	75,26	76,72	78,54

Water – Triethylene glycol ($C_6H_{14}O_4$)

[531]

φ_2	x_2	ε at t °C			φ_2	x_2	ε at t °C		
		15	25	35			15	25	35
100	100	24,0	23,0	21,8	50	10,7	59,8	57,5	54,5
97,5	82,1	26,8	25,4	24,2	40	7,4	64,9	62,4	59,1
95	69,3	29,2	27,5	26,2	30	4,9	69,5	66,8	63,4
90	51,8	33,5	31,8	30,3	20	2,9	73,9	71,0	67,4
80	32,4	41,4	39,4	37,6	10	1,3	78,0	74,8	71,2
70	21,8	48,2	46,0	43,9	0	0	81,9	78,3	74,80
60	15,2	54,3	52,1	49,3	—	—	—	—	—

Water – Mannitol ($C_6H_{14}O_6$)

w_2	ε at *t* °C [500], ν=2 MHz					
	20	25	30	40	50	60
20	—	75,47	73,64	70,10	66,74	63,61
15	78,07	76,30	74,47	70,87	67,57	64,38
10	78,84	77,12	75,18	71,57	68,33	65,08
5	79,66	77,83	76,06	72,40	69,16	65,85
0	80,37	78,54	76,73	73,12	69,85	66,62

Water – Sucrose ($C_{12}H_{22}O_{11}$)

[538]

w_2=40		w_2=30		w_2=20	
t °C	ε	*t* °C	ε	*t* °C	ε
—3,5	78,00	—2,1	80,66	—1,5	87,26
—3,2	77,93	—1,7	80,50	0	82,77
—1,2	77,19	0	79,69	1,7	81,99
0,4	76,60	2,6	78,70	3,3	81,33
1,1	76,19	5,0	77,71	5,3	80,43
4,1	74,93	6,9	76,89	7,9	79,36
6,6	73,96	10,1	75,63	10,4	78,40
9,6	72,67	13,2	74,37	13,5	77,23
12,9	71,43	15,5	73,53	16,4	76,03
14,9	70,71	15,7	73,47	17,1	75,83
15,5	70,46	20,7	71,50	17,4	75,47
17,4	69,81	26,5	69,43	19,9	74,63
19,1	69,26	29,7	68,29	24,9	72,81
22,4	65,07	34,1	66,76	25,0	72,74
27,4	66,31	40,2	65,13	30,1	70,84
33,1	64,54	43,3	63,77	34	69,54
36,9	63,24	49,6	62,04	38	67,84
41,9	61,75	56,2	60,19	44	66,23
48,2	60,00	60,5	58,89	50	64,37
52,2	58,90	60,8	58,79	54	62,81
57,5	57,37	64,6	57,69	56	62,41
63,2	55,76	69,0	56,43	60	61,37
67,7	54,44	75,4	54,67	65	60,00
72,4	53,07	79,5	53,93	69	58,00
76,3	51,89	84,7	51,89	74	57,50
81,6	50,34	89,5	50,47	79	55,90
86,0	49,09	93,7	49,13	85	54,39
90,4	47,89	94,3	49,21	89	53,04
95,6	46,36	99,5	47,31	94	51,50
		99,8	47,16	99	50,06

$w_2=10$		$w_2=5$		$w_2=5$		$w_2=0$	
t °C	ε	t °C	ε	t °C	ε	t °C	ε
0	85,49	0	86,64	7,6	83,36	0	87,90
2,2	84,54	0,7	86,21	9,2	82,77	3,95	86,29
3,3	84,04	1,3	86,16	11,8	81,59	9,8	83,96
5,4	83,13	1,6	85,81	13,3	81,01	15,0	81,99
8,2	81,97	2,7	85,51	13,5	80,94	20,3	79,67
10,7	81,11	3,3	85,17	13,9	80,81	25,0	77,81
12,9	80,17	4,2	84,80	15,6	80,17	30,2	76,10
14,0	79,66	5,2	84,39	17,5	79,14	39,0	72,51
14,6	79,41	5,6	84,14	18,8	78,94	50,1	68,93
18,4	77,87	6,4	83,74	19,4	78,73	59,4	65,73
19,5	77,50	7,0	83,49	22,3	77,81	70,3	62,77
20,9	76,97	8,1	83,19	26,9	75,91	79,0	60,80
25,3	75,23	8,1	83,09	30,0	74,66	89,5	58,31
29,0	73,83	8,65	82,81	37,7	71,84	98,7	55,51
34,0	71,97	9,4	82,69	43,5	69,86	—	—
39,5	70,10	10,4	82,20	49,5	67,96	—	—
44,6	68,37	11,4	81,80	53,6	66,76	—	—
49,6	66,76	12,3	81,46	56,5	65,73	—	—
50,5	66,49	13,5	81,03	60,6	64,56	—	—
55,2	65,04	14,4	80,66	66,7	62,80	—	—
60,0	63,56	—	—	76,7	60,14	—	—
64,9	62,14	—	—	79,8	59,30	—	—
69,5	60,91	—	—	86,8	57,37	—	—
84,5	56,80	—	—	91,6	56,11	—	—
85,1	56,64	—	—	92,1	56,11	—	—
89,1	55,40	—	—	92,3	55,29	—	—
94,3	54,10	—	—	95,5	55,17	—	—
99,0	52,59	—	—	97,0	54,66	—	—
—	—	—	—	97,6	54,53	—	—
—	—	—	—	99,0	54,33	—	—

w_2	ε at t °C [500], ν=2 MHz									
	10	20	30	40	50	60	70	80	90	100
70	—	55,0	52,5	50,1	47,6	45,1	42,7	40,3	38,0	35,7
60	—	61,2	58,2	55,5	52,5	49,8	47,1	44,3	—	—
50	—	65,7	62,4	59,4	56,3	53,4	50,5	47,6	—	—
40	72,8	69,8	66,1	63,0	59,8	56,8	53,6	50,5	47,7	44,8
30	72,8	72,6	68,9	65,7	62,5	59,4	56,2	53,2	50,4	47,6
20	78,8	75,4	71,7	68,3	65,0	62,0	58,8	55,8	52,9	50,2
10	81,5	78,0	74,2	70,7	67,5	64,4	61,2	58,2	55,4	52,7
0	84,2	80,4	76,7	73,1	69,9	66,6	63,5	60,6	57,8	55,1

w_2	ε at *t* °C [537] *)								
	20	30	40	50	60	70	80	90	95
90	14,5	19,1	24,5	29,6	32,3	32,6	32,2	31,5	31,0
80	43,5	44,4	43,2	42,9	41,6	39,8	38,5	36,6	35,6
60	59,8	57,4	54,7	52,7	50,9	47,8	45,6	43,4	42,5
40	70,0	66,7	63,7	60,7	58,0	55,2	52,7	49,9	48,8
20	75,4	72,05	68,55	65,95	62,8	60,2	57,7	54,7	53,6
0	80,4	76,7	73,5	70,2	67,4	64,4	62,5	60,5	—

w_2	ε at *t* °C [536], ν=24 kHz			w_2	ε at *t* °C		
	20	25	30		20	25	30
60	61,80	60,19	58,64	20	75,45	73,63	71,90
50	65,88	64,20	62,57	10	78,04	76,19	74,43
40	69,45	67,72	66,05	0	80,38	78,54	76,76
30	72,64	70,86	69,13	—	—	—	—

*) See also [498, 492, 490, 491, 539].

Supplementary information on the dielectric permittivity of aqueous solutions of organic compounds

Water – Oxalic acid ($C_2H_2O_4$), fig. 54.
Water – Malonic acid ($C_3H_4O_4$), fig. 54.
Water – Succinic acid ($C_4H_6O_4$), fig. 54.
Water – Cresol (C_7H_8O), fig. 56.

§ 3. Dielectric dispersion parameters of aqueous solutions of inorganic compounds

Water – Aluminium chloride ($AlCl_3$)
[540], $t = 20$ °C, $\lambda = 3,5$ cm

N	ε′	ε″	*N*	ε′	ε″	*N*	ε′	ε″
0,06	62,0	30,4	0,36	51,1	39,2	1,60	56,6	34,2
0,09	61,4	32,5	0,39	52,1	42,6	1,73	55,5	34,8
0,12	58,6	32,8	0,49	48,6	41,2	2,16	54,7	36,7
0,16	58,7	33,9	0,60	47,4	40,6	2,7	52,4	40,0
0,23	55,6	34,0	0,94	57,0	34,8	—	—	—
0,30	53,3	36,0	1,30	56,6	34,2	—	—	—

Water – Aluminium nitrate $Al(NO_3)_3$ *)

[541, 540], $t = 25$ °C, $N = 1$

$\lambda = 10$ cm		$\lambda_1 = 3,12$ cm		$\lambda = 1,17$ cm	
ε′	ε″	ε′	ε″	ε′	ε″
5,25	5,0	38,3	16,8	24,7	16,8

*) $\varepsilon = 55,7$, $\varepsilon_\infty = 5,5$, $\lambda_m = 1,51$, $\alpha = 0,22$, $g = 1,92$.

Water – Aluminium sulphate ($Al_2(SO_4)_3$)

[540], $t = 24$ °C, $\lambda = 3,5$ cm

N	ε′	ε″	N	ε′	ε″	N	ε′	ε″
0,01	61,6	30,8	0,19	60,0	32,0	0,6	50,6	43,1
0,02	61,6	30,8	0,24	58,9	35,6	0,77	49,1	44,4
0,04	61,0	31,4	0,3	58,0	36,2	0,97	47,0	46,5
0,08	61,0	31,4	0,4	52,0	41,6	—	—	—
0,15	60,0	32,0	0,5	52,0	41,6	—	—	—

Water – Barium chloride ($BaCl_2$)

[543, 540], $t = 25$ °C *)

N	$\lambda = 10$ cm		$\lambda = 3$ cm		ε	λ_m
	ε′	ε″	ε′	ε″		
1,0	64,1	8,8	52,3	22,2	64,0	1,50
2,0	—	—	44,1	16,9	51,0	1,42

*) See. [544].

[541], $t = 25$ °C

N	$\lambda = 10,0$ cm		$\lambda = 3,12$ cm		$\lambda = 1,17$ cm		ε	λ_m	α	g
	ε′	ε″	ε′	ε″	ε′	ε″				
0,5	—	—	53,9	23,9	29,0	28,2	68,9	1,6	0,04	2,31
1,0	—	—	49,2	20,3	27,6	23,9	63,4	1,52	0,09	2,13
1,5	54,5	4,2	44,2	14,9	26,1	17,6	58,2	1,3	0,22	1,96

Water – Calcium bromide ($CaBr_2$) *)
[541], $t=25$ °C

N	$\lambda=3,12$ cm		$\lambda=1,17$ cm		ε	λ_m	α
	ε'	ε''	ε'	ε''			
0,5	52,2	22,7	31,0	27,3	65,0	1,45	0,03
1,0	45,9	20,1	26,3	22,8	60,2	1,55	0,09

*) См. [544].

Water – Calcium chloride ($CaCl_2$) *)
[541, 540], $t=25$ °C, fig. 59

N	$\lambda=10,0$ cm		$\lambda=3,12$ cm		$\lambda=1,17$ cm		ε	ε_∞	λ_m	α	g
	ε'	ε''	ε'	ε''	ε'	ε''					
0,5	63,6	4,0	52,6	23,2	29,0	28,4	64,3	5,5	1,41	0,01	2,15
1,0	56,5	2,7	46,1	19,0	26,1	23,4	57,5	5,2	1,43	0,07	1,92
2,0	53,0	3,0	35,8	14,1	22,4	14,0	54,0	5,0	1,55	0,29	1,83
4,0	—	—	24,5	8,0	16,3	6,1	49	5,0	5,6	0,53	1,73

*) See also [544].

Water – Cadmium nitrate ($Cd(NO_3)_2$)
[546], $t=28$ °C, $\lambda=3,14$; 4,4 cm

N	0,25	0,50	0,75	1,0	1,50	2,0
λ_m	1,235	1,175	1,096	1,072	1,047	1,023
ε	70,5	68,0	63,0	60,0	56,0	51,0

Water – Cobalt(II)chloride ($CoCl_2$)
[161], $t=28$ °C, $\lambda=1,276$; 3,14; 4,4 cm

N	0,25	0,5	0,75	1,0
ε	67,0	64,5	62,5	60,0
λ_m	1,349	1,334	1,307	1,3
α	0	0	0,01	0,017

[547, 540], $t=28$ °C, $\lambda=1,276$; 314; 10 cm

N	ε	ε расчет		λ	α
		with α	without α		
0,25	67,0	70,7	67,6	1,349	0,0
0,50	64,5	67,4	61,0	1,334	0,0
0,75	62,5	64,4	54,1	1,307	0,011
1,0	60,0	62,2	49,9	1,3	0,017

Water – Cobalt(II)nitrate ($Co(NO_3)_2$)

[546], $t=28$ °C, $\lambda=3,14$; 4,4 cm

N	0,25	0,50	0,75	1,0	2,0
λ_m	1,215	1,160	1,096	1,065	1,021
ε	71,0	68,0	61,0	57,5	52,0

Water – Cobalt(II)sulphate ($CoSO_4$)

[548], $t=28$ °C, $\lambda=1,267$; 3,14 cm, fig. 60

N	0,25	0,50	0,75	1,00
ε	68,8	66,5	64,5	62,0
λ_m	1,32	1,31	1,30	1,25
α	0,008	0,022	0,027	0,055

[549], $n=1,04$

ν, GHz	ε′ at t °C						ε″ at t °C					
	5	15	25	35	45	55	5	15	25	35	45	55
0,34	90,5	88,9	84,8	80,8	78,9	77,9	11,0	8,6	8,0	6,8	4,0	3,8
0,5	85,7	84,9	83,6	82,1	79,7	77,4	14,1	12,1	9,0	8,4	7,6	7,1
0,75	84,4	83,3	81,9	80,8	78,5	76,1	15,3	13,7	12,4	10,2	9,8	7,1
1,15	79,3	78,5	77,4	76,0	75,3	73,8	16,6	15,1	13,1	13,5	9,9	8,3
1,7	75,7	75,3	74,6	74,0	72,5	71,9	18,2	17,2	16,3	15,0	13,3	11,6
2,5	71,8	72,3	72,1	72,0	70,2	68,2	20,3	19,0	17,5	16,6	14,2	12,7

[549], $t=5$ °C, $n=1,04$

ν, GHz	4,74	5,86	8,5	12,07	18,37	26,00
ε′	63,7	58,8	49,5	37,2	23,9	14,8
ε″	29,6	30,5	35,1	35,7	29,2	23,3

Water – Chromium(III)nitrate ($Cr(NO_3)_3$)
[546], $t = 28$ °C, $\lambda = 3,4;\ 4,4$ cm

N	0,25	0,50	0,75	1,0	2,0
λ_m	1,210	1,168	1,144	1,069	1,013
ε	70,5	64,0	57,5	54,5	46,0

Water – Caesium chloride ($CsCl$) *)
[551], $t = 24,96$ °C

N	λ=3,6 cm		λ=3,0 cm		λ=2,5 cm	
	ε′	ε″	ε′	ε″	ε′	ε″
0,5	61,59	24,3	58,98	27,3	54,94	29,9
1,0	58,43	22,1	56,11	25,1	52,39	27,4
1,5	55,65	20,5	53,21	22,9	49,92	25,2
2,0	52,32	18,4	50,58	20,9	47,55	23,2
3,0	46,73	14,4	48,45	16,8	43,46	19,0

*) See also [550, 540], $t=20$ °C, $\nu=5-38$ GHz

Water – Caesium iodide (CsI)
[552], fig. 61, 62

N	t °C	ε′	ε″	α	$10^{-11}\tau$
0,5	5	6,1	80,4	0	1,37
	25	5,1	74,0	0,007	0,79
1,0	5	5,7	77,3	0,017	1,26
1,5	5	6,0	73,3	0,024	1,17

Water – Caesium nitrate ($CsNO_3$)
[551], $t = 24,96$ °C

N	λ=3,6 cm		λ=3,0 cm		λ=2,5 cm	
	ε′	ε″	ε′	ε″	ε′	ε″
0,5	62,26	24,3	59,66	27,3	55,59	29,8
1,0	60,08	22,4	57,41	25,5	53,72	28,0

Water – Copper(II)chloride ($CuCl_2$)

[161, 547], $t=28$ °C, $\lambda=1,267;\ 3,14;\ 10$ cm

N	0,25	0,50	0,75	1,0
ε	70,0	67,0	63,5	61,0
λ_m	1,365	1,349	1,295	1,291
α	0	0	0,005	0,011

Water – Copper(II)nitrate ($Cu(N)_3)_2$)

[546], $t=28$ °C, $\lambda=3,14;\ 4,4$ cm

N	0,25	0,50	0,75	1,0	2,0
λ_m	1,235	1,175	1,148	1,072	1,047
ε	69,5	67,0	61,5	57,0	54,0

Water – Copper(II)sulphate ($CuSO_4$) *)

[548], $t=28$ °C, $\lambda=1,267;\ 3,14;\ 4,4$ cm

N	0,25	0,50	0,75
ε	69,5	67,5	63,8
λ_m	1,35	1,34	1,30
α	0,011	0,017	0,028

*) See also [553, 554, 545].

Water – Hydrogen chloride (HCl)

[377], fig. 63

N	t °C	$\lambda=9,95$ cm		$\lambda=3,22$ cm		α	ε	λ_m
		ε'	ε''	ε'	ε''			
0,0625	3	77,6	21,6	47,0	38,6	0,02	85,0	2,96
	10	78,0	17,4	53,6	35,5	0,018	82,6	2,35
0,125	3	76,5	20,8	47,0	37,9	0,015	83,0	2,90
	10	76,3	16,7	54,0	35,2	0,015	81,0	2,31
0,25	3	69,6	17,8	45,0	34,0	0,007	77,7	2,75
	10	70,5	15,1	51,3	32,4	—	74,0	2,25
0,5	3	66,0	15,7	44,6	31,7	—	70,0	2,57
	10	63,2	12,7	47,6	28,0	0,007	66,7	2,12

[559, 543], $t=21$ °C

$10^{-3}w$	0,0	0,1	0,2	0,3	0,4
σ_{10}^{-4}	—	9,0	18,0	25,0	33,0
ε	80	73,2	71,9	74,7	80,0

Water – Sulphuric acid (H_2SO_4)
[377], fig. 63, 64

N	t °C	λ=9,95 cm		λ=3,22 cm		α	ε	λ_m
		ε′	ε″	ε′	ε″			
0,0625	3	78,7	22,3	47,5	39,4	0,015	85,2	3,02
	10	79,0	17,5	54,6	36,2	0,02	83,8	2,35
0,125	3	77,5	22,1	46,3	39,2	0,012	84,4	3,03
	10	77,3	17,4	53,0	36,4	0,015	81,5	2,42
0,25	3	75,4	21,6	43,6	37,8	0,02	81,2	3,11
	10	76,2	17,3	51,0	35,7	0,025	79,4	2,47
0,5	3	71,5	21,3	39,7	35,0	0,03	78,0	3,25
	10	72,6	16,8	48,6	33,8	0,025	76,4	2,50

Water – Phosphoric acid (H_3PO_4)
[555], t = 25 °C

N	λ=17 cm		λ=10 cm		λ=3 cm		ε	λ_m
	ε′	ε″	ε′	ε″	ε′	ε″		
0,00	46,61	36,53	63,06	29,88	76,6	11,04	78,3	1,53
0,01	44,9	36,4	62,7	29,7	75,9	11,6	77,8	1,54
0,10	43,9	35,9	61,7	29,4	74,3	10,7	76,3	1,57
0,25	42,8	35,3	60,0	29,4	73,0	10,9	75,1	1,61
0,50	41,1	34,4	57,7	29,5	71,0	11,9	73,5	1,66
0,75	39,6	33,5	55,5	29,1	69,2	12,5	71,8	1,70
1,00	38,3	32,5	53,5	28,8	67,6	12,8	70,2	1,73
1,25	37,0	31,6	51,6	28,3	65,6	12,3	68,2	1,76
1,50	35,9	30,7	49,7	27,8	63,7	12,1	66,3	1,78
2,00	33,8	28,8	46,9	26,5	59,8	10,8	62,4	1,80
3,0	30,2	25,4	41,8	23,8	52,9	9,1	55,4	1,80
5,0	24,8	19,8	34,1	18,9	41,0	5,3	43,5	1,76
7,5	22,1	15,0	28,6	15,4	32,9	2,1	35,0	1,69
10,0	20,9	12,8	25,9	13,4	30,1	0,5	31,5	1,63
15,0	15,2	10,3	23,5	11,2	28,8	1,1	28,7	1,52
20,0	18,1	8,7	22,1	9,7	26,6	1,8	26,3	1,43
30,0	16,8	6,9	20,1	7,9	23,9	0,5	23,2	1,29
50,0	15,8	5,6	18,0	5,8	21,6	0,3	20,5	1,14
75,0	15,0	4,9	17,1	5,3	20,2	1,5	19,2	1,14
95,0	14,6	4,9	16,9	5,6	20,3	4,0	19,6	1,21

Water – Potassium bromide (KBr)
[541], fig. 59

N	λ=3,21 cm		λ=1,17 cm		ε	λ_m	α
	ε′	ε″	ε′	ε″			
1	55,4	22,4	30,0	26,1	69,3	1,46	0,09
2	51,9	17,8	30,6	20,7	67,5	1,36	0,23

[544], $t = 25$ °C, $\lambda = 10$ cm

N	0,3	0,5	1,0
ε′	—	71,6	67,1
ε″	32,0	43,6	75,3

Water – Potassium chloride (KCl)
[556], fig. 59, 65, 66

N	t °C	λ=20 cm			λ=10 cm			ε
		ε′	ε″	ε''_{corr} *)	ε′	ε″	ε''_{corr} *)	
0	15	80,75	8,03	—	78,20	15,89	—	81,95
	25	77,71	5,79	—	76,39	11,42	—	78,30
	35	74,58	4,88	—	73,76	8,89	—	74,83
0,05	15	79,82	14,67	8,18	77,46	17,86	14,62	80,6
	25	77,02	14,08	6,08	75,65	15,20	11,20	77,5
	35	73,80	14,61	5,08	73,04	12,75	7,99	74,1
0,1	15	79,08	20,64	8,08	76,82	21,58	15,30	80,0
	25	76,38	21,70	6,22	74,70	19,11	11,37	76,8
	35	73,39	23,39	4,87	72,32	17,85	8,59	73,5
0,3	15	77,13	44,13	8,42	73,95	32,15	14,30	77,5
	25	73,70	49,13	5,53	72,69	32,02	10,22	74,2
	35	71,62	55,78	3,87	69,97	34,20	8,24	71,2
0,5	25	—	—	—	70,5	44,85	9,68	71,9
	35	—	—	—	67,56	49,23	7,74	68,5

*) ε''_{corr} – is the value of ϵ'' with allowance for the ionic conductivity.

[551, 609] $t = 24,96$ °C

N	λ=3,6 cm		λ=3,0 cm		λ=2,5 cm	
	ε′	ε″	ε′	ε″	ε′	ε″
0,5	61,42	24,0	58,42	27,1	54,35	29,7
1,0	57,65	21,6	54,87	24,3	51,32	26,3
1,5	54,27	19,4	51,55	21,7	48,67	23,7
2,0	51,11	17,4	48,61	19,7	45,86	21,6
3,0	45,00	13,0	43,34	15,8	41,21	17,9
4,0	39,75	11,5	38,45	13,7	36,92	14,4

[557], $t = 18$ °C, $\nu = 9400$ MHz

N	0,05	0,2	0,5	1	2	3
ε	79,65	76,9	74,3	69	61,9	56,4
σ	115,5	107,7	102,2	98,1	92,3	88,1
ε	32,3	30,8	29,2	25,7	20,5	15,4
$10^{-12}\tau$	9,78	9,66	9,33	8,55	7,2	5,7

[544], $t = 25$ °C, $\lambda = 10$ cm

N	0,1	0,2	0,3	0,6	0,8	1,0	1,5	2,0
ε′	74,7	73,9	72,9	70,3	68,3	66,8	62,9	58,9
ε″	18,6	26,5	32,1	51,3	63,5	75,7	103,5	133,0

[558], $t = 25$ °C

N	λ=3,175 cm		λ=1,264 cm		ε	λ_m
	ε′	ε″	ε′	ε″		
0,5	59,7	25,1	33,5	32,5	71,3	1,47
1,0	55,6	22,0	33,5	29,7	64,9	1,36

[541], $t = 25$ °C

N	λ=10,0 cm		λ=3,12 cm		λ=1,17 cm		ε	λ_m	α
	ε′	ε″	ε′	ε″	ε′	ε″			
1	68,5	7,5	54,4	21,7	30,8	26,6	70,5	1,52	0,11
2	—	—	50,7	18,2	30,7	20,9	66,3	1,36	0,22
4	—	—	44,0	10,2	30,1	11,2	62,0	1,10	0,56

[543, 559, 545, 186], $t=25$ °C

N	$\lambda=10$ cm		$\lambda=3$ cm		ε	λ_m
	ε'	ε''	ε'	ε''		
0,5	70,3	10,5	60,0	25,9	73,5	1,54
1,0	64,3	6,8	57,4	23,6	68,5	1,50
1,5	—	—	54,4	21,8	63,5	1,47
2,0	—	—	49,0	18,4	58,5	1,43

[560], $\nu=9400$ MHz

N	$t=0$ °C		$t=6$ °C		$t=10$ °C		$t=20$ °C		$t=40$ °C		$10^{-12}\tau$
	ε'	ε''	ε'	ε''	ε'	ε''	ε'	ε''	ε'	ε''	
0,05	45,5	42,9	53,2	40,6	55,6	39,0	62,3	31,4	64,4	22,0	17,8
0,2	45,4	43,7	52,7	41,5	55,1	40,1	61,2	33,5	62,4	25,4	17,0
0,5	45,4	45,0	52,4	43,7	54,7	42,1	59,6	37,5	60,9	33,0	15,9
1,0	45,3	47,6	51,5	46,9	53,6	46,0	57,2	44,1	58,3	43,8	15,0
1,5	44,8	51,3	50,2	50,8	52,0	51,5	55,4	49,7	56,6	53,5	—
2,0	44,3	53,8	48,6	54,8	51,2	55,9	53,9	57,0	55,7	63,5	12,6
2,5	43,7	57,0	48,2	58,8	50,0	59,1	52,9	62,3	54,9	72,0	—
3,0	43,2	60,0	47,2	62,9	48,8	64,4	52,0	67,9	54,0	79,6	11,0

Water – Potassium fluoride (KF)

[541], $t=25$ °C, fig. 59

N	$\lambda=10,0$ cm		$\lambda=3,12$ cm		$\lambda=1,17$ cm		ε	λ_m	α
	ε'	ε''	ε'	ε''	ε'	ε''			
1	68,5	7,8	56,0	25,0	31,0	30,0	70,0	1,60	0,00
4	—	—	42,4	16,1	26,8	15,8	64,0	1,96	0,32
8	—	—	29,2	12,1	18,1	8,1	60,0	6,7	0,47
12	48,5	4	9,5	9,5	16,2	5,6	57,0	19,0	0,58

[558], $t=25$ °C

N	$\lambda=9,22$ cm		$\lambda=3,175$ cm		$\lambda=1,264$ cm		ε	λ_m
	ε'	ε''	ε'	ε''	ε'	ε''		
0,33	72,0	11,7	60,7	26,7	33,6	33,7	74,0	1,53
0,66	—	—	58,0	25,3	32,4	32,0	70,1	1,52
1	—	—	56,3	24,2	31,7	30,2	67,0	1,47

[543], $t = 25$ °C

N	λ=10 cm		λ=3 cm		ε	λ_m
	ε′	ε″	ε′	ε″		
0,242	73,6	8,8	61,5	27,9	75,0	1,58
0,484	72,1	7,3	58,9	25,7	72,0	1,56

[544], $t = 25$ °C, λ = 10 cm

N	0,3	0,5	1,0
ε′	—	69,2	66,2
ε″	28,2	38,7	64,9

Water – Potassium iodide (KI)

[560], ν = 9400 MHz, fig. 59

N	t=0 °C		t=6 °C		t=10 °C		t=20 °C		t=40 °C		$10^{-12}\tau$
	ε′	ε″	ε′	ε″	ε′	ε″	ε′	ε″	ε′	ε″	
0,05	45,7	43,1	53,6	40,7	55,8	39,2	62,2	31,4	64,5	22,1	18,0
0,2	45,8	43,8	53,1	41,5	55,4	40,0	61,3	33,6	62,1	25,7	17,0
0,5	46,3	44,8	52,9	43,6	55,0	42,5	59,4	37,7	60,5	32,7	15,7
1	46,6	47,9	52,3	47,2	54,0	46,6	57,0	44,0	57,5	43,5	14,3
1,5	45,8	51,5	50,8	51,0	52,6	52,0	55,3	50,5	55,4	54,0	—
2	45,2	54,5	48,9	55,8	51,2	56,8	53,7	57,6	53,5	64,8	11,5
2,5	44,4	58,7	48,0	60,0	49,6	61,6	52,1	63,8	52,8	73,5	—
3	43,6	62,5	46,8	64,8	48,0	67,2	50,7	70,0	51,4	81,7	10,0
3,5	42 5	65,2	45,5	68,4	46,9	70,4	49,5	75,5	50,8	88,3	—
4	41,5	68,3	44,6	72,0	45,7	74,4	48,5	80,3	49,6	94,3	8,1
4,5	40,0	70,4	43,3	75,1	44,6	77,6	47,4	84,0	48,8	98,9	—
5	39,3	72,8	42,0	78,2	43,7	80,4	46,4	87,2	48,0	103,0	6,9

[557], $t = 18$ °C, ν = 9400 MHz

N	0,05	0,2	0,5	1	2	3	4	5
ε	79,8	76,6	73,9	68,3	60,5	55,5	49,6	47,4
σ	117,0	110,4	106	103	99,6	92,8	89,4	80
ε″	32,3	30,5	28,8	25	18,7	15,4	9,8	9,2
$10^{-12}\tau$	9,84	9,5	9,2	8,33	6,67	5,3	3,96	3,87

[543], $t = 25$ °C

N	λ=10 cm		λ=3 cm		ε	λ_m
	ε′	ε″	ε′	ε″		
0,198	73,1	10,4	61,0	27,2	75,0	1,56
0,396	73,3	11,1	58,7	25,1	72,0	1,52

[541], $t = 25$ °C

N	$\lambda = 3,12$ cm		$\lambda = 1,17$ cm		ε	λ_m	α
	ε'	ε''	ε'	ε''			
1	55,4	21,7	30,1	26,1	69,5	1,45	0,11
2	51,8	17,4	35,9	20,8	67,6	1,35	0,25
4	44,0	5,4	29,8	6,3	57,5	1,07	0,68

[544], $t = 25$ °C, $\lambda = 10$ cm

N	0,3	0,47	0,94
ε'	—	70,1	65,7
ε''	32,9	46,4	80,1

Water – Potassium nitrate (KNO_3)

[551], $t = 24,96$ °C

N	$\lambda = 3,6$ cm		$\lambda = 3,0$ cm		$\lambda = 2,5$ cm	
	ε'	ε''	ε'	ε''	ε'	ε''
0,5	61,45	23,6	59,12	26,7	54,92	29,4
1,0	58,95	22,0	56,89	24,8	53,06	27,03
1,5	56,67	20,8	54,58	23,4	51,09	25,7
2,0	54,65	19,4	52,67	22,1	49,40	24,2

[377]

N	t° C	$\lambda = 51,5$ cm		$\lambda = 9,95$ cm		$\lambda = 3,22$ cm		ε	α	λ_m
		ε'	ε''	ε'	ε''	ε'	ε''			
0,0625	3	85,4	5,2	79,4	21,0	47,2	39,2	86,0	0,024	2,99
	10	—	—	78,5	18,2	54,2	36,6	83,1	0,015	2,39
0,125	3	84,3	5,2	78,5	20,2	47,2	38,8	84,9	0,017	2,97
	10	—	—	78,0	17,8	53,7	36,3	82,4	0,012	2,38
0,25	3	82,0	5,4	77,0	21,0	46,1	37,8	82,2	0,012	2,96
	10	80,4	2,6	76,2	17,0	52,4	35,3	81,0	0,022	2,37
0,5	3	80,7	3	74,5	19,0	45,0	36,4	80,2	0,018	2,93
	10	—	—	74,0	16,5	51,2	33,3	78,3	0,025	2,35
1,0	3	—	—	70,0	18,5	42,9	34,0	75,8	0,023	2,89
	10	—	—	69,5	15,5	48,2	31,8	73,8	0,029	2,34

Water – Lanthanum chloride ($LaCl_3$)

[543], $t = 25$ °C

N	$\lambda = 10$ cm		$\lambda = 3$ cm		ε	λ_m
	ε'	ε''	ε'	ε''		
0,52	70,9	8,9	57,3	25,2	71,0	1,54
1,04	66,1	7,8	52,6	22,0	64,0	1,50

Water – Lithium bromide (LiBr)

[552], $t = 25$ °C $N = 1$, fig. 67

ε'	ε''	α	$10^{-11}\tau$
5,6	71,0	0	1,34

[541], $t = 25$ °C

N	λ=3,12 cm		λ=1,17 cm		ε	$λ_m$	α	g
	ε′	ε″	ε′	ε″				
2	42,8	16,2	26,3	23,4	51,8	1,34	0,00	1,78
4	35,2	12,1	22,3	12,5	49,3	1,61	0,29	1,76
10	20,8	6,0	13,4	4,5	39	7,2	0,53	1,61

Water – Lithium chloride (LiCl) *)

[557], $t = 18$ °C, $\nu = 9400$ MHz, fig. 59, 68

N	0,05	0,2	0,5	1	2	3	4	5
ε	79	76,5	71,5	66,5	56,8	49,2	—	39,1
σ	85,9	77,7	70,5	63,18	53	45,2	—	33,3
ε″	32,2	30,9	28,3	25,4	20,8	17,5	14,9	13,2
$10^{-12}\tau$	10	9,83	9,46	9,0	8,6	8,4	—	8,1

*) See also [550], $t=25$ °C, $\nu=5$—38GHz; [561], $\nu=0,2$—2000kHz, $N=6$—20.

[543, 559, 544], $t = 25$ °C

N	λ=10 cm		λ=3 cm		ε	$λ_m$
	ε′	ε″	ε′	ε″		
0,5	69,0	10,7	58,3	26,0	71,2	1,55
1,0	60,6	8,0	53,9	23,3	64,2	1,52
1,5	52,5	5,0	49,0	20,2	57,0	1,49
2,0	—	—	44,2	17,0	51,0	1,45

[558], $t = 25$ °C

N	λ=9,22 cm		λ=3,175 cm		λ=1,264 cm		ε	$λ_m$
	ε′	ε″	ε′	ε″	ε′	ε″		
0,31	71,6	12,7	60,4	26,5	33,4	33,0	73,2	1,55
0,62	65,8	13,1	57,1	24,7	32,1	30,7	68,4	1,52
0,93	—	—	53,7	22,7	30,9	28,6	64,0	1,48

[551], $t = 24,96$ °C

N	λ=3,6 cm		λ=3,0 cm		λ=2,5 cm	
	ε′	ε″	ε′	ε″	ε′	ε″
0	66,83	27,0	62,67	30,4	57,60	33,3
0,5	59,68	23,3	56,51	26,2	52,26	28,9
1,0	54,26	20,7	52,74	23,3	48,31	25,8
1,5	50,02	18,8	48,0	21,2	44,97	23,3
2,0	45,65	16,4	43,75	18,4	41,15	20,2
3,0	38,58	13,0	37,19	14,9	35,12	16,2
4,0	33,17	10,1	32,23	11,6	30,58	13,1
6,0	25,34	6,5	24,95	7,6	24,31	8,5
8,0	21,29	5,1	21,00	5,8	20,34	6,3
10,0	19,12	4,4	18,80	4,9	18,20	5,5
13	16,94	4,1	16,56	4,4	16,06	4,7

[541], $t = 25$ °C

N	λ=10,0 cm		λ=3,12 cm		λ=1,17 cm		ε	ε_∞	λ_m	α	g
	ε′	ε″	ε′	ε″	ε′	ε″					
2	52,0	6,0	42,8	17,6	25,8	24,1	52,5	5,5	1,43	0,00	1,75
4	—	—	35,0	12,9	21,2	13,1	50	—	1,75	0,28	1,74
8	36,0	1,9	22,0	8,6	14,4	5,8	39	5,5	4,3	0,50	1,45
12	34,0	1,3	17,6	7,8	11,8	5,4	35	4,2	8,0	0,57	1,42

[560], $\nu = 9400$ MHz

N	t=0 °C		t=6 °C		t=10 °C		t=20 °C		t=40 °C		$10^{-12}\tau$
	ε′	ε″	ε′	ε″	ε′	ε″	ε′	ε″	ε′	ε″	
0,05	45	42,6	52,6	40,0	55,0	37,7	61,7	31,3	64,5	22,3	18,4
0,2	44,8	42,7	50,8	40,4	53,7	38,0	60,0	33,3	62,1	24,6	17,3
0,5	43,4	42,6	48,6	41,0	51,2	39,1	57,7	34,5	59,3	28,7	16,9
1	41,4	42,5	45,5	41,4	48,0	39,9	54,0	37,5	55,3	34,5	16,2
1,5	39,1	42,0	42,9	41,7	45,2	41,1	50,1	40,0	51,3	39,3	—
2	36,4	41,2	40,2	42,0	42,0	42,0	47,1	41,6	48,4	43,7	15,2
2,5	34,8	40,6	38,0	41,7	39,6	42,5	44,2	43,0	45,6	47,0	—
3,0	32,9	40,3	35,5	41,3	37,0	42,7	41,3	44,8	43,2	50,1	14,5
3,5	31,2	39,1	33,4	40,7	34,6	42,2	39,1	45,5	40,9	52,7	—
4,0	29,1	38,6	31,2	39,8	32,5	41,8	36,6	46,2	38,6	53,8	13,8
4,5	27,8	37,2	29,5	38,8	31,2	40,9	35,0	46,5	37,0	54,5	—
5,0	26,3	36,0	27,8	38,0	29,3	40,1	33,2	46,5	35,5	55,2	13,0

[552], $t = 25$ °C, $N = 0{,}5$, fig. 62

ε′	ε″	α	10^{-11} τ
4,9	71,1	0	0,787

*) See also [544, 550].

[377]

t °C	N	λ=51,5 cm		λ=9,95 cm		λ=3,22 cm		λ=1,25 cm		ε	α	λ_m
		ε′	ε″	ε′	ε″	ε′	ε″	ε′	ε″			
3	0,0625	85,3	4,7	78,0	22,0	47,7	39,3	—	—	86,0	0,02	2,96
	0,125	84,2	4,6	77,5	21,6	47,6	38,7	17,1	27,9	86,0	0,015	2,93
	0,188	82,1	4,4	75,4	29,9	—	—	16,7	26,7	82,5	0,019	2,92
	0,25	81,9	4,4	75,3	20,7	48,0	37,8	—	—	81,8	0,012	2,90
	0,50	—	—	—	—	45,0	35,6	—	—	—	—	2,86
10	0,0625	—	—	—	—	55,6	37,2	22,7	32,4	83,6	0,01	2,37
	0,125	—	—	77,8	17,4	54,3	36,1	—	—	82,4	0,015	2,36
	0,188	79,5	3,7	76,0	16,8	54,9	35,0	22,0	30,8	80,2	0,017	2,31
	0,25	—	—	74,7	16,4	53,8	34,8	21,9	30,3	80,0	0,02	2,30

[541], $t = 25$ °C

N	λ=10,0 cm		λ=3,12 cm		λ=1,17 cm		ε	λ_m	α
	ε′	ε″	ε′	ε″	ε′	ε″			
1	56,4	5,0	46,2	20,7	26,6	25,0	57,1	1,51	0,00
2	—	—	35,0	15,0	25,1	16,1	49,0	1,20	0,19
3	41,0	3,0	28,0	10,6	16,8	9,6	44,3	1,70	0,37
4	—	—	24,1	9,9	13,8	7,5	42,0	—	0,39

[543], $t = 25$ °C

N	λ=10 cm		λ=3 cm		ε	λ_m
	ε′	ε″	ε′	ε″		
0,468	70,0	13,0	57,4	25,2	71,0	1,56
0,936	64,9	8,1	53,8	23,0	64,5	1,53

Water – Lithium nitrate ($LiNO_3$)

[551], $t=24,96$ °C

N	λ=3,6 cm		λ=3,0 cm		λ=2,5 cm	
	ε′	ε″	ε′	ε″	ε′	ε″
0,5	59,79	23,0	57,23	25,9	53,15	28,6
1,0	54,83	20,1	52,85	22,7	49,27	25,4
1,5	50,55	18,0	48,48	20,1	45,64	22,5
2,0	46,40	15,9	44,93	18,0	42,31	20,0
3,0	39,91	12,8	38,76	14,3	37,03	16,2
4,0	35,25	10,5	34,26	11,9	32,65	13,1
6,0	27,68	7,4	27,19	8,4	26,18	9,3

Water – Magnesium chloride ($MgCl_2$)

[541], $t=25$ °C, fig. 69

N	λ=10,0 cm		λ=3,12 cm		λ=1,17 cm		ε	$ε_∞$	$λ_m$	α	g
	ε′	ε″	ε′	ε″	ε′	ε″					
1	56,4	5,0	46,2	20,7	26,6	25,0	57,1	5,5	1,51	0,00	1,91
2	—	—	35,0	15,0	25,1	16,1	49,0	—	1,20	0,19	1,64
3	41,0	3,0	28,0	10,6	16,8	9,6	44,3	3,3	1,70	0,37	1,52
4	—	—	24,1	9,9	13,8	7,5	42,0	—	—	0,39	1,46

Water – Manganese (II)nitrate ($Mn(NO_3)_2$)

[161, 547], $t=28$ °C, λ=1,276; 3,14; 4,4 cm

N	0,25	0,5	0,75	1,0
ε	68,5	65,0	63,0	61,0
$λ_m$	1,288	1,23	1,217	1,185
α	0	0	0,005	0,001

Water – Manganese (II)chloride ($MnCl_2$)

[546], $t=28$ °C, λ=3,14; 4,4 cm

N	0,25	0,50	0,75	1,0	2,0
$λ_m$	1,210	1,130	1,083	1,047	1,012
ε	68,5	65,5	60,0	57,5	54,0

Water – Sodium bromide (NaBr)

[548], $t=28$ °C, λ=1,267; 3,14; 4,4 cm

N	0,25	0,50	0,75	1,00	2,00
ε	67,5	65,3	62,5	60,0	55,0
$λ_m$	1,33	1,32	1,31	1,30	1,24
α	0,011	0,013	0,022	0,028	0,039

[558], $t = 25$ °C

N	$\lambda = 9{,}22$ cm		$\lambda = 3{,}175$ cm		$\lambda = 1{,}264$ cm		ε	λ_m
	ε'	ε''	ε'	ε''	ε'	ε''		
0,33	69,9	11,00	60,3	26,0	33,9	32,8	72,0	1,49
0,66	—	—	57,0	22,7	33,4	30,7	66,8	1,40
1,00	—	—	54,0	21,0	33,0	29,4	63,3	1,36

[541], $t = 25$ °C

N	$\lambda = 10{,}0$ cm		$\lambda = 3{,}12$ cm		$\lambda = 1{,}17$ cm		ε	λ_m	α	g
	ε'	ε''	ε'	ε''	ε'	ε''				
2	60,0	7,2	48,2	18,9	27,8	22,1	62,4	1,47	0,13	2,16
4	—	—	38,2	11,5	23,6	14,8	47,2	1,24	0,19	1,68

[562], $N = 0{,}5$

t °C	$\lambda = 9{,}22$ cm		$\lambda = 3{,}282$ cm		λ_m	ε
	ε'	ε''	ε'	ε''		
5	74,2	18,4	54,2	35,0	2,37	79,0
10	73,0	16,1	55,5	32,0	2,18	76,5
20	69,2	11,7	57,8	26,7	1,67	71,5
30	67,0	8,6	59,7	22,0	1,30	68,0
40	65,3	7,1	59,5	18,5	1,10	66,0
50	63,5	6,4	58,5	16,5	0,95	63,6
60	62,2	6,0	57,7	14,2	0,84	62,0

$N = 1$

t °C	$\lambda = 9{,}22$ cm		$\lambda = 3{,}282$ cm		$\lambda = 1{,}267$ cm		λ_m	ε
	ε'	ε''	ε'	ε''	ε'	ε''		
5	67,6	13,0	50,0	29,5	20,0	27,3	2,22	70,0
10	66,6	11,2	51,8	27,2	23,7	28,0	1,94	68,2
20	65,0	8,0	54,0	23,5	29,7	28,5	1,54	65,2
30	63,0	7,0	55,4	19,8	35,0	28,5	1,25	64,4
40	61,0	6,5	55,7	17,5	38,5	27,5	1,08	62,6
50	59,0	5,5	54,3	15,5	40,2	27,5	0,94	58,7
60	56,5	5,0	52,8	14,5	41,0	23,5	0,84	56,5

Water – Sodium chloride (NaCl)

[377], fig. 70

N	t °C	λ=9,95 cm		λ=3,22 cm		α	ε	λ_m
		ε′	ε″	ε′	ε″			
0,125	3	78,4	22,3	47,6	39,8	0,013	85,8	3,0
	25	75,3	11,0	63,7	27,6	0,025	78,0	1,54
	40	70,6	7,3	64,3	19,5	0,015	71,5	1,09
0,33	3	75,3	20,4	46,6	37,2	0,012	81,9	2,87
	25	—	—	60,3	26,0	0,022	73,0	1,52
	40	68,2	7,3	61,7	19,9	0,024	69,7	1,15
0,66	3	71,8	18,8	45,7	35,5	0,004	77,2	2,82
	25	67,5	8,8	—	—	0,013	68,7	1,42
	40	65,8	7,2	59,7	19,1	0,04	67,6	1,19
1,0	3	69	17,8	44,1	33,7	0,007	74,0	2,74
	25	63,6	8,2	54,4	21,3	0,017	65,0	1,39
	40	64,4	8,0	56,8	21,2	0,035	65,3	1,35

[543]

N	t °C	λ=10 cm		λ=3 cm		λ=1,27 cm		ε	λ_m
		ε′	ε″	ε′	ε″	ε′	ε″		
0,5	1,5	—	—	44,7	36,1	—	—	77,6	2,95
	21	72,8	12,6	58,6	28,3	—	—	73,9	1,71
0,66	0	70,7	22,9	45,5	36,5	—	—	79,1	2,96
	10	70,4	19,1	51,9	32,1	—	—	75,5	2,17
	20	69,1	15,1	56,3	27,2	—	—	72,0	1,70
	30	67,7	12,0	55,7	21,8	—	—	66,6	1,35
	40	62,8	10,0	54,3	16,9	—	—	63,0	1,10
1,0	1,5	—	—	43,0	32,8	—	—	71,0	2,80
	21	67,4	13,1	56,2	25,0	30,1	30,4	69,1	1,63
1,5	21	61,5	12,7	52,5	22,3	—	—	64,3	1,55
2,0	1,5	—	—	39,4	26,4	—	—	60,0	2,49
	21	56,2	11,8	49,6	20,7	27,8	26,0	59,0	1,51
3,0	1,5	—	—	35,9	21,4	—	—	51,0	2,26
	21	—	—	45,8	18,5	—	—	54,0	1,47
5,0	1,5	—	—	28,8	15,1	—	—	38,3	2,07
	21	—	—	41,8	15,7	—	—	48,4	1,39

[557, 553, 554], $t = 18$ °C, $\nu = 9400$ MHz

N	0,05	0,2	0,5	1	2	3
ε	79,2	76,5	73,3	67,7	58,5	52,3
σ	95,3	87,5	80,76	74,2	64,66	56,4
ε″	31,8	30,45	28,6	25,5	20	17,3
$10^{-12}\tau$	9,65	9,5	9,2	8,76	7,7	7,47

[562], $N = 0,5$

t °C	λ=9,22 cm		λ=3,282 cm		λ=1,267 cm		λ_m	ε
	ε″	ε″	ε′	ε″	ε′	ε″		
0	73,0	23,0	44,5	38,0	—	—	3,2	80,0
10	72,0	14,5	52,5	33,5	22,5	29,5	2,29	75,8
20	70,7	9,5	57,5	27,5	29,0	31,7	1,7	71,0
30	68,3	7,5	59,5	22,5	35,0	31,5	1,37	69,5
40	66,2	5,0	58,5	18,7	38,0	30,5	1,10	66,8
50	62,8	3,8	58,2	15,0	40,8	28,6	0,98	63,0
60	58,0	3,5	55,5	12,3	41,8	24,5	0,87	58,7

$N = 1$

t °C	λ=9,22 cm		λ=3,282 cm		λ=1,267 cm		λ_m	ε
	ε′	ε″	ε′	ε″	ε′	ε″		
0	76,2	23,0	45,5	39,0	—	—	3,18	83,0
10	76,0	16,7	55,5	34,0	23,5	32,0	2,22	79,0
20	75,0	12,3	60,5	29,5	31,0	34,7	1,71	77,0
30	72,7	8,7	62,3	25,2	36,8	34,5	1,37	73,8
40	70,2	6,3	62,3	21,0	40,5	33,0	1,18	70,7
50	66,7	4,5	60,5	16,8	43,0	29,5	0,98	66,5
60	63,2	3,7	58,3	15,0	44,5	26,0	0,87	63,0

[555], $t = 25$ °C

N	ν=3 GHz		ν=10 *Гец*		ν=17 *Гец*		ε	λ_m
	ε′	ε″	ε′	ε″	ε′	ε″		
1,020	60,5	7,0	52,7	23,0	40,4	29,6	62,8	1,41
2,083	50,2	3,1	45,7	1,61	35,9	23,7	52,2	1,29
4,366	35,3	—3,6	34,9	11,2	29,0	15,7	37,6	1,07
5,619	29,3	—2,4	30,9	8,9	26,3	12,7	32,1	0,96

[563, 376], λ=0,6; 1,25; 3,22 cm

N	t=0 °C			t=10 °C			t=20 °C		
	ε	$10^{-12}\tau$	$10^{-11}\sigma$	ε	$10^{-12}\tau$	$10^{-11}\sigma$	ε	$10^{-12}\tau$	$10^{-11}\sigma$
0	88	18,7	0	84	13,6	0	80	10,1	0
0,5	77	17,1	0,22	74	12,2	0,31	71	9,2	0,40
1,0	69	16,4	0,43	66	11,8	0,56	63	9,0	0,69
1,5	62	15,3	0,55	60	11,3	0,78	57	8,7	0,97
2,0	56	14,4	0,74	54	10,9	0,98	51	8,3	1,24
2,5	51	13,6	0,86	48	10,6	1,15	46	7,9	1,46
3,0	46	13,0	0,98	43	10,3	1,28	41	7,6	1,60

N	t=30 °C			t=40 °C		
	ε	$10^{-12}\tau$	$10^{-11}\sigma$	ε	$10^{-12}\tau$	$10^{-11}\sigma$
0	77	7,5	0	73	5,9	0
0,5	68	7,2	0,47	65	5,7	0,56
1,0	60	7,1	0,83	58	5,6	0,97
1,5	55	6,9	1,17	52	5,5	1,36
2,0	49	6,7	1,48	47	5,4	1,72
2,5	44	6,4	1,74	42	5,3	2,00
3,0	39	6,2	1,93	37	5,2	2,26

[541, 545, 559, 564], t=25 °C

N	λ=3,12 cm		λ=1,17 cm		ε	λ_m	α
	ε′	ε″	ε′	ε″			
2	49,2	17,2	28,1	23,0	59,5	1,33	0,10
4	38,2	12,7	23,3	15,8	46,0	1,22	0,14
5	33,7	11,5	21,2	13,6	42,0	1,32	0,18

[558], t=25 °C

N	λ=9,22 cm		λ=3,175 cm		λ=1,264 cm		ε	λ_m
	ε′	ε″	ε′	ε″	ε′	ε″		
0,33	72,4	14,3	59,5	25,9	33,8	32,9	71,8	1,49
0,66	—	—	—	—	33,1	30,9	67,2	1,42
1,00	—	—	54,6	21,7	32,4	28,7	63,0	1,37

[560]. ν = 9400 MHz

N	t=0 °C		t=6 °C		t=10 °C		t=20 °C		t=40 °C		$10^{-12}\tau$
	ε′	ε″	ε′	ε″	ε′	ε″	ε′	ε″	ε′	ε″	
0,05	45,4	42,4	52,7	40,1	55,5	38,5	61,9	31,5	64,6	22,8	17,4
0,2	44,9	42,9	52,1	40,6	54,5	39,3	60,7	32,7	—	—	16,8
0,5	44,0	43,2	51,3	41,8	53,4	40,8	58,4	35,6	59,8	29,9	16,1
1	43,5	44,0	49,5	43,0	50,5	42,5	55,2	38,8	56,1	37,8	15,5
1,5	41,8	44,2	46,7	43,7	48,1	44,5	52,4	42,5	53,3	44,6	—
2	40,2	45,2	44,5	45,4	45,9	46,1	49,8	45,8	50,6	51,5	14,0
2,5	38,3	44,9	42,3	46,0	43,7	47,8	47,5	48,5	48,5	57,2	—
3	36,5	44,6	40,4	46,4	41,2	48,4	45,2	51,0	46,4	60,8	13,2
3,5	35,1	44,6	38,4	46,4	39,7	48,9	42,9	53,0	44,7	64,6	—
4	33,5	44,3	36,3	46,9	37,6	49,1	40,5	53,5	43,0	67,2	11,7
4,5	32,1	43,2	34,8	46,6	35,8	49,3	38,5	54,2	41,4	68,4	—
5	30,5	42,3	33,3	45,6	33,8	48,8	36,6	55,4	39,8	70,7	11,0

[551], t = 24,96 °C

N	λ=3,6 cm		λ=3,0 cm		λ=2,5 cm	
	ε′	ε″	ε′	ε″	ε′	ε″
0	66,83	27,0	62,67	30,4	57,60	33,3
0,5	60,58	23,5	57,68	26,4	53,51	28,7
1,0	55,55	20,5	53,21	23,1	49,72	25,5
1,5	51,44	18,6	49,48	20,7	46,47	23,1
2,0	48,01	16,5	46,06	18,4	43,28	20,5
3,0	41,52	12,4	40,04	14,6	38,04	16,5
4,0	36,38	10,0	34,78	11,8	33,56	13,4

Water – Sodium perchlorate ($NaClO_4$)

[541], t = 25 °C

N	λ=3,12 cm		λ=1,17 cm		ε	λ_m	α
	ε′	ε″	ε′	ε″			
2	48,4	18,2	28,1	24,5	57,8	1,31	0,05
4	34,5	10,5	24,2	12,3	45,2	1,30	0,30

Water – Ammonium chloride (NH_4Cl)

[545], $t = 25$ °C

N	λ=3,175 cm		λ=1,264 cm		ε	$λ_m$
	ε′	ε″	ε′	ε″		
0,33	60,6	26,3	34,3	33,3	73,0	1,50
0,66	58,5	24,6	33,4	31,7	69,4	1,46
1,00	55,8	22,8	33,0	30,1	66,0	1,41

Water – Sodium fluoride (NaF)

[543], $t = 25$ °C

N	λ=10 cm		λ=3 cm		ε	$λ_m$
	ε′	ε″	ε′	ε″		
0,415	71,3	7,4	58,3	25,0	73,0	1,55
0,830	72,4	8,4	56,2	24,0	69,0	1,52

[545], $t = 25$ °C

N	λ=9,22 cm		λ=3,175 cm		λ=1,264 cm		ε	$λ_m$
	ε′	ε″	ε′	ε″	ε′	ε″		
0,33	71,9	14,7	61,2	27,3	33,7	33,5	74,5	1,55
0,66	69,2	14,3	59,0	26,3	32,8	31,8	71,3	1,54
1,00	—	—	56,9	25,5	32,3	30,9	68,8	1,53

Water – Sodium iodide (NaI)

[557, 541], $t = 18$ °C, $ν = 9400$ MHz

N	0,05	0,2	0,5	1	2	3	4
ε	79,1	75,9	73,1	67,5	58,3	50,9	45
σ	97,6	90,1	84	78,5	70,1	61,8	53,9
ε″	31,9	30,2	28,4	25,2	20,1	17,1	14,7
$10^{-12}τ$	9,7	9,5	9,13	8,6	7,75	7,45	7,3

[558], $t = 25$ °C

N	λ=9,22 cm		λ=3,175 cm		λ=1,264 cm		ε	λ_m
	ε′	ε″	ε′	ε″	ε′	ε″		
0,33	72,1	11,4	60,6	25,5	34,1	32,7	72,0	1,47
0,66	—	—	57,4	22,7	33,7	30,6	67,0	1,39
1,00	—	—	55,1	20,7	33,3	28,4	62,7	1,30

[541], $t = 25$ °C

N	λ=10,0 cm		λ=3,12 cm		λ=1,17 cm		ε	λ_m	α
	ε′	ε″	ε′	ε″	ε′	ε″			
2	—	2	46,7	17,4	29,0	22,5	55,5	1,28	0,06
4	—	—	37,8	10,5	23,8	14,2	45,7	1,16	0,20
8	35,0	2	23,5	7,3	16,0	7,0	37,0	2,8	0,40

[543], $t = 25$ °C

N	λ=10 cm		λ=3 cm		ε	λ_m
	ε′	ε″	ε′	ε″		
0,428	71,5	8,3	59,4	25,1	71,0	1,50
0,856	—	—	54,5	20,8	64,0	1,42

[560], ν = 9400 MHz

N	t=0 °C		t=6 °C		t=10 °C		t=20 °C		t=40 °C	
	ε′	ε″	ε′	ε″	ε′	ε″	ε′	ε″	ε′	ε″
0,05	45,3	42,6	52,7	40,2	55,6	38,2	61,9	31,4	64,5	21,9
0,2	45,4	42,8	52,6	40,8	54,8	39,4	61,0	32,7	62,5	25,5
0,5	44,9	44,0	52,2	42,0	54,1	40,9	58,8	35,2	59,8	31,4
1	44,5	44,8	50,1	43,4	51,6	42,7	55,6	39,4	55,7	39,3
1,5	43,7	45,2	48,1	45,0	49,3	45,3	52,2	43,5	52,6	47,1
2	42,2	45,7	45,4	46,1	46,6	47,2	49,5	47,5	49,6	54,0
2,5	40,5	45,9	43,5	47,4	44,3	49,0	46,8	50,4	47,2	60,1
3	38,5	46,7	41,3	48,0	42,0	50,5	44,0	53,3	45,0	64,9
3,5	36,4	46,4	39,0	48,8	39,9	51,6	41,5	55,7	43,2	68,4
4	34,9	46,4	36,8	49,4	37,5	52,5	39,6	57,6	41,3	72,0
4,5	32,9	45,9	34,5	49,5	35,3	52,8	37,4	58,3	39,5	73,1
5	31,4	45,6	32,6	49,2	33,2	52,8	35,5	59,0	38,4	74,6

Water – Sodium nitrate ($NaNO_3$)

[541], $t=25$ °C

N	$\lambda=3,12$ cm		$\lambda=1,17$ cm		ε	λ_m	α
	ε'	ε''	ε'	ε''			
2	49,8	19,5	28,4	25,0	60,3	1,37	0,05
4	38,7	12,0	26,6	15,1	48,4	1,15	0,21

[551], $t=24,96$ °C

N	$\lambda=3,6$ cm		$\lambda=3,0$ cm		$\lambda=2,6$ cm	
	ε'	ε''	ε'	ε''	ε'	ε''
0,5	60,61	28,2	57,88	26,0	53,83	28,8
1,0	56,78	20,4	54,40	23,5	50,86	25,9
1,5	53,04	18,9	51,11	21,3	48,14	23,4
2,0	49,61	17,1	48,07	19,3	45,60	21,8
3,0	44,81	14,7	43,24	16,4	41,27	18,0
4,0	40,23	12,8	39,13	14,3	37,25	15,8

Water – Sodium hydroxide ($NaOH$)

[543], $t=25$ °C

N	$\lambda=10$ cm		$\lambda=3$ cm		ε	λ_m
	ε'	ε''	ε'	ε''		
0,25	75,7	10,3	58,9	26,6	73,0	1,58
0,5	66,5	9,5	57,0	24,7	68,0	1,58

Water – Sodium sulphate (Na_2SO_4)

[543], $t=25$ °C, $\lambda=3$ cm

N	ε'	ε''	ε	λ_m
0,5	57,3	26,4	73,0	1,54
1,0	55,0	25,1	67,0	1,48
20	51,2	20,8	60,5	1,40

[541], $t=25$ °C, $N=1$

λ=3,12 cm		λ=1,17 cm		ε	λ_m	α
ε′	ε″	ε′	ε″			
50,3	22,4	27,1	26,4	62,5	1,46	0,04

Water – Nickel chloride ($NiCl_2$)
[161, 547], $t=28$ °C,
$\lambda=1{,}276$; 3,14; 4,4 cm

N	0,25	0,5	0,75	1,0
ε	68,0	63,5	60,0	56,8
λ_m	1,28	1,27	1,22	1,20
α	0	0	0,005	0,011

Water – Nickel nitrate ($Ni(NO_3)_2$)
[546], $t=28$ °C
$\lambda=3{,}14$; 4,4 cm

N	0,25	0,50	0,75	1,0	2,0
λ_m	1,190	1,072	1,045	1,021	1,016
ε	70,5	65,0	60,5	56,0	50,5

Water – Nickel sulphate ($NiSO_4$)
[548], $t=28$ °C, $\lambda=1{,}267$; 3,14; 4,4 cm

N	0,25	0,50	0,75	1,00	2,00
ε	67,5	64,0	62,5	60,0	57,0
λ_m	1,28	1,26	1,25	1,10	1,16
α	0,022	0,028	0,033	0,040	0,051

Water – Rubidium chloride (RbCl)
[377]

N	t °C	λ=16,7 cm		λ=9,95 cm		λ=3,22 cm		ε	α	λ_m
		ε′	ε″	ε′	ε″	ε′	ε″			
0,125	3	82,0	13,5	77,0	21,3	47,4	38,5	84,0	0,016	2,92
	10	—	—	78,4	17,4	55,3	36,7	83,3	0,17	2,35
0,25	3	80,0	12,7	76,5	20,4	48,1	37,7	83,0	0,016	2,82
	10	78,0	9,9	75,8	16,1	54,7	34,8	80,3	0,01	2,25
0,375	3	—	—	73,3	18,6	48,2	36,1	78,4	0	2,69
	10	—	—	73,8	15,3	54,6	33,8	77,7	0,002	2,20
0,5	3	—	—	72,5	18,1	48,1	35,5	77,5	0	2,66
	10	—	—	73,2	14,9	54,0	33,1	76,3	0	2,17
1,0	3	—	—	68,8	16	47,5	33,0	73,3	0	2,50
	10	—	—	68,5	13,4	51,4	30,0	71,4	0,002	2,09

[543], $t = 25$ °C

N	λ=10 cm		λ=3 cm		ε	$λ_m$
	ε′	ε″	ε′	ε″		
0,5	72,1	8,9	59,5	25,0	73,5	1,54
1,0	71,0	8,4	56,1	23,2	68,5	1,50
1,5	—	—	52,5	22,0	63,5	1,47
2,0	—	—	49,2	20,2	58,5	1,43

[558], $t = 25$ °C

N	λ=9,22 cm		λ=3,175 cm		λ=1,264 cm		ε	$λ_m$
	ε′	ε″	ε′	ε″	ε′	ε″		
0,33	72,6	15,3	60,5	27,1	34,1	32,9	74,0	1,54
0,66	—	—	59,7	25,1	34,1	31,8	—	1,46
1,00	—	—	56,3	22,2	33,7	29,5	65,2	1,38

[551], $t = 24,96$ °C

N	λ=3,6 cm		λ=3,0 cm		λ=2,5 cm	
	ε′	ε″	ε′	ε″	ε′	ε″
0,5	61,41	23,9	58,80	26,8	54,22	29,4
1,0	57,64	21,4	55,47	24,3	51,97	27,0
1,5	55,01	19,7	52,81	22,1	49,46	24,7
2,0	51,85	17,0	50,13	19,5	47,57	21,9
3,0	46,09	13,8	45,15	15,8	42,74	17,7

Water – Rubidium nitrate ($RbNO_3$)

[551], $t = 24,96$ °C

N	λ=3,6 cm		λ=3,0 cm		λ=2,5 cm	
	ε′	ε″	ε′	ε″	ε′	ε″
0,5	62,26	24,1	59,29	27,0	55,29	30,0
1,0	59,59	22,4	56,93	25,1	53,48	27,6
1,5	57,66	21,1	55,38	23,8	51,99	26,0

Supplementary information on the dielectric dispersion parameters of aqueous solutions of inorganic compounds

Water – Aluminium oxide (γ-Al_2O_3), fig. 57.
Water – Beryllium sulphate ($BeSO_4$), fig. 58.
Water – Cadmiun sulphate ($CdSO_4$) [545].
Water – Caesium bromide (CsBr) [550], t = 25°C, ν = 5 – 38 GHz.
Water – Caesium fluoride (CsF) [550], t = 25°C, ν = 5 – 38 GHz.
Water – Caesium sulphate ($CsSO_4$) [540], t = 24°C, λ = 3.5 cm, N = 0.5 – 3.
Water – Iron(III)chloride ($FeCl_3$) [540], t = 20°C, λ = 3.5 cm, N = 0.05 – 3.4.
Water –Iron(III)nitrate ($Fe(NO_3)_3$) [540], t = 20°C, λ = 3.5 cm, N = 0.05 – 0.8.
Water – Potassium sulphate (K_2SO_4) [540], t = 20°C, λ = 3.5 cm, N = 0.01 – 0.6.
Water – Lithium sulphate (Li_2SO_4) [540] t = 20°C, λ = 3.5 cm, N = 0.01 – 2.5.
Water – Sodium tungstate (Na_2WO_4) [540], t = 25°C, λ = 3.5 cm, N = 0.36 – 2.2.
Water – Nickel sulphate ($NiSO_4$), fig. 63, 71.
Water – Rubidium bromide (RbBr) [550], t = 25°C, ν = 5 – 38 GHz.
Water – Rubidium fluoride (RbF) [550]. t = 25°C, ν = 5 – 38 GHz.
Water – Rubidium iodide (RbI) [550], t = 25°C, ν = 5 – 38 GHz.
Water – Strontium chloride ($SrCl_2$) [540], t = 22°C, λ = 3.5 cm, N = 0.01 – 3.
Water –Zinc chloride ($ZnCl_2$) [544], t = 25°C, λ = 10 cm.
Water – Zinc sulphate ($ZnSO_4$) [553].

§ 4. Dielectric dispersion parameters of aqueous solutions of organic compounds

Water – Formamide (CH_3NO)

[377, 565], fig. 72

N	t °C	λ=51,5 cm		λ=16,7 cm		λ=9,95 cm		λ=3,22 cm		α	ε	λ_m
		ε'	ε''	ε'	ε''	ε'	ε''	ε'	ε''			
0,125	3	84,2	5,2	82,4	13,5	77,6	21,3	48,2	39,2	0,01	85,4	2,92
	10	—	—	81,2	10,8	78,7	17,4	54,6	36,2	0,018	83,0	2,35
	25	—	—	—	—	75,7	10,9	63,7	27,9	0,01	77,6	1,53
	40	—	—	—	—	71,2	7,3	64,6	20,2	0,017	72,2	1,09

N	t °C	λ=51,5 cm ε′	λ=51,5 cm ε″	λ=16,7 cm ε′	λ=16,7 cm ε″	λ=9,95 cm ε′	λ=9,95 cm ε″	λ=3,22 cm ε′	λ=3,22 cm ε″	α	ε	λ_m
0,25	3	81,4	5,2	79,8	12,7	75,0	20,1	47,8	37,9	0,005	83,6	2,85
	10	80,2	4,0	78,8	10,3	75,7	16,6	54,0	35,4	0,01	80,7	2,33
	25	—	—	—	—	74,0	10,5	62,4	26,9	0,008	75,8	1,51
	40	—	—	—	—	69,4	7,0	63,0	19,4	0,023	70,4	1,08
0,5	3	—	—	77,2	12	73,3	19,2	47,0	36,9	0,007	79,1	2,79
	10	—	—	75,7	9,5	73,5	15,8	53,2	33,9	0,01	77,8	2,26
	25	—	—	—	—	71,5	10,4	60,0	26,5	0,01	73,2	1,55
	40	—	—	—	—	67,2	7,2	60,4	19,8	0,02	68,3	1,15
1,0	3	—	—	—	—	68,6	18,0	45,7	33,7	0,002	73,7	2,66
	10	—	—	70,0	8,4	67,7	13,6	50,8	30,6	0	72,1	2,15
	25	—	—	—	—	67,4	9,9	55,5	24,6	0,04	69,0	1,57
	40	—	—	—	—	64,2	7,2	56,8	19,3	0,03	65,4	1,20

Water – Methanol (CH_4O) *)

[566], fig. 73

φ_2	t °C	λ=10,67 cm ε′	λ=10,67 cm ε″	λ=3,2 cm ε′	λ=3,2 cm ε″	λ=1,14 cm ε′	λ=1,14 cm ε″	λ=0,818 cm ε′	λ=0,818 cm ε″	λ=0,423 cm ε′	λ=0,423 cm ε″
90	10	73,8	20,4	40,5	37,4	—	—	11,8	16,95	7,4	11,0
	20	73,1	14,4	50,2	34,2	22,2	32,0	14,0	23,9	8,0	14,0
	30	71,3	10,9	56,4	29,3	28,6	34,0	18,5	27,0	9,9	16,4
	40	68,6	8,4	58,5	24,5	33,5	32,5	23,2	29,5	11,8	18,6
	50	65,9	6,6	59,0	19,6	38,5	29,2	28,0	29,6	14,2	20,4
70	10	60,6	24,3	—	—	—	—	7,75	11,9	6,10	6;70
	20	62,5	18,7	34,0	30,7	12,4	19,0	9,50	15,1	6 75	8,65
	30	61,9	14,3	40,2	29,1	15,8	21,9	11,7	18,2	7,25	10,6
	40	60,5	10,8	44,4	26,3	19,8	23,6	14,4	21,4	8,45	12,5
	50	58,2	8,2	48,6	21,6	23,5	24,7	17,0	22,4	9,57	14,3
50	10	46,5	24,4	17,8	22,4	—	—	5,77	8,86	5,10	4,68
	20	49,9	20,3	21,5	24,5	8,5	12,5	7,07	10,0	5,52	5,85
	30	50,8	16,4	26,6	24,8	10,3	14,8	8,30	12,0	5,85	7,14
	40	50,7	12,5	31,0	24,0	12,8	16,6	9,70	13,9	6,34	8,56
	50	49,4	9,6	35,5	21,7	15,7	18,2	11,4	15,9	6,79	10,1
30	10	33,5	22,2	11,3	14,8	—	—	—	—	—	—
	20	37,2	20,1	13,9	16,5	6,85	8,4	5,45	4,17	4,76	3,84
	30	39,3	17,0	17,0	18,0	7,75	9,9	5,68	4,66	4,94	4,62
	40	39,7	13,8	20,7	18,9	8,9	11,2	5,85	5,17	5,25	5,3
	50	39,7	10,8	13,2	18,2	9,9	12,2	6,00	5,65	5,60	6,05
10	10	21,1	17,2	8,1	9,1	—	—	5,40	3,30	4,49	2,48
	20	23,9	16,6	9,4	10,4	6,00	5,10	5,45	4,17	4,51	2,74
	30	26,2	15,1	10,8	11,6	6,30	5,75	5,68	4,66	4,56	3,05
	40	27,5	13,3	12,2	12,6	6,65	6,40	5,85	5,17	4,63	3,32
	50	28,6	10,9	13,9	12,8	6,90	7,00	6,00	5,65	4,87	3,62

*) See also [421], ν = 5 GHz [533], λ = 14 cm; [568], λ = 1 – 30 cm.

[567], $\lambda = 10,6;\ 3,2;\ 1,14;\ 0,818;\ 0,423$ cm

φ_2	ε at t °C					$\varepsilon_{\infty 1}$ at t °C					$\varepsilon_{\infty 2}$ at t °C				
	10	20	30	40	50	10	20	30	40	50	10	20	30	40	50
100	35,7	33,6	31,6	29,7	28,0	5,9	5,8	5,6	5,5	5,4	3,5	3,4	3,4	3,4	3,4
90	42,2	39,9	37,8	35,6	33,8	6,7	6,5	6,4	6,2	6,0	3,7	3,6	3,5	3,5	3,6
70	53,7	50,8	48,4	45,2	42,9	7,0	7,3	7,5	7,8	7,5	4,0	3,9	3,7	3,8	3,6
50	62,8	60,6	56,9	54,6	52,1	8,0	8,6	9,3	10,0	11,1	4,4	4,4	4,2	4,2	4,2
30	72,9	69,6	66,2	63,1	60,2	8,9	10,6	12,2	14,1	13,2	5,2	5,1	4,8	4,9	5,0
10	80,8	77,2	73,9	70,6	67,9	5,8	5,5	5,8	5,8	6,0	—	—	—	—	—

φ_2	$10^{-11}\tau_1$ at t °C					$10^{-12}\tau_2$ at t °C				
	10	20	30	40	50	10	20	30	40	50
100	6,75	5,44	4,40	3,63	3,02	2,6	2,5	2,4	2,2	2,1
90	7,0	5,35	4,2	3,3	2,66	4,3	3,7	3,2	3,0	2,6
70	5,0	3,82	3,0	2,33	1,86	5,5	4,9	4,3	3,8	3,2
50	3,63	2,78	2,16	1,72	1,38	6,0	5,3	4,5	4,0	3,6
30	2,58	2,0	1,55	1,25	0,97	6,6	5,7	5,0	4,5	3,9
10	1,55	1,16	0,91	0,735	0,595	—	—	—	—	—

Water – Urea (CH_4N_2O)

[544], $\lambda = 10$ cm, fig. 65, 74, 75

N	0,5	1,0	2,0
ε'	77,6	78,7	79,7
ε''	12,15	13,04	14,73

Water – Acetic acid ($C_2H_4O_2$) *)

[571], $t = 30$ °C, $\nu = 1200$ MHz

φ_2	ε'	ε''	ε	λ_m	$10^{-11}\tau$
96,06	28,820	3,001	29,10	2,807	1,490
92,28	4,204	3,348	42,30	2,116	1,123
73,86	45,51	4,437	55,86	2,104	1,116
66,6	61,27	4,810	61,63	2,063	1,094
59,21	75,63	5,948	76,11	2,053	1,089
54,00	73,01	5,684	73,48	2,041	1,082
50,00	65,14	9,92	66,74	4,023	2,134
41,5	73,94	8,762	75,01	3,121	1,656
36,00	78,72	8,139	79,60	2,720	1,443
21,75	63,64	5,674	64,14	2,392	1,269

*) See also [569, 570, 415], $t = 10$—40°C, $\nu = 9,37$ GHz [610], $t = 25$ °C, $\nu = 600$—3600 MHz

Water – Glycine ($C_2H_5O_2$)

[572, 186], $N=1$

t °C	ν, MHz	ε′	ε″	ε	λ_m
0	450	107,5	10,92	109,6	22,0
	600	107,7	14,62	111,4	21,8
	760	107,5	18,74	113,5	21,9
10	450	105,3	8,47	106,7	17,1
	600	106,2	11,36	108,6	16,5
	760	103,4	13,77	107,2	17,3
20	450	101,0	6,34	101,9	13,5
	600	101,4	8,51	102,9	13,4
	760	100,4	10,49	102,8	13,3
	900	99,35	12,25	102,7	13,9

t °C	ν, MHz	ε′	ε″	ε	λ_m
30	450	98,7	4,94	99,2	10,1
	600	99,0	6,45	99,8	9,6
	760	98,1	8,18	99,5	9,9
	900	98,05	9,39	99,8	9,6
40	450	94,7	3,89	95,0	8,3
	600	94,9	4,99	95,4	7,7
	760	94,2	6,53	95,1	8,3
	900	94,6	7,21	95,7	7,8
50	450	91,3	3,34	91,6	7,4
	600	91,2	4,05	91,6	6,8
	760	91,1	6,05	91,9	7,6

[573], $\lambda=3,2$ cm

N	t °C	ε′	ε″	ε	ε_∞	$10^{11}\tau$	ΔF	ΔH
0,0	20	62,8	31,5	80,4	6,4	0,949	2,36	3,94
	25	64,5	28,4	78,5	6,9	0,837	2,34	—
	30	65,7	25,7	76,8	6,2	0,734	2,31	—
	40	66,2	20,9	73,3	4,7	0,577	2,26	—
	50	65,6	16,9	70,0	—	0,442	—	—
0,25	20	61,9	33,2	86,1	16,4	1,24	2,52	1,92
	25	63,8	30,7	84,1	17,4	1,12	2,51	—
	30	65,3	28,3	82,4	18,5	1,03	2,51	—
	40	66,0	23,8	78,8	21,7	0,914	2,54	—
	50	65,9	20,2	75,4	22,9	0,799	2,56	—
0,50	20	61,1	34,7	91,8	21,9	1,50	2,63	1,46
	25	63,3	32,6	89,8	23,2	1,38	2,64	—
	30	64,7	30,5	88,0	24,8	1,30	2,65	—
	40	65,7	26,3	84,3	28,5	1,20	2,71	—
	50	66,3	23,2	80,8	29,2	1,06	2,74	—
0,75	20	60,0	36,0	97,5	25,4	1,77	2,73	0,83
	25	62,2	34,1	95,4	27,2	1,65	2,74	—
	30	64,2	32,6	93,6	28,1	1,53	2,75	—
	40	65,8	28,8	89,8	31,2	1,42	2,82	—
	50	66,2	25,7	86,2	33,2	1,32	2,88	—
1,00	20	59,2	37,4	103,2	27,5	2,00	2,80	1,23
	25	61,8	36,0	101,1	28,8	1,85	2,81	—
	30	63,4	34,2	99,2	30,7	1,78	2,84	—
	40	65,8	31,2	95,3	32,8	1,61	2,80	—
	50	66,8	28,5	91,7	34,2	1,48	2,95	—

Water – Ethanol (C_2H_6O) *)

[110], ν=5 Hz—5 MHz, fig. 68, 69, 73

t °K	ε	ε_∞	$\varepsilon_{\infty 2}$	ν_{m1}, kHz	ν_{m2}, kHz
w_2=95					
116,8	104	10,6	4,5	0,029	3,1
126,0	96,1	9,2	4,7	1,84	460
133,7	90,1	8,6	4,9	14,6	4900
141,9	83,7	8,2	—	76	—
153,1	75,4	7	—	425	—
162,3	70,0	—	—	1240	—
w_2=99					
110,1	102,0	8,6	4,1	0,0168	0,52
115,3	98,2	8,6	4,7	0,203	14,0
120,4	91,7	8,5	4,7	2,67	225
126,0	87,4	8,5	5,1	11,4	1500
138,9	77,8	7,6	—	125	—
153,7	68,7	—	—	780	—

*) Temperature dependence $\varepsilon=-A+\frac{B}{T}$, where A=24,20; B=15,26·10^{-3} (w_2=95) and A=21,09; B=13,68·10^{-3} (w_2=99) at 110—300 °K with accuracy 1%.

See also [568, 533, 421].

Water – Ethylamine (C_2H_7N)

[558], t=25 °C

N	λ=9,22 cm		λ=3,175 cm		λ=1,264 cm		ε	λ_m
	ε′	ε″	ε′	ε″	ε′	ε″		
0,6	72,4	12,9	57,6	30,0	30,1	33,1	75,0	1,77
1,16	—	—	55,2	30,8	27,9	31,7	72,5	1,84

Water – 1, 2–Diaminoethane ($C_2H_8N_2$)

[558], t=25 °C

N	λ=9,22 cm		λ=3,175 cm		λ=1,264 cm		ε	λ_m
	ε′	ε″	ε′	ε″	ε′	ε″		
0,525	74,3	12,0	59,7	29,2	31,3	33,0	75,5	1,71
1,05	71,0	12,5	—	—	—	—	73,0	—
1,57	68,5	13,5	51,5	30,2	24,7	29,5	71,0	2,06

Water – Acetone (C_3H_6O)

[574, 392, 393, 416], $\lambda = 3{,}21$ cm, fig. 76

x_1	$t=0$ °C		$t=10$ °C		$t=20$ °C		$t=30$ °C		$t=40$ °C	
	ε′	ε″	ε′	ε″	ε′	ε″	ε′	ε″	ε′	ε″
0	22,5	4,7	21,4	4,0	20,9	3,5	19,8	2,9	18,8	2,7
20	22,7	8,8	22,5	7,5	22,5	6,3	22,2	5,2	21,9	4,1
40	21,5	12,5	22,6	11,4	23,7	10,3	24,4	8,8	24,9	7,3
60	18,5	15,8	22,0	16,4	24,4	15,6	26,5	14,4	28,3	12,8
80	18,3	22,7	23,7	24,9	28,5	24,8	32,1	23,3	35,6	21,0
90	21,5	28,5	29,1	30,5	36,0	30,7	41,3	27,9	46,1	24,3
95	26,1	33,5	37,1	34,4	44,6	32,9	50,6	29,8	54,0	24,7
98	34,5	38,2	45,6	36,4	53,9	33,3	58,0	29,0	60,4	27,8
100	43,4	41,1	54,8	37,4	62,3	31,9	65,3	26,1	65,9	21,1

x_1	ε at t °C					$10^{-12}\tau$ at t °C				
	0	10	20	30	40	0	10	20	30	40
0	23,5	22,4	21,3	20,2	19,2	3,7	3,2	3,0	2,8	2,6
20	28,8	27,1	25,5	23,9	22,6	7,2	5,9	4,8	4,1	3,4
40	36,3	34,2	32,3	30,3	28,5	14,2	11,0	8,6	6,7	5,3
60	45,6	42,5	40,0	38,0	35,6	25,7	18,4	14,0	10,9	8,4
80	62,4	58,9	55,5	52,2	49,3	33,1	23,9	17,5	17,7	10,6
90	73,6	70,0	66,4	63,0	59,8	31,0	22,5	16,4	12,7	9,7
95	80,4	76,4	73,3	69,9	66,3	27,5	18,8	14,1	10,7	8,4
98	83,4	80,8	77,4	73,9	70,4	22,1	15,4	11,6	9,1	7,0
100	87,6	84,0	80,3	76,8	73,3	18,2	13,0	9,4	7,4	5,8

x_1	$t=0$ °C		$t=10$ °C		$t=20$ °C		$t=30$ °C		$t=40$ °C	
	Δ*F*	Δ*H*	Δ*F*	Δ*H*	Δ*F*	Δ*H*	Δ*F*	Δ*H*	Δ*F*	Δ*H*
0	1,47	0,64	1,46	0,62	1,49	0,60	1,52	0,58	1,55	0,56
20	1,83	2,62	1,80	2,60	1,76	2,58	1,75	2,56	1,71	2,54
40	2,19	3,66	2,14	3,64	2,09	3,62	2,04	3,60	1,99	3,58
60	2,52	4,26	2,44	4,24	2,38	4,22	2,34	4,20	2,27	4,18
80	2,65	4,50	2,59	4,48	2,51	4,46	2,48	4,44	2,42	4,42
90	2,62	4,58	2,55	4,56	2,47	4,54	2,43	4,52	2,36	4,50
95	2,55	4,58	2,45	4,56	2,38	4,54	2,33	4,52	2,27	4,50
98	2,43	4,50	2,34	4,48	2,26	4,46	2,23	4,44	2,16	4,42
100	2,33	4,42	2,24	4,40	2,14	4,38	2,10	4,36	2,04	4,34

x_1	α at t °C					x_1	α at t °C				
	0	10	20	30	40		0	10	20	30	40
0	0,040	0,036	0,023	0	0	90	0,039	0,040	0,028	0,017	0
20	0,140	0,120	0,095	0,060	0,111	95	0,028	0,039	0,033	0	0
40	0,200	0,183	0,167	0,139	0,094	98	0,016	0,022	0,017	0	0
60	0,167	0,140	0,117	0,083	0,054	100	0,011	0,017	0,011	0	0
80	0,78	0,63	0,52	0,033	0,011						

[575]

x_2	t °C	λ=10,67 cm ε'	λ=10,67 cm ε''	λ=0,818 cm ε'	λ=0,818 cm ε''	λ=0,423 cm ε'	λ=0,423 cm ε''
90	1	28,4	4,14	10,8	9,93	6,31	7,37
	20	27,4	3,38	11,7	10,45	6,86	7,82
	30	26,35	2,68	12,6	10,2	7,40	7,85
	40	25,6	2,22	13,4	9,76	7,90	7,75
70	10	38,6	12,5	7,80	9,47	5,50	5,85
	20	38,1	9,55	9,0	10,5	5,81	6,68
	30	37,7	7,44	10,25	11,45	6,12	7,53
	40	36,7	5,74	11,45	12,55	6,48	8,26
50	10	49,9	19,8	7,20	10,0	5,63	6,03
	20	50,8	14,6	9,10	11,4	6,30	7,70
	30	49,2	11,3	11,1	13,3	7,0	9,44
	40	47,8	8,77	13,0	15,1	7,69	11,24
30	10	62,7	21,4	8,60	13,1	6,87	8,76
	20	62,45	15,6	11,8	16,8	7,78	11,0
	30	61,7	11,55	15,0	19,45	8,70	13,0
	40	59,7	8,68	18,4	22,05	9,70	15,1
10	10	73,9	18,3	12,4	18,8	7,92	12,15
	20	72,8	13,6	16,4	23,8	9,0	14,7
	30	70,6	10,15	21,2	27,55	10,6	17,6
	40	68,0	7,83	26,4	29,0	12,4	20,4
0	10	80,1	16,2	15,0	24,5	8,94	14,9
	20	77,1	12,6	20,1	29,7	10,7	18,4
	30	75,2	9,05	25,2	32,9	12,25	21,8
	40	72,8	7,10	30,3	34,4	14,4	25,4

φ_2 *)	$\frac{\tau_{max}}{10^{11}}$	$\frac{\tau_0}{10^{11}}$	β	α	ε_∞	φ_2 *)	$\frac{\tau_{max}}{10^{11}}$	$\frac{\tau_0}{10^{11}}$	β	α	ε_∞
100	—	0,34	—	0,0	2,5	30	1,72	—	0,85	—	4,3
90	—	0,66	—	0,103	2,7	10	1,27	—	0,87	—	4,2
70	—	1,50	—	0,092	4,0	0	0,99	—	0,92	—	4,2
50	1,90	—	0,90	—	4,2						

*) τ_{max} determined by the Davidson–Cole formula, τ_0 determined by the Cole–Cole formula.

[576], λ=10,67; 3,2; 0,813; 0,423 cm

φ_2	x_2	t=10 °C ε	$\varepsilon_{\infty 1}$	$\varepsilon_{\infty 2}$	$\frac{\tau_1}{10^{11}}$	$\frac{\tau_2}{10^{12}}$	t=20 °C ε	$\varepsilon_{\infty 1}$	$\varepsilon_{\infty 2}$	$\frac{\tau_1}{10^{11}}$	$\frac{\tau_2}{10^{12}}$
100	100	22,3	2,5	—	0,370	—	21,2	2,5	—	0,340	—
90	68,8	29,1	21,6	3,6	1,95	5,5	28,3	20,8	3,6	1,55	4,8
70	36,3	45,4	18,0	4,3	2,80	7,8	42,2	17,4	4,2	2,10	6,7

φ_2	x_2	t=10 °C					t=20 °C				
		ε	$\varepsilon_{\infty 1}$	$\varepsilon_{\infty 2}$	$\frac{\tau_1}{10^{11}}$	$\frac{\tau_2}{10^{12}}$	ε	$\varepsilon_{\infty 1}$	$\varepsilon_{\infty 2}$	$\frac{\tau_1}{10^{11}}$	$\frac{\tau_2}{10_{12}}$
50	19,7	59,1	11,6	4,5	—	—	55,8	14,8	4,5	—	—
30	9,5	70,0	12,0	5,0	2,33	5,4	67,0	13,2	5,2	1,75	4,6
10	2,6	80,0	17,0	5,4	1,72	6,4	76,2	18,0	5,1	1,30	5,0
0	0	83,8	35,8	5,6	1,57	8,4	80,2	36,2	5,4	1,20	6,2

φ_2	x_2	t=30 °C					t=40 °C				
		ε	$\varepsilon_{\infty 1}$	$\varepsilon_{\infty 2}$	$\frac{\tau_1}{10^{11}}$	$\frac{\tau_2}{10^{12}}$	ε	$\varepsilon_{\infty 1}$	$\varepsilon_{\infty 2}$	$\frac{\tau_1}{10^{11}}$	$\frac{\tau_2}{10^{12}}$
100	100	20,2	2,6	—	0,317	—	19,2	2,8	—	0,293	—
90	68,8	27,0	20,0	3,6	1,25	4,8	25,7	18,2	3,8	1,03	3,7
70	36,3	39,6	16,9	4,2	1,64	5,7	37,3	17,0	3,8	1,30	5,1
50	19,7	52,7	15,7	4,3	—	—	49,6	15,6	4,2	—	—
30	9,5	64,2	14,2	5,3	1,34	4,1	61,2	17,2	3,9	1,05	3,5
10	2,6	72,8	18,8	4,2	1,01	3,8	69,6	19,6	3,6	0,80	3,1
0	0	76,7	36,7	3,6	0,925	4,8	73,2	35,2	2,7	0,730	3,8

t °C	10	20	30	40
C_1	0,60	0,59	0,55	0,53
C_2	0,40	0,41	0,45	0,47

Water – Propionic acid ($C_3H_6O_2$)

[558], $t=25$ °C

N	λ=9,22 cm		λ=3,17 cm		λ=1,264 cm		ε	λ_m
	ε'	ε''	ε'	ε''	ε'	ε''		
0,5	73,3	12,7	59,1	28,5	31,4	33,3	75,0	1,66
1,0	69,4	13,4	55,4	28,5	28,9	31,4	71,5	1,77
1,5	65,7	13,0	51,4	28,7	26,4	29,6	6,4	1,91

[571], $t=30$ °C, $\nu=1200$ MHz

φ_2	ε'	ε''	ε	λ_m	$\frac{\tau}{10^{11}}$	φ_2	ε'	ε''	ε	λ_m	$\frac{\tau}{10^{11}}$
99,00	3,626	0,9891	4,615	17,84	9,466	50,00	23,90	12,17	31,16	14,91	7,912
80,00	18,26	5,807	20,34	9,268	4,929	41,38	39,3	9,963	42,11	7,015	3,638
64,28	34,97	10,82	38,63	8,484	4,502	32,43	56,10	13,58	59,65	6,518	3,458
56,26	32,47	3,525	34,96	7,312	3,880	26,08	54,98	10,17	57,01	5,012	2,660

Water – α–Alanine ($C_3H_7NO_2$)

[558], $t=25$ °C

N	λ=9,22 cm		λ=1,264 cm		ε	$λ_m$
	ε′	ε″	ε′	ε″		
0,5	70,9	71,0	31,5	32,9	72,5	1,57
1,0	64,4	65,0	29,0	30,1	66,5	1,59
1,5	57,7	59,3	26,7	27,4	60,2	1,60

Water – β–Alanine ($C_3H_7NO_2$)

[558], $t=25$ °C

N	λ=9,22 cm		λ=1,264 cm		ε	$λ_m$
	ε′	ε″	ε′	ε″		
0,5	71,3	10,9	31,9	33,3	73,0	1,56
1,0	64,8	10,0	29,9	30,5	67,0	1,55
1,5	—	—	28,1	28,0	61,1	1,54

Water – Propan–1–ol (C_3H_8O)

[562], $N=0,41$, fig. 73

t °C	λ=9,22 cm		λ=3,282 cm		λ=1,267 cm		$λ_m$	ε
	ε′	ε″	ε′	ε″	ε′	ε″		
5	76,0	24,0	48,0	38,0	17,0	28,0	—	82,8
10	75,5	20,0	53,7	36,0	20,5	90,0	2,50	80,3
20	74,5	14,0	58,5	31,3	28,0	33,0	1,89	76,7
30	73,0	10,6	62,2	26,5	34,0	34,0	1,51	74,5
40	71,3	8,5	63,0	22,0	38,5	33,5	1,28	71,7
50	68,8	6,2	62,6	18,0	42,5	32,0	1,09	68,3
60	65,8	4,5	62,0	15,0	45,0	29,0	0,93	65,7

$N=0,82$

t °C	λ=9,22 cm		λ=3,282 cm		λ=1,267 cm		$λ_m$	ε
	ε′	ε″	ε′	ε″	ε′	ε″		
5	72,8	24,3	45,0	37,5	15,0	25,0	—	80,0
10	72,3	21,4	49,0	36,0	18,5	28,0	2,67	78,5
20	71,7	14,7	55,0	32,0	25	31,5	2,03	75,0
30	70,7	10,4	58,8	27,2	31	33,0	1,60	72,2
40	69,2	7,6	60,3	22,5	36,0	32,5	1,37	69,5
50	67,4	6,0	60,5	19,0	39,5	31,7	1,14	67,8
60	65,6	4,5	60,0	16,0	43,0	28,7	0,97	64,8

[558, 553], $t=25$ °C

N	$\lambda=9,22$ cm		$\lambda=3,175$ cm		$\lambda=1,264$ cm		ε	λ_m
	ε'	ε''	ε'	ε''	ε'	ε''		
0,33	72,6	12,5	61,3	29,3	31,6	33,3	75,5	1,69
0,66	70,7	13,0	58,2	30,2	29,5	32,0	73,9	1,81
1,00	70,0	14,6	54,9	30,3	27,1	31,3	72,3	1,94

Water – Propan–2–ol (C_3H_8O) *)

[558], $t=25$ °C

N	$\lambda=9,22$ cm		$\lambda=1,264$ cm		ε	λ_m
	ε'	ε''	ε'	ε''		
0,33	74,8	11,5	31,6	33,8	75,9	1,63
0,66	72,3	11,0	28,9	32,3	73,4	1,73
1,00	—	—	26,2	30,7	70,9	1,85

*) See also [568, 421]; [110], $w_2=95$, $\nu=5$ Hz – 5 MHz. Temperature dependence ϵ with accuracy 1% is given by the expression

$$\varepsilon=-A+\frac{B}{t},$$

where $A=23,65$, $B=12,78\cdot10^{-3}$ at $t=123-300$ °K.

Water – Glycerine ($C_3H_8O_3$) *)

[578, 579], $\nu=0,5$—250 MHz

x_1	t °C	ε	β	$10^{-9}\tau$	x_1	t °C	ε	β	$10^{-9}\tau$
5,04	—7,5	51,6	0,63	33,2	30,0	—7,5	56,4	0,63	8,14
	—15,3	53,2	0,63	83,0		—15,3	57,2	0,63	17,8
	—19,5	54,3	0,60	158		—19,5	60,0	0,60	30,7
12,8	—7,5	52,8	0,63	21,2	35,8	—7,5	57,6	0,63	5,60
	—15,3	54,1	0,63	44,3		—15,3	58,8	0,63	12,3
	—19,5	56,0	0,60	89,5		—19,5	61,4	0,60	20,4
15,7	—7,5	53,6	0,63	18,2	41,7	—7,5	59,2	0,63	3,71
	—15,3	54,9	0,63	39,9		—15,3	61,3	0,63	8,06
	—19,5	56,6	0,60	79,6		—19,5	63,2	0,60	14,9
20,02	—7,5	54,1	0,63	14,5	47,5	—7,5	61,6	0,63	2,66
	—15,3	55,6	0,63	30,1		—15,3	63,0	0,63	5,17
	—19,5	57,2	0,60	63,6		—19,5	65,3	0,60	9,6

*) $\varepsilon_\infty=4,0$ for all x_1 and t °C.

[580], $\nu=30$ MHz $w_1=5$

t °C	10,4	5,0	2,6	0,5	—3,1	—7,4	—13,5	—29,5
ε'	43,2	38,9	35,9	34,6	28,7	27,0	20,5	10,6
ε''	12,2	14,4	15,8	16,3	16,8	16,6	14,0	5,8

$t=-28$ °C, $w_1=5$

ν, MHz	8	15	25	37,5
ε'	19,2	14,9	12,5	11,1
ε''	11,0	9,3	7,7	6,2

Water – Propylamine (C_3H_9N)
[558], $t=25$ °C, $\lambda=1,264$ cm

N	ε'	ε''
0,33	31,8	33,7
0,66	28,8	31,6

Water – Pyrazine ($C_4H_4N_2$)
[581], $t=25$ °C, $\nu=1,5$—38 GHz

N	ε_∞	ε	ε'_m	α	$10^{-11}\tau$
0	5,3	78,4	41,9	0,006	0,82
0,5	5,8	75,0	40,4	0,012	0,88
1,0	6,4	71,8	39,1	0,013	0,94
1,5	5,4	68,5	37	0,016	0,99

Water – Pioxan ($C_4H_8O_2$)
[375, 585], $t=25$ °C, fig. 77

x_1	λ=10,0 cm		λ=3,22 cm		λ=1,25 cm		λ=0,435 cm	
	ε'	ε''	ε'	ε''	ε'	ε''	ε'	ε''
5,01	2,51	0,024	2,48	0,060	2,39	0,059	—	—
9,11	2,82	0,062	2,73	0,22	2,58	0,11	—	—
19,9	3,76	0,24	3,47	0,50	—	—	2,57	0,36
28,3	4,77	0,50	4,09	0,94	3,27	0,87	2,73	0,51
38,8	6,50	1,06	5,02	1,70	3,79	1,37	3,05	0,78
49,6	—	—	6,30	2,88	4,34	2,18	—	—
58,5	—	—	7,67	4,46	4,93	3,03	—	—

x_1	ε	ε_∞	$\varepsilon_{\infty 2}$	$10^{-12}\tau_1$	$10^{-12}\tau_2$	$10^{-12}\tau$	C_1
5,01	2,51	—	2,34	—	—	—	8,5
9,11	2,80	—	2,50	—	—	—	9,9
19,9	3,81	2,94	2,46	13,0	4,0	0,65	—
28,3	4,90	3,21	2,57	16,0	4,2	0,72	—
38,8	6,80	3,63	2,74	18,5	3,8	0,78	—
49,6	9,75	4,25	2,87	22,0	4,0	0,80	—
58,5	13,37	4,66	3,00	23,0	4,0	0,84	—

[582], $t = 20$—50 °C, $\nu = 10^5$—10^7 Hz

x_1	ΔF	x_1	ΔF	x_1	ΔF	x_1	ΔF
22,81	0,82	59,10	1,01	83,46	1,12	95,08	1,05
27,41	0,84	70,38	1,08	87,91	1,10	97,86	1,03
41,87	0,90	80,99	1,13	91,94	1,07	100	0,99

[583]

x_1	t °C	λ=3,193 cm		λ=1,223 cm		λ=1,217 cm	
		ε′	ε″	ε′	ε″	ε′	ε″
19,97	10	3,390	0,680	2,948	0,529	2,387	0,260
	25	3,478	0,502	3,020	0,532	2,401	0,284
	40	3,471	0,385	3,074	0,509	2,447	0,328
	55	3,412	0,280	3,097	0,426	2,481	0,340
30,01	10	4,003	1,137	3,261	0,852	2,584	0,319
	25	4,240	1,055	3,401	0,937	2,601	0,344
	40	4,336	0,848	3,581	0,967	2,642	0,379
	55	4,329	0,647	3,724	0,910	2,694	0,457
40,0	10	4,621	1,791	3,592	1,328	2,790	0,547
	25	5,107	1,770	3,873	1,323	2,873	0,590
	40	5,382	1,621	4,184	1,559	2,878	0,658
	55	5,540	1,396	4,376	1,505	2,885	0,755
50,0	10	5,549	2,878	4,054	1,793	3,025	0,706
	25	6,320	2,765	4,460	2,046	3,095	0,847
	40	6,942	2,598	4,806	2,212	3,129	0,922
	55	7,232	2,140	5,278	2,222	3,236	1,270
60,0	10	6,765	4,666	4,488	2,568	3,295	0,903
	25	7,966	4,921	5,137	2,955	3,335	1,149
	40	8,989	4,528	5,663	3,185	3,336	1,341
	55	9,694	3,970	6,439	3,380	3,405	1,454
69,49	1	7,096	6,264	4,850	3,284	3,603	1,063
	10	8,569	6,847	5,336	3,706	3,565	1,296
	25	10,62	7,653	6,009	4,678	3,558	1,396
	40	12,77	7,489	7,200	5,487	3,778	1,812
79,56	1	9,735	9,808	5,716	5,587	4,094	1,550
	10	12,108	11,059	6,534	6,429	4,350	1,910
	25	14,368	11,497	8,457	7,610	5,350	3,240
	40	—	—	—	—	—	—

x_1	t °C	ε	ε_∞	$\varepsilon_{\infty 2}$	$10^{-12}\tau_1$	$10^{-12}\tau_2$
19,97	10	3,97	3,00	2,32	20,4	3,75
	25	3,84	3,10	2,34	15,6	3,25
	40	3,69	3,18	2,32	11,7	2,92
	55	3,55	3,19	2,34	9,5	2,54

x_1	t °C	ε	ε_∞	$\varepsilon_{\infty 2}$	$10^{-12}\tau_1$	$10^{-12}\tau_2$
30,0	10	5,44	3,31	2,56	23,9	3,60
	25	5,22	3,42	2,57	18,5	3,08
	40	4,95	3,67	2,58	13,8	2,60
	55	4,66	3,85	2,58	11,0	2,29
40,0	10	7,46	3,73	2,68	27,8	3,45
	25	7,12	3,83	1,68	20,7	3,13
	40	6,75	4,00	2,64	16,1	2,60
	55	6,45	4,19	2,62	13,0	2,34
50,0	10	10,73	4,05	2,78	31,8	3,13
	25	10,06	4,40	2,81	23,6	2,68
	40	9,45	4,58	2,78	16,7	2,28
	55	8,80	4,67	2,72	13,0	1,91
60,0	10	15,70	4,39	2,99	33,7	3,32
	25	14,20	4,75	2,95	23,5	2,85
	40	13,45	4,94	2,94	17,8	2,39
	55	12,55	5,20	2,90	11,9	2,14
69,75	1	24,57	4,70	3,20	49,0	3,60
	10	23,08	4,97	3,18	35,8	3,18
	25	21,44	5,19	3,18	24,0	2,65
	40	19,93	5,36	3,18	16,2	2,23
79,56	1	35,64	4,85	3,57	43,7	3,18
	10	33,55	5,08	3,58	31,5	2,74
	25	31,19	5,52	3,58	21,0	2,28

x_1	t °C	C_1	ΔF_1	ΔF_2	ΔH_1	ΔH_2	ΔS_1	ΔS_2
19,97	10	0,59	2,71	1,75	2,5	1,1	—0,6	—2,3
	25	0,49	2,72	1,79				
	40	0,37	2,71	1,80				
	55	0,30	2,73	1,84				
30,00	10	0,74	2,80	1,73	2,6	1,3	—0,8	—1,6
	25	0,68	2,83	1,76				
	40	0,54	2,82	1,77				
	55	0,39	2,83	1,80				
40,0	10	0,78	2,89	1,71	2,7	1,2	—0,8	—1,9
	25	0,74	2,89	1,77				
	40	0,67	2,91	1,77				
	55	0,59	2,94	1,82				
50,0	10	0,84	2,96	1,65	3,2	1,4	0,6	—0,9
	25	0,78	2,97	1,67				
	40	0,73	2,94	1,69				
	55	0,68	2,94	1,69				
60,0	10	0,89	2,99	1,68	3,4	1,3	1,4	—1,4
	25	0,84	2,97	1,71				
	40	0,81	2,98	1,72				
	55	0,76	2,89	1,76				

x_1	t °C	C_1	ΔF_1	ΔF_2	ΔH_1	ΔH_2	ΔS_1	ΔS_2
69,75	1	0,93						
	10	0,91	3,09	1,66	4,2	1,5	4,1	—0,6
	25	0,89	2,98	1,67				
	40	0,87	2,92	1,68				
79,56	1	0,96	—	—				
	10	0,95	3,02	1,59	4,7	1,6	6,2	0,0
	25	0,93	2,91	1,58				

[584], $\nu = 300—400$ *кгц*

x_1	$\varepsilon - 1/\varepsilon + 2$ at t °C			$10^{-11}\tau$ at t °C		
	30	40	50	30	40	50
4,1288	0,32239	0,31815	0,31373	1,09	0,89	0,73
10,034	0,37237	0,36647	0,36036	1,11	0,90	0,73
14,391	0,40757	0,40071	0,39356	1,13	0,91	0,77
22,802	0,47186	0,46368	0,45494	1,19	0,95	0,77
27,410	0,50519	0,49645	0,48716	1,22	1,03	0,78
41,869	0,65166	0,65139	0,63532	1,27	1,04	0,84
59,104	0,78846	0,78007	0,76926	1,55	1,24	1,01
70,384	0,84989	0,84696	0,84395	1,68	1,31	1,06
80,985	0,90643	0,90261	0,89742	1,70	1,32	1,06
83,459	0,91609	0,91178	0,90684	1,61	1,24	0,98
100,00	0,96190	0,96014	0,95830	0,807	0,638	0,516

[194], $t = 25$ °C, $w_2 = 33,2$

t °C	$\lambda = 9,22$ cm		$\lambda = 1,262$ cm		ε	λ_m	$10^{-11}\tau$
	ε'	ε''	ε'	ε''			
25	45,3	11,3	15,0	18,3	47,9	2,4	1,27
35	44,75	8,93	18,3	19,4	46,1	1,9	1,01
45	42,21	7,07	21,7	20,5	44,0	1,6	0,85
55	40,41	5,62	—	—	41,2	1,4	0,74
65	40,89	4,83	—	—	41,2	1,2	0,635

[584], $t = 25$ °C

w_2	$\lambda = 9,22$ cm		$\lambda = 1,262$ cm		ε	λ_m	$10^{-11}\tau$
	ε'	ε''	ε'	ε''			
45,2	36,0	10,8	11,3	13,9	39,1	2,7	1,43
33,2	45,3	11,3	15,0	18,3	47,0	2,4	1,27
24,9	56,2	12,4	18,6	21,9	58,0	2,1	1,11
14,2	67,2	12,5	24,3	27,7	68,7	1,8	0,95

Water – *tert*–Butyl alcohol ($C_4H_{10}O$)
[558], $t=25$ °C

N	λ=9,22 cm		λ=3,175 cm		λ=1,264 cm		ε	λ_m
	ε′	ε″	ε′	ε″	ε′	ε″		
0,33	73,7	12,9	59,9	29,6	31,6	33,3	75,9	1,74
0,66	71,6	14,0	55,1	30,2	—	—	73,5	1,90
1	69,9	15,3	50,4	30,2	—	—	71,5	2,06

Water – Tetramethylammonium bromide ($C_4H_{12}BrN$)
[562], $N=0,25$, fig. 67, 78

t °C	λ=9,22 cm		λ=3,282 cm		λ=1,267 cm		λ_m	ε
	ε′	ε″	ε′	ε″	ε′	ε″		
5	75,5	20,0	49,5	38,0	18,0	29,0	2,85	82
10	74,5	18,5	53,0	35,0	21,2	30,0	2,42	78,2
20	73,0	16,5	57,5	30,5	28,0	32,5	1,83	75,0
30	71,0	14,0	60,0	25,5	34,2	33,0	1,45	72,2
40	68,5	11,0	61,0	21,0	39,2	31,5	1,20	70,0
50	66,0	8,5	60,4	17,0	43,5	30,0	0,99	66,0
60	63,5	5,5	59,0	14,0	45,7	27,5	0,86	64,0

$N=0,5$

t °C	λ=9,22 cm		λ=3,282 cm		λ=1,267 cm		λ_m	ε
	ε′	ε″	ε′	ε″	ε′	ε″		
0	72,9	24,5	41,5	37,5	13,8	24,0	3,46	81,5
10	71,5	18,5	50,0	33,7	20,5	28,5	2,47	76,0
20	70,0	15,2	55,5	30,5	26,3	31,0	1,00	73,5
30	68,3	11,5	58,2	25,0	32,0	31,9	1,50	70,0
40	66,0	9,0	59,5	21,0	36,3	30,2	1,23	67,0
50	64,5	7,5	59,0	17,7	40,2	28,2	1,02	64,5
60	62,0	6,5	56,8	15,8	42,5	25,5	0,89	61,5

[552]

N	t=5 °C				t=25 °C			
	ε′	ε″	α	$10^{-11}\tau$	ε′	ε″	α	$10^{-11}\tau$
0	86,2	5,8	0,006	1,48	78,6	5,3	0,006	0,823
0,5	79,4	5,6	0	1,56	72,9	4,8	0	0,885
1,0	74,0	5,7	0,019	1,68	67,0	5,0	0,011	0,95
1,5	68,2	5,6	0,012	1,80	61,7	5,2	0,02	1,035
2,0	62,2	5,6	0,015	1,91	55,7	5,6	0,019	1,11

Water – Pyridine (C_5H_5N)
[581], $t=25$ °C, $\nu=1,5$—38 GHz

N	ε_∞	ε	ε'_m	α	$10^{-11}\tau$
0,0	5,3	78,4	41,9	0,006	0,82
0,5	4,4	76,9	40,7	0,022	0,90
1,0	5,1	74,3	39,7	0,031	0,99
2,0	4,2	70,6	37,4	0,06	1,20
4,0	4,2	62,3	33,2	0,09	1,77

Water – 2-Methylpyrazine ($C_5H_6N_2$)
[581], $t=25$ °C, $\nu=1,5$—38 GHz

N	ε_∞	ε	ε'_m	α	$10^{-11}\tau$
0	5,3	78,4	41,9	0,006	0,82
0,5	6,9	74,8	40,8	0,02	0,92
1,0	5,7	70,8	38,2	0,021	1,00
2,0	4,6	63,4	34,0	0,041	1,19
4,0	4,8	48,4	26,6	0,06	1,83

Water – Glutaric acid ($C_5H_8O_4$)
[558], $t=25$ °C

N	$\lambda=9,22$ cm		$\lambda=3,175$ cm		$\lambda=1,264$ cm		ε	λ_m
	ε'	ε''	ε'	ε''	ε'	ε''		
0,33	73,6	13,1	60,6	28,4	32,5	33,1	74,9	1,63
1,00	65,6	14,8	53,5	26,2	28,6	28,8	68,0	1,71

Water – Proline ($C_5H_9NO_2$) *)
[586], $N=1$ (pH=6,5)

ν, MHz	$t=0$ °C		$t=10$ °C		$t=20$ °C	
	ε'	ε''	ε'	ε''	ε'	ε''
300	107,9	12,0	101,5	8,1	100,1	6,4
600	102,6	20,6	99,7	15,3	99,5	11,8
900	95,3	25,8	97,9	20,1	96,4	15,7
1200	91,3	28,8	94,4	23,7	96,3	19,3
1500	86,3	31,5	89,2	26,4	91,5	21,6
1800	82,0	32,7	83,9	28,4	87,3	23,2
2000	74,5	31,8	80,7	28,4	84,4	24,3

*) See also [587], $\nu=9455$ MHz.

ν, MHz	t=30 °C		t=40 °C		t=50 °C	
	ε′	ε″	ε′	ε″	ε′	ε″
300	96,7	4,7	92,4	3,6	88,1	3,0
600	95,7	9,3	92,1	7,4	87,9	6,2
900	92,8	13,2	90,9	10,2	87,7	8,8
1200	92,0	15,7	90,2	13,1	86,9	10,4
1500	91,0	17,8	88,6	14,9	86,5	12,6
1800	87,5	20,2	86,7	16,7	84,3	14,0
2000	84,1	21,1	83,1	17,9	82,0	15,7

Water – Oxyproline ($C_5H_9O_3N$)
[586], $N=1$ (pH=5,65)

ν, MHz	t=10 °C		t=20 °C		t=30 °C		t=40 °C		t=50 °C	
	ε′	ε″	ε′	ε″	ε′	ε″	ε′	ε″	ε′	ε″
300	102,3	10,9	98,8	8,6	94,8	5,5	91,4	4,4	87,7	2,7
600	96,6	18,1	95,4	14,5	91,4	11,1	90,9	8,4	86,8	7,2
900	90,5	22,6	91,5	17,5	89,7	15,3	88,6	12,7	85,9	10,5
1200	86,5	23,5	86,9	20,2	87,0	17,4	87,1	14,5	84,6	12,0
1500	82,9	24,0	83,9	21,9	83,6	19,7	83,9	16,4	83,8	14,2
1800	79,5	24,4	82,0	22,5	82,7	19,4	80,8	17,6	80,3	16,2
2000	78,5	25,3	78,5	22,7	78,1	19,9	79,1	18,3	77,6	16,5

Water – Pentan-3-one ($C_5H_{10}O$)
[558], $t=25$ °C

N	λ=9,22 cm		λ=3,175 cm		λ=1,264 cm		ε	λ_m
	ε′	ε″	ε′	ε″	ε′	ε″		
0,17	75,2	10,6	62,2	29,0	33,2	35,0	77,0	1,62
0,33	74,6	10,8	61,1	29,3	31,6	34,1	76,5	1,67

Water – Phenol (C_6H_6O)
[558], $t=25$ °C

N	λ=9,22 cm		λ=3,175 cm		λ=1,264 cm		ε	λ
	ε′	ε″	ε′	ε″	ε′	ε″		
0,25	74,6	11,5	61,3	28,2	32,1	33,6	75,6	1,62
0,5	71,3	12,1	58,4	27,9	29,1	31,2	73,0	1,67

Water – Aniline (C_6H_7N)

[558], $t = 25$ °C

N	λ=9,22 cm		λ=3,175 cm		λ=1,264 cm		ε	λ_m
	ε′	ε″	ε′	ε″	ε′	ε″		
0,125	75,5	10,3	62,9	28,4	33,6	35,2	77,1	1,58
0,25	74,1	10,4	61,9	28,3	32,5	34,2	76,0	1,61

Water – 2–6–Dimethylpyrazine ($C_6H_8N_2$)

[581], $t = 25$ °C, ν = 1,5—38 GHz

N	ε_∞	ε	ε'_m	α	$10^{-11}\tau$
0	5,3	78,4	41,9	0,006	0,82
1,0	4,7	70,0	37,4	0,045	1,05
1,5	4,8	65,1	34,9	0,06	1,22

Water – Quinoxaline ($C_8H_6N_2$)

[581], $t = 25$ °C, ν = 1,5—38 GHz

N	ε_∞	ε	ε'_m	α	$10^{-11}\tau$
0	5,3	78,4	41,9	0,006	0,82
0,5	5,4	73,6	39,5	0,022	0,91
1,0	4,3	68,9	36,6	0,039	0,97
2,0	4,3	58,7	31,5	0,07	1,22
4,0	4,9	40,5	22,7	0,1	2,11

Water – Tetraethylammonium bromide ($C_8H_{20}BrN$)

[544], $t = 25$ °C, λ = 10 cm, fig. 65

N	0,25	0,5	0,75	1,00
ε′	72,2	67,6	63,7	59,6
ε″	23,2	31,0	37,6	42,0

[552], $t = 25$ °C, λ = 3,2 cm

N	ε′	ε″	α	$10^{-11}\tau$
1,0	4,7	61,7	0,031	1,18

Water – 2–Methylquinoxaline ($C_9H_8N_2$)
[581], $t = 25$ °C, $\nu = 1{,}5$—38 GHz

N	ε_∞	ε	ε'_m	α	$10^{-11}\tau$
0	5,3	78,4	41,9	0,006	0,82
1,0	4,1	67,2	35,6	0,07	1,05
1,5	4,6	61,6	33,1	0,09	1,26

Water – Sucrose ($C_{12}H_{22}O_{11}$)
[519], $t = 19$ °C, fig. 79

λ	$w_2=65$		$w_2=50$		$w_2=30$	
	ε'	ε''	ε'	ε''	ε'	ε''
150 *км*	67,2	—	67,2	—	73,2	—
91 *м*	58,7	—	65,6	—	72,6	—
36 *м*	56,8	—	65,0	—	72,5	—
12 *м*	54,4	—	64,7	—	72,4	—
2,43 *м*	51,6	9,96	63,4	6,0	71,1	2,96
58,4 *см*	39,3	12,0	61,7	12,0	70,9	7,03
16,3 *см*	23,1	15,0	46,5	19,5	64	17,5
10,59 *см*	17,7	15,4	35,0	20,5	55,6	20,4

Water – Tetrapropylammonium bromide ($C_{12}H_{28}BrN$)
[552], $t = 25$ °C, $N = 1$

ε'	ε''	α	$10^{-11}\tau$
4,8	55,4	0,056	1,52

Water – Tetrabutylammonium bromide ($C_{16}H_{36}BrN$)
[552], $t = 25$ °C, fig. 62

N	ε'	ε''	α	$10^{-11}\tau$
0,5	4,5	64,0	0,047	1,22
1,0	5,2	50,8	0,073	1,83

Water – 1, 4–Diazobicyclo–[2, 2, 2] –octane ($C_6H_{12}N_2$)
[581], $t=25$ °C, $\nu=1,5-38$ GHz

N	ε_∞	ε	ε'_m	α	$10^{-11}\tau$
0,0	5,3	78,4	41,9	0,006	0,82
1,0	4,6	70,7	37,6	0,05	1,21
2,0	4,3	63,6	34,0	0,10	1,85
3,0	4,7	56,0	30,3	0,12	3,24

Supplementary information on the dielectric dispersion parameters of aqueous solutions of organic compounds

Water – Thiourea (CH_4N_2S), fig. 74.
Water – Guanidine (CH_5N_3), fig. 74.
Water – Acetamide (C_2H_5NO), fig. 74.
Water – Dimethylsulphoxide (C_2H_6OS) [611].
Water – Propylene oxide (C_3H_6O) [577].
Water – Tetrahydrofuran (C_4H_8O) [577].
Water – Butan–1–ol ($C_4H_{10}O$) [568].
Water – Starch ($(C_6H_{10}O_5)_x$), fig. 79.
Water – Glucose ($C_6H_{12}O_6$), fig. 79.
Water – Aminohexanoic acid ($C_6H_{13}NO_2$) [588], $t = 10 - 40$°C, $\nu = 70 - 2000$ MHz. [586], [612].
Water – Tributylphosphate ($C_{12}H_{27}PO_4$) [379].
Water – Gelatine, fig. 79.

CHAPTER IV

DIELECTRIC DATA PRESENTED IN GRAPHICAL FORM

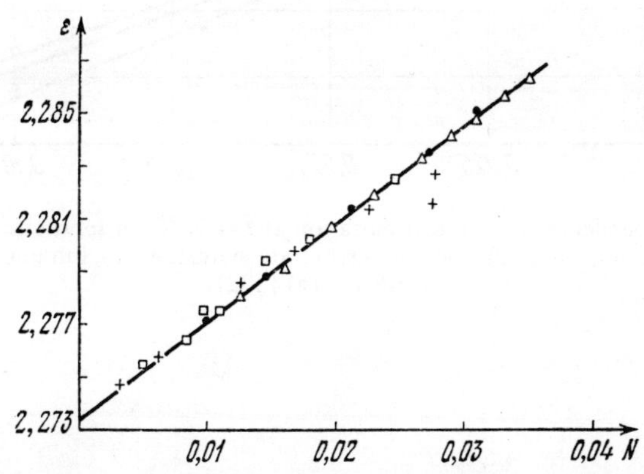

Fig. 1. Dependence of ϵ on concentration for $\nu = 2$MHz, $t = 25°$C for solutions: potassium chloride-benzene (Δ), sulphuric acid-benzene (•) [589]. □, + indicate values for the same solutions, respectively, from [590].

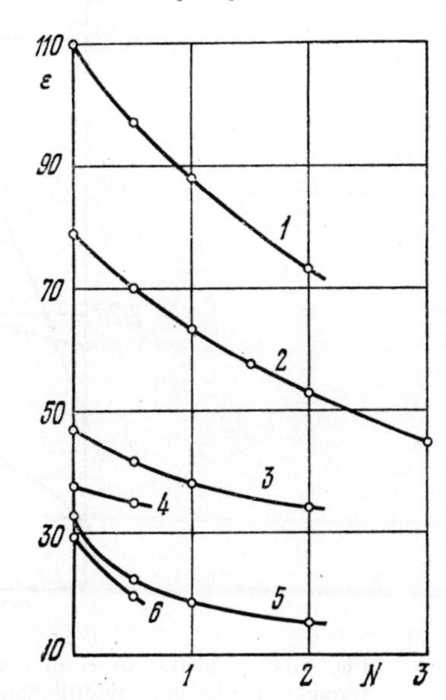

Fig. 2. Dependence of ϵ on concentration at $t = 25°$C for lithium nitrate in formamide (1), in water (2), in dimethylsylphoxide (3), in *N,N*-Dimethylformamide (4), in methanol (5), in hexamethylphosphorictriamide (6) [591].

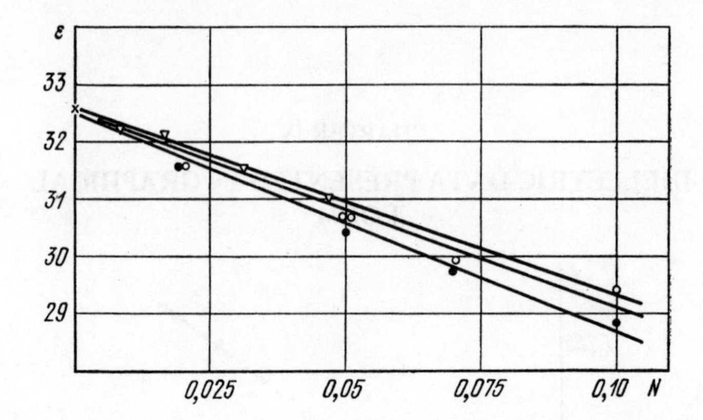

Fig. 3. Dependence of ϵ on concentration at t = 25°C for solutions: magnesium perchlorate-methanol (▽), sodium perchlorate-methanol (○), lithium perchlorate-methanol (●) [592].

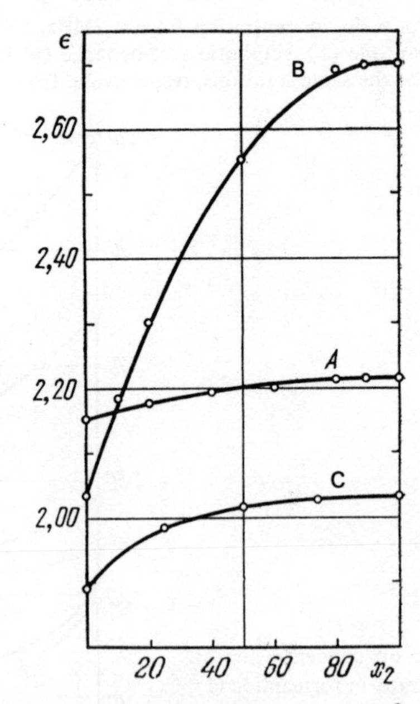

Fig. 4. Dependence of ϵ on concentration at t = 20°C for solutions: carbon tetrachloride-hexamethyldisiloxane (A), cyclohexane-pentamethylcyclopentasiloxane (B), cyclohexane-hexadecane (C) [241].

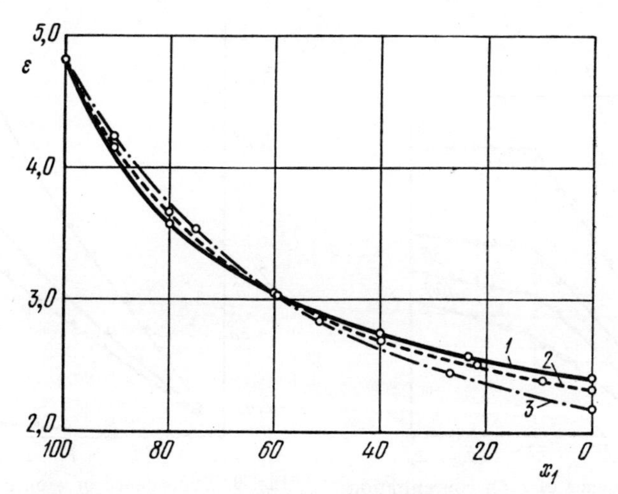

Fig. 5. Dependence of ϵ on concentration at $t = 20°C$ for solutions: chloroform-octamethyltrisiloxane (1), chloroform-octamethyltrisiloxane (2), chloroform-hexamethyldisiloxane (3) [241].

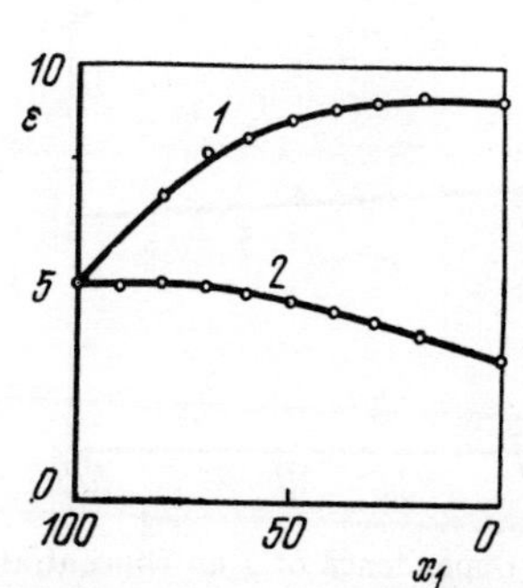

Fig. 6. Dependence of ϵ on concentration at $t = 20°C$ for solutions: chloroform-quinoline (1) and chloroform-dibutyl ether (2) [165].

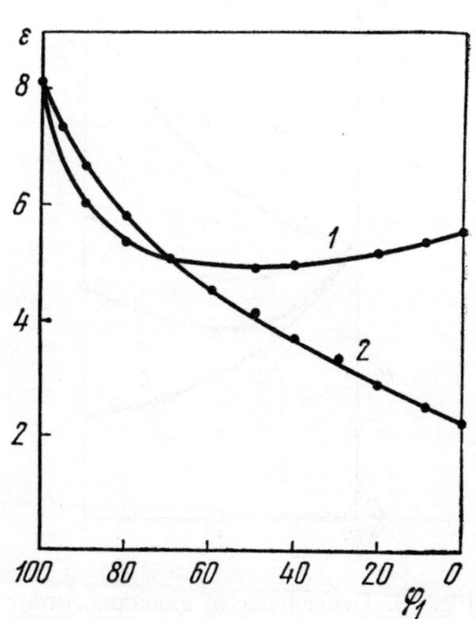

Fig. 7. Dependence of ϵ on concentration at $t = 25°C$ for solutions of trifluoroacetic acid in chlorobenzene (1) and in benzene (2) [593].

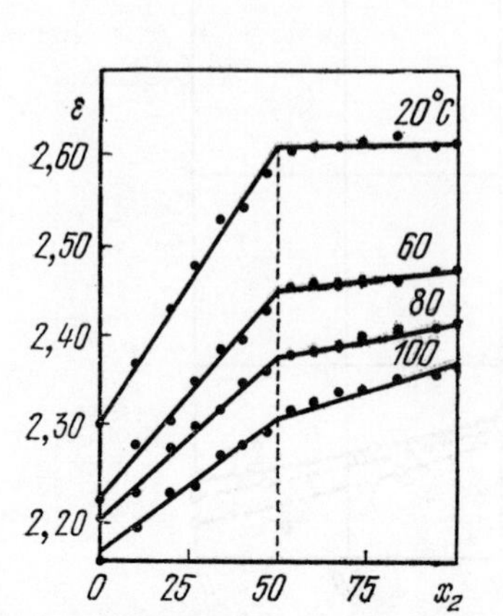

Fig. 8. Dependence of ε on concentration at various temperatures of solutions of acetic acid-pyridine [276].

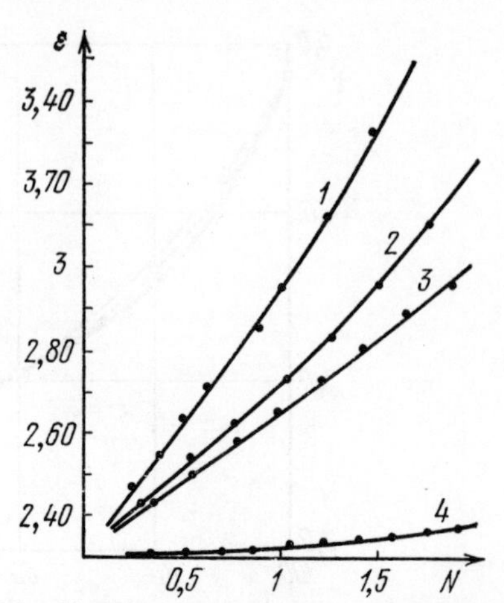

Fig. 9. Dependence of ε on concentration at $t = 20°C$ of solutions: pyridine- benzene (1), phenol-benzene (2), aniline-benzene (3), acetic acid-benzene (4)

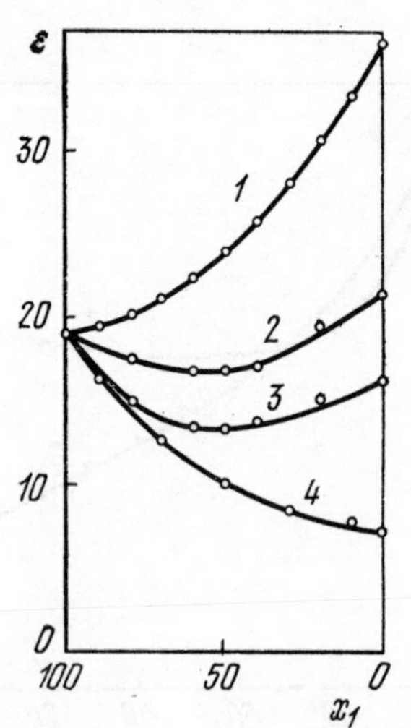

Fig. 10. Dependence of ε on concentration at $t = 20°C$ for solutions: isobutanol-nitrobenzene (1), isobutanol-acetone (2), isobutanol-cyclohexanone (3), iso- butanol-aniline (4) [165].

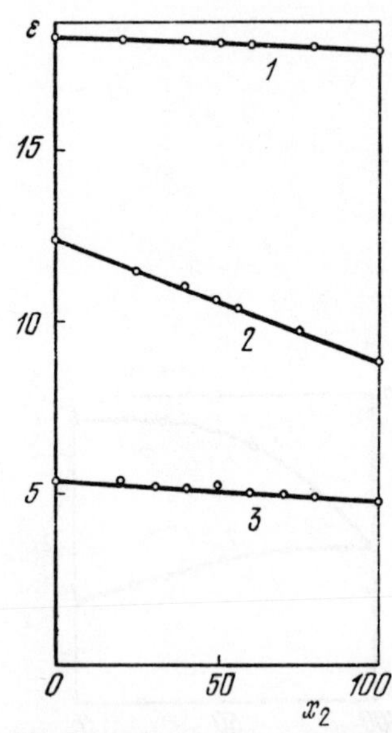

Fig. 11. Dependence of ε on concentration at $t = 20°C$ for solutions: butanone-benzaldehyde (1), pyridine-quinoline (2), methyl hexanoate-diethylaniline (3) [165].

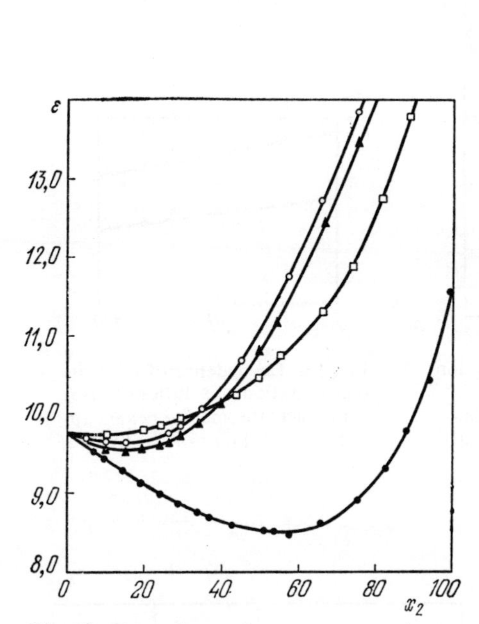

Fig. 12. Dependence of ϵ on concentration at $t = 30°C$, $\nu = 2$ MHz for solutions of *o*-dichlorobenzene in the four isomeric butyl alcohols: butan-1-ol (○), isobutanol (▲), butan-2-ol (□), *tert*-butanol (●) [317].

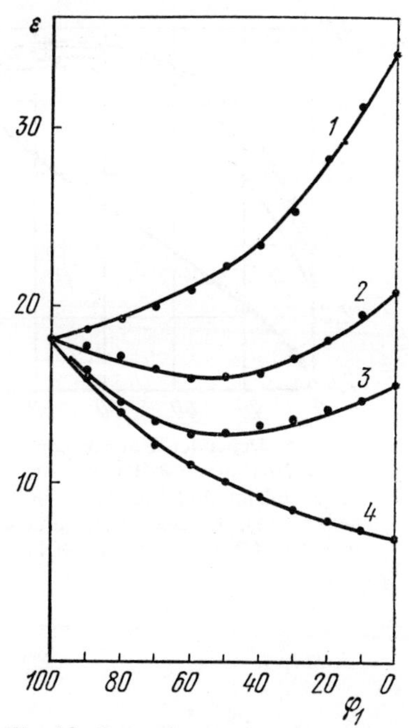

Fig. 13. Dependence of ϵ on concentration at various temperatures of solutions of isobutanol-cyclohexanone: $t = 10°C$ (1), $t = 20°C$ (2), $t = 30°C$ (3), $t = 40°C$ (4) [165, 594].

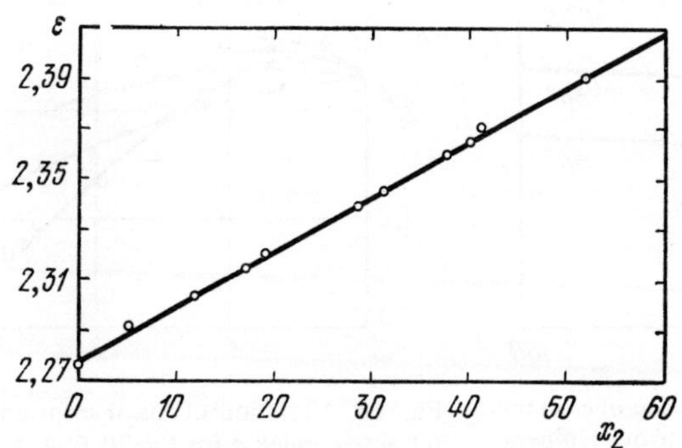

Fig. 14. Dependence of ϵ on concentration for $t = 25°C$, $\nu = 2$MHz for solutions of benzene-cyclohexylamine [589].

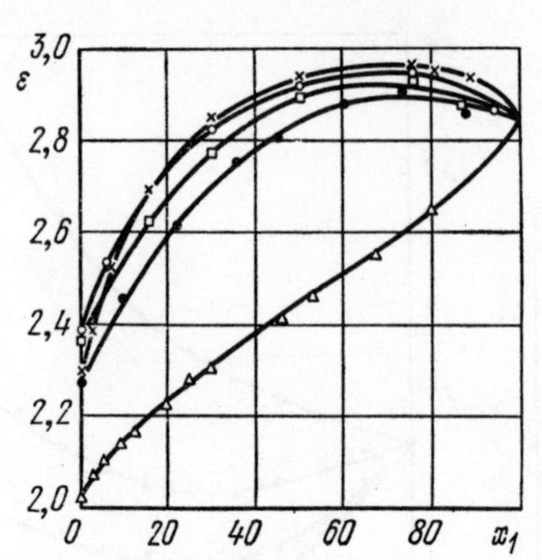

Fig. 15. Dependence of ϵ on concentration at $t = 20°C$, $\nu = 1.8$ MHz for solutions of 2, 6-Dimethyl-4-pentanol in cyclohexane (Δ), benzene (X), toluene (O), *m*-xylene (□), *p*-xylene (•) [401].

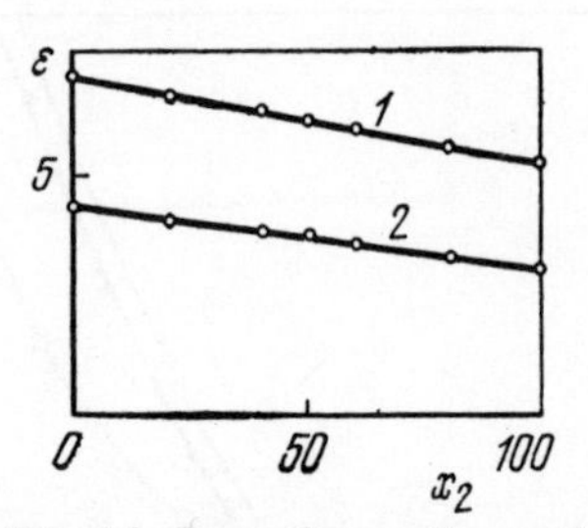

Fig. 16. Dependence of ϵ on the concentration of solutions of iso pentyl acetate- methyl hexanoate at $t = 10°C$ (1), $t = 40°C$ (2) [165].

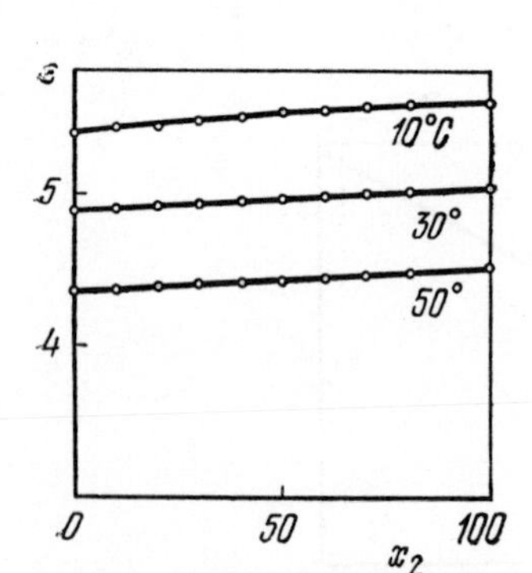

Fig. 17. Dependence of ϵ on the concentration at various temperatures of solutions of dimethylaniline-diethylaniline [165].

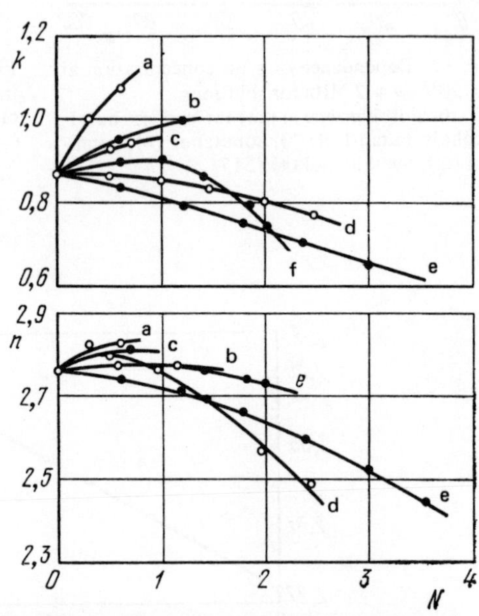

Fig. 18. Variation of the absorption coefficient k refractive index n for $t = 20°C$, $\lambda = 1.24$ cm, for solutions in methanol of KI (a). NaBr (b), $BaBr_2$ (c), $Ca(NO_3)_2$ (d), LiCl (e), $SrBr_2$ (f) ($\epsilon' = n^2 - k^2$, $\epsilon'' = 2nk$)

(The concentration variable N is defined in Symbols p. 7)

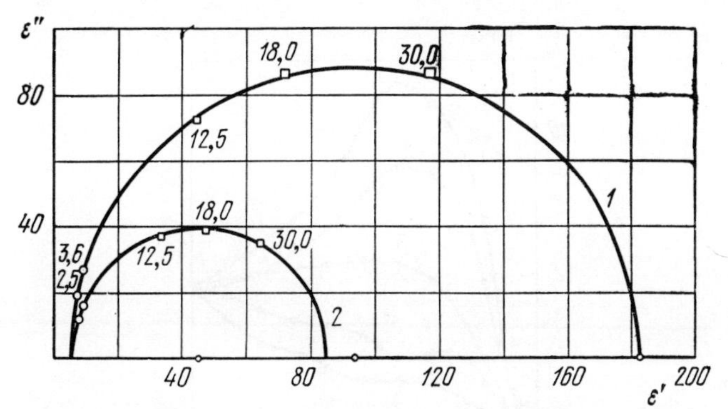

Fig. 19. Plot of ϵ'' against ϵ' for t = 25°C for pure *N*-methylformamide (1) and for solutions of *N*-methylformamide-lithium nitrate $N = 1$ (2) [591].

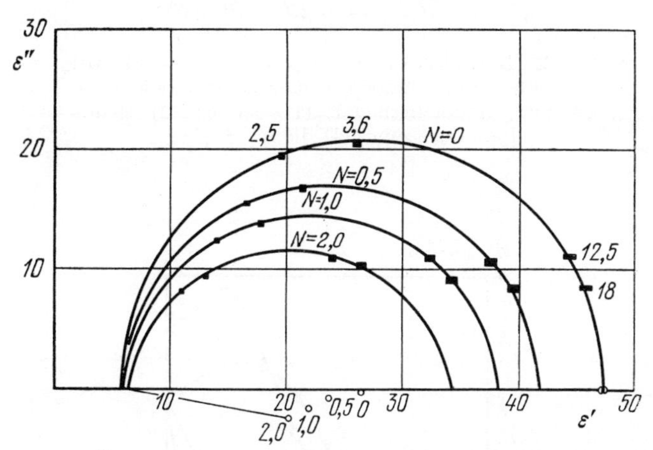

Fig. 20. Plot of ϵ'' against ϵ' for t = 25°C for pure dimethylsulphoxide and for solutions of dimethylsulphoxide-lithium nitrate (wave lengths given in cm) [591].

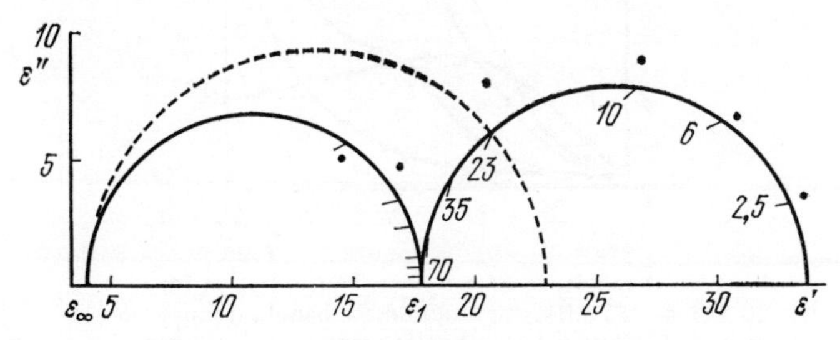

Fig. 21. Plot of ϵ'' against ϵ' for t = –40°C for pure ammonia (broken line) and for a solution of sodium in ammonia at x_1 = 50 (frequencies given in Ghz, • are experimental points) [384].

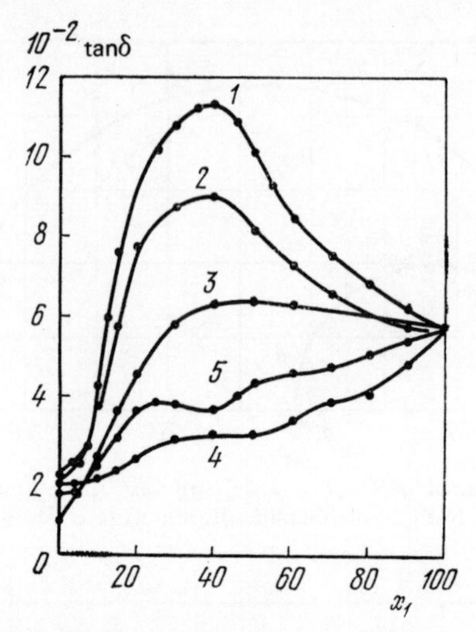

Fig. 22. Dependence of tan δ at $t = 20°C, \nu = 137$ MHz for solutions of methanol on concentration in various solvents: carbontetrachloride (1), benzene (2), chloroform (3), acetone (4), dioxan (5) [389].

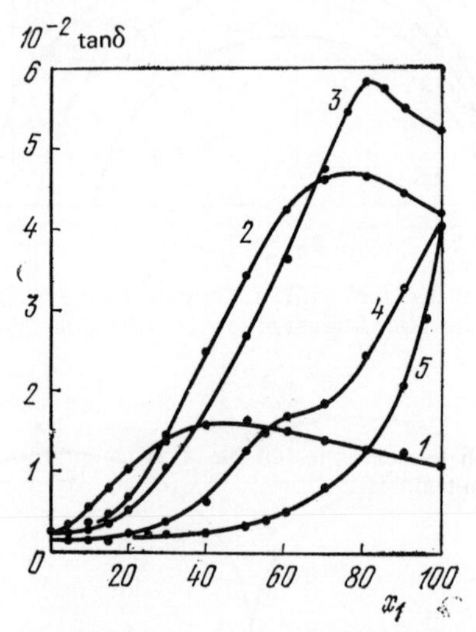

Fig. 23. Dependence of tan δ on concentration for $t = 20°C, \nu = 137$ MHz for solutions: ethanol-carbon tetrachloride (1), butan-1-ol carbon tetrachloride (2), isopentyl alcohol-carbon tetrachloride (3), butan-1-ol-chloroform (4), butan-1-ol-acetone (5) [389].

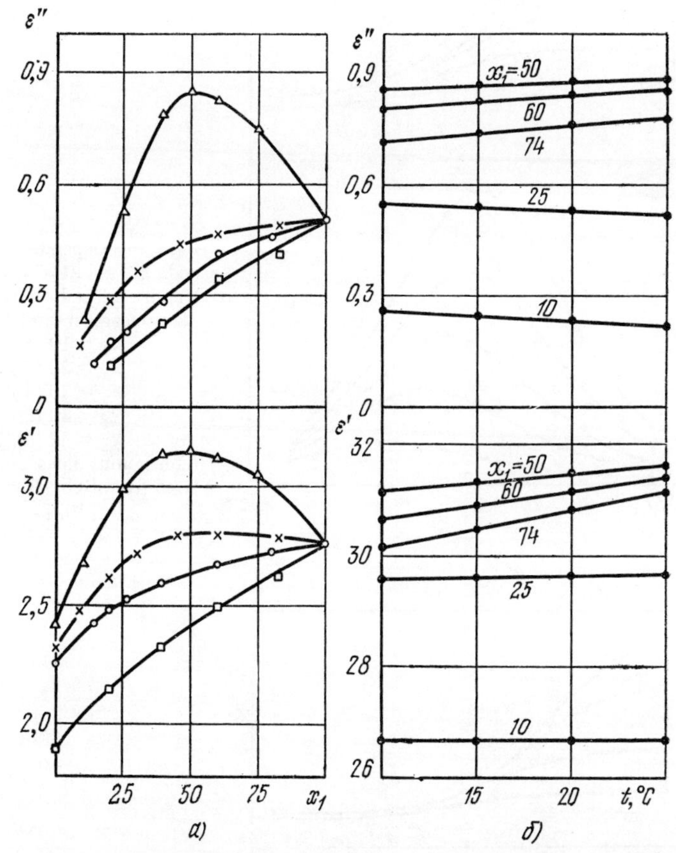

Fig. 24. a) Dependence of ϵ' and ϵ'' on concentration for $t = 20°C$, $\lambda = 3$ cm for solutions: hexanol-dioxan (△), hexanol-benzene (×), hexanol-carbon tetrachloride (○), hexanol-hexane (□). b) Dependence of ϵ' and ϵ'' on temperature for solutions of hexanol-dioxan [595].

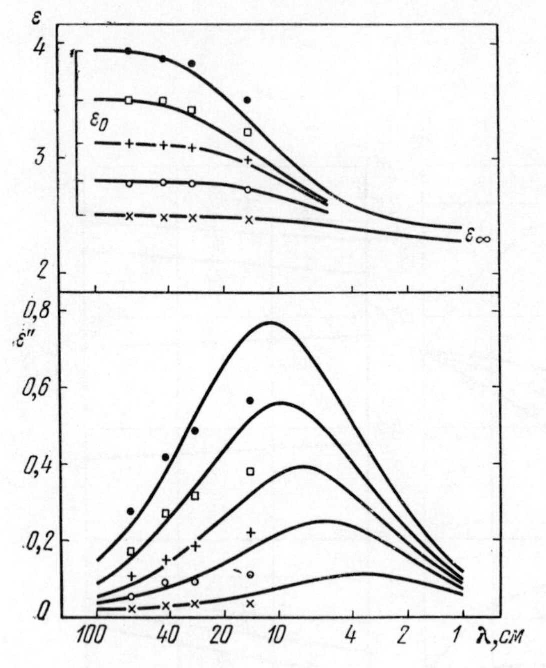

Fig. 25. Dependence of ϵ' and ϵ'' at 21°C on wave length for solutions of *o*-cresol-carbon tetrachloride
$x_2 = 10$ (×)
$x_2 = 20$ (○)
$x_2 = 30$ (+)
$x_2 = 40$ (□)
$x_2 = 50$ (●)
(continuous lines are calculated from Debye's equation) [402].

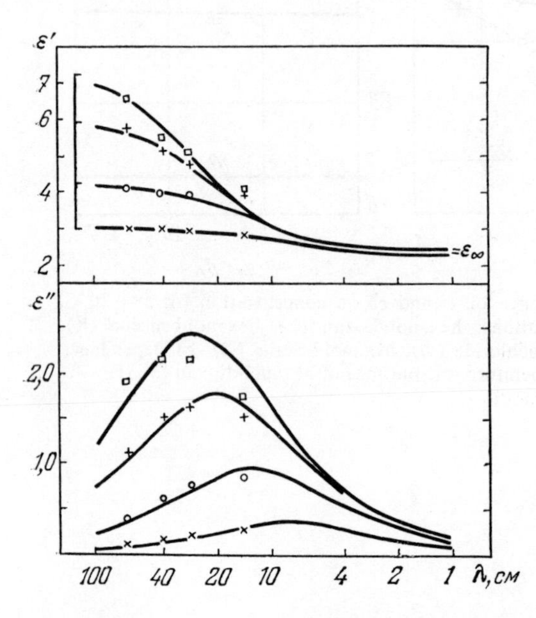

Fig 26. Dependence of ϵ' and ϵ'' at 21°C on wave length for solutions of eugenol-carbon tetrachloride
$x_2 = 10$ (×)
$x_2 = 25$ (○)
$x_2 = 50$ (+)
$x_2 = 70$ (●)
(continuous lines are calculated from Debye's equation) [402].

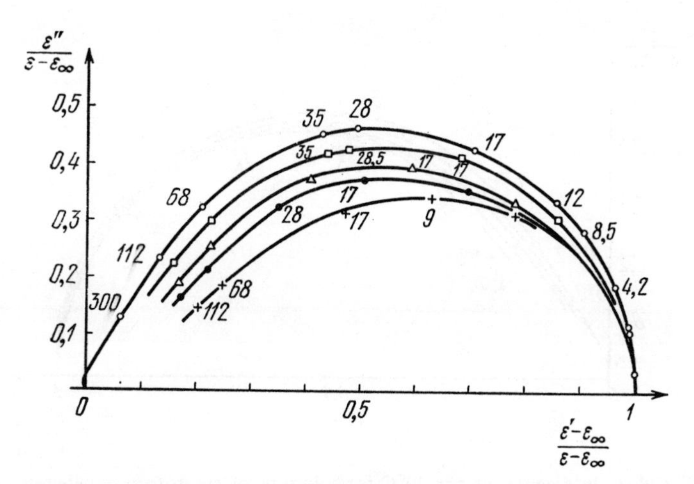

Fig. 27. Cole-Cole diagram at $t = 20°C$ for solutions of chloroform-ketene: $\phi_1 = 100$ (○), $\phi_1 = 80$ (□), $\phi_1 = 60$ (Δ), $\phi_1 = 40$ (●), $\phi_1 = 20$ (+). (Frequencies are expressed in GHz. [596].

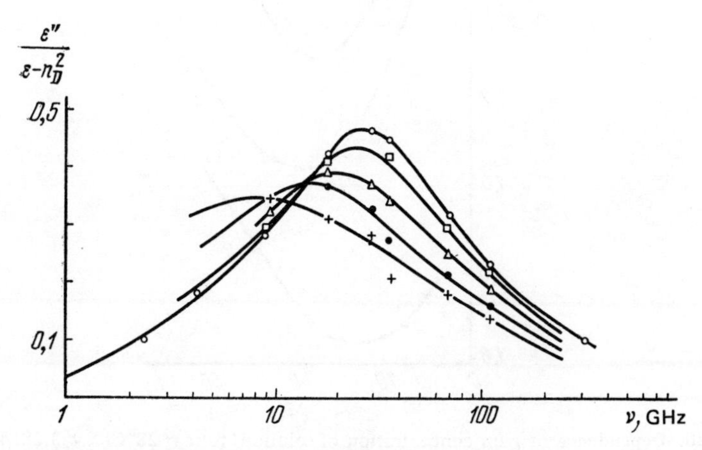

Fig. 28. Dependence of dielectric loss on frequency at $t = 20°C$ for solutions of chloroform-ketene: $\phi_1 = 100$ (○), $\phi_1 = 80$ (□), $\phi_1 = 60$ (Δ), $\phi_1 = 40$ (●), $\phi_1 = 20$ (+) [596]. ν in GHz.

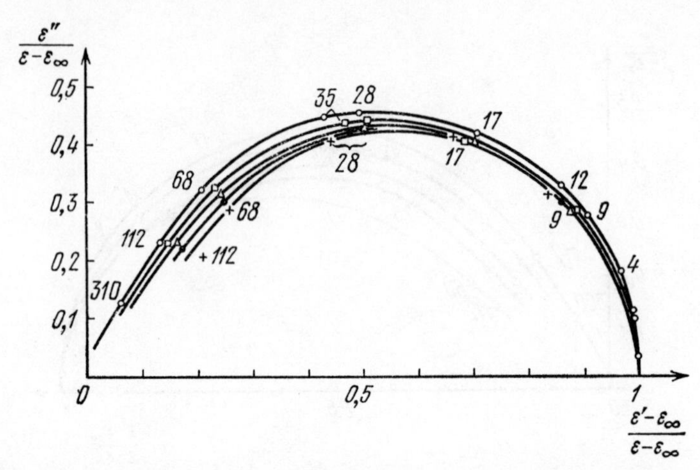

Fig. 29. Cole-Cole diagram at t = 20°C for solutions of chloroform-p-xylene: ϕ_1 = 100 (○), ϕ_1 = 80 (□), ϕ_1 = 60 (Δ), ϕ_1 = 40 (•), ϕ_1 = 20 (+) (frequencies given in GHz) [596].

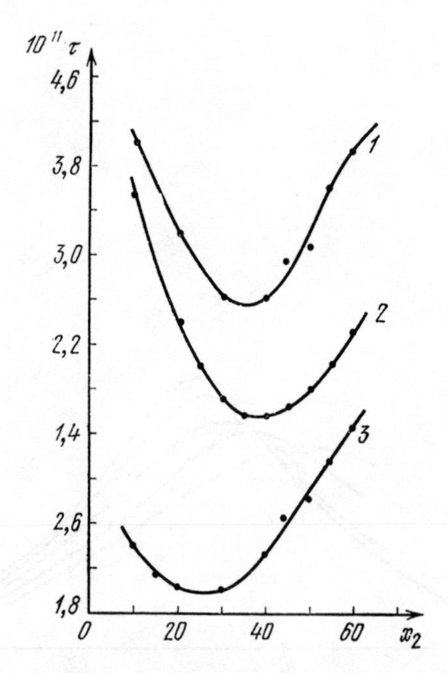

Fig. 30. Dependence of τ on concentration of solutions for t = 28°C, λ = 3.28; 4.38; 15; 20; 30 cm: 2-propanol-dioxan (1), methanol-dioxan (2), butan-1-ol-dioxan (3) [114].

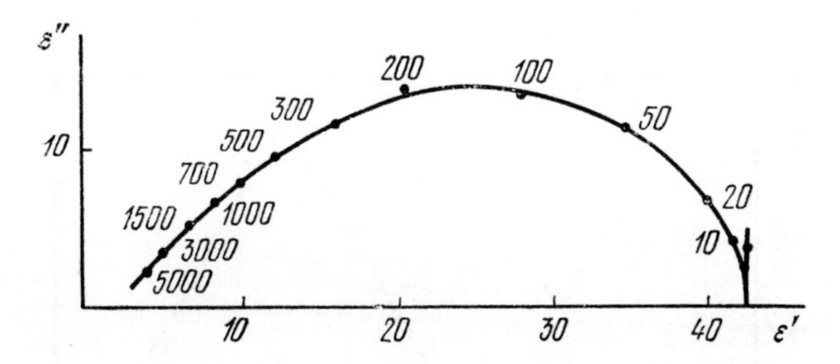

Fig. 31. Plot of ϵ'' against ϵ' at t = –110°C, x_1 = 25, for solutions of methanol-2-ethylhexan-1-ol (frequencies given in kHz) [599].

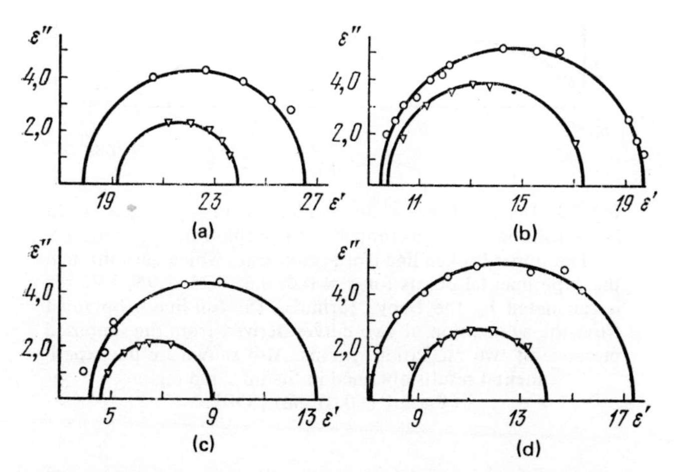

Fig. 32. Plot of ϵ'' against ϵ' for t = 25°C, ν = 200–3000 MHz of solutions: (a) tributylammonium iodide-acetone, N = 0.4 (○), N = 0.2 (▽); (b) tributylammonium picrate-1,2-dichloroethane, N = 0.4 (○), tributylammonium iodide-1,2-dichloroethane, N = 0.4 (▽); (c) tributylammonium picrate-trichloroethylene, N = 0.4 (○), tributylammonium iodide-trichloroethylene, N = 0.2 (▽); (d) tributylammonium picrate-tetrahydrofuran, N = 0.4 (○), tributylammonium picrate-dichloroethylene, N = 0.2 (▽) [600].

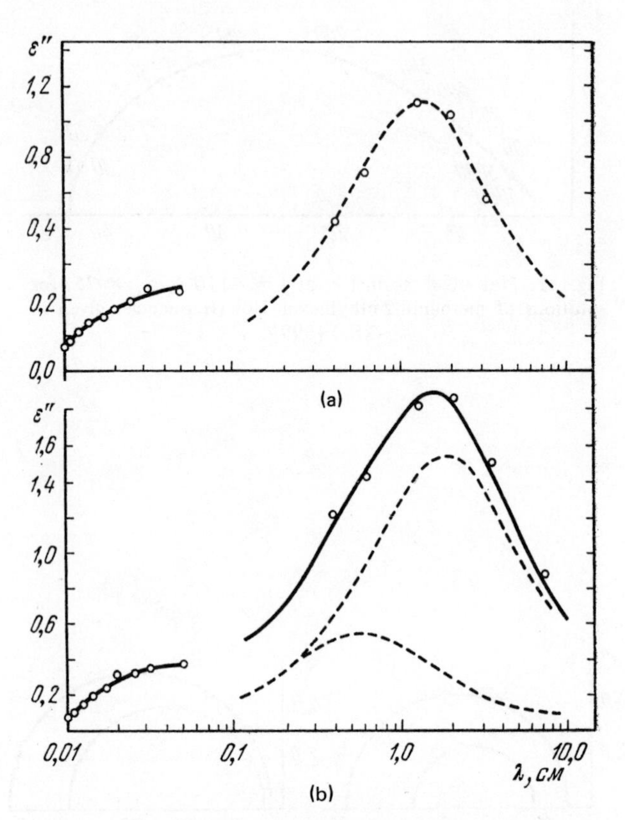

Fig. 33. Dependence of ϵ'' on wave length at $t = 40°C$, $x_1 = 25$ for solutions of (a) acetonitrile (b) propionitrile in benzene. The upper broken line (for acetonitrile) which goes through the experimental points for $\lambda = 0.4; 0.6; 1.25; 1.95; 3.23$ cm is calculated by the Debye formula. The full line is obtained from the summation of two curves derived from the supposed presence of two relaxation regions. Also shown are the experimental results obtained in the infra-red region ($\lambda = 0.1 - 0.01$ cm) [390].

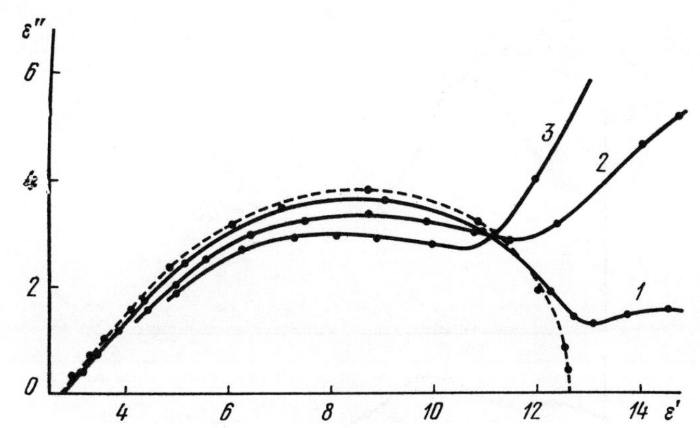

Fig. 34. Plot of ϵ'' against ϵ' for the high-frequency range for solutions of isopentyl bromide-ethanol: $x_2 = 20$ (1), $x_2 = 50$ (2), $x_2 = 67$ (3); the broken line is for pure isopentyl bromide. [328].

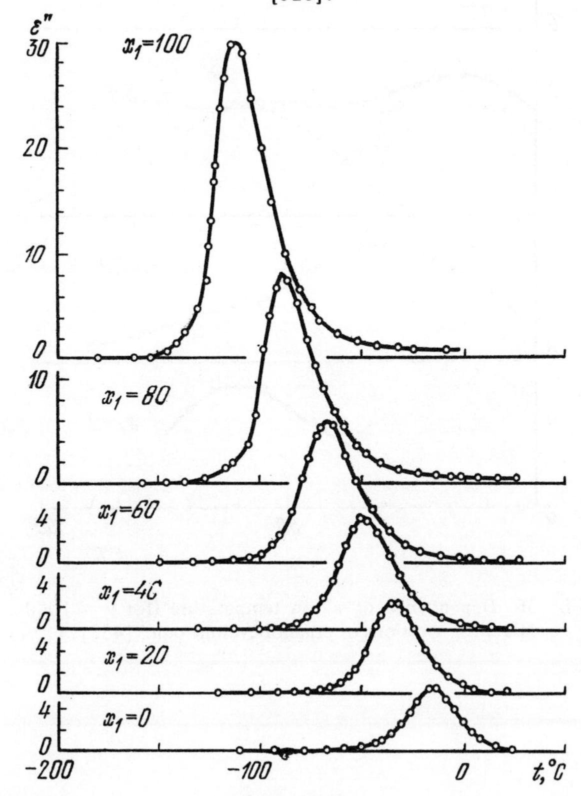

Fig. 35. Dependence of ϵ'' on the temperature (for $\nu = 2$ MHz) for solutions of ethanol-cyclohexanol [433].

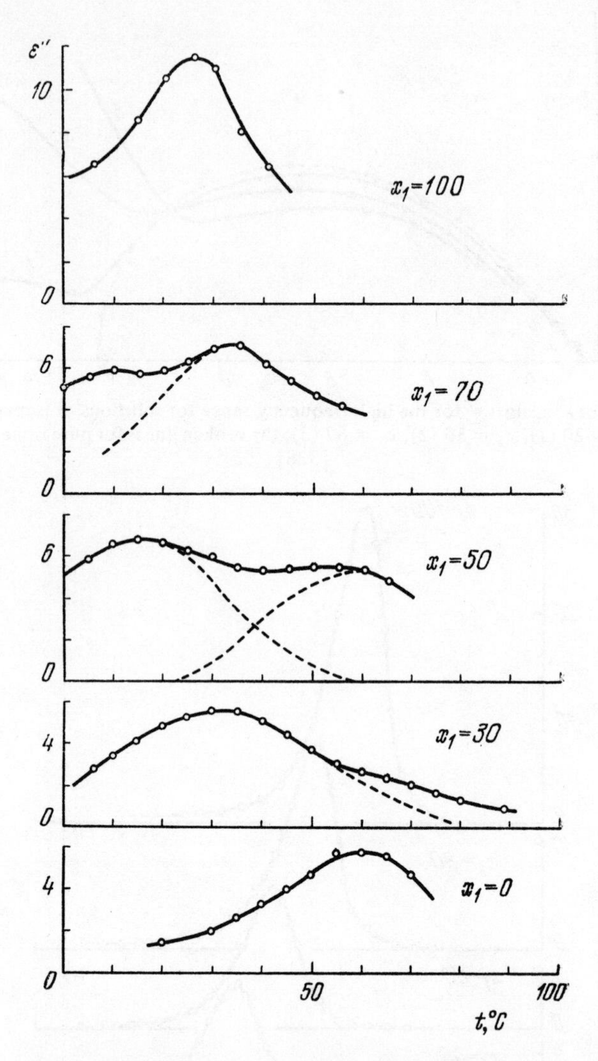

Fig. 36. Dependence of ϵ'' on temperature (for ν = 1000 MHz) for solutions of ethanol-cyclohexanol [433].

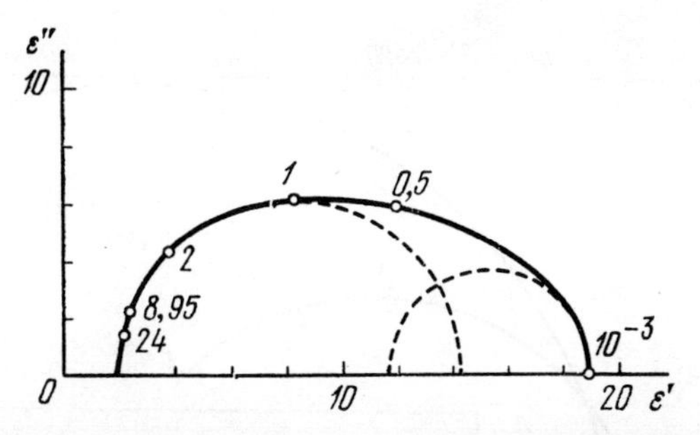

Fig. 37. Plot of ϵ'' against ϵ' for t = 40°C, x_1 = 70 for solutions of ethanol-cyclohexanol (adjacent to the experimental points the frequencies are shown in GHz) [433].

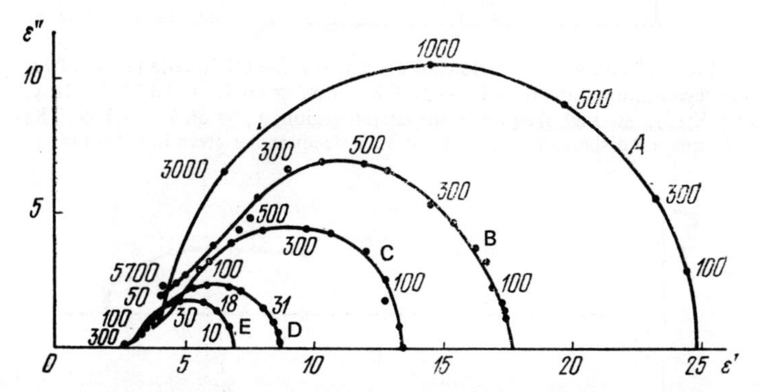

Fig. 38. Plot of ϵ'' against ϵ' for t = 25°C for solutions: pure ethanol (A); ethanol-*o*-tolylphosphate, x_1 = 88.8 (B), x_1 = 75 (C), x_1 = 33.3 (D); pure tolylphosphate (E) (frequencies given in MHz) [599].

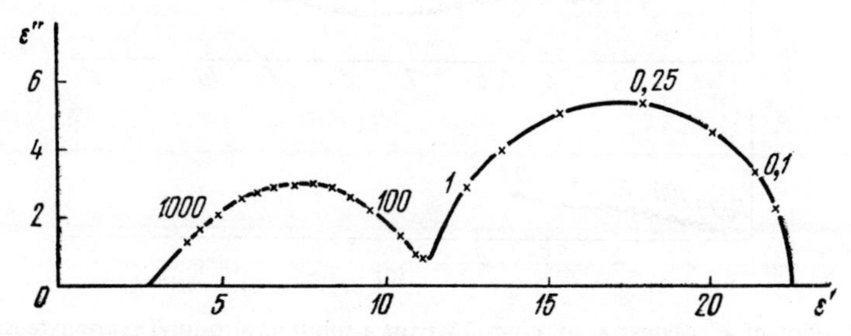

Fig. 39. Plot of ϵ'' against ϵ' at t = 135°K, x_1 = 33, for solution of isopentyl bromide-propan-1-ol (frequencies expressed in kHz) [330].

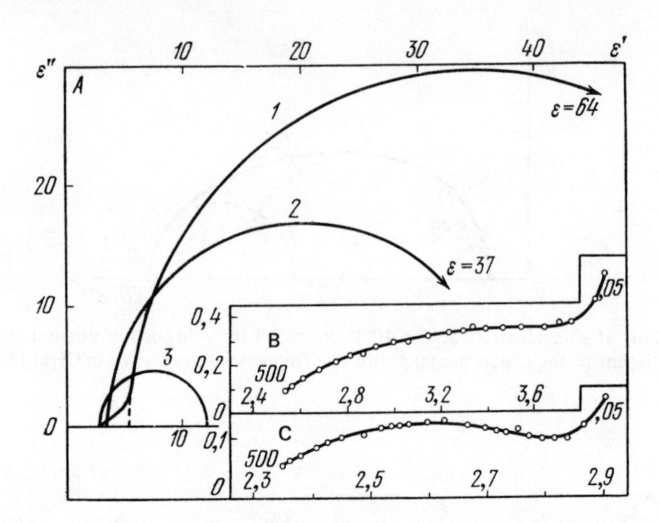

Fig. 40. Plot of ϵ'' and ϵ' for solutions of propan-1-ol-2-methylpentane: A–low-frequency dispersion region, pure propan-1-ol, $t = 133°$K (1); $x_1 = 66.7$, $t = 132.2°$K (2); $x_1 = 33.3$, $t = 130.8°$K (3); B–high-frequency dispersion region, $x_1 = 66.7$, $t = 109.7°$K; C–high-frequency dispersion region, $x_1 = 33.3$ (frequencies given in kHz) [444].

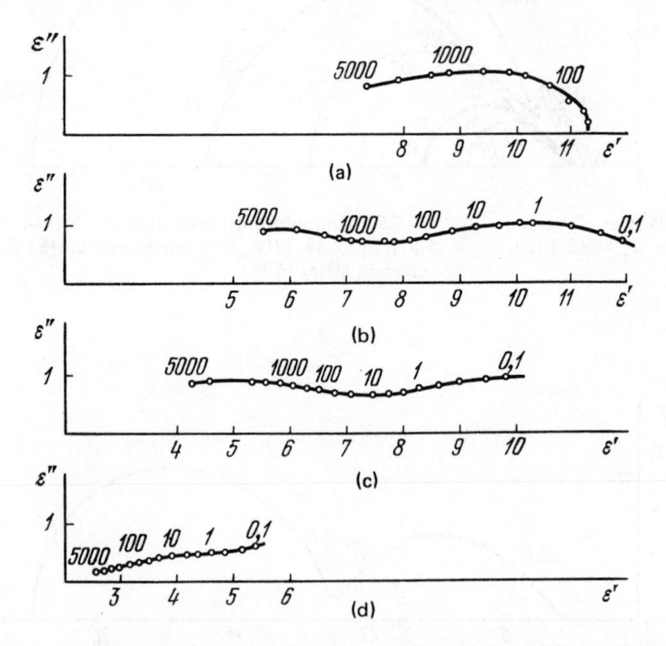

Fig. 41. Plot of ϵ'' against ϵ' at $x_1 = 16.7$ for a solution of o-tolyl phosphate-isobutyl chloride: $t = -98°$C (a), $t = -119°$C (b), $t = -129°$C (c), $t = -148°$C (d) (frequencies in kHz are shown above the experimental points) [601].

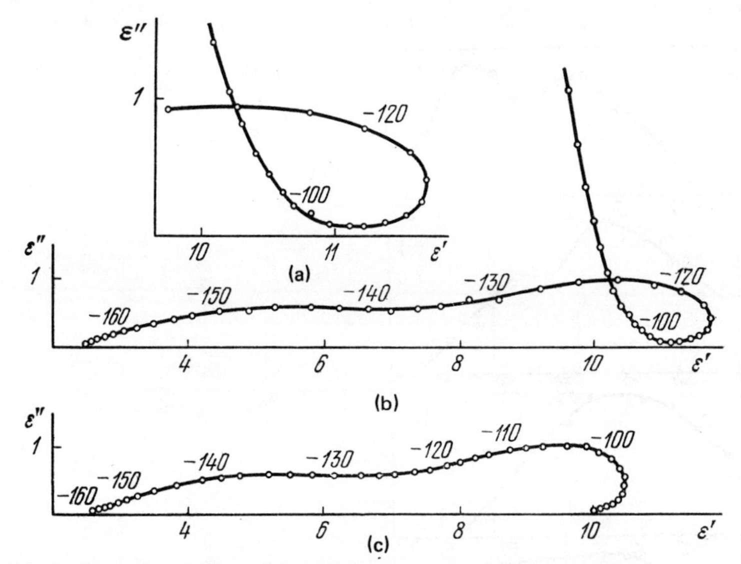

Fig. 42. Dependence of ϵ'' on ϵ' at $x = 16.7$ for an *o*-tolylphosphate-isobutyl chloride solution: $\nu = 0.1$ kHz (a, b); $\nu = 100$ kHz (c) [601].

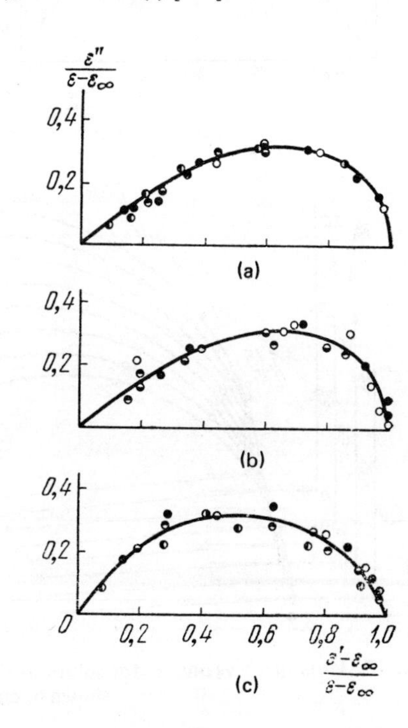

Fig. 43. Diagrams for solutions of chlorobenzene-butan-1-ol in the liquidus-solidus interval: $x_2 = 27$ (a), $x_2 = 52.6$ (b), $x_2 = 76.9$ (c), ($t = -65°C$ (○), $t = -70°C$ (●), $t = -75°C$ (◒), $t = -80°C$ (◓), $t = -85°C$ (◐) [451].

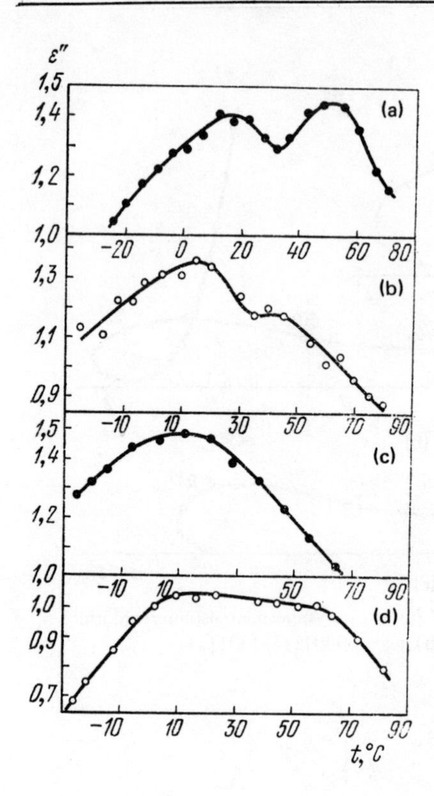

Fig. 44. Dependence of ϵ'' on temperature for $\nu = 16.2$ kHz for solutions of chlorobenzene-bromobenzene $\phi_2 = 28.6$ (a), $\phi_2 = 37.5$ (b), $\phi_2 = 100$ (c), $\phi_2 = 0$ (d) [459].

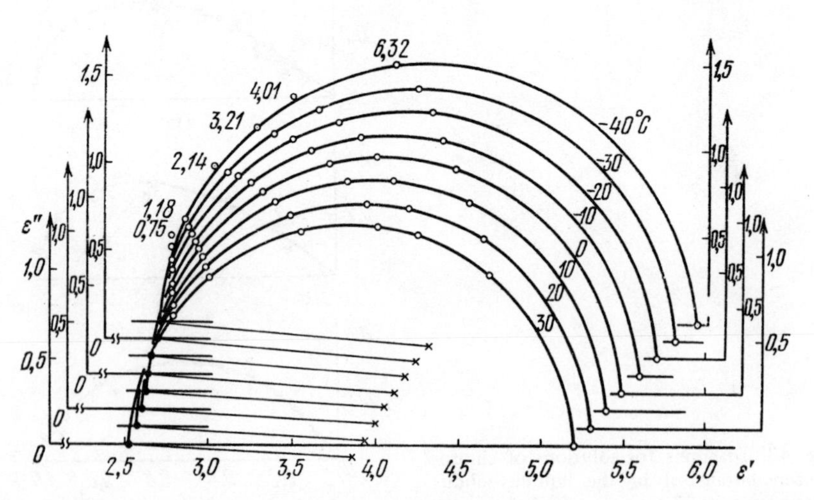

Fig. 45. Plots of ϵ'' against ϵ' for solutions of chlorobenzene-iodobenzene (wave lengths shown in cm) [464].

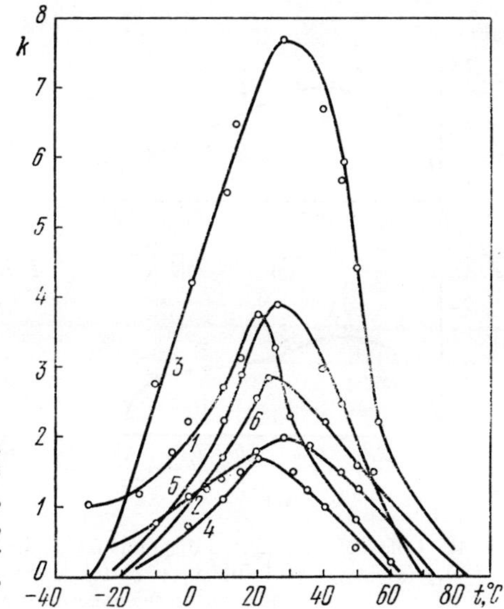

Fig. 46. Dependence of absorption coefficient k on temperature at $\lambda = 3.18$ cm, $x_1 = 30$ for solutions of aniline (1), *o*-chloroaniline (2), *m*-chloroaniline (3), *o*-toluidine (4), *p*-toluidine (5), *m*-toluidine (6) in diphenyl ether, $\epsilon' = n^2 - k^2$, $\epsilon'' = 2nk$ [475].

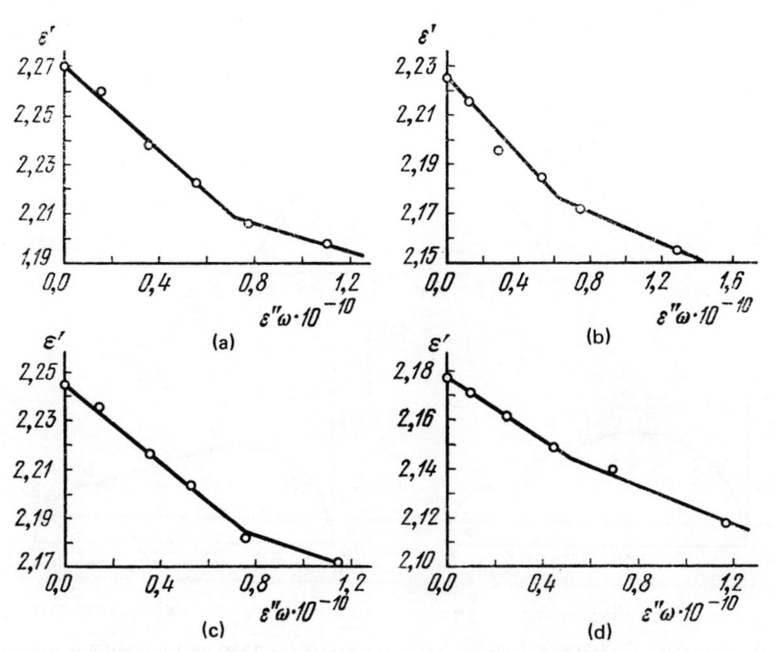

Fig. 47. Plots of ϵ' against $\epsilon''\omega$ for solutions of toluene-cyclohexane, $w_1 = 69.32$: $t = 15°C$ (a), $t = 25°C$ (b), $t = 40°C$ (c), $t = 60°C$ (d) [478].

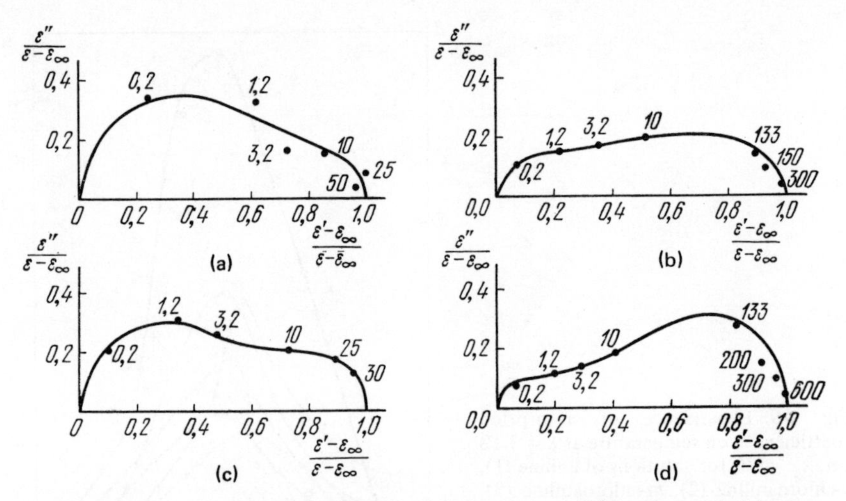

Fig. 48. Cole-Cole normalized diagrams for solutions at t = 25°C of hexan-1-ol-heptane: x_1 = 4.5 (a), x_1 = 10.1 (b), x_1 = 23 (c), x_1 = 28.6 (d) (wave lengths in cm. are shown adjacent to the experimental points) [482].

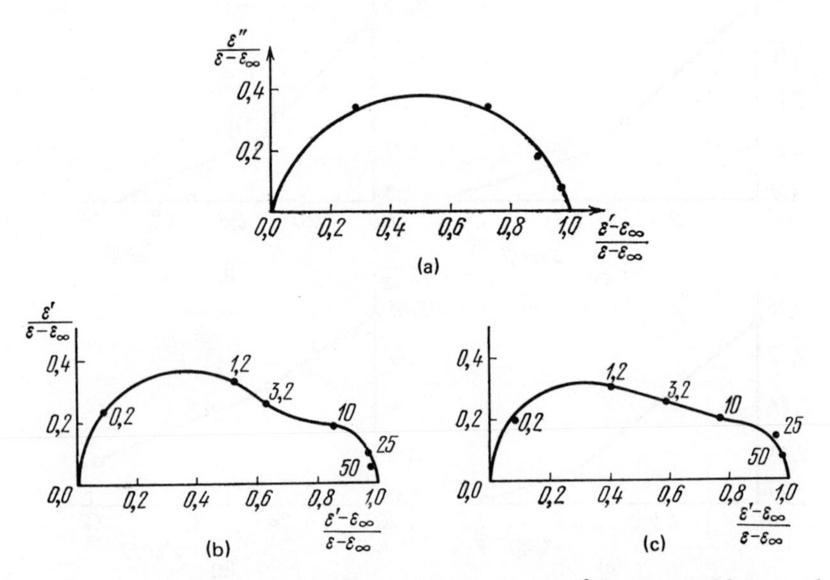

Fig. 49. Cole-Cole normalized diagrams for solutions at t = 25°C of 2-methyl-3-pentanol-heptane: x_1 = 3.4 (a), x_1 = 20.6 (b), x_1 = 34.7 (c) (wave lengths in cm are shown adjacent to the experimental points) [482].

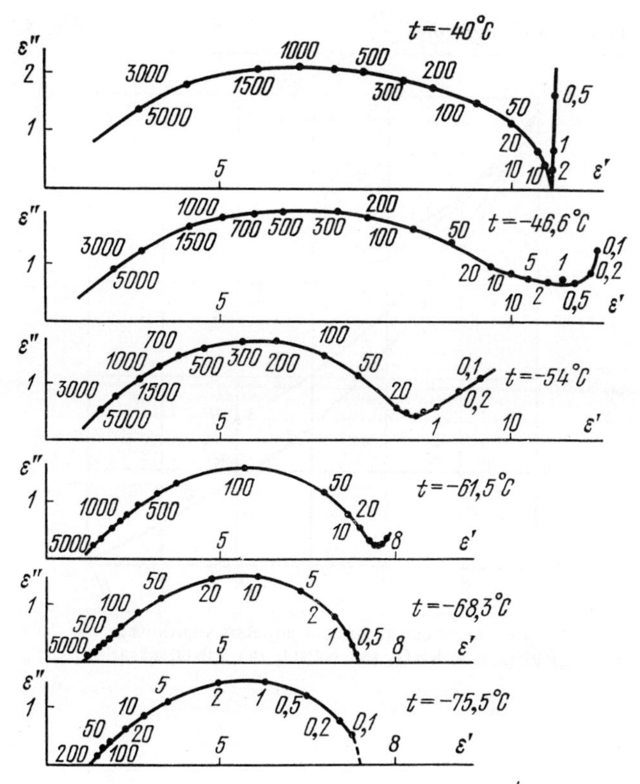

Fig. 50. Plots of ϵ'' against ϵ' for solutions $x_1 = 33.3$ of 3-methyl-2,2-heptandiol-dibutyl phthalate for various temperatures (frequencies expressed in kHz) [605].

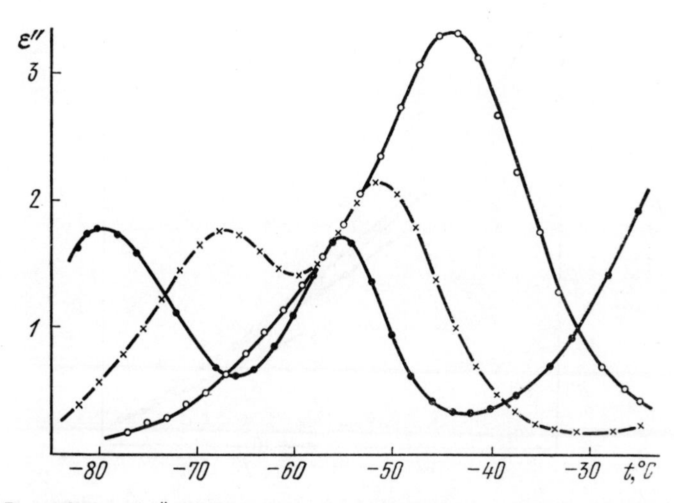

Fig. 51. Dependence of ϵ'' on temperature for a solution $x_1 = 66.6$ of 2-ethylhexan-1-ol-o-tolyl phosphate: $\nu = 1$ kHz (●), $\nu = 10$ kHz (×), $\nu = 100$ kHz (○) [605].

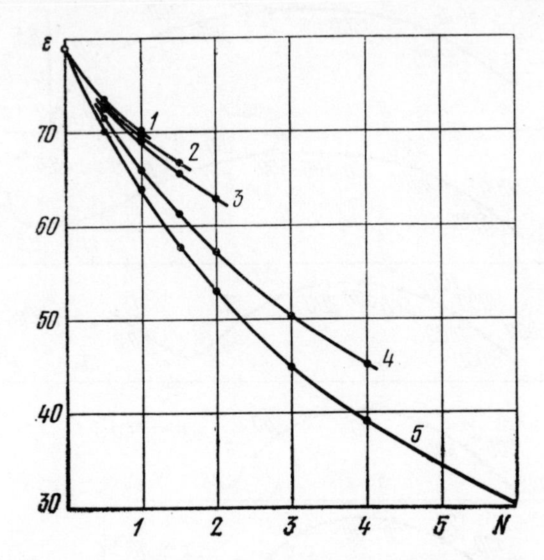

Fig. 52. Dependence of ϵ on concentration for aqueous solutions at t = 25°C of $CsNO_3$ (1), $RbNO_3$ (2), KNO_3 (3), $NaNO_3$ (4), $LiNO_3$ (5) [591].

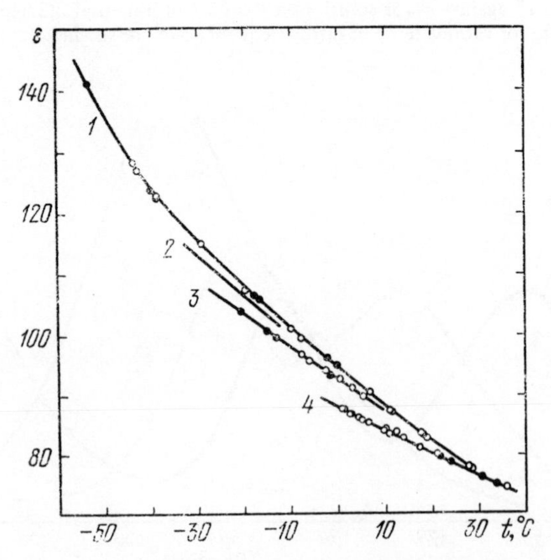

Fig. 53. Dependence of ϵ on temperature for aqueous solutions of hydrogen peroxide: w_2 = 46.5 (1), w_2 = 35.3 (2), w_2 = 26.2 (3), water (4) (for increase of temperature ○, for decrease of temperature ● [606].

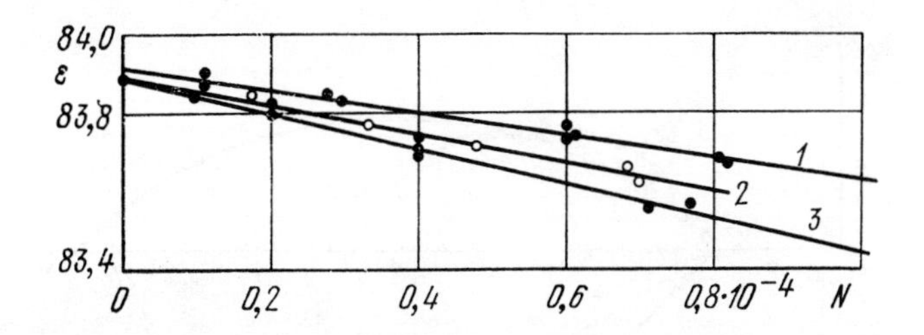

Fig. 54. Dependence of ϵ on concentration of aqueous solutions, (t = 12.3°C), of oxalic acid (1), malonic acid (2), succinic acid (3) [607].

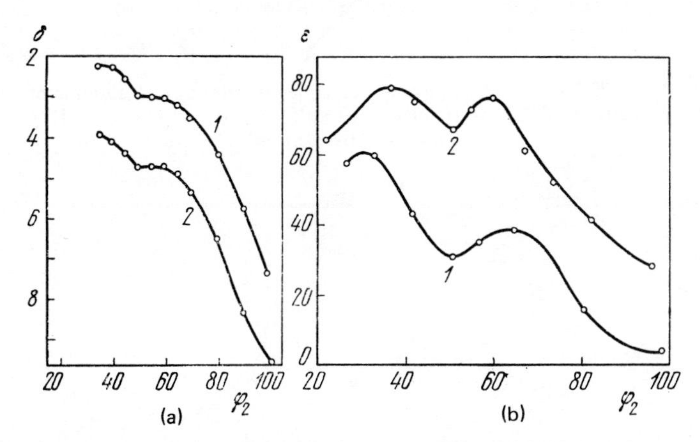

Fig. 55. Proton resonance shift, obtained by means of NMR (a) [608] and the variation with concentration for aqueous solutions (t = 30°C, ν = 1200 MHz) (b). Results for acetic acid (1) and propionic acid (2) [571].

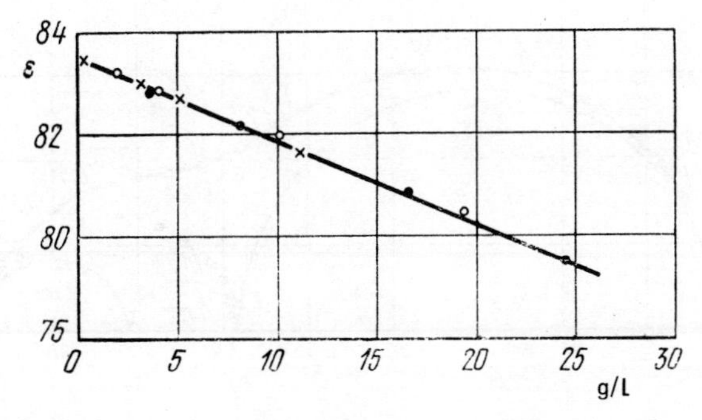

Fig. 56. Dependence of ϵ on concentration for aqueous solutions (t = 14°C) of cresols, ν = 1091 kHz: o-cresol (○), m-cresol (×), p-cresol (●) [607].

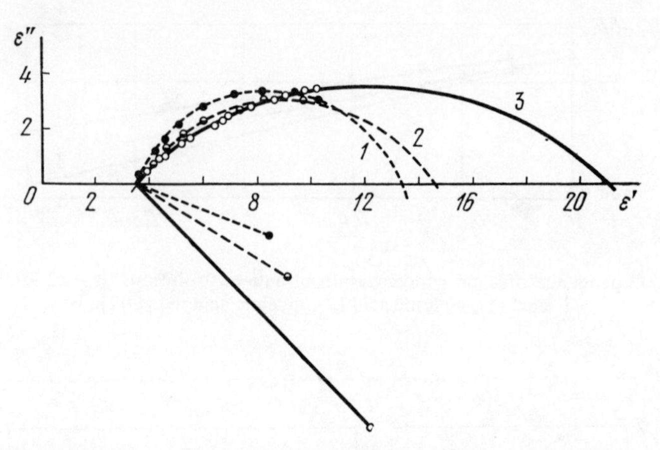

Fig. 57. Plot of ϵ'' against ϵ' for aqueous solutions of aluminium oxide (t = 50°C, ν = 10 kHz – 10 MHz); w_1 = 3.61 (1), w_1 = 4.87 (2), w_1 = 7.67 (3) [542].

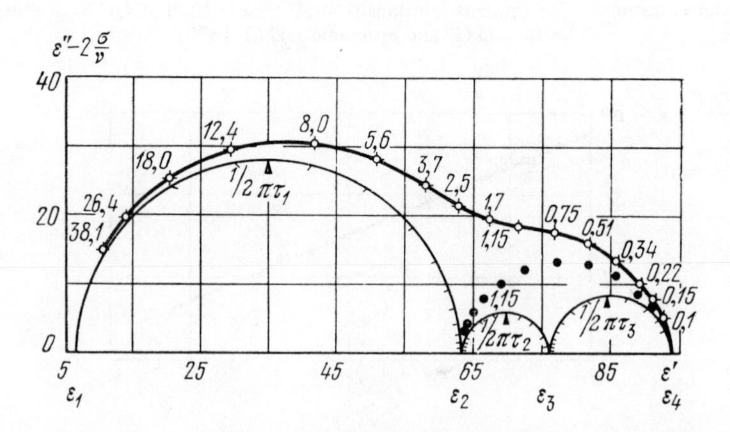

Fig. 58. Plot of ϵ'' against ϵ' for an aqueous solution of beryllium sulphate, t = 5°C, N = 1. Frequencies are given in GHz. [608].

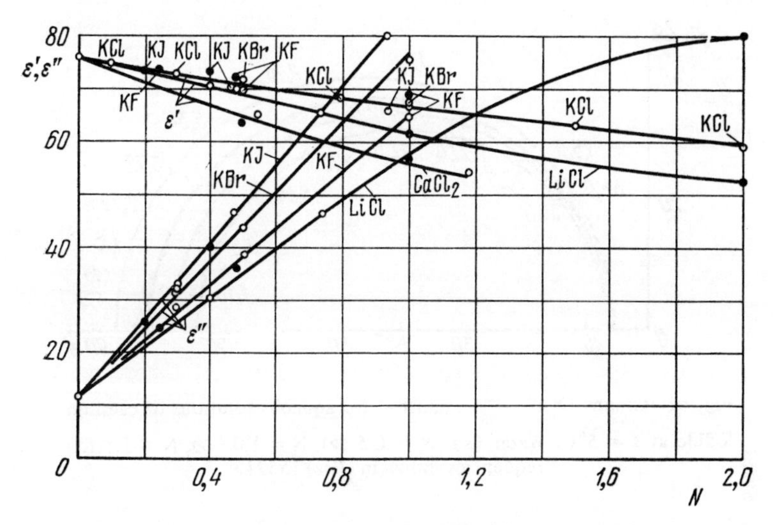

Fig. 59. Dependence of ϵ' and ϵ'' on concentration at $t = 25°C$ and $\lambda = 10$ cm for aqueous solutions of the halides of potassium, lithium and calcium (●–values from [441, 443], ○–values from [544]).

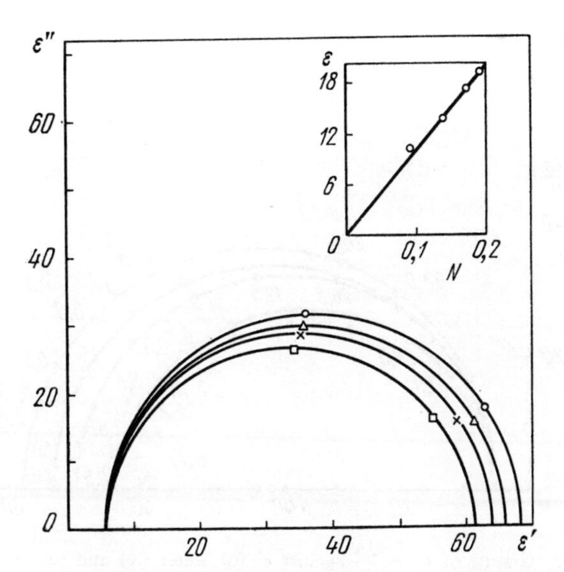

Fig. 60. Plot of ϵ'' against ϵ' for aqueous solutions of cobalt sulphate at $t = 28°C$ with $\lambda = 1.276; 3.14$ cm: N = 0.25 (□), N = 0.5 (Δ), N = 0.75 (X), N = 1.0 (O) (in the top corner the graph shows the variation of ϵ with the concentration [548].

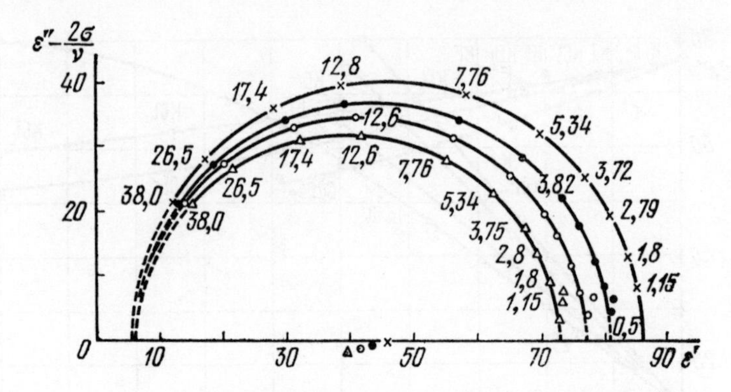

Fig. 61. Graphs of $\epsilon'' - \frac{2\sigma}{\nu}$ against ϵ' for aqueous solutions of caesium iodide at t = 5°C: Water (×), N = 0.5 (•), N = 1.0 (○), N = 1.5 (Δ) (frequencies shown in GHz) [552].

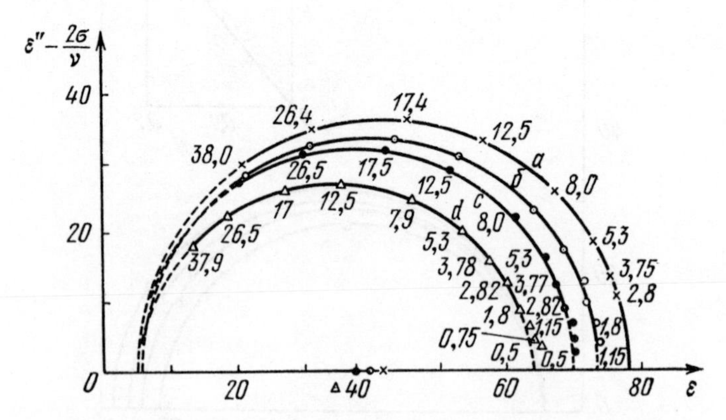

Fig. 62. Graphs of $\epsilon'' - \frac{2\sigma}{\nu}$ against ϵ' for water (×) and for N = 0.5 aqueous solutions: caesium iodide (○), lithium iodide (•), tetrabutyl-ammonium bromide (Δ) (frequencies shown in GHz) [552].

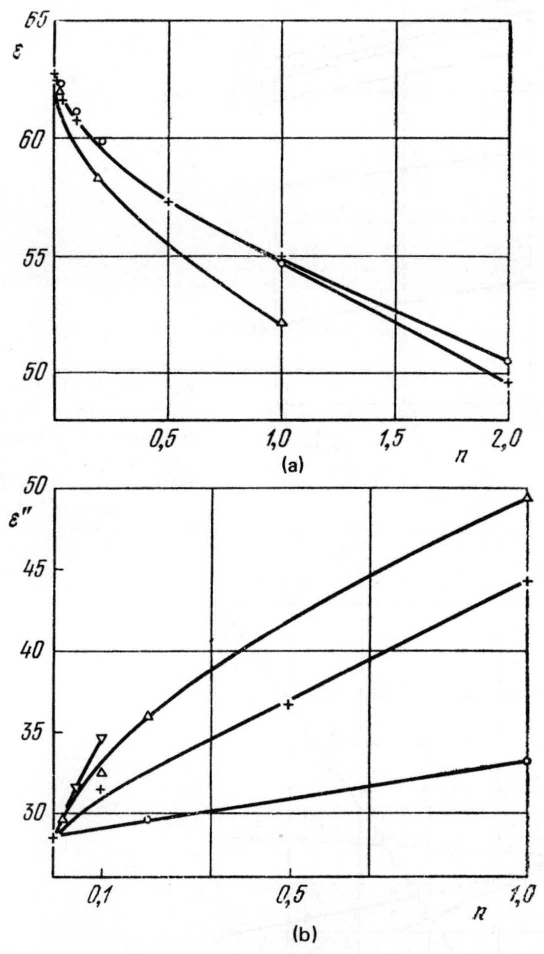

Fig. 63. Plots of ϵ' (a) and ϵ'' (b) at $t = 25°C$, $\lambda = 3.2$ cm, against concentration of aqueous solutions: KCL(+), $NiSO_4$ (○), H_2SO_4 (Δ), HCl (▽) (Concentrations in gram-equivalents/litre) [609].

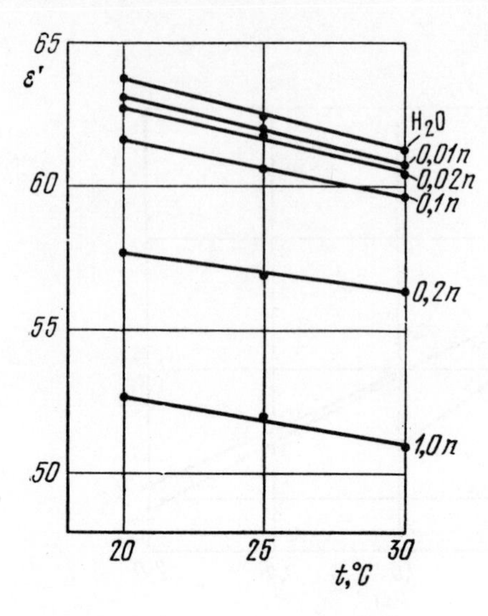

Fig. 64. Dependence of ϵ' on temperature for aqueous solutions of sulphuric acid with λ = 3.2 cm (concentrations in gram-equivalents/litre) [609].

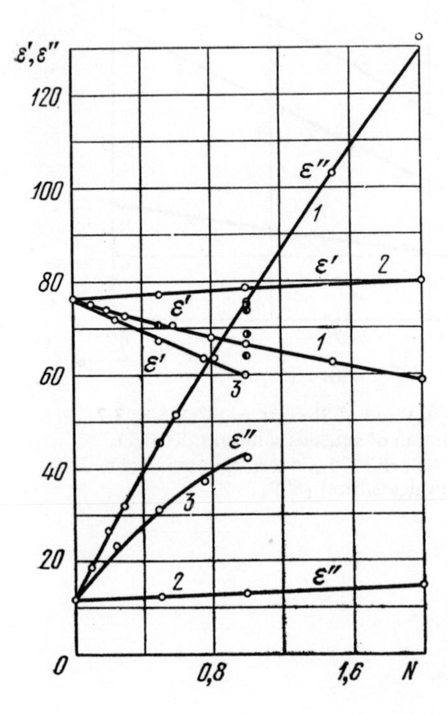

Fig. 65. Dependence of ϵ' and ϵ'' at t = 20°C and λ = 10 cm on concentrations of aqueous solutions: potassium chloride (1), urea (2), tetraethylammonium bromide (3) (◐ – [543], ◑ – [541], -○ [541]).

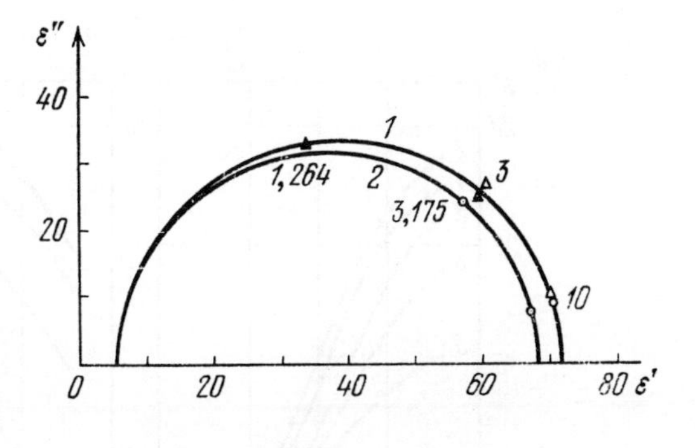

Fig. 66. Plot of ϵ'' against ϵ' for an N = 0.5 aqueous solution of potassium chloride at t = 25°C (1), t = 35°C (2) (○–[556], Δ–[543], ▲–[558], wave lengths given in cm).

Fig. 67. Dependence of ϵ' and $\epsilon'' - \frac{2\sigma}{\nu}$ on frequency at $t = 25$°C for water (Δ) and for N = 1 aqueous solutions: tetramethylammonium bromide (●), lithium bromide (○) [552].

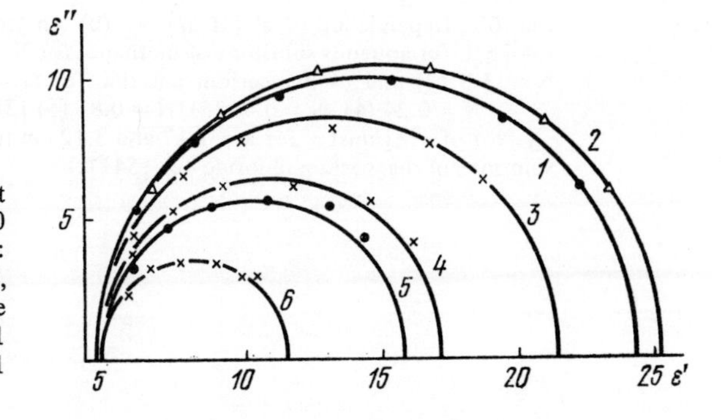

Fig. 68. Plots of ϵ'' against ϵ' ($t = 25$°C, $\nu = 330-3300$ MHz) for aqueous solutions: ethanol at N = 0.22 (1), N = 1.3 (2), lithium chloride at N = 0.125 (3), N = 0.31 (4), N = 0.48 (5), N = 0.81 (6) [381].

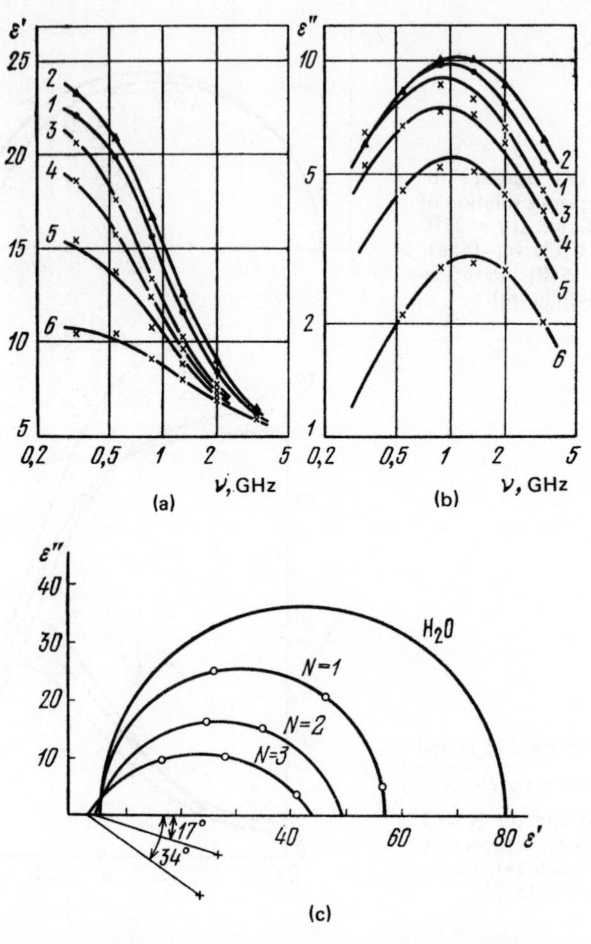

Fig. 69. Dependence of ϵ' (a) and ϵ'' (b) on frequency at $t = 25°C$ for aqueous solutions of methanol for N = 0.22 (1), N = 1.3 (2) and of magnesium chloride for N = 0.12 (3), N = 0.24 (4), N = 0.40 (5), N = 0.81 (6) [381].

Plot of ϵ'' against ϵ' for $\lambda = 1.17$ and 3.12 cm for aqueous solutions of magnesium chloride (c) [541].

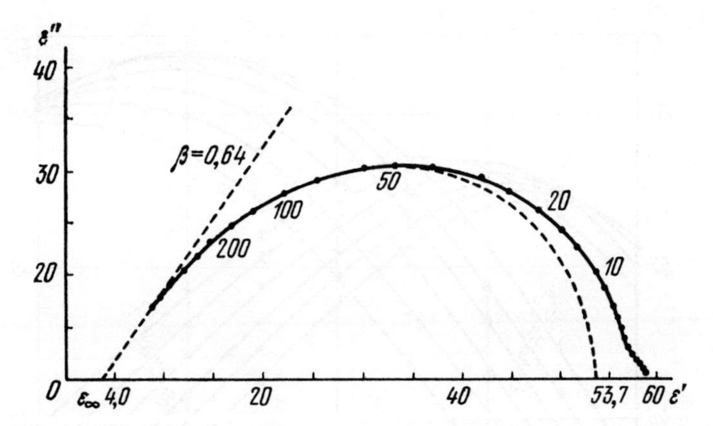

Fig. 70. Plot of ϵ'' against ϵ' for $t = 35.6°C$, N = 1 for aqueous solution of sodium chloride (frequencies in kHz) [386].

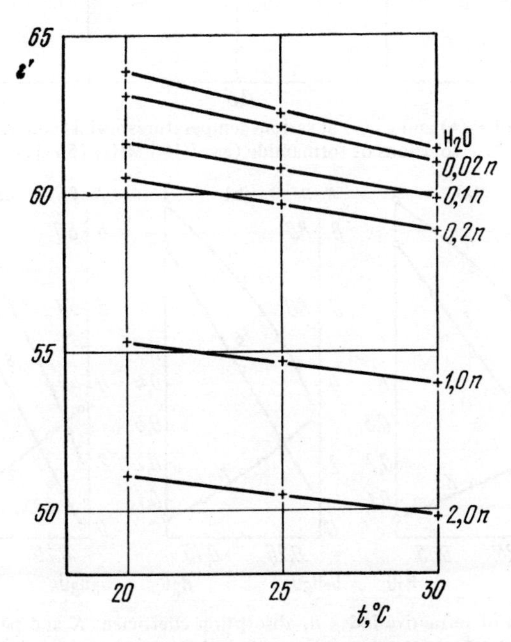

Fig. 71. Dependence of ϵ' on temperature for aqueous solutions of nickel sulphate for $\lambda = 3.2$ cm. (concentrations in gram-equivalents/litre) [603].

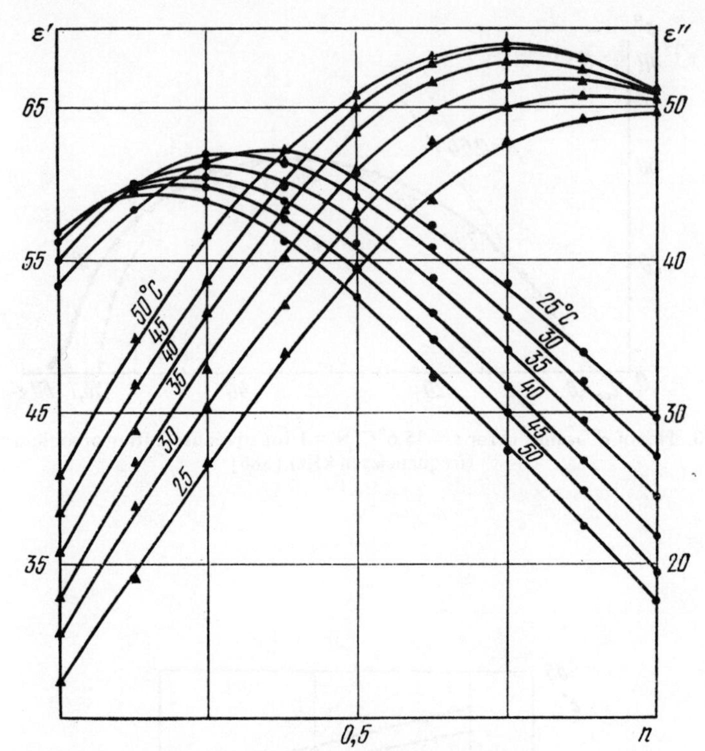

Fig. 72. Graphs of ϵ' (▲) and ϵ'' (●) at various temperatures against concentrations of aqueous solutions of formamide ($\nu = 9465$ MHz) [565].

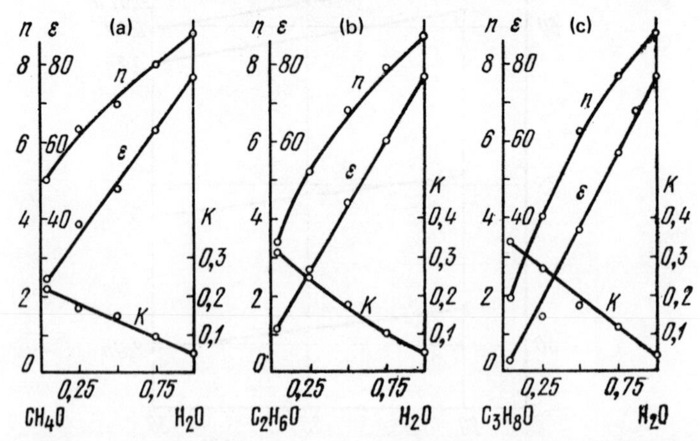

Fig. 73. Variation of refractive index n, absorption coefficient K and permittivity ϵ at $t =$ 18°C and $\lambda = 14$ cm for aqueous solutions of alcohols: methanol (a), ethanol (b), propan-1-ol (c) (concentration as volume fraction of water), $\epsilon^1 = n^2 - k^2$, $\epsilon'' = 2nK$. [553].

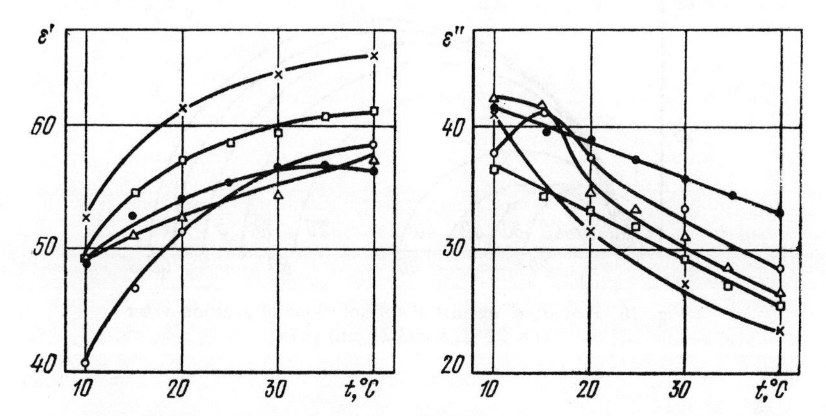

Fig. 74. Dependence of ϵ' and ϵ'' for $\nu = 9730$ MHz and N = 3 on temperature for pure water (×) and for aqueous solutions: urea (Δ), thiourea (□), acetamide (○), guanidine (•) [569].

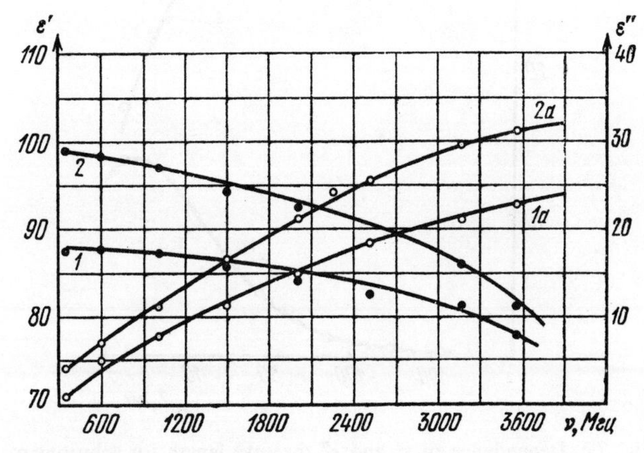

Fig. 75. Dependence of ϵ' (•) and ϵ'' (○) at t = 25°C on frequency for aqueous solutions of urea for N = 4 (1, 1a), N = 8 (2, 2a) [610].

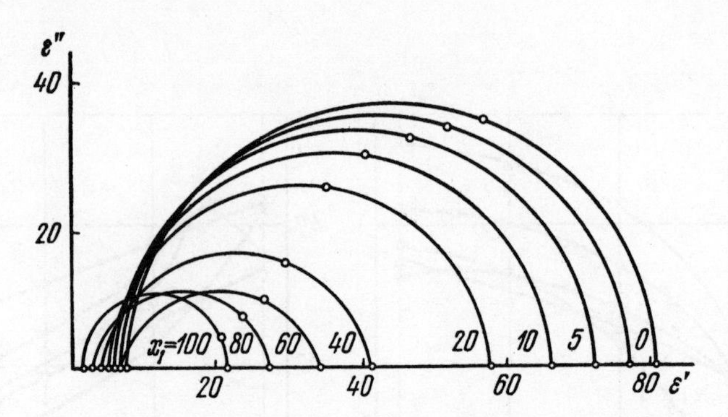

Fig. 76. Plot of ϵ'' against ϵ' for solutions of acetone-water ($t = 20°C, \lambda = 3.21$ cm) [574].

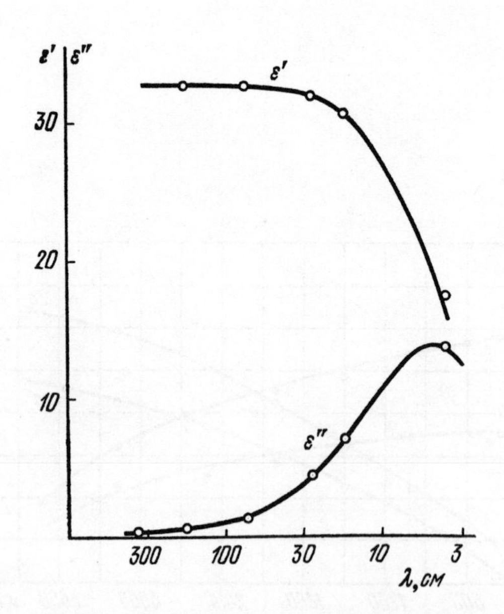

Fig. 77. Dependence of ϵ' and ϵ'' on wave length for solutions of water-dioxan ($t = 25°C, x_2 = 50$) [592].

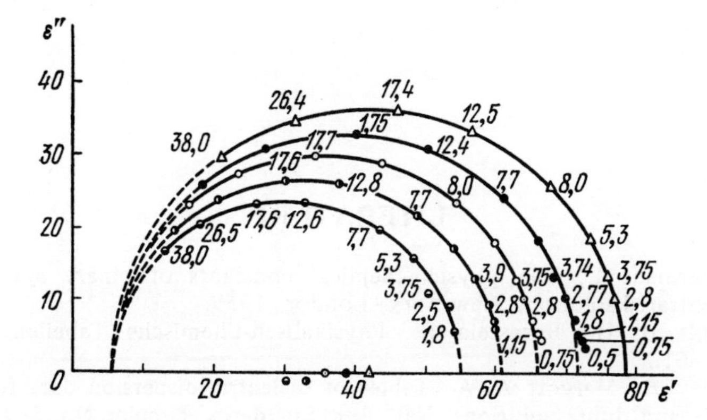

Fig. 78. Dependence of ϵ'' on ϵ' at $t = 25°C$ for a water-tetramethylammonium bromide solution: water (Δ), N = 0.5 (●), N = 1.0 (○), N = 1.5 (○), N = 2 (○) (frequencies given in GHz) [552].

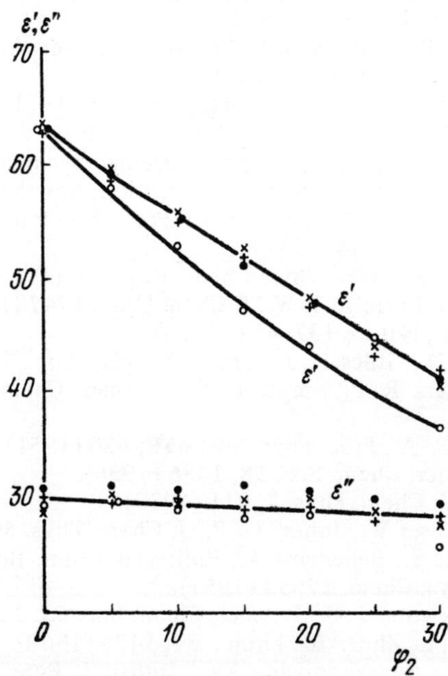

Fig. 79. Dependence of ϵ' and ϵ'' at $t = 25°C$, $\lambda = 3.2$ cm on the concentration of water solutions: glucose (●), saccharose (X), starch (+), gelatine (○) [545].

LITERATURE

1. Timmermans I., The physico-chemical constants of binary systems in concentrated solutions, New-York–London, 1959.
2. Landolt H. H., Bornstein R., Physikalisch-Chemische Tabellen, Berlin 1950–61.
3. Buckley F., Maryott A. A., Tables of dielectric dispersion data for pure liquids and dilute solutions, Nat. Bur. Standards, circular No. 589, 1958.
4. Akhadov Y.Y., Dielektricheskie svoistva chistykh zhidkostei, Izd. standartov, 1972.
5. Debye P., Polar Molecules, 1931.
6. Skanavi G. I., Fizika dielektrikov, Gostekhizdat, 1949.
7. Fröhlich H., The Theory of Dielectrics, 1960.
8. Hippel A. R., Dielectrics and Waves, 1960.
9. Braun V. F., Dielektriki, IL, 1961.
10. Shakhparonov M. I., Metody issledovaniya teplogo dvizheniya molekul i stroenie zhidkostei, Izd, MGU, 1963.
11. Bogoroditsky N. P., Volakobinsky Y. M., Vorob'ev A. A., Tareev B. M., Teoriya dielektrov, 'Energiya', 1965.
12. Gubkin A. N., Fizika dielektrov, 'Vysshya shkola', 1971.
13. Smyth C. P., Dielectric behaviour and structure, New York, 1955.
14. Hill N. E., Dielectric properties and molecular behaviour, London, 1969.
15. Bottcher C. I. F., Theory of electric polarisation, New-York, 1952.
16. Daniel V. V., Dielectric relaxation, Acad. Press, London, 1967.
17. Müller F. H., Physik. Z. **38**, 283 (1937).
18. Higasi K., Sci. Papers. Inst. Phys. Chem. Research (Tokyo) **28**, 284 (1936).
19. Le Fevre C. G., Le Fevre R. J. W., J. Chem. Soc., 1747 (1955).
20. Guggenheim E. A., Nature **137**, 459 (1936).
21. Buckingham A. D., Stiles P. J., Trans. Faraday Soc. **67** (3), 577 (1971).
22. Conner W. P., Clark R. P., Smyth C. P., J. Amer. Chem. Soc. **64** (1), 1379 (1942).
23. Ross J. G., Sack R. A., Proc. Phys. Soc. **64B**, 620 (1951).
24. Onsager L., J. Amer. Chem. Soc. **58**, 1486 (1936).
25. Kirkwood J. G., J. Chem. Phys. **7**, 911 (1939).
26. Chen T., Dannhauser W., Jonori G. P., J. Chem. Phys. **50** (5), 2046 (1969).
27. Loon R. V., Fuks S., Bellemans A., Bull. Soc. Chim. Belg. **76**, 202 (1967).
28. Cole R. H., J. Phys. Chem. **27**, 33 (1957).
29. Fouss R. M., Kirkwood J. C., J. Amer. Chem. Soc. **63**, 385 (1941).
30. Shakhparonov M. I., Zhur. fiz. khim., **34**, 1478 (1960).
31. Dashpande D. K., Suryanaruyana Rao, Indian J. Pure Appl. Phys. **7**, 439 (1969).
32. Trench I. A., Suaya R., Westerkamp J. F., Anal. Soc. Cient. Argent. **188**, 161 (1969).

33. Adamski P. Kryszewski M., Acta Physica Polonica **33**, 307 (1968).
34. Kotake K., Nakamura N., and Chihara H., Bull. Chem. Soc. (Japan) **43**, 2429 (1970).
35. Rizk H. A., Youssef N., Grace H., Canad. J. Chem. **47**, 3767 (1969).
36. Scholte Th. G., De Vos F. C., Rec. Trav. Chim. **72**, 625 (1953).
37. Rataichak G., Sobchik L., Zhurnal strukturnoi khimii **6**, 262 (1965).
38. Scholte Th. G., Physica **13**, 437 (1949).
39. Poley J. P., J. Chem. Phys. **22**, 1466 (1954).
40. Buckingham A. D., Trans. Faraday Soc. **49**, 881 (1953).
41. Buckingham A. D., Austr. J. Chem. **6**, 93 (1953).
42. Wilson J. N., Chem. Rev. **25**, 377 (1939).
43. Abbott J. A., Bolton H. C., Trans. Faraday Soc. **48**, 428 (1952).
44. Poley J. P., Appl. Sci. Res. **B4**, 337 (1955).
45. Klug D. D., Waughan W. E., J. Chem. Phys. **56** (10), 5005 (1972).
46. Syrkin V. K., Dokl. Akad. Nauk. **35**, 43, (1942).
47. Smith J. W., Walshaw S., J. Chem. Soc., 3217, 4727 (1957).
48. Smith J. W., Walshaw S., J. Chem. Soc., 3784 (1959).
49. Khare B. N., J. Chem. Phys. **47**, 5173 (1967).
50. Shakhparonov M.I., Goremykina V.V., Vestnik MGU, ser. khimiya **1**, 18 (1971).
51. Loon R. V., Dauchot J. P., Bellemans A., Bull. Soc. Chim. Belg. **77**, 397 (1968).
52. Weisbecker A., J. Chim. Phys. Biol. **66**, 2, 226 (1969).
53. Colderwood J. H., Smyth C. P., J. Amer. Chem. Soc **78**, 1295 (1956).
54. Bauer P., Diguet R., Comp. Rend. Acad. Sci. Paris **272C** (15), 1341 (1971).
55. Scaife B. K. P., Proc. Phys. Soc. **81**, 194 (1963).
56. Chandra S., Prakash J., Indian J. Pure Appl. Phys. **6**, 116 (1968).
57. Cole R. H., J. Chem. Phys. **23**, 493 (1955).
58. Brot C., Compt. Rend. Acad. Sci. Paris **19**, 397 (1959).
59. Gopala Krishna K. V., Trans. Faraday Soc. **53**, 769 (1957).
60. Higasi K., Uchiyana K., Bull. Res. Inst. Appl. Elect. **17**, 164 (1965).
61. Prakashi J., J. Phys. Soc. Japan **34 (1)**, 134 (1973).
62. Heston W. M., Hannelly E. I., Smyth C. P., J. Amer. Chem. Soc. **70**, 4093 (1948).
63. Laquer H. L., Smyth C. P., J. Amer. Chem. Soc. **70**, 4097 (1948).
64. Hennelly E. I., Heston W. M., Smyth C. P., J. Amer. Chem. Soc. **70**, 4102 (1948).
65. Pitt D. A., Smyth C. P., J. Amer. Chem. Soc. **80**, 1061 (1958).
66. Clark D. E., Kumar S. N., Brit. J. Appl. Phys. **7**, 282 (1956).
67. Curtis A. J., McGeer P. L., Rathman G. B., Smyth C. P., J. Amer. Chem. Soc. **74**, 645 (1952).
68. Petro A. J., Smyth C. P., J. Amer. Chem. Soc. **79**, 6142 (1957).
69. Grossley J., Advances molecul. relax. process. **2** (1), **69** (1970).
70. Buchanan T. J., Grant E. H., Brit. J. Appl. Phys. **6**, 64 (1955).
71. Girard P., Abadie P., Trans. Faraday Soc. **42A**, 40 (1946).
72. Yager W. A., Physica **7**, 434 (1936).
73. Cole K. S., Cole R. H., J. Chem. Phys. **9**, 341 (1949).
74. Grant E. H., J. Phys. Chem. **73**, 4386 (1969).
75. Williams G., J. Phys. Chem. **63**, 537 (1959).
76. Davies M., J. Chem. Educ. **46**, 17 (1969).

77. Clark G. L., J. Chem. Phys. **25** (1), 125 (1956).
78. Higasi K., Bull. Chem. Soc. Japan **39**, 2157 (1966).
79. Srivastava K. S. C., Nath D., Bull. Chem. Soc. Japan **43** (9), 2805 (1970).
80. Davidson D. W., Cole R. H., J. Chem. Phys. **19**, 1484 (1951).
81. Davidson D. W., Canad. J. Chem. **39** (3), 571 (1961).
82. Cole R. H., J. Chem. Phys. **6**, 385 (1938).
83. Powles J. G., J. Chem. Phys. **21**, 633 (1953).
84. O'Dwyer J. J., Sack R. A., Austral. J. Sci. Res. **A5**, 647 (1952).
85. Hill N. E., Proc. Phys. Soc. **72**, 1532 (1958).
86. Srivastava G. P., Mathur P. C., Tripathi K. N., J. Phys. Soc. Japan **27** (2), 460 (1969).
87. Jackson W., Powles J., Trans. Faraday Soc. **42A**, 101 (1946).
88. Higasi K., Bergman K., Smyth C. P., J. Phys. Chem. **64** (7), 880 (1960).
89. Govande S. G., Garg S. K., Kadaba P. K., Mater. Sci. Engrs. **4** (4), 206 (1969).
90. Fuoss R. M., Kirkwood J. G., J. Amer. Chem. Soc. **63**, 385 (1941).
91. Muralidhara Rao V., Indian J. Pure Appl. Phys. 1 (1), 33 (1963).
92. Budo A., Physik Z. **39**, 706 (1938).
93. Bergman K., Roberti D. M., Smyth C. P., J. Phys. Chem. **5**, 665 (1960).
94. Magee M. D., Walker S., Trans. Faraday Soc. **62**, 3093 (1966).
95. Bhattacharyya J., Hasan A., Roy S. B., Kastha G. S., J. Phys. Soc. Japan **28** (1), 204 (1970).
96. Barriol J., Boule P., Diguet R., Comp. Rend. **268C** (23), 1977 (1969).
97. Boule P., J. Chim. Phys. France **66** (11–12), 1987 (1969).
98. Sabench J., Layer G., Regnier J., Comp. Ren. Acad. Sci. Paris **274C** (10), 973 (1972).
99. Higasi K., Koga Y., Makamura M., Bull. Chem. Soc. Japan **44** (4), 988 (1971).
100. Turner E. M., Ehrhardt W. W., Leone G., Vaughan E., J. Phys. Chem. **74** (19), 3543 (1970).
101. Magee M. D., Walker S., J. Chem. Phys **50** (2), 1019 (1969).
102. Klanbuehl D. E., Klug D. D., Vaughan W. E., J. Chem. Phys. **50** (12), 5266 (1969).
103. Birnbaym G., Cohen E. R., J. Chem. Phys. **53** (7), 2885 (1970).
104. G. Williams, Chem. Rev. **72** (1), 55 (1972).
105. Magat M., Izv. AN SSSR **24** (1), 10 (1960).
106. Cole R. H., Davidson D. W., J. Chem. Phys. **20**, 1388 (1952).
107. Dannhauser W. J., J. Chem. Phys. **55**, 629 (1971).
108. Danney D. J., J. Chem. Phys. **30**, 1019 (1959).
109. Dalbert M., Magat M., Surdut A., Bull. Soc. Chim. France **7–8**, D 345, 1949.
110. Hassion F. X., Cole R. H., J. Chem. Phys. **23**, 1756 (1955).
111. Lane A., Saxton J. A., Proc. Roy. Soc. **211A**, 530 (1953).
112. Grant E. M., Proc. Phys. Soc. **70**, 937 (1957).
113. Rampolla R. W., Miller R. S., Smyth C. P., J. Chem. Phys. **30**, 566, (1959).
114. Sarojini V., Trans. Faraday Soc. **57**, 425 (1961).
115. Garg S. K., Smyth C. P., J. Phys. Chem. **69**, 1294 (1965).
116. Johari G. P., Smyth C. P., J. Amer. Chem. Soc. **91**, 6215 (1969).
117. Crossley J., Glasser L., Smyth C. P., J. Chem. Phys. **52** 6203, (1970).
118. Meakins R. J., Trans. Faraday Soc. **54**, 1160 (1958).

119. Pitt A. J., Smyth C. P., J. Phys. Chem. **63**, 582 (1959).
120. Miller R. C., Smyth C. P., J. Am. Chem. Soc. **79**, 3310 (1957).
121. Perrin F. J., Phys. Radium (Paris) **5**, 497 (1934).
122. Fischer E., Phys. Z. **40**, 645 (1939).
123. Gierer A., Wirtz K., Z. Naturf. **8a**, 532 (1953).
124. Andrade E. N., Phil. Mag. **17**, 497 (1934).
125. Hill N. E., Proc. Phys. Soc. **67B**, 149 (1954).
126. Hayghan W. E., Russell W. P., Smyth C. P., J. Am. Chem. Soc. **83** (3), 571 (1961).
127. Magee M. D., Walker S., J. Chem. Phys. **50**, 2580 (1969).
128. Magee M. D., Canad. J. Chem. **49**, 1106 (1971).
129. Higasi K., Dielectric relaxation and molecular structure, Sapporo, Japan, 1961.
130. Chitoku K., Higasi K., Bull. Chem. Soc. Japan **36**, 1064 (1963).
131. Kalman O. F., Smyth C. P., J. Amer Chem. Soc. **82**, 783 (1960).
132. Krishnaji, Srivastava S. L., Nath D., J. Chem. Phys. **52**, 940 (1970).
133. Suamalamba K., Premaswarup D., Indian J. Pure Appl. Phys. **3**, 389 (1965).
134. Suamalamba K., Premaswarup D., Indian J. Pure Appl. Phys. **4**, 84 (1966).
135. Deshpande D. K. Suryanarayana Rao, Indian J. Pure Appl. Phys. **7**, 71 (1969).
136. Fairweather A., Frast E. J., Molecular Relaxation Processes, Chem. Society spec. pub. No. 20, 45, 1966.
137. Glasstone S., Laidler K. J., Eyring H., The Theory of Rate Processes, 1941.
138. Panchenkov G. M., Lebedev V. P., Khimicheskaya kinetika i kataliz, Izd. MGU, 1961.
139. Bhandari R. C., Sisodia M. L., Indian J. Pure Appl. Phys. **2**, 132 (1964).
140. Krishnaji and Abhai Mansingh, Indian J. Pure Appl. Phys. **2**, 176 (1964).
141. Shukla J. P., Shukla D. D., Saxena M. C., J. Phys. Chem. **73**, 2187 (1969).
142. Srivastava G. P., Mattur P. C., Tripathi K. N., Indian J. Pure Appl. Phys. **6**, 561 (1968).
143. Shakhparonov M. I., Zhur. fiz. khim. **35**, 5 (1961).
144. Lomova N. N., Shakhparonov M. I., Dokl. Akad. Nauk **134** (3), 632 (1960).
145. Shakhparonov M. I. in a review Kriticheskie yavleniya i flyuktuatsii v rastvorakh, Izd. MGU **151**, 1960.
146. Deich A. Y., Borovikov Y. Y., Izv. AN Latviiskoi SSR, ser. khim. **1**, 45 (1969).
147. Borovikov Y. Y., Deich A. Y., Izv. AN Latviiskoi SSR, ser. khim. **1**, 49 (1969).
148. Harig H. J., Michel W., Wiss. Z., Techn. Hochsch., Leuna-Merseburg, **8**, 168 (1966).
149. Winkelman J., Quitzsch K., Z. Phys. Chemie **250** (5/6), 355 (1972).
150. Kasimov R. M., Shakhparonov M. I. in a review Kriticheskie yavleniya i flyuktatsii v rastvorakh, Izd. MGU **37**, 1960.
151. Kasimov R. M., Shakhparonov M. I., Akhadov Y. Y., Zhur. Strukt. khim. **2** (1), 13 (1961).
152. Plotnikov V. A., Shcheka I. A., Yankevich Z. A., Zhur. org. khim., **9** (10), 868, (1939).
153. Lowry T. M., Jessop G. J., Chem. Soc., 782 (1930).
154. Tourky A. R., Rizk H. A., Girgis Y. M., Z. Phys. Chem. **32** (1–2), 102 (1962).

155. Graffunder W., Hayman E., Z. Phys. Chem. **15B**, 377 (1931).
156. Harris F., Haycock E. W., Alder H. J., J. Chem. Phys. **21** (11), 1943 (1953).
157. Borovikov Y. Y., Fialkov Y. Y., Elektrokhimiya **1**, 1106 (1965).
158. Hersh C. K., Platz G. M., Swehla R. J., J. Phys. Chem. **63**, 1968 (1959).
159. Eggers H. S., J. Phys. Chem. **8**, 14 (1904).
160. Rosental S., Bull. Inter. Acad. Polanaise Sci. **8A**, 377 (1928).
161. Dobinski M. D., Bull. Inter. Acad. Polanaise Sci. **8–9A**, 239 (1932).
162. Furth R., Keller R., Biochem. Z. **141** (1–3), 187 (1923).
163. Manduit Y., Weinreich H., J. Chim. Phys. **68** (2), 267 (1971).
164. Osipov O. A., Lysenko Y. A., Zhur. obshch. khim. **30** (12), 3866 (1960).
165. Osipov O. A., Kashireninov O. E., Nemirov G. V., Shelemov I. K., Zhur. obshch. khim. **31**, 3153 (1961).
166. Earp D. P., Glasstone S., J. Chem. Soc., 1709, 1720 (1935).
167. Lewis G. L., Smyth C. P., J. Amer. Chem. Soc. **61** 3067 (1939).
168. Briegleb G., Z. Phys. Chem. **16B**, 249 (1932).
169. Briegleb G., Z. Phys. Chem. **16B**, 276 (1932).
170. Linebarger C. E., Z. Phys. Chem. **20** (1), 131 (1896).
171. Bergholm S., Drudes Ann. **53**, 169 (1918).
172. Dobroserdov D. K., Zhur. Russ. fiz. khim. Obshch. Chast khimicheskaya **44**, 679 (1912).
173. Rolinski J., Phys. Z. **29** (18), 658 (1928).
174. Grutzmacher M., Z. Phys. **28** (6), 342 (1924).
175. Decroocq D., Bull. Soc. Chim. France **1**, 127 (1964).
176. Goss F. R., J. Chem. Soc., 752 (1933).
177. Sayce L. A., Briscoe H. V. A., J. Chem. Soc., 2663 (1926).
178. Krchma I. J., Williams J. W., J. Amer. Chem. Soc. **49** (10), 2408 (1927).
179. Morgan S. O., Lowry H. H., J. Phys. Chem. **34** , 2385 (1930).
180. Audsley A., Goss F. R., J. Chem. Soc., 864 (1941).
181. Huyskens P., Gillerot G., Zeegers-Huyskens Th., Bull. Soc. Chim. Belg. **72**, 666 (1963).
182. Harms R., Thesis Wurzburg 1938; 2, 286 [1].
183. Sadek H., Fuoss R. M., J. Amer. Chem. Soc. **76** 5897 (1954).
184. Gold P. I., Perrine R. L., J. Chem. Eng. Data **12**, 4 (1967).
185. Brandt A. A., Shakhparonov M. I., Vestnik MGU, ser. khimiya **9**, 45 (1954).
186. Gerdes E., Kraeft W. D., Monatsber. Dtsch. Akad. Wissen. Ber. **10**, 781 (1964).
187. Berns D. S., Fuoss R. M., J. Amer. Chem. Soc. **82**, 5585 (1960).
188. Accascina F., Petrucci S., Fuoss R. M., J. Amer. Chem. Soc. **81**, 1301 (1959).
189. Berns D. S., Fuoss R. M., J. Amer. Chem. Soc. **83**, 1321 (1961).
190 Fujishiro R., Kimura K., Bull. Chem. Soc. Japan **32**, 1237 (1959).
191. Timmermans I., The physico-chemical constants of binary systems in concentrated solutions, New York–London, 1959, p. 549.
192. Drutman Z. S., Zhur. fiz. khim. **32**, 769 (1958).
193. Sadek H., Fuoss R. M., J. Amer. Chem. Soc. **76** , 5902 (1954).
194. Graffunder M., Heyman E., Z. Phys. **72** 744 (1931).
195. King J. F., Patruk W. A., J. Amer. Chem. Soc. **43**, 1835 (1921).
196. Hederstrand G., Z. Phys. Chem. **B2**, 428 (1929).

197. Suryanarayana C. V., Somasundaram K. M., Bull. Chem. Soc. Japan **32**, 666 (1959).
198. Loiseleur H., Merlin J. G., Paris R. A., J. Chim. Phys. **64** (4), 634 (1967).
199. Kolling O. W., Trans. Kansas Acad. Sci. **69** (2), 123 (1966).
200. Huyskens P., Cracco F., Bull. Soc. Chim. Belg. **69** (7), 422 (1960).
201. Das L. M., Roy S. C., Indian J. Phys. **5**, 441 (1930).
202. Fuoss R. M., Hirsch E., J. Amer. Chem. Soc. **82**, 1013 (1960).
203. Kimura K., Fujishira R., Bull. Chem. Soc. Japan. **35**, 85, (1962).
204. Pal N. N., Philos. Mag. **11**, 265 (1930).
205. Jenkins H. C., Sutton L. E., J. Chem. Soc., 609 (1935).
206. Goss T. R., J. Chem. Soc., 1789 (1939).
207. Williams J. W., Krchma I. J., J. Amer. Chem. Soc. **48**, 1888 (1926).
208. Schulze A., Z. Phys. Chem. **86**, 309 (1914).
209. Williams J. W., Krchma I. J., J. Amer. Chem. Soc. **49**, 1676 (1927).
210. Pilpel N., J. Amer. Chem. Soc. **77**, 2949 (1955).
211. Hückel W., Wenzke U., Z. Phys. Chem. **193A**, 132 (1944).
212. Krasil'nikov V. Y., Zhur. fiz khim. **18**, 174 (1944).
213. Malecki J., Nowakowa J., Poznan Tow. Przyj. Nauk. Fiz. Dielektrykow Radiospek, **4**, 33 (1968).
214. Michalczuk J., Roczn. Chem. Ann. Soc. Chim. Polonorum **38**, 694 (1964).
215. Philip J. C., Z. Phys. Chem. **24**, 18 (1897).
216. Williams J. W., Ogg E. F., J. Amer. Chem. Soc. **50**, 94 (1928).
217. Le Fevre C. G., Le Fevre R. J. W., J. Chem. Soc., 487 (1936).
218. Lange P., Z. Phys. Chem. **A 33**, 169 (1925).
219. Golubkov P. W., Rapp. Soc. Rech. Nat. Saratow **1**, 37 (1925).
220. Müller F. H., Phys. Z. **38** (8), 283 (1937).
221. Briegleb G., Z. Phys. Chem. **14**, 97 (1931).
222. Jatkar S. K. K., Deshpande C. M., J. Univ, Poona **12**, 31 (1957).
223. Smyth C. P., Rogers H. E., J. Amer. Chem. Soc. **52**, 2227 (1930).
224. Fialkov Y. Y., Borovikov Y. Y., Ukrainsky khimichesky zhurnal **32**, 590 (1966).
225. Campbell A. N., Gieskes J. M. T. M., Canad. J. Chem. **42**, 1379 (1963).
226. Drude P., Z. Phys. Chem. **23A**, 267 (1897).
227. Kerr R. N., J. Chem. Soc., 2796 (1926).
228. Dulitskaya K. A., Trudy Moskovskogo aviatsionnogo instituta **52**, 56 (1955).
229. Hurwic J., Michalczuk J., Roczn. chem. **34**, 1423 (1960).
230. Neale S. M., Weyl D. A., Proc. Roy. Soc. **291**, 368 (1966).
231. Osipov O. A., Shelomov I. K., Zhur. fiz. khim. **30**, 608 (1956).
232. Osipov O. A., Uch. zap. univ. Rostov-na-Donu **41**, 3 (1958).
233. Dolezalek F., Schulze A., Z. Phys. Chem. **83A**, 45 (1913).
234. Coop I. E., Trans. Faraday Soc. **33**, 583 (1937).
235. Goremykina V. V., Vakalov I. A., Shakhparonov M. I., Vestnik MGU, ser. khimiya **19**, 33 (1964).
236. Le Fevre R. J. W., Russell P., J. Chem. Soc. **391** (1936).
237. Schulze A., Z. Phys. Chem. **97A**, 338 (1921).
238. Smyth C. P., Morgan S. O., J. Amer. Chem. Soc. **50**, 1547 (1928).
239. Plucnett W. K., Dowd T., J. Chem. Eng. Data 8, 207 (1963).
240. Corfield G., Davies M., Trans. Faraday Soc. **60**, 10 (1964).

241. Heyden M. J., Markgraf H. G., Nikuradse A., Ulbrich R., Z. Naturforsh. **10a**, 10 (1955).
242. Mahanti P. C., Das-Gupta R. N., J. Indian Chem. Soc. **6**, 441 (1929).
243. Tewari P. H., Johari G. P., J. Phys. Chem. **69**, 2857 (1965).
244. Fialkov Y. Y., Borovikov Y. Y., Vestnik Kievskogo politekhnicheskogo inst., ser. khimiya **1**, 73 (1965).
245. Kolling O. W., McMillan D. J., Trans Can. Acad. Sci. **67**, 470 (1964).
246. Cavell A. S., Jerrara H. G., J. Phys. Chem. **69**, 3657 (1965).
247. Cunningham G. P., Vidulich G. A., Kay P. L., J. Chem. Eng. Data **12**, 336 (1967).
248. Celiano A. V., J. Chem. Eng. Data **7**, 391 (1962).
249. Thwing C. B., Z. Phys. Chem. **14**, 286 (1894).
250. Campbell A. N., Anand S. C., Canad. J. Chem. **50** (8), 1109 (1972).
251. Reynaud R., C. R. Sean. Acad. Sci., Paris **270C** (22), 1813 (1970).
252. Winkelman J., Quitzsch K., Z. Phys. Chem. **250**, 355 (1972).
253. Watanabi A., Rep. Gov. Ind. Res. Inst. Nagoya, Japan **17** (6), 142 (1968).
254. D'Aprano A., Triolo R., J. Phys. Chem. **71** (11), 3474 (1967).
255. Sadek H., Fuoss R. M., J. Amer. Chem. Soc. **72**, 301 (1950).
256. Barthel J., Krener M., Engel G., Z. Phys. Chem. **69**, 283 (1970).
257. Starobinets G. L., Starobinets K. S., Ryzhikova L. A., Zhur. fiz. khim. **25**, 1186 (1951).
258. La Rochelle J. H., Vernon A. A., J. Amer. Chem. Soc. **72**, 3293 (1950).
259. Romanow W. I., Eltzin I. A., Phys. Z. Sovjetunion **11**, 526 (1937).
260. Suryanarayana C. V., Somasundaram K. M., J. Sci. Industry. Res. **19**, 375 (1960).
261. Velasco M., Anal. Fis. Quim., Madrid **28**, 1228 (1930).
262. Starobinets G. L., Uch. zap. Belorus. univ., khimiya **14**, 82 (1953).
263. Colmant P., Bull. Soc. Chim. Belg. **63**, 5 (1954).
264. Noyes R. M., Dickinson R. G., J. Amer. Chem. Soc. **65**, 1427 (1943).
265. D'Aprano A., Fuoss R. M., J. Phys. Chem. **67** 1871 (1963).
266. Smyth C. P., Dornte R. W., Wilson E. B., J. Amer. Chem. Soc. **53**, 4242 (1931).
267. Higasi K., Sci. Papers. Inst. Phys. Chem. Research, Tokyo **24**, 57 (1934).
268. Vernon A. A., Wyman J., Avery R. A., J. Amer. Chem. Soc. **67**, 1422 (1945).
269. Dahl W., Muller F. H., Kolloid Z. **198**, 27 (1964).
270. Fialkov Y. Y., Tarasenkov Y. A., Borovikov Y. Y., Zhur. org. khim. **36**, 981 (1966).
271. Gore R. S., Briscoe H. T., J. Phys. Chem. **40**, 619 (1936).
272. Kolling O. W., Van Arsdale C. R., Trans. Kansas Acad. Sci. **68**, 65 (1965).
273. Smyth C. P., Rogers H. E., J. Amer. Chem. Soc. **52**, 1824 (1930).
274. Fialkov Y. Y., Borovikov Y. Y., Zhur. org. khim. **36**, 1554 (1966).
275. Briegleb G., Z. Phys. Chem. **10**, 205 (1930).
276. Abgrall C., Barre R., C. R. Seances Acad. Sci., Paris **253**, 439 (1961).
277. needs translating
278. Smyth C. P., Stops W. N., J. Amer. Chem. Soc. **51**, 3312 (1929).
279. Starobinets G. L., Starobinets K. S., Zhur. fiz. khim. **25**, 753 (1951).
280. Clerbaux Th., Zeegers-Huyskens Th.,Bull. Soc. Chim. Belg. **75**, 366 (1966).
281. Martin A. R., Brown A. C., Trans. Faraday Soc. **34**, 742 (1938).

282. Cohn L., Arons L., Wiedemann's Ann. **33**, 13 (1888).
283. Wang Y. L., Z. Phys. Chem. **45**, 323 (1940).
284. Lindberg J. J., Kenttämaa J., Nissema A., Suomen kem. **34**, 156 (1961).
285. Walls W. S., Smyth C. P., Chem. Phys. **1**, 337 (1933).
286. Miller C. G., Maass O., Canad. J. Chem. **38**, 1606 (1960).
287. Vakalov I. A., Shakhparonov M. I., Zhur. strukt. khim. **5**, 620 (1964).
288. Drude P., Ann. Phys. Chem. **61**, 466 (1897).
289. Akerlof G., Short O. A., J. Amer. Chem. Soc. **58**, 1241 (1936).
290. Shidlovskaya A. N., Syrkin Y. S., Zhur. fiz. khim. **22**, 913 (1948).
291. Wehrle J. A., Phys Rev. **37**, 1135 (1931).
292. Karhi M F , Suikkanen Ofversigt Helsingfors **19**, 54 (1912).
293. Ehrenhaft F., Sitzungsb. Wien **11** (111), 1549 (1902).
294. Andersin G., Hirn E., Overs. Finska Vetensk. Soc. Forh. **11**, 51 (1909).
295. Borovikov Y. Y., Zhur. org. khim. **38** (6), 1216 (1968).
296. Smyth C. P., Walls W. S., J. Amer. Chem. Soc. **53**, 527 (1931).
297. Dawson L. R., Kim K. H., Eckstrom H. C., J. Phys. Chem. **70**, 775 (1966).
298. Samosundaram K. M., Suryanarayana C. V., Bull. Chem. Soc. Japan **31**, 838 (1958).
299. Byattacharrya T. J., Indian J. Phys. **34**, 358 (1960).
300. Mahanti P. C., J. Indian Chem. Soc. **8**, 743 (1929).
301. Philippe R., Bull. Classe Sci. Acad. Roy. Belg. **40**, 544 (1954).
302. Smyth C. P., Walls W. S., J. Phys. Chem. **1**, 200 (1933).
303. Myers R. T., Sun V. M. L., J. Phys. Chem. **70**, 3217 (1966).
304. Moelwyn-Hughes E. A., Thorre P. L., Proc. Roy. Soc. **277**, 423 (1964).
305. Bywater S., Worsfold D. J., J. Phys. chem. **70**, 162 (1966).
306. Rojala G. E., Crossley J., Canad. J. Chem. **49** (22), 3616 (1971).
307. Pursel W. P., Singer J. A., J. Phys. Chem. **69**, 4097 (1965).
308. Jatkar S. K. K., Phansalkar V. K., J. Univ. Poona **22**, 43 (1962).
309. Yasumi M., Shirai M., Bull. Chem. Soc. Japan. **28**, 193 (1955).
310. Jatkar S. K. K., Phansalkar V. K., J. Univ. Poona **22**, 27 (1962).
311. Jatkar S. K. K., Phansalkar V. K., J. Univ. Poona **22**, 57 (1962).
312. Jatkar S. K. K., Phansalkar V. K., J. Univ. Poona **22**, 35 (1962).
313. Phansalkar V. K., J. Univ. Poona **22**, 69 (1962).
314. Jatkar S. K. K., Phansalkar V. K., J. Univ. Poona **22**, 65 (1962).
315. Grishko N. I., Gur'yanova E. N., Zhur. fiz. khim. **32**, 2725 (1958).
316. Smyth C. P., Dornte R. W., J. Amer. Chem. Soc. **53**, 545 (1931).
317. Combs L. L., J. Phys. Chem. **75** (14), 2133 (1971).
318. McKey R. B., Hillson P. J., Trans. Faraday Soc. **62**, 1439 (1966).
319. Huyskens P., Henry R., Gillerot G., Bull. Soc. Chim. France **4**, 720 (1962).
320. John H. Möller G., Z. Phys. Chem. **13**, 385 (1894).
321. Jagielski A., Wesolowski J., Bull. Inter. Acad. Polon. Sci. **A**, 260 (1935).
322. Williams J. W., Phys. Z. **29**, 174 (1928).
323. Bannet P., Barriol J., J. Chim. Phys. Biol. **68**, 1703 (1971).
324. Gur'yanova E. N., Zhur. fiz. khim. **24**, 479 (1950).
325. Langguth U., Bittrich H. J., Z. Phys. Chem. **244** (5/6), 327 (1970).
326. Janelli L., Orsini P. G., Gazz. Chim. Ital. **86**, 1104 (1956).
327. Jatkar S. K. K., Desphande C. M., J. Indian. Chem. Soc. **37**, 11 (1960).
328. Kovalenko K. N., Osipov O. A., Trifonov N. A., Zhur. fiz. khim. **29**, 685 (1955).

329. Farmer E. H., Warren F. L., J. Chem. Soc., 1297 (1933).
330. Daumezon P., Heitz R., J. Chem. Phys. **55** (12), 5704 (1971).
331. Hassel O., Naeshagen E., Z. Phys. Chem. **19**, 434 (1932).
332. Rudolfi E., Z. Phys. Chem. **66**, 705 (1909).
333. Hrynakowski K., Jeske J., Ber. Dtsch. Chem. Ges. **71**, 1415 (1938).
334. Wood R. F., Dickinson R. G., J. Amer. Chem. Soc. **61**, 3259 (1939).
335. Errera J., Phys. Z. **27**, 134 (1941).
336. Gilkerson W. P. , Xalmin E., J. Amer. Chem. Soc. **82**, 5295 (1960).
337. Smyth C. P., Morgan S. O., Boyse J. C., J. Amer. Chem. Soc. **50**, 1536 (1928).
338. Bodensen H. K., Ramsey J. B., J. Phys. Chem. **67**, 140 (1963).
339. Maktin A. R., Trans. Faraday Soc. **33**, 191 (1937).
340. Spengler H., Phys. Z. **42**, 134 (1941).
341. Mecke R., Rosswog K., Z. Electrochem. **60**, 47 (1956).
342. Nespital W., Z. Phys. Chem. **16B**, 153 (1932).
343. Thomson G., J. Chem. Soc., 1051 (1937).
344. Jatkar S. K. K., Phansalkar V. K., J. Univ. Poona **22**, 49 (1962).
345. Schadow E., Steiner R., Z. Phys. Chem. **66**, 105 (1969).
346. Kahlenberg L., Anthony R. B., J. Phys. Chem. **4**, 358 (1906).
347. Lomova N. N., Shakhparonov M. I., Vestnik MGU, ser. khimiya **3**, 11 (1960).
348. Hanai T., Koizumi N., Gotoh R., Nippon Kagaku Zasshi **80**, 17 (1959).
349. Semenchenko V. K., Azimov M., Zhur. fiz. khim. **30**, 1821, (1956).
350. Pickara A., Bull. Inter. Acad. Polon. Sci. **9A**, 319 (1933).
351. Lomova N. N., Shakhparonov M. I., Vestnik MGU, ser. khimiya **3**, 33 (1961).
352. Semenchenko V. K., Azimov M., Zhur. fiz. khim. **30**, 2228, (1956).
353. Fairbrother F., J. Chem. Soc., 1846 (1934).
354. Williams J. W., Allgeir R. J., J. Amer. Chem. Soc. **49**, 2416 (1927).
355. Philip J. C., Haynes D., J. Chem. Soc. **87**, 998 (1905).
356. Hassel O. Nashagen E. Z., Phys. Chem. **15B**, 373 (1931).
357. Wulff N., Tokashima E., Z. Phys. Chem. **A398**, 322 (1938).
358. Hrynakowski K., Zochowski A., Ber. Dtsch. Chem. Ges. **70**, 1739 (1937).
359. Mohler H., Helv. Chim. Acta **20**, 1447 (1937).
360. Williams J. W., Schwingel C. H., J. Amer. Chem. Soc. **50**, 362 (1928).
361. Fialkov Y. Y., Fenerli G. N., Ukr. khim. zhur. **31**, 258 (1965).
362. Silberstein L., Ann. Phys. und Chem. **56**, 661 (1895).
363. Kalinowski K., Roczn. Chem. **13**, 384 (1933).
364. Gur'yanova E. N., Kuzina L. S. Zhur. fiz. khim. **30**, 616 (1956).
365. Smyth C. P., Stops W. N., J. Amer. Chem. Soc. **51**, 3330 (1929).
366. Coopock J. B. M., Goss F. R., J. Chem. Soc., 1789 (1939).
367. Jatkar S. K. K., Deshpande C. M., J. Indian Chem. Soc. **37**, 15 (1960).
368. Wolorowitsch M. P., Stepanenko N. N., Acta Physicochem. URSS **13**, 647 (1940).
369. Howell O. R., Jackson W., Proc. Roy. Soc. **A145**, 539 (1934).
370. Winstein A., Wood R. E., J. Amer. Chem. Soc. **62**, 548 (1940).
371. Brot C., Soulard A., Arch. Sci. Phys. Nat. **12**, 9 (1959).
372. Suryanarayana C. V., Somasundaram K. M., Acta Chim. Hungar. **20**, 231 (1959).

373. Tschamler H., Krischai H., Monatish. Chem. **82**, 259 (1951).
374. Fisher E., Phys. Z. **36**, 585 (1935).
375. Clemett C. J., Forest E., Smyth C. P., J. Chem. Phys. **40**, 2123 (1964).
376. Lane J. A., Saxton J. A., Proc. Roy. Soc. **214A**, 531 (1952).
377. Hasted J. B., Roderick G. W., J. Chem. Phys. **29**, 17 (1958).
378. Sastry P. S., Premaswarup D., Current Sol. **39** (11), 253 (1970).
379. Miles J. H., J. Inorg. and Nucl. Chem. **26**, 2308 (1964).
380. Apfeel H., Ber. Bungesges. Phys. Chem. **72** (7), 1059 (1968).
381. Cachet H., Lestrade C., Epelboin I., C. R. Seances Acad. Sci. (Paris) **261**, 678 (1965).
382. Badiali J. P., Cached H. C., Lestrade J. C., Berich. Bunsen. Phys. Chem. **75**, 297 (1971).
383. Hasted J. B., Tirmazi S. H., J. Chem. Phys. **50**, 4116 (1969).
384. Breitschwerdt K. G., Radscheit H., Phys. letters **29A**, 381 (1969).
385. Mahaffey D. W., Jarde D. A., Rev. Modern. Phys. **40**, 710 (1968).
386. Bartoli F. J., Birch J. N., Nguyen-Huu-Toan, Duffie G. E., J. Chem. Phys. **49**, 1916 (1958).
387. Miller R. C., Smyth C. P., J. Chem. Phys. **24**, 814 (1956).
388. Antony A. A., Smyth C. P., J. Amer. Chem. Soc. **86**, 152 (1964).
389. Sapenko V. V., Zhur. fiz. khim. **37**, 1172 (1963).
390. Eloranta J. K., Kadaba P. K., Mater. Sci. Eng. **8**, 203 (1971).
391. Akhadov Y. Y., Candidate's dissertation, Investigation of the Dielectric Properties of Pure Liquids and Solutions in the Centimetre Range of Electromagnetic Waves, Moscow State University, 1962.
392. Shakhparonov M. I., Kasimov R. M., Akhadov Y. Y., Ukr. fiz. zhur. **7**, 874 (1962).
393. Akhadov Y. Y., Shakhparonov M. I., Kasimov R. M. in a review Primenenie ultraakustiki k issledovaniyu veshchestva, Izd. MOPI **15**, 29 (1961).
394. Kasimov R. M., Akhadov Y. Y., Shakhparonov M. I. Dokl. AN Azerbaidzhanskoi SSR **17**, 207 (1961).
395. Kasimov R. M., Dissertation, Dielectric Properties of concentrated solutions in the millimetre range of radio-waves and their relationship to molecular structure, Moscow State University, 1962.
396. Sen S. N., Indian J. Phys. **25**, 187 (1951).
397. Bhattacharrya T. J., Indian J. Pure Appl. Phys. **2**, 33 (1964).
398. Mitteilungen K., Chimia **23**, (3), 105 (1969).
399. Eloranta J. K., Kadaba P. K., Trans. Faraday Soc. **67**, 1355 (1971).
400. Dieringer F., Z. Physik. **145**, 184 (1956).
401. Szurkowski B., Danielewicz-Ferchmin I., Acta Phys. Polon. **A39**, 227 (1971).
402. Fischer E., Zengin N., Z. Physik **147**, 113 (1959).
403. Price A. H., Arhiv. Sci., 12 fas. spec., 31, (1959).
404. Boule P., J. Chem. Phys. **65**, 777 (1969).
405. Bandopadhyaya T. K., Indian J. Pure Appl. Phys. **6**, 286 (1968).
406. Klages G., Roth D., Z. Naturforsc. **14A**, 628 (1959).
407. Kranbuehl D., Klug D., Vayghan W., Ann. Rep. Confer. Electrical Insulation and Dielectric Phenomena, USA, Nat. Acad. Sci., Washington, 1970.
408. Shakhparonov M. I., Vakalov I. A., Zhur. fiz. khim. **38**, 1978 (1964).
409. Yadava R. S., Dube D. C., Parshad R., Indian J. Pure Appl. Phys. **6**, (1), 10 (1968).

410. Savchenko B. B., Levin V. V., Shakhparonov M. I. Zhur. strukt. khim. **12**, 1086 (1971).
411. Savchenko V. V., Vakalov I. A., Shakhparonov M. I., Vestnik MGU, ser. khimiya **6**, 732 (1970).
412. Akhmamet'ev M. A., Kazakov S. M., Izv. Vusov, seriya fizika **10**, 15 (1967).
413. Levin V. V., Kazantseva S. I., Zhur strukt. khim. **13**, 148 (1972).
414. Quinn R. G., Smyth C. P., J. Chem. Phys. **41**, (20), 37 (1964).
415. Borsukova G. A., Bogdanov L. I., Uch. zap. MOPI **165** (7), 237 (1968).
416. Kasimov R. M., Akhadov Y. Y., Shakhparonov M. I., Dokl. AN Azerbaidzhanskoi SSR **17**, 687 (1961).
the Azerbaijan S. S. R. **17**, 687 (1961).
417. Denney D. J., Cole R. H., J. Chem. Phys. **23**, 1767 (1955).
418. Sarojini V., J. Sci. Indust. Res. **21B**, 158 (1962).
419. Yasumi M., Nukazawa K., Mizushima S., Bull. Chem. Soc. Japan **24**, 60 (1951).
420. Jakel T., Ann. Physic. **17**, 42 (1955).
421. Brescia G., Grossetti E., Atti Acad. Naz. Lincei **48**, 619 (1970).
422. Turner E. M., Anderson D. W., Reich L. A., Vaughan W. E., J. Phys. Chem. **74**, 1275 (1970).
423. Constany E., Leroy Y., Barois J., Desplanques P., C. R. Sean. Sci., Paris **264**, 228 (1967).
424. Mansingh A., Indian J. Pure Appl. Phys. **2**, 33 (1964).
425. Eloranta J. K., Kadaba P. K., Frenzel C. F. 1969 Ann. Report. Confer. Electrical Insulation and Dielectric Phenomena, USA, Nat. Acad. Sci., Washington, 38, 1970.
426. Eloranta J. K., Kaadaba P. K., Trans. Faraday Soc. **66**, 817 (1970).
427. Imanov L. M., Mirzoev F. G., Izv. AN Azerbaidzhanskoi SSR, ser. Fiz.-mat. **4**, 37 (1963).
428. Heston W. M., Franklin A. D., Hennelly E. J., Smyth C. P., J. Amer. Chem. Soc. **72**, 3443 (1950).
429. Heitz R., Daumezon P., J. Chem. Phys. **68**, 1 (1971).
430. Panchenkov G. M., Davtyan O. K., Zhur. fiz. khim. **13**, 651 (1959).
431. Sagal M. W., J. Chem. Phys. **36**, 2437 (1962).
432. Kamiyoshi K., Fujimura T., J. Phys. Radium **23**, 337 (1962).
433. Baba K., Fujimura T., Kamiyoshi K., J. Phys. Chem. **73**, 1146 (1969).
434. Sarojini V., J. Sci. Indust. Res. **19B**, 91 (1960).
435. Sakellaridis P., Hinopoulos B., J. Phys. D. 5, 1815 (1972).
436. Kasimov R. M., Akhadov Y. Y., Shakhparonov M. I., Vestnik MGU, ser. khimiya **5**, 22 (1963).
437. Akhadov Y. Y., Shakhparonov M. I., Kasimov R. M., Kornilova N. B., Zhur. strukt. khim. **2**, 13 (1961).
438. Akhadov Y. Y., Shakhparonov M. I. in a review Kriticheskie yavleniya i flyuktuatsii v rastvorakh, Izd. MGU 14, 1961.
439. Vakalov I. A., Shakhparonov M. I., Vestnik MGU, ser. khimiya **2**, 16 (1964).
440. Cavell E. A. S., Trans. Faraday Soc. **61**, 1578 (1965).
441. McDuffie G. E., La Macchia J. T., Conord A. E., J. Chem. Phys. **39**, 1878 (1963).
442. Kono R., Litovitz T. A., McDuffie G. E., J. Chem. Phys. **45**, 1790 (1966).

443. Sarojini V., J. Sci. Industr. Res. **19**, 52 (1960).
444. Denney D. J., J. Chem.Phys. **30**, 1019 (1959).
445. Rizk H. A., Elanwar I. M., Z. Phys. Chem. **245**, 289 (1970).
446. Rizk H. A., Elanwar I. M., Z. Phys. Chem. **245**, 299 (1970).
447. Muralidhara R. V., Trans. Faraday Soc. **58**, 2139 (1962).
448. Muralidhara R. V., J. Sci. Industr. Res. **18B**, 103 (1959).
449. Kilp H., Garg S. K., Smyth C. P., J. Chem. Phys. **45**, 2799 (1966).
450. Tucker S. W., Walker S., J. Chem. Phys. **45**, 1302 (1966).
451. Imanov L. M., Zul'figarzade K. E., Bafadarov Z. A., Zhur. fiz. khim. **42**, 2450 (1968).
452. Rizk H. A., Youssef N., Z. Phys. Chem. **244**, 165 (1970).
453. Hafelin J., Arch. Sci. Phys. Nat. **28**, 19 (1946).
454. Vakalov I. A., Shakhparonov M. I., Zhur. strukt. khim. **7**, 505 (1966).
455. Denney D. J., Ring J. W., J. Chem. Phys. **44**, 4621 (1966).
456. Phillips C. S. E., Nature **166**, 866 (1950).
457. Price A. H., J. Phys. Chem. **64**, 1442 (1960).
458. Forest E., Smyth C. P., J. Phys. Chem. **69**, 1303 (1965).
459. Garg S. K., Kadaba P. K., J. Phys. Chem. **69**, 674 (1965).
460. Bhattacharyya T. J., Mathur M. S., Gopalan T. V., Kadaba P. K., Material Sci. Eng. **1**, 183 (1966).
461. Imanov L. M., Zul'fugarzade K. E., Izv. AN Azerbaidzhansloi SSR, ser. fiz.-mat. **6**, 75 (1962).
462. Imanov L. M., Zul'fugarzade K. E., Zhur. fiz. khim. **38**, 2437 (1964).
463. Imanov L. M., Zul'fugarzade K. E., Izv. AN Azerbaidzhansloi SSR, ser. fiz.-mat. **5**, 39 (1962).
464. Imanow L. M., Zülfügarzade K. E., Hadshijew H. A., Z. Phys. Chem. **238**, 133 (1968).
465. Daumezon P., Moziamez M., Arnoult R., Magnet. electr. reson. relaxation. 235, 1962.
466. Schunzel M., Stokhausen F., Z. angew. Physik **21**, 508 (1966).
467. Chamberlain J. E., Werner E. B. C., Gebbie H. A., Trans. Faraday Soc. **63**, 2605 (1967).
468. Kasimov R. M., Shakhparonov M. I., Azerbaidzhansky khimichesky zhur. **4**, 101 (1961).
469. Arkhangel'sky K. V. in a review Fizika dielektrov, Izd AN SSSR, 48, 1960. Sciences of U.S.S.R., 48, (1960).
470. Choudhury A., Indian J. Phys. **24**, 507 (1950).
471. Chitoku K., Higasi K., Bull. Chem. Soc. Japan. **40**, 773 (1960).
472. Semenchenko V. K., Arkhangel'sky K. V., Zhur. fiz. khim. **35** (4), 927 (1961).
473. Kraubuehl D. E., Klug D. E., Vaughan W. E., J. Chem. Phys. **50**, 5266 (1969).
474. Kastha G. S., Indian J. Phys. **26**, 103 (1952).
475. Bhattacharyya T. J., Indian J. Phys. **36**, 533 (1962).
476. Glushkova N. V., Levin V. V., Ovchinnikov V. I., Vestnik MGU, ser. khimiya **1**, 26 (1971).
477. Arnoult R., Lebrun A., Moriamez C., Moriamez M., Wemelle R., Arch. Sci. Phys. et nat. fas. spec. **10**, 48 (1957).
478. Hassell W. F., Walker S., Trans. Faraday Soc. **62**, 861 (1966).

479. Crossley J., Hassell W. F., Walker S., J. Chem. Phys. **48**, 1261 (1968).
480. Crossley J., Walker S., Canad. J. Chem. **46**, 841 (1968).
481. Loveluck G., J. Phys. Chem. **64**, 385 (1960).
482. Grossley J., Classer L., Smyth C. P., J. Chem. Phys. **52** (12), 6203 (1970).
483. Crossley J., Glasser L., Smyth C. P., J. Chem. Phys. **55**, 2197 (1971).
484. Bhattacharrya T. J., Gopalan T. V., Kadaba P. K., Mater. Sci. Eng. **1**, 257 (1967).
485. Crossley J., Walker S., Canad. J. Chem. **46**, 847 (1968).
486. Milicka O., Slama A., Ann. Physik. **8**, 663 (1931).
487. Sack H., Phys. Z. **28**, 199 (1927).
488. Carman A. P., Smith K. O., Phys. Rev. **34**, 1042 (1929).
489. Krishna P. S., Rao M., Premaswarup D., Indian J. Pure Appl. Phys. **4**, 322 (1966).
490. Schmidt C. C., Phys. Rev. **30**, 925 (1927).
491. Deubner A., Ann. Physik. **84**, 429 (1927).
492. Lattey R. T., Phil. mag. **41**, 829 (1921).
493. Gross P. M., Toylor R. C., J. Amer. Chem. Soc. **72**, 2075 (1950).
494. Cuthbertson A. C., Maass O., J. Amer. Chem. Soc. **92**, 489 (1970).
495. Grubb H. M., Hunt H., J. Amer. Chem. Soc. **61**, 565 (1939).
496. Drake F. H., Pierce G. W., Dow M. T., Phys. Rev. **35**, 613 (1930).
497. Errara J., J. Phys. Radium. **6**, 390 (1925).
498. Harrington, E. A. Phys. Rev. **8**, 581 (1916).
499. Jones T. T., Davies R. M., Philos Mag. **28** (188), 307 (1939).
500. Akerlof G., J. Amer. Chem. Soc. **54**, 4125 (1932).
501. Albright P. S., Gosting H. J., J. Amer. Chem. Soc. **68**, 1061 (1946).
502. Travers F., Douzon P., J. Phys. Chem. **74**, 2243 (1970).
503. Salazar G. G., Anal. fis. quim. Madrid. **22**, 275 (1924).
504. Havestadt L., Fricke R. Z., anorgan. allgem. Chem. **188**, 357 (1930).
505. Fürth R., Ann. Physik **70**, 63 (1923).
506. Wyman J., J. Amer. Chem. Soc. **55**, 4116 (1933).
507. Douheret G., Morenas M., C. R. Seances Acad. Sci. Paris **264C** 729 (1967).
508. Wuman J., J. Amer. Chem. Soc. **56**, 536 (1934).
509. Hederstrand G., Z. Phys. Chem. **135**, 36 (1928).
510. Nernst W., Cottinger Nachrichten **20**, 12 (1893).
511. Rock G. D., Klosky S., J. Phys. Chem. **33**, 143 (1929).
512. Remesow J., Tavaststjerna N., Z. Biolog. **218**, 147 (1930).
513. Amis E. S., Potts J. E., J. Amer. Chem. Soc. **63**, 2883 (1941).
514. Heurou M. L., Guerillot C. R., C. R. Seances Acad. Sci. **258**, 2549 (1964).
515. Wyman J., J. Amer. Chem. Soc. 53, 3292 (1931).
516. Lindberg J. J. Kenttamaa J., Soumen Kemistilch. **33B**, 104 (1960).
517. Humbel T. F., Scherrer J. P., Helv. Phys. Acta **26**, 17 (1953).
518. Morenas M., Douheret G., C. R. Acad. Sci. Paris **270C**, 2197 (1970).
519. Slevogt K. E., Ann. Phys. **36**, 141 (1939).
520. Bouchard J., J. Chim. Phys. **33**, 127 (1936).
521. Albright P. S., J. Amer. Chem. Soc. **59**, 2098 (1937).
522. Reynaud R., C. R. Acad. Sci. Paris **266C**, 8, 489 (1968).
523. Goffredi M., Shedlovsky T., J. Phys. Chem. **71**, 2176 (1967).
524. Blumenschine R. L., Sears P. G., J. Chem. Eng. Data **11**, 141 (1966).
525. Critchfield F. E., Gibson J. A., Hall J. L., J. Amer. Chem. Soc. **75**, 6044 (1953).

526. Kunze R. W., Fuoss R. M., J. Phys. Chem. **67**, 911 (1963).
527. Accascina F., D'Aprano A., Triolo R., J. Phys. Chem. **71**, 3469 (1967).
528. Farkas P., Z. Electrochem. Angew. Phys. Chem. **38**, 654 (1932).
529. Harned H. S., Morrison J. O., J. Amer. Chem. Soc. **58**, 1908 (1936).
530. Baturin A. N., Nauch. zap. Dnepropetrov. univ. **77** (9), 170 (1962).
77 (9), 170 (1962).
531. Morenas M., Douheret G., C. R. Seances Acad. Sci. **273C**, 1060 (1971).
532. Oehme F., Angewandte Chem. **71**, 572 (1959).
533. Ghosh J. C., J. Chem. Soc. **117**, 1390 (1920).
534. Hurwic J., Maczynski S., Starczewski W., Rocz. Chem. **42**, 1335 (1968).
535. Sheka I. A., Kotorlenko L. A., Lugovaya L. N., Ukr. khim. zhur. **25**, 602 (1959).
536. Malmberg C. G., Maryott A. L., J. Res. Nat. Bur. Stand. **45**, 299 (1950).
537. Sheka I. A., Kaban A. P., Trudy Kievskogo tekhnoloicheskogo instituta pishchevoi prom. **17**, 179 (1957).
538. Kockel J., Ann. Physik. **77**, 417 (1925).
539. Furth R., Physical. Z. **25**, 676 (1924).
540. Uzbekov V. A., Yermakov V. I., Scherbakov V. V., Zhur. fiz. khim. **45**, 481 (1971).
540a. Gorpnichenko I. M., Vestnik MGU, ser. khimiya **5**, 110 (1969).
541. Harris F. E., O'Konski C. T., J. Phys. Chem. **61**, 310 (1957).
542. Ebert G., Z. Electrochem. **65**, 672 (1961).
543. Hasted J. B., Riston D. M., Collie C. H., J. Chem. Phys. **16**, 1 (1948).
544. Kraeft W. D., Gerdes E., Z. Phys. Chem. **228**, 331 (1965).
545. Shevchik F., Vetterl' V., Biofizika **10** (3), 441 (1965).
546. Sastry P. S., Preemaswarup D., Indian J. Pure Appl. Phys. **8**, 675 (1970).
547. Monana Rao P. S. K., Premaswarup D., Indian J. Pure Appl. Phys. **5**, 581 (1967).
548. Monana Rao P. S. K., Premaswarup D., Indian J. Pure Appl. Phys. **7**, 68 (1969).
549. Tamm K., Schneider M., Z. angew. Phys. **20**, 544 (1966).
550. Giese K., Kaatze U., Pottel R., J. Phys. Chem. **74**, 3718 (1970).
551. Barthel J., Schmithals F., Behret H., Z. Phys. Chem. **71**, 115 (1970).
552. Pottel R., Lossen O., Ber. Bunsenges. Phys. Chem. **71**, 135 (1967).
553. Seeberger M., Ann. Phys. **16**, 77 (1933).
554. Satyanarayana R., Khastgir S. R., Indian J. Phys. **25**, 163, (1951).
555. Christensen J. H., Smith A. J., Reed R. B., Elmore K. L., J. Chem. Eng. Data **11**, 60 (1966).
556. Wei B. E., Gerdes E., Hoffman H. J., Z. Phys. Chem. **227**, 51, (1965).
557. Yastremsky P. S., Uch. zap. Leningradsk. ped. inst. **207**, 227 (1961).
207, 227 (1961).
558. Haggis G. H., Hasted J. B., Buchanan T. J., J. Chem. Phys., **20**, 1453 (1952).
559. Pechhold R., Ann. Physics **83**, 427 (1927).
560. Yastremsky P. S., Zhur. strukt. khim. **2**, 268 (1961).
561. Moynihan C. T., Bressel R. D., Angell C. A., J. Chem. Phys. **55**, 4414 (1971).
562. Hasted J. B., Elsabeh S. H. M., Trans. Faraday Soc. **49**, 1003 (1953).
563. Saxton J. A., Lane J. A., Wireless engineer. **29**, 269 (1952).
564. Yastremsky P. S., Uch. zap. Stalingradsk. ped. inst. **11**, 92 (1959).
Institute **11**, 92 (1959).

565. Demau M. G., C. R. Seances Acad. Sci. **266B**, 707 (1968).
566. Chekalin N. V., Zhur. fiz. khim. **44**, 3090 (1970).
567. Chekalin N. V., Shakhparonov M. I., Zhur. fiz. khim. **45**, 452 (1971).
568. Sarojini V., Trans. Faraday Soc. **57**, 1534 (1961).
569. Barsukova G. N., Zhur. strukt. khim. **9**, 552 (1968).
570. Barsukova G. N., Uch. zap. Omsk. ped. inst. **33**, 15 (1969).
33, 15 (1968).
571. Ramakrishna Rao K. S., Indian J. Pure Appl. Phys. **6**, 430 (1968).
572. Aaron M. W., Grant E. H., Trans. Faraday Soc. **59**, 185 (1963).
573. Sandus O., Lubitz B. B., J. Phys. Chem. **65**, 881 (1961).
574. Shakhparonov M. I., Akhadov Y. Y., Zhur. strukt. khim. **6**, 21 (1965).
575. Chekalin N. V., Kazantseva S. I., Vestnik MGU, ser. khimiya **4**, 96 (1969).
576. Shakhparonov M. I., Chekalin N. V., Zhur. strukt. khim. **11**, 599 (1970).
577. Hawkins R. E., Davidson D. W., J. Phys. Chem. **70**, 1889 (1966).
578. McDuffie G. E., J. Chem. Phys. **37**, 239 (1962).
579. Kauther E., Annal. Physics **27** (5), 29 (1936).
580. Litovitz T. A., J. Chem. Phys. **21**, 17 (1953).
581. Pottel R., Kaatze U., Ber. Bunsenges Phys. Chem. **73**, 437 (1969).
582. Rizk H. A., Z. Phys. Chem. **245**, 413 (1970).
583. Garg S. K., Smyth C. P., J. Chem. Phys. **43**, 2959 (1965).
584. Tourky A. R., Rizk H. A., Girgis Y. M., J. Phys. Chem. **65**, 40 (1961).
585. Cook H. F., Trans. Faraday Soc. **47**, 751 (1951).
586. Shepherd J. C. W., Grant E. H., Proc. Roy. Soc. **A307**, 335, 345 (1968).
587. Delbos G., C. R. Acad. Sci. Paris **273B** (14), 608 (1971).
588. Shepherd J. C. W., Grant E. H., Proc. Roy. Soc. **A307**, 335 (1968).
589. Gregory M. d., Affsprung H. E., Christian S. D., J. Phys. Chem. **72**, 1748 (1968).
590. Risbourg A., Liebaert R., C. R. Acad. Sci. Paris **264C**, 237 (1967).
591. Barthel J., Behret H., Schmithals F., Berich. Bunsen. Phys. Chem. **75**, 305 (1971).
592. Badiali J. P., Cachet H., Lestrade J. G., J. Chim. Phys. Biol. **64**, 1350 (1967).
593. Kudra O. K., Fialkov Y. Y., Tarasenko Y. A., Ukr. khim. zhur. **30** (4), 347 (1964).
594. Osipov O. A., Panina M. A., Lempert L. Y., Zhur. org. khim. **25** (4), 662 (1955).
595. Szczygielski M., Ziolo J., Acta Phys. Polonica **38A** (5), 799 (1970).
596. Goulon J., Rivail J. L., C. R. Acad Sci. Paris. **275C** (13), 641 (1972).
597. Krishnaji and Mansingh A., J. Chem. Phys. **41**, 827 (1964).
598. Sarojini V., Trans. Faraday Soc. **58** (9), 1729 (1962).
599. Soxou P., Daumezon P., Dansas P., J. Chim. Phys. Biol. **64** (5), 825 (1967).
600. Cavell E. A. S., Knight P. C., Sheikh M. A., Trans. Faraday Soc. **67** (8), 2225 (1971).
601. Sixou P., Dansas P., Daumezon P., Comp. Rend. Acad. Sci. Paris **264B** (15), 1119 (1967).
602. Arkhangel'sky K. V., Semenchenko V. K. Zhur. fiz. khim. **36** (11), 2501 (1962).
603. Ripley B. D., McIntosh R., Canad J. Chem. **39**, 526 (1961).

604. Lubezky I., McIntosh R., Canad J. Chem. **51** (4), 545 (1973).
605. Sixou P., Dansas P., Gillot D., J. Chim. Phys. Biol. **64** (5), 834 (1967).
606. Gross P. M., Taylor R. C., J. Amer. Chem. Soc. **72** (5), 2075 (1950).
607. Weber R., Z. Phys. **70**, 711 (1931).
608. Pottel R., Rerich. Bunseng. Phys. Chem. **69** (5), 363 (1965).
609. Kraeft G. W. D., Zecha M., Z. Phys. Chem. **241** (1–2), 25 (1969).
610. Hartman H., Jaenicke R., Lertes E., Z. Naturforsch. **22A** (10), 1652 (1967).
611. Doucet M. Y., Calmes-Perrault F. C., Durant T. C., Comp. Rend. Acad. Sci. Paris. **260B** (7), 1878 (1965).
612. Delbos G., Salefran J. L., Comp. Rend. Acad. Sci. Paris. **274B** (4), 259 (1972).

INDEX

All compounds mentioned in the book, including all forms of isomers, are given in a primary alphabetic list, and the alphabetic list under each entry gives all the binary solutions in which the compound is involved. In a few cases where the substance is not exactly defined the naming is necessarily imprecise. This accounts for such index entries as *Xylene* which precedes *m-Xylene.*

CHAPTER V

FURTHER DATA FROM PUBLICATIONS APPEARING FROM 1973–1978

§ 1. Permittivity of non-aqueous solutions of inorganic and organic compounds

Carbon tetrachloride (CCl_4) – Nitromethane (CH_3NO_2)

[615]

X_2	ϵ at t°C			X_2	ϵ at t°C		
	20	30	40		20	30	40
100,00	38,5700	36,3920	34,3530	45,19	10,8060	10,3010	9,8210
81,57	25,6800	24,2620	22,9340	28,14	6,7544	6,4905	6,2434
63,68	17,1060	16,2160	15,3830	0,00	2,2362	2,2178	2,2006
54,44	13,6650	12,9860	12,3490	-	-	-	-

Carbon tetrachloride (CCl_4) – Acetic acid ($C_2H_4O_2$)

[616], $\nu = 300$ kHz, $t = 35$°C

N	2,5	2,0	1,8	1,6	1,4	1,0
ϵ	2,3655	2,3180	2,3105	2,2955	2,2755	2,3630

Carbon tetrachloride (CCl_4) – Acetone (C_3H_6O)

[615]

X_2	ϵ at t°C			X_2	ϵ at t°C		
	20	30	40		20	30	40
100,00	21,248	20,069	18,965	40,93	7,8340	7,5060	7,1960
80,13	15,872	15,035	14,252	22,53	4,8715	4,7136	4,5659
59,94	11,350	10,849	10,294	0,00	2,2362	2,2178	2,2006
50,21	9,199	8,786	9,398	-	-	-	-

Carbon tetrachloride (CCl_4) – Butan–1–ol ($C_4H_{10}O$)
[617], ν = 2 MHz, t = 25°C

X_2	ϵ	X_2	ϵ	X_2	ϵ
100,000	17,49	65,561	9,858	30,284	4,1360
97,568	16,66	60,829	9,110	26,271	3,5121
89,107	14,59	57,583	7,323	17,952	2,8970
76,038	12,303	49,711	6,4412	13,707	2,6803
68,205	10,462	37,171	4,3841	0,000	2,2590

Carbon disulphide (CS_2) – Benzene (C_6H_6)
[620]

X_1	ϵ at t°C			X_1	ϵ at t°C		
	20	30	40		20	30	40
0	2,2827	2,2632	2,2431	38,91	2,3891	2,3681	2,3471
5,58	2,3058	2,2866	2,2682	59,75	2,4606	2,4379	2,4159
9,32	2,3236	2,3043	2,2846	79,94	2,5361	2,5120	2,4890
20,52	2,3615	2,3408	2,3195	100,00	2,6329	2,5728	2,5487

Carbon disulphide (CS_2) – Toluene (C_7H_8)
[620]

X_1	ϵ at t°C			X_1	ϵ at t°C		
	20	30	40		20	30	40
0	2,3898	2,3663	2,3454	39,88	2,4597	2,4361	2,4136
4,99	2,4012	2,3798	2,3583	59,99	2,5130	2,4882	2,4646
10,06	2,4040	2,3828	2,3613	79,99	2,5617	2,5347	2,5108
19,77	2,4227	2,4003	2,3787	100,00	2,6329	2,5728	2,5487

Chloroform ($CHCl_3$) – Benzene (C_6H_6)
[620]

X_1	ϵ at t°C			X_1	ϵ at t°C		
	20	30	40		20	30	40
0	2,2827	2,2632	2,2431	50,50	3,4079	3,3530	3,2590
5,54	2,4403	2,4103	2,3912	59,50	3,6447	3,5584	3,4696
10,01	2,5189	2,4783	2,4630	80,02	4,2868	4,1437	4,0276
19,73	2,7093	2,6985	2,6345	100,00	4,8069	4,6932	4,6680
40,39	3,1668	3,1300	3,0659	-	-	-	-

Methanol (CH_4O) – Ethanol (C_2H_6O)
[621]

X_2	ϵ at t°C			X_2	ϵ at t°C		
	15	25	35		15	25	35
100,00	26,83	25,33	23,75	10,07	34,52	32,85	31,21
99,82	26,76	25,11	23,57	6,44	34,88	33,26	31,65
99,56	26,24	24,84	23,20	0,992	35,71	34,12	32,37
98,83	26,32	25,04	23,35	0,501	35,80	34,26	32,43
97,86	27,00	25,38	23,73	0,289	35,82	34,12	32,41
80,94	27,75	26,22	24,63	0,076	35,44	33,86	32,03
60,22	28,76	27,40	25,89	0,00	34,86	33,00	31,19
40,20	30,26	28,72	27,22	-	-	-	-

Methanol (CH_4O) – Benzene (C_6H_6)
[622]

X_1	ϵ at t°C				g at t°C			
	20	30	40	50	20	30	40	50
0,00	2,2825	2,2646	2,2463	2,2279	-	-	-	-
7,45	2,6787	2,5812	2,5332	2,4812	0,72	0,62	0,61	0,58
18,29	3,5717	3,4607	3,3385	3,2165	0,97	0,95	0,93	0,90
34,98	5,8026	5,4814	5,1693	4,8379	1,53	1,48	1,44	1,37
55,52	11,73	10,96	10,23	9,59	2,85	2,79	2,72	2,65
76,62	20,16	18,84	17,69	16,60	4,54	4,46	4,38	4,31
88,07	26,05	24,49	22,99	21,53	5,67	5,63	5,36	5,44
100,00	33,38	31,38	29,05	27,03	7,21	7,09	6,89	6,73

Tetrachloroethylene (C_2Cl_4) – Benzene (C_6H_6)
[614], $t = 30$°C

X_1	ϵ	X_1	ϵ	X_1	ϵ
0,00	2,268	37,79	2,281	91,22	2,300
13,62	2,271	50,90	2,286	100	2,301
19,86	2,274	58,93	2,288	-	-
30,20	2,278	87,80	2,300	-	-

Tetrachloroethylene (C_2Cl_4) – Toluene (C_7H_8)

[614], $t = 30°C$

X_1	0,00	16,17	31,73	41,11	51,55	65,09	70,88	91,75	100,00
ϵ	2,369	2,359	2,349	2,343	2,336	2,326	2,322	2,307	2,301

Tetrachloroethylene (C_2Cl_4) – *p*–Xylene (C_8H_{10})

[614], $t = 30°C$

X_1	0,00	10,63	36,42	57,45	66,76	75,32	84,92	92,57	100,00
ϵ	2,256	2,263	2,274	2,287	2,292	2,294	2,298	2,300	2,301

1,2–Dibromoethane ($C_2H_4Br_2$) – Benzene (C_6H_6)

[623], $t = 25°C$

X_1	ϵ	X_1	ϵ	X_1	ϵ	X_1	ϵ
0,00	2,2626	33,87	3,1398	67,40	3,8246	100,00	4,7503
11,38	2,5397	46,81	3,4009	77,91	4,0823	-	-
20,69	2,7806	56,31	3,4607	89,05	4,6011	-	-

[620]

X_1	ϵ at t°C			X_1	ϵ at t°C		
	20	30	40		20	30	40
0	2,2827	2,2632	2,2431	40,401	3,3255	3,2768	3,2385
4,976	2,4655	2,4288	2,4216	60,224	3,8443	3,7919	3,7563
10,107	2,5766	2,5375	2,5217	81,297	4,4267	4,3791	4,3160
20,252	2,8167	2,7779	2,7549	100,00	4,9612	4,9003	4,8416

1,2–Dibromoethane ($C_2H_4Br_2$) – Cyclohexane (C_6H_{12})

[623], $t = 30°C$

X_1	ϵ	X_1	ϵ	X_1	ϵ	X_1	ϵ
0,00	2,0107	23,80	2,2800	55,94	2,9790	83,21	3,8601
7,85	2,1189	34,71	2,5806	65,28	3,2396	91,72	4,2814
12,26	2,1816	45,39	2,7214	74,48	3,6020	100,00	4,7503

1,2–Dibromoethane ($C_2H_4Br_2$) – Toluene (C_7H_8)
[623], $t = 30°C$

X_1	ϵ	X_1	ϵ	X_1	ϵ	X_1	ϵ
0,00	2,3522	34,68	2,9206	64,94	3,4409	91,41	4,6018
10,33	2,3431	45,06	3,0799	73,91	3,8212	100,00	4,7503
23,78	2,6000	55,40	3,2188	82,99	4,1406	-	-

1,2–Dibromoethane ($C_2H_4Br_2$) – *o*–Xylene (C_8H_{10})
[623], $t = 30°C$

X_1	ϵ	X_1	ϵ	X_1	ϵ	X_1	ϵ
0,00	2,5403	25,96	2,8224	58,60	3,4413	84,74	4,0214
5,85	2,6204	36,36	2,9810	67,87	3,7007	93,39	4,2806
13,62	2,6818	48,17	3,1816	76,79	3,8220	100,00	4,7503

1,2–Dibromoethane ($C_2H_4Br_2$) – *m*–Xylene (C_8H_{10})
[623], $t = 30°C$

X_1	ϵ	X_1	ϵ	X_1	ϵ	X_1	ϵ
0,00	2,3503	37,71	2,9414	68,45	3,5006	92,30	4,4424
11,84	2,6011	48,86	3,0400	81,02	3,7628	100,00	4,7503
25,66	2,7198	58,84	3,3794	85,22	3,8221	-	-

1,2–Dibromoethane ($C_2H_4Br_2$) – *p*–Xylene (C_8H_{10})
[623], $t = 30°C$

X_1	ϵ	X_1	ϵ	X_1	ϵ	X_1	ϵ
0,00	2,2502	40,44	2,8810	68,01	3,5022	100,00	4,7503
13,62	2,4006	49,19	3,0812	85,48	4,0802	-	-
26,22	2,6633	58,89	3,3016	92,15	4,4000	-	-

1,2–Dichloroethane ($C_2H_4Cl_2$) – Benzene (C_6H_6)
[620]

X_1	ε at t°C			X_1	ε at t°C		
	20	30	40		20	30	40
0	2,2827	2,2632	2,2431	40,33	4,5195	4,3783	4,2107
6,24	2,6402	2,5295	2,4968	61,98	6,6563	5,8015	5,5579
9,91	2,8292	2,6863	2,6455	77,93	8,1598	7,0501	6,7253
20,65	3,4230	3,1861	3,1177	100,00	10,6497	9,4487	9,0298

1,2–Dichloroethane ($C_2H_4Cl_2$) – Toluene (C_7H_8)
[620]

X_1	ε at t°C			X_1	ε at t°C		
	20	30	40		20	30	40
0	2,3898	2,3663	2,3454	40,09	4,2844	4,1666	4,0503
4,81	2,6052	2,5694	2,5285	61,12	5,6589	5,4290	5,2133
10,31	2,8291	2,7814	2,7339	80,11	7,6834	7,3430	7,0456
20,12	3,2453	3,1854	3,1127	100,00	10,6497	9,4487	9,0298

Acetic acid ($C_2H_4O_2$) – Acetic anhydride ($C_4H_6O_3$)
[624], ν = 1,5 MHz, t = 25°C

W_2	ε	W_2	ε	W_2	ε	W_2	ε
100,0	21,70	70,0	16,59	40,0	11,85	10,0	7,51
90,0	19,96	60,0	14,99	30,0	10,30	0,0	6,19
80,0	18,25	50,0	13,40	20,0	8,88	-	-

Acetic acid ($C_2H_4O_2$) – 1,4-Dioxan ($C_4H_8O_2$)
[616], ν = 300 kHz, t = 35°C

N	1,0	1,4	1,6	1,8	2,0	2,5
ε	2,5980	2,6530	2,7005	1,8118	2,8280	3,0155

Acetic acid ($C_2H_4O_2$) – Benzene (C_6H_6)

[616], ν = 300 kHz, t = 35°C

N	1,0	1,4	1,6	1,8	2,0	2,5
ϵ	2,3030	2,3255	2,3355	2,3455	2,3630	2,4030

Acetic acid ($C_2H_4O_2$) – Cyclohexane (C_6H_{12})

[616], ν = 300 kHz, t = 35°C

N	1,0	1,4	1,6	1,8	2,0	2,5
ϵ	2,0505	2,0955	2,1155	2,1480	2,1755	2,2130

Acetic acid ($C_2H_4O_2$) – Hexane (C_6H_{14})

[616], ν = 300 kHz, t = 35°C

N	1,0	1,4	1,6	1,8	2,0	2,5
ϵ	1,9480	1,9930	2,0330	2,0805	2,1030	2,1530

Ethanol (C_2H_6O) – Benzene (C_6H_6)

[622]

X_1	ϵ at t°C				g at t°C			
	20	30	40	50	20	30	40	50
5,34	2,5004	2,4851	2,4525	2,4287	1,06	1,13	1,14	1,17
7,75	2,6442	2,6116	2,5770	2,5299	1,13	1,15	1,18	1,18
9,34	2,7159	2,6790	2,6420	2,5966	1,17	1,19	1,21	1,19
11,49	2,8457	2,8137	2,7662	2,7183	1,22	1,26	1,27	1,27
20,18	3,5779	3,4165	3,3598	3,2522	1,47	1,46	1,42	1,40
40,47	6,979	6,556	6,165	5,822	2,20	2,14	2,08	2,01
70,73	15,646	14,545	13,482	12,520	2,93	2,86	2,77	2,67
100,00	25,29	23,59	22,52	21,12	3,03	2,97	2,96	2,91

Acetone (C_3H_6O) – Cyclohexane (C_6H_{12})

[615]

X_1	ϵ at t°C			X_1	ϵ at t°C		
	20	30	40		20	30	40
0	2,0229	2,0078	1,9934	60,24	10,4390	9,9390	9,4610
19,70	3,9178	3,8018	3,6932	80,05	15,1540	14,3540	13,6050
40,45	6,4867	6,2190	5,9684	100,00	21,2480	20,0690	18,9650
50,19	8,2230	7,8500	7,5020	-	-	-	-

Propan–1–ol (C_3H_8O) – Benzene (C_6H_6)
[622]

X_1	ε at t°C				ε at t° C			
	20	30	40	50	20	30	40	50
0,00	2,2825	2,2645	2,2463	2,2279	-	-	-	-
2,48	2,3877	2,3636	2,3378	2,3124	1,06	1,07	1,08	1,10
5,01	2,4787	2,4488	2,4194	2,3819	1,07	1,09	1,09	1,12
7,60	2,5758	2,5577	2,5233	2,4896	1,08	1,10	1,11	1,13
9,97	2,7474	2,7097	2,6633	2,6197	1,20	1,22	1,21	1,20
19,95	3,4297	3,3233	3,2265	3,1319	1,38	1,35	1,33	1,30
39,98	6,458	5,9556	5,511	5,124	2,15	2,02	1,91	1,80
70,15	13,677	12,608	11,582	10,649	2,95	2,83	2,71	2,58
100,00	21,24	19,86	18,52	17,70	3,30	3,24	3,15	3,07

Propan-1-ol (C_3H_8O) – Cyclohexane (C_6H_{12})
[622]

X_1	ε at t°C				ε at t° C			
	20	30	40	50	20	30	40	50
0,00	2,0240	2,0119	1,9981	1,9787	-	-	-	-
4,74	2,1045	2,0933	2,0787	2,0659	0,60	0,63	0,67	0,76
10,08	2,2265	2,2145	2,1968	2,1825	0,68	0,72	0,74	0,80
20,34	2,7457	2,6783	2,6125	2,5615	1,09	1,06	1,04	1,04
30,46	3,8153	3,6068	3,4150	3,2510	1,61	1,53	1,44	1,38
49,63	7,6000	6,9200	6,3700	5,8000	2,57	2,41	2,28	2,12
79,64	15,41	14,30	13,21	12,19	3,21	3,12	3,00	2,89
100,00	21,24	19,8	18,52	17,27	3,30	3,24	3,15	3,07

Ethyl acetate ($C_4H_8O_2$) – Benzene (C_6H_6)
[620]

X_1	ε at t°C			X_1	ε at t°C		
	20	30	40		20	30	40
0	2,2827	2,2632	2,2431	39,651	4,0171	3,9480	3,8384
5,156	2,5097	2,4802	2,4591	60,981	4,7423	4,7301	4,5902
9,817	2,6867	2,6488	2,6106	79,892	5,3219	5,1759	4,9968
19,788	3,0612	3,0042	2,9487	100,00	6,0814	5,7320	5,6074

Butan–1–ol ($C_4H_{10}O$) – Benzene (C_6H_6)

[622]

X_1	ϵ at t°C				ϵ at t°C			
	20	30	40	50	20	30	40	50
0,00	2,2825	2,2645	2,2463	2,2279	-	-	-	-
4,76	2,4547	2,4334	2,4031	2,3644	1,17	1,15	1,13	1,15
9,05	2,6428	2,6083	2,5734	2,5288	1,21	1,19	1,23	1,22
19,21	3,2480	3,1608	3,0851	2,9956	1,41	1,39	1,38	1,35
30,47	4,4376	4,1631	3,9424	3,7416	1,84	1,72	1,65	1,57
49,39	7,58	6,97	6,46	5,93	2,56	2,42	2,30	2,16
74,36	12,86	11,79	10,84	9,95	3,23	3,09	2,95	2,80
100,00	18,27	16,96	15,73	14,59	3,58	3,47	3,36	3,24

[625], $\nu = 1$ MHz, $t = 25$°C

X_1	1,96	3,45	6,54	13,45	21,12	40,55	60,06	82,22	100,00
ϵ	2,347	2,408	2,542	2,895	3,438	5,799	9,434	13,901	17,203
ϵ_∞	2,269	2,265	2,256	2,237	2,216	2,162	2,106	2,042	1,989
n_D^2	2,238	2,234	2,222	2,200	2,175	2,115	2,059	1,999	1,953
μ	1,68	1,71	1,75	1,82	2,07	2,33	2,67	2,82	3,00
g	1,00	1,04	1,08	1,17	1,17	1,52	2,52	2,98	3,19

Butan–1–ol ($C_4H_{10}O$) – Aniline (C_6H_7N)

[617], $\nu = 2$ MHz, $t = 25$°C

X_1	ϵ	X_1	ϵ	X_1	ϵ	X_1	ϵ
100,000	17,49	70,493	12,09	51,784	10,305	20,833	7,843
90,624	15,32	62,353	11,243	43,214	9,408	10,092	7,180
80,570	13,58	57,140	10,706	32,333	8,579	0,000	7,014

Butan–1–ol ($C_4H_{10}O$) – Cyclohexane (C_6H_{12})
[622]

X_1	ε at t°C				ε at t°C			
	20	30	40	50	20	30	40	50
0,00	2,0240	2,0119	1,9981	1,9787	-	-	-	-
5,37	2,1179	2,1106	2,0888	2,0823	0,65	0,71	0,79	0,83
10,78	2,2409	2,2261	2,2207	2,1997	0,72	0,75	0,82	0,85
19,25	2,5819	2,5482	2,5181	2,4757	0,98	0,99	1,01	1,01
32,69	3,7760	3,5286	3,3394	3,2098	1,61	1,48	1,39	1,35
58,23	8,97	8,19	7,84	6,83	3,00	2,83	2,82	2,50
86,47	15,22	14,09	12,96	11,91	3,49	3,37	3,22	3,08
100,00	18,27	16,96	15,73	14,59	3,58	3,47	3,36	3,24

Diethyl ether ($C_4H_{10}O$) – Benzene (C_6H_6)
[620]

X_1	ε at t°C		X_1	ε at t°C	
	20	30		20	30
0	2,2827	2,2632	39,87	3,0839	2,9976
5,09	2,3885	2,3598	60,31	3,4755	3,3966
8,92	2,4731	2,4290	80,75	3,9602	3,7975
17,46	2,6338	2,5945	100,00	4,4291	4,1807

Diethyl ether ($C_4H_{10}O$) – Hexane (C_6H_{14})
[622]

X_1	ε at t°C			ε at t°C		
	10	18	22	10	18	22
0,00	1,9627	1,8899	1,8838	-	-	-
13,96	2,1624	2,0598	2,0702	1,21	1,07	1.19
24,55	2,3252	2,2277	2,2364	1,19	1,15	1,22
35,72	1,5265	2,4824	2,4132	1,21	1,31	1,20
43,74	2,6958	2,6329	2,5659	1,24	1,29	1,25
54,77	2,9613	2,8851	2,8221	1,28	1,31	1,27
64,43	3,2187	3,1508	3,0667	1,33	1,35	1,30
74,73	3,4768	3,4226	3,3519	1,35	1,36	1,33
80,15	3,7121	3,5949	3,5322	1,37	1,37	1,36
91,17	4,1131	3,9640	3,9029	1,41	1,40	1,39
100,00	4,4692	4,2970	4,2294	1,68	1,42	1,41

Diethyl ether ($C_4H_{10}O$) – **Hexane** (C_6H_{14})
[615]

X_1	ε at t°C			X_1	ε at t°C		
	15	20	25		15	20	25
0	-	1,8865	1,8795	62,15	3,0718	3,0275	2,9863
16,11	2,1319	2,1181	2,1050	74,02	3,4093	3,3512	3,2981
29,80	2,3545	2,3354	2,7172	79,53	3,5783	3,5155	3,4539
37,54	2,5086	2,4832	2,4602	92,23	4,0283	3,9485	3,8725
50,33	2,7749	2,7406	2,7088	100,00	4,3614	4,2666	4,1751
53,07	2,8350	2,7982	2,7635	-	-	-	-

Diethyl ether ($C_4H_{10}O$) – **Heptane** (C_7H_{16})
[615]

X_1	ε at t°C			X_1	ε at t°C		
	15	20	25		15	20	25
0	-	1,9206	1,9141	62,54	2,9853	2,9438	2,9055
12,72	2,0556	2,0445	2,0344	77,07	3,4564	3,3992	3,3520
18,45	2,1697	2,1565	2,1437	88,34	3,8607	3,7838	3,7192
33,41	2,4047	2,3857	2,3615	89,88	3,9211	3,8439	3,7757
41,52	2,5510	2,5281	2,5007	100,00	4,3614	4,2666	4,1751
55,55	2,8527	2,8195	2,7851	-	-	-	-

Diethyl ether ($C_4H_{10}O$) – **Octane** (C_8H_{18})
[615]

X_1	ε at t°C			X_1	ε at t°C		
	15	20	25		15	20	25
0	-	1,9546	1,9392	64,06	3,0222	2,9814	2,9419
18,31	2,1745	2,1610	2,1487	75,84	3,3742	3,3232	3,2723
27,58	2,3192	2,3009	2,2832	87,41	3,7881	3,7201	3,6528
37,25	2,4599	2,4386	2,4168	91,42	3,9589	3,8841	3,8037
40,83	2,6598	2,6333	2,6045	100,00	4,3614	4,2666	4,1751
60,34	2,9391	2,9008	2,8650	-	-	-	-

Diethylamine ($C_4H_{11}N$) – Benzene (C_6H_6)

[629], $\nu = 50$ kHz

X_1	ϵ at t°C		X_1	ϵ at t°C	
	20	50		20	50
0,00	2,283	2,226	50,60	2,997	2,799
1,19	2,299	2,239	60,13	3,158	2,907
5,76	2,358	-	79,97	3,518	3,131
10,79	2,426	2,348	90,00	3,724	3,245
12,25	2,442	-	95,07	3,830	3,302
20,24	2,549	2,455	97,97	3,855	3,335
40,79	2,848	2,688	100,00	3,881	3,358

Diethylamine ($C_4H_{11}N$) – Cyclohexane (C_6H_{12})

[629], $\nu = 50$ kHz

X_1	ϵ at t°C		X_1	ϵ at t°C	
	20	40		20	40
0,00	2,027	2,000	60,96	2,944	2,758
2,10	2,050	2,020	80,73	3,382	3,116
5,44	2,086	2,051	90,30	3,622	3,302
10,08	2,141	2,097	95,17	3,775	3,408
20,69	2,271	2,207	98,08	3,839	3,466
40,99	2,577	2,461	100,00	3,885	3,516
50,90	2,753	2,606	-	-	-

Pyridine (C_5H_5N) – Benzene (C_6H_6)

[630], $t = 20$°C

ψ_1	10,3	20,0	30,1	40,2	50,2	60,2	70,2	80,1	90,1	100,0
ϵ	2,82	3,73	4,79	5,92	7,19	8,46	9,75	10,99	12,28	13,52

Methyl methacrylate ($C_5H_8O_2$) – Benzene (C_6H_6)
[632]

X_1	ε at t°C			X_1	ε at t°C		
	20	30	40		20	30	40
100,00	2,293	2,280	2,264	69,61	3,560	3,494	3,361
97,13	2,445	2,422	2,390	50,42	4,393	4,260	4,065
95,14	2,518	2,490	2,457	29,93	5,142	4,962	4,771
91,90	2,650	2,618	2,578	10,02	5,862	5,662	5,451
90,17	2,722	2,685	2,641	0,00	6,160	5,943	5,742

Methyl methacrylate ($C_5H_8O_2$) – Toluene (C_7H_8)
[632]

X_1	ε at t°C			X_1	ε at t°C		
	20	30	40		20	30	40
100,00	2,425	2,371	2,360	30,98	4,880	4,733	4,592
88,80	2,847	2,762	2,729	9,73	5,730	5,539	5,355
69,53	3,524	3,441	3,361	0,00	6,160	5,943	5,742
49,46	4,323	4,183	4,077	-	-	-	-

Methyl methacrylate ($C_5H_8O_2$) – Styrene (C_8H_8)
[632]

X_2	ε at t°c			X_2	ε at t°C		
	20	30	40		20	30	40
100,00	2,442	2,431	2,420	70,29	3,111	3,249	3,186
96,04	2,552	2,528	2,503	50,89	4,111	4,009	3,907
94,90	2,672	2,603	2,574	29,53	4,958	4,811	4,670
92,31	2,707	2,673	2,632	9,87	5,760	5,574	5,387
89,97	2,780	2,743	2,707	0,00	6,160	5,943	5,742

Methyl methacrylate ($C_5H_8O_2$) – *o*–Xylene (C_8H_{10})
[632]

X_2	ε at t°C			X_2	ε at t°C		
	20	30	40		20	30	40
100,00	2,581	2,560	2,541	69,68	3,477	3,394	3,329
97,51	2,671	2,634	2,608	49,83	4,158	4,031	3,930
94,92	2,737	2,702	2,672	30,01	4,880	4,733	4,597
92,12	2,817	2,782	2,747	10,03	5,773	5,566	5,377
89,44	2,895	2,851	2,814	0,00	6,160	5,943	5,742

Methyl methacrylate ($C_5H_8O_2$) – *m*–Xylene (C_8H_{10})
[632]

X_2	ε at t°C			X_2	ε at t°C		
	20	30	40		20	30	40
100,00	2,391	2,379	2,362	73,46	3,173	3,109	3,054
97,13	2,471	2,456	2,429	50,51	3,972	3,863	3,764
94,70	2,546	2,519	2,491	29,83	4,786	4,656	4,342
91,83	2,617	2,590	2,562	10,04	5,693	5,510	5,319
89,85	2,679	2,643	2,613	0,00	6,160	5,943	5,742

Methyl methacrylate ($C_5H_8O_2$) – *p*–Xylene (C_8H_{10})
[632]

X_2	ε at t°C			X_2	ε at t°C		
	20	30	40		20	30	40
100,00	2,365	2,344	2,310	29,51	4,763	4,623	4,493
89,23	2,665	2,647	2,607	9,47	5,666	5,488	5,308
67,62	3,370	3,313	3,235	0,00	6,160	5,943	5,742
49,55	4,084	3,978	3,883	-	-	-	-

Benzene (C_6H_6) – Triethylamine ($C_6H_{15}N$)

[629], $\nu = 50$ kHz

X_1	ϵ at t°C		X_1	ϵ at t°C	
	20	40		20	40
100,00	2,425	2,349	28,89	2,363	2,305
96,95	2,426	2,350	14,36	2,328	2,285
92,28	2,425	2,353	7,00	2,309	2,266
85,11	2,421	2,346	3,47	2,299	2,262
72,01	2,412	2,344	1,82	2,293	2,253
49,13	2,391	2,325	0,00	2,284	2,245
38,98	2,377	2,320	-	-	-

Benzene (C_6H_6) – Toluene (C_7H_8)

[620]

X_2	ϵ at t°C			X_2	ϵ at t°C		
	20	30	40		20	30	40
100,00	2,3898	2,3663	2,3454	20,06	2,3162	2,2974	2,2784
79,99	2,3734	2,3529	2,3322	9,96	2,3041	2,2853	2,2669
60,30	2,3532	2,3326	2,3130	5,01	2,2969	2,2784	2,2605
36,73	2,3361	2,3159	2,2968	0	2,2827	2,2632	2,2431

Benzene (C_6H_6) – Heptan-1-ol ($C_7H_{16}O$)

[625], $\nu = 1$ MHz, $t = 25$°C

X_2	100,00	79,00	60,51	39,41	24,37	7,68	3,17	1,29	0,54
ϵ	11,013	8,820	6,801	4,414	3,304	2,571	2,398	2,314	2,290
ϵ_∞	2,074	2,104	2,135	2,176	2,209	2,252	2,265	2,270	2,272
n^2	2,023	2,054	2,085	2,123	2,158	2,215	2,232	2,237	2,242
μ	2,84	2,68	2,48	2,08	1,83	1,73	1,74	1,71	1,54
g	2,88	2,54	2,18	1,54	1,19	1,07	1,07	1,04	0,84

Benzene (C_6H_6) – 3–Methylhexan–2–ol ($C_7H_{16}O$)

[625], $\nu = 1$ MHz, $t = 25$°C

X_2	100,00	80,56	59,90	31,69	11,19	5,13	2,69	1,69	0,44
ϵ	4,476	3,989	3,569	3,010	2,640	2,450	2,372	2,333	2,289
ϵ_∞	2,058	2,088	2,125	2,188	2,240	2,258	2,233	2,269	2,273
n^2_D	2,022	2,044	2,075	2,139	2,201	2,224	2,265	2,236	2,242
μ	1,58	1,51	1,47	1,44	1,62	1,64	1,69	1,65	1,62
g	0,88	0,81	0,76	0,74	0,93	0,96	1,01	0,96	0,93

Bromobenzene (C_6H_5Br) – Benzene (C_6H_6)
[615]

X_1	ε at t°C			X_1	ε at t°C		
	20	30	40		20	30	40
0	2,2827	2,2645	2,2456	50,65	4,0144	3,9429	3,8570
10,56	2,6763	2,6437	2,6046	59,66	4,2927	4,2018	4,1084
19,91	3,0083	2,9591	2,9106	70,56	4,6303	4,5194	4,4185
30,03	3,3598	3,3039	3,2397	80,06	4,9047	4,7910	4,6783
40,22	3,6896	3,6223	3,5512	100,00	5,4585	5,3138	5,1785

Chlorobenzene (C_6H_5Cl) – Benzene (C_6H_6)
[615]

X_1	ε at t°C			X_1	ε at t°C		
	20	30	40		20	30	40
0	2,2827	2,2645	2,2456	59,87	4,3825	4,2726	4,1698
9,01	2,6107	2,5775	2,5420	69,18	4,6458	4,5663	4,4528
19,93	3,0052	2,9540	2,9032	79,10	5,0253	4,8909	4,7570
31,47	3,4048	3,3412	3,2689	89,83	5,3745	5,2274	5,0832
40,58	3,7385	3,6539	3,5740	100,00	5,6895	5,5312	5,3728
49,96	4,0538	3,9584	3,8726	-	-	-	-

Benzene (C_6H_6) – Hexan–1–ol ($C_6H_{14}O$)
[622]

X_2	ε at t°C				g at t°C			
	20	30	40	50	20	30	40	50
100,00	13,647	12,576	11,494	10,588	2,95	2,82	2,67	2,55
69,47	9,374	8,480	7,715	7,040	2,51	2,33	2,20	2,04
39,81	4,8712	4,5643	4,3118	4,0851	1,60	1,51	1,46	1,39
20,75	3,1936	3,1374	3,0821	3,0266	1,10	1,13	1,14	1,15
10,42	2,6721	2,6422	2,6120	2,5807	0,97	0,98	1,02	1,02
7,48	2,5399	2,5176	2,5089	2,4815	0,96	0,93	1,01	1,01
5,05	2,4702	2,4459	2,4186	2,3904	0,90	0,92	0,98	0,98
2,42	2,3605	2,3446	2,3317	2,3065	0,85	0,91	0,97	0,99
0,00	2,2825	2,2645	2,2463	2,2279	-	-	-	-

Benzene (C_6H_6) – 3–Methylhexan–3–ol ($C_7H_{16}O$)

[625], $\nu = 1$ MHz, $t = 25°C$

X_2	100,00	81,79	57,17	30,85	11,98	6,57	3,27	1,81	0,99[illegible]	0,37
ϵ	3,248	3,242	3,225	2,996	2,627	2,487	2,381	2,324	2,313	2,285
ϵ_∞	2,034	2,065	2,114	2,178	2,234	2,251	2,263	2,268	2,270	2,273
n^2	2,019	2,040	2,082	2,139	2,197	2,216	2,230	2,236	2,240	2,243
μ	1,16	1,21	1,33	1,46	1,56	1,62	1,61	1,57	1,76	1,52
g	0,48	0,52	0,63	0,75	0,86	0,92	0,92	0,87	1,10	0,82

Benzene (C_6H_6) – Octan–1–ol ($C_8H_{18}O$)

[622]

X_2	ϵ at t°C				g at t°C			
	20	30	40	50	20	30	40	50
100,00	10,090	9,223	8,453	7,808	2,63	2,48	2,33	2,22
73,01	7,353	6,700	6,221	5,8204	2,14	1,99	1,90	1,82
40,18	4,2730	4,0339	3,8460	3,6890	1,42	1,34	1,28	1,24
19,90	3,0119	2,9637	2,9144	2,8708	1,01	1,02	1,03	1,04
10,19	2,6285	2,6040	2,5790	2,5513	0,94	0,99	0,99	1,01
7,58	2,5481	2,5269	2,5010	2,4738	0,92	0,98	0,98	1,00
5,005	2,4544	2,4359	2,4091	2,3820	0,91	0,97	0,95	0,95
2,44	2,3644	2,3513	2,3279	2,3062	0,92	0,99	0,98	0,99
0,00	2,2825	2,2645	2,2463	2,2279	-	-	-	-

Aniline (C_6H_7N) – Toluene (C_7H_8)

[617], $\nu = 2$ MHz, $t = 25°C$

X_1	ϵ	X_1	ϵ	X_1	ϵ
100,000	7,014	56,692	4,6384	12,597	2,6969
87,869	6,290	47,224	4,1494	5,237	2,5268
79,471	5,808	37,574	3,7827	0,000	2,3884
69,884	5,243	25,533	3,1801	-	-
61,139	5,0110	19,752	2,9287	-	-

Cyclohexene (C_6H_{10}) – Cyclohexane (C_6H_{12})
[615]

X_1	ε at t°C			X_1	ε at t°C		
	20	30	40		20	30	40
0	2,0228	2,0077	1,9928	59,37	2,1326	2,1139	2,0964
10,18	2,0408	2,0252	2,0106	69,98	2,1540	2,1346	2,1164
20,12	2,0585	2,0421	2,0268	79,77	2,1739	2,1540	2,1354
29,74	2,0760	2,0590	2,0431	90,57	2,1965	2,1760	2,1569
39,25	2,0938	2,0763	2,0598	100,00	2,2176	2,1973	2,1772
50,35	2,1150	2,0968	2,0798	-	-	-	-

Cyclohexanone ($C_6H_{10}O$) – Cyclohexanol ($C_6H_{12}O$)
[635]

t°C	ε at X_1			
	20	40	60	80
20	15,32	14,45	14,05	14,72
30	14,67	13,76	13,38	13,81
40	14,02	13,06	12,47	12,80
50	13,40	12,42	11,77	11,77
60	12,77	11,80	11,08	10,87
70	12,16	11,21	10,42	10,05
80	11,63	10,67	9,82	9,30
90	11,24	10,30	9,45	8,75
100	10,80	9,80	9,05	8,20
110	10,50	9,55	8,70	7,70
120	9,83	9,10	8,32	7,42
130	9,53	8,80	7,87	7,05
140	9,20	8,50	7,55	6,55
150	8,81	8,20	7,35	6,22

Cyclohexane (C_6H_{12}) – Cyclohexanol ($C_6H_{12}O$)
[635]

t°C	ε at φ_2				
	20	40	60	80	100
20	2,71	4,72	9,06	12,80	16,40
30	2,68	4,20	7,97	11,65	15,29
40	2,64	3,99	7,12	10,35	14,12
50	2,60	3,77	6,47	9,13	12,80
60	2,56	3,64	5,78	8,11	11,51
70	2,52	3,43	5,34	7.31	10,33

Cyclohexane (C_6H_{12}) – Triethylamine ($C_6H_{15}N$)

[629], ν = 50 kHz

X_1	ϵ at t°C		X_1	ϵ at t°C	
	20	40		20	40
100,00	2,422	2,348	34,34	2,182	2,133
98,74	2,422	2,347	16,12	2,104	2,061
93,66	2,404	2,330	8,44	2,070	2,032
87,48	2,383	2,312	3,74	2,047	2,012
76,00	2,342	2,280	0,98	2,033	1,998
53,37	2,260	2,204	0,00	2,028	1,995
44,12	2,222	2,171	-	-	-

Cyclohexane (C_6H_{12}) – Toluene (C_7H_8)

[615]

X_2	ϵ at t°C			X_2	ϵ at t°C		
	20	30	40		20	30	40
100,00	2,3820	2,3571	2,3338	47,49	2,1788	2,1593	2,1421
81,98	2,3080	2,2866	2,2665	15,47	2,1130	2,0953	2,0787
64,28	2,2396	2,2195	2,2006	0	2,0229	2,0078	1,9934
55,37	2,2066	2,1871	2,1689	-	-	-	-

Toluene (C_7H_8) – *N,N*–Dimethylaniline ($C_8H_{11}N$)

[617], ν = 2MHz, t = 25°C

X_1	ϵ	X_1	ϵ	X_1	ϵ
100,000	5,0202	62,115	4,2029	23,893	3,0567
89,452	4,8440	53,184	3,8343	10,613	2,7519
80,526	4,5926	40,347	3,5178	0,000	2,3884
74,390	4,4308	34,505	3,3426	-	-

Supplementary information on the permittivity of aqueous solutions of inorganic and organic substances.

Argon (Ar) – Nitrogen (N_2) [613], $t = 91 - 115°K$, P = 1.04 – 14 atm.
Argon (Ar) – Methane (CH_4) [613], $t = 91 - 115°K$, P = 1.04 – 14 atm.
Nitrogen (N_2) – Methane (CH_4) [613], $t = 91 - 115°K$, P = 1.04 – 14 atm.
Nitrogen (N_2) – Ethane (C_2H_6) [613], $t = 91 - 115°K$, P = 1.04 – 14 atm.
Carbon tetrachloride (CCl_4) – Tetrachloroethylene (C_2Cl_4) [614], $t = 30°C$.
Carbon tetrachloride (CCl_4) – 2–Methylbutan–2–ol ($C_5H_{12}O$) [618], $t = 30°C, x = 0 - 100$.
Carbon tetrachloride (CCl_4) – Octan–1–ol ($C_8H_{18}O$) [619], $\nu = 2$ MHz, $t = 25°C$.
Methane (CH_4) – Ethane (C_2H_6) [613], $t = 91 - 115°K$, P = 1.4 – 14 atm.
Methane (CH_4) – Propane (C_3H_8) [613], $t = 91 - 115°K$, P = 1.04 – 14 atm.
Methane (CH_4) – Butane (C_4H_{10}) [613], $t = 91 - 115°K$, P = 1.04 – 14 atm.
1,1,1–Trichloroethane ($C_2H_3Cl_3$) – 2–Methylbutan–2–ol ($C_5H_{12}O$) [618].
Propan–1–ol (C_3H_8O) – Benzene (C_6H_6) [625], $\nu = 1$ MHz, $t = 25°C$.
Propan–2–ol (C_3H_8O) – Benzene (C_6H_6) [625], $\nu = 1$ MHz, $t = 25°C$.
2–Pyrrolidone (C_4H_7NO) – Benzene (C_6H_6) [626], $\nu = 1.5$ kHz, $t = 20 - 50°C$.
Isobutanol ($C_4H_{10}O$) – Benzene (C_6H_6) [625], $\nu = 1$ MHz, $t = 25°C$.
tert–Butyl alcohol ($C_4H_{10}O$) – Hexane (C_6H_{14}) [627], $\nu = 2$ MHz, $t = 20°C$.
tert–Butyl alcohol ($C_4H_{10}O$) – Dodecan–1–ol ($C_{12}H_{26}O$) [628], $t = 30°C$.
Cyclopentane (C_5H_{10}) – Benzonitrile (C_7H_5N) [631], $\nu = 2$ MHz, $t = 10 - 50°C$.
2–Methylbutan–2–ol ($C_5H_{12}O$) – Benzene (C_6H_6) [618].
2–Methylbutan–2–ol ($C_5H_{12}O$) – Cyclohexane (C_6H_{12}) [618].
2–Methylbutan–2–ol ($C_5H_{12}O$) – Hexane (C_6H_{14}) [618].
2–Methylbutan–2–ol ($C_5H_{12}O$) – Xylene (C_8H_{10}) [618].
2–Methylbutan-2-ol ($C_5H_{12}O$) – Triethylamine ($C_6H_{15}N$) [618].
Pentan–1–ol ($C_5H_{12}O$) – Benzene (C_6H_6) [625], $\nu = 1$MHz, $t = 25°C$.
Pentan–3–ol ($C_5H_{12}O$) – Dodecan–1–ol ($C_{12}H_{26}O$) [628], $t = 40–50°C$.
Nitrobenzene ($C_6H_5NO_2$) – Benzene (C_6H_6) [633], $t = 220 - 330°K$, $x = 30 - 100$.
Nitrobenzene ($C_6H_5NO_2$) – Hexane (C_6H_{14}) [644], $t = 19.65 - 19.85$.
Benzene (C_6H_6) – Hexan–1–ol ($C_6H_{14}O$) [625], $\nu = 1$ MHz, $t = 25°C$.

Benzene (C_6H_6) – Octan–1–ol ($C_8H_{18}O$) [619], $\nu = 2$ MHz, $t = 25°C$.

Cyclohexane (C_6H_{12}) – Octan–1–ol ($C_8H_{18}O$) [619], $\nu = 2$ MHz, $t = 25°C$.

Hexane (C_6H_{14}) – *o*–Nitrotoluene ($C_7H_7NO_2$) [636], $t = 253 - 293°K$, $x = 30 - 70$.

Benzonitrile (C_7H_5N) – 2–Methylheptane (C_8H_{18}) [631], $\nu = 2$MHz, $t = 0 - 50°C, x = 30 - 50$.

§2. Dielectric dispersion parameters of non-aqueous solutions of inorganic and organic compounds.

Deuterium oxide (D_2O) – Potassium chloride (KCl)

[637], $\nu = 9370$ MHz

X_2	ϵ' at t°C					
	4,6	8,5	15	25	35	50
0,00	37,3	42,4	50,3	58,0	60,5	62,2
0,18	37,7	42,5	50,1	57,3	59,8	60,7
0,45	37,9	42,6	49,7	56,4	58,9	59,7
0,89	38,7	43,0	49,2	55,1	57,3	57,6
1,33	39,0	43,2	48,7	54,0	55,6	55,8
1,77	39,3	43,0	48,3	52,9	54,7	54,4
3,48	39,7	42,4	46,2	49,0	50,4	50,3

Deuterium oxide (D_2O) – Lithium chloride (LiCl)

[637], $\nu = 9370$ MHz

X_2	ϵ' at t°C					
	4,6	8,5	15	25	35	50
0,00	37,2	42,4	50,4	58,0	60,5	62,2
0,18	37,0	42,0	49,6	56,9	59,8	60,7
0,45	36,4	41,5	48,6	55,2	58,1	58,9
0,88	35,7	40,4	46,9	53,3	55,4	56,1
1,32	35,0	39,3	45,3	51,3	53,5	54,0
1,74	34,3	38,3	43,9	49,6	51,4	52,0
3,36	31,7	34,7	39,2	48,6	44,7	45,1

Deuterium oxide (D_2O) – Sodium chloride (NaCl)

[637], $\nu = 9370$ MHz

X_2	ϵ' at t°C					
	4,6	8,5	15	25	35	50
0,00	37,3	42,4	50,4	58,0	60,5	62,2
0,18	37,4	42,2	50,0	56,9	59,9	60,6
0,45	37,5	42,3	49,5	56,1	58,4	59,4
0,89	37,3	42,1	48,5	54,2	56,3	56,9
1,33	36,8	41,3	47,4	52,7	54,3	54,9
1,77	36,5	40,9	46,5	51,5	52,7	52,8
3,48	35,2	38,4	42,6	45,9	46,6	46,6

Deuterium oxide (D_2O) – 1,6–Diaminohexane ($C_6H_{16}N_2$)

[638], $\nu = 1{,}8$–40 GHz, $t = 25$°C

N	φ	ϵ	ϵ_∞	$10^{-12}\ \tau$	a
0	0	76,8	4,7	10,41	0,016
0,3288	3,599	75,2	4,7	11,81	0,034
0,6576	6,829	73,0	4,6	13,39	0,042
1,005	10,75	71,3	4,2	15,37	0,074

Lithium perchlorate ($LiClO_4$) – Tetrahydrofuran (C_4H_8O)

[642], $\nu = 0{,}137$–34 GHz, $t = 25$°C, $N = 0{,}59$

ν, MHz	ϵ'	ϵ''	ν, MHz	ϵ'	ϵ''
137	16,6	22,1	625	14,6	8,3
	16,9	22,5	1060	12,7	7,08
260	16,2	13,1	1666	10,9	6,05
	16,2	13,1	2825	9,3	4,72
375	15,8	10,4	9347	7,18	2,66
383	15,7	10,3	34720	5,80	2,55
	15,6	10,2	-	-	-

Lithium perchlorate ($LiClO_4$) – Ethyl acetate ($C_6H_{10}O_4$)

[643], $N = 0{,}6$, $t = 25$°C

ν, MHz	ϵ'	ϵ''	ν, MHz	ϵ'	ϵ''
0,12802+	13,94	20,27	0,47106	12,67	8,66
0,12802	14,27	20,61	0,89508	10,66	6,50
0,18022	13,96	15,34	1,28507	9,43	5,68
0,18022	13,95	15,63	2,05143	8,10	4,52
0,26004	13,63	11,96	3,23497	7,13	3,50
0,26004	13,66	11,95	9,369	5,82	2,29
0,47106	12,58	8,62	-	-	-

+) Measurements carried out in a different condenser

Sodium (Ha) – Ammonia (NH_3)

[644]

ν, GHz	t°C	X_1 = 0.7		X_1 = 0.7		t°C	X_1 = 0.7		X_1 = 1.2	
		ε′	ε″	ε′	ε″		ε′	ε″	ε′	ε″
2,5	–40	32,3	3,6	33,6	3,6	–75	41,0	2,8	42,8	2,5
6		30,7	7,3	30,8	6,8		34,9	7,5	39,9	8,5
10		26,0	6,6	26,7	8,9		31,8	8,8	34,3	11,4
23		21,8	7,1	20,4	8,0		26,7	10,6	26,9	12,0
35		20,4	8,5	16,9	4,7		24,6	11,3	22,5	11,2
70		16,4	10,2	14,5	5,1		17,5	11,7	18,4	8,4

Sodium perchlorate ($NaClO_4$) – Tetrahydrofuran (C_4H_8O)

[645], ν = 0,2–8,5 GHz, t = 25°C

ν, GHz	N = 0,05		N = 0,1		ν, GHz	N = 0,05		N = 0,1	
	ε′	ε″	ε′	ε″		ε′	ε″	ε′	ε″
0,30	8,21	0,28	8,71	0,42	1,50	7,99	0,52	8,36	0,78
0,50	-	-	8,68	0,40	2,00	7,85	0,58	8,20	0,80
0,60	8,26	0,38	8,65	0,43	3,00	7,77	0,74	7,88	0,90
0,90	8,15	0,45	8,59	0,59	4,00	7,56	0,76	7,73	1,03
1,20	8,14	0,50	8,42	0,63	8,52	7,21	0,96	7,24	1,03

Carbon tetrachloride (CCl_4) – Methanol (CH_4O)

[646]

ϕ_2	t°C	λ = 10 cm		λ = 3,2 cm		λ = 0,408 cm	
		ε′	ε″	ε′	ε″	ε′	ε″
20	10	3,35	0,90	3,00	0,50	2,62	0,26
	20	3,43	1,02	3,03	0,58	2,61	0,28
	30	3,58	1,13	3,05	0,65	2,59	0,29
	40	3,80	1,20	3,06	0,73	2,65	0,32
	50	4,05	1,20	-	-	2,66	0,38
30	10	4,05	1,73	3,35	0,87	2,80	0,39
	20	4,21	2,00	3,45	0,96	2,81	0,42
	30	4,50	2,26	3,57	1,14	2,81	0,44
	40	4,85	2,43	3,57	1,25	2,83	0,48
	50	5,35	2,45	-	-	2,80	0,51
50	10	5,65	4,15	4,25	1,90	3,26	0,78
	20	6,30	4,75	4,26	2,11	3,26	0,81
	30	6,85	5,00	4,32	2,44	3,30	0,85
	40	7,68	5,07	4,35	2,82	3,29	0,90
	50	8,62	5,00	-	-	3,28	0,94
70	10	8,35	7,65	4,52	3,86	3,74	1,21
	20	9,59	8,40	5,06	3,91	3,74	1,24
	30	10,8	8,32	5,31	4,45	3,69	1,32
	40	12,1	8,03	5,45	4,93	3,75	1,39
	50	13,1	7,45	-	-	3,72	1,46
90	10	12,5	12,0	6,61	5,72	4,24	1,71
	20	14,6	11,8	6,57	6,42	4,25	1,87
	30	16,5	11,5	6,81	7,18	4,22	1,98
	40	17,9	10,5	7,32	7,72	4,17	2,08
	50	18,5	9,67	-	-	4,18	2,21

Chloroform ($CHCl_3$) – Methanol (CH_4O)

[654)

ϕ_2	t°C	λ = 10 cm		λ = 3,2		λ = 0,408 cm	
		ε'	ε''	ε'	ε''	ε'	ε''
0	10	-	-	4,57	0,959	2,56	0,794
	20	-	-	4,48	0,832	2,58	0,817
	30	-	-	4,39	0,726	2,62	0,838
	40	-	-	4,26	0,632	2,65	0,864
	50	-	-	-	-	-	-
10	10	6,50	1,33	5,14	1,71	2,70	0,96
	20	6,46	1,14	5,21	1,60	2,73	1,00
	30	6,33	0,95	5,20	1,47	2,77	1,03
	40	6,11	0,79	5,10	1,33	2,81	1,07
	50	5,86	0,63	4,90	1,17	2,84	1,10
30	10	7,90	3,65	5,43	2,91	2,90	1,06
	20	8,35	3,65	5,69	2,92	2,93	1,13
	30	8,65	3,40	5,90	2,88	2,96	1,21
	40	8,88	3,02	6,06	2,93	3,00	1,28
	50	8,92	2,53	6,15	2,90	3,03	1,36
50	10	9,17	6,50	5,67	3,94	3,15	1,16
	20	10,1	6,70	5,93	4,20	3,18	1,26
	30	11,0	6,49	6,24	4,46	3,21	1,37
	40	11,9	5,97	6,39	4,57	3,23	1,47
	50	5,20	5,20	6,69	4,55	3,25	1,52
70	10	10,9	9,65	5,93	4,95	3,56	1,38
	20	12,4	9,70	6,39	5,38	3,57	1,51
	30	14,0	9,39	6,80	6,00	3,58	1,64
	40	15,2	8,71	7,38	6,57	3,58	1,77
	50	16,1	7,68	8,04	7,01	3,60	1,89
90	10	14,1	12,5	6,81	6,34	4,16	1,94
	20	16,0	12,4	7,30	7,07	4,16	2,04
	30	17,6	11,9	7,77	7,80	4,16	2,14
	40	18,9	11,0	8,44	8,42	4,16	2,24
	50	19,8	9,47	9,15	8,96	4,16	2,33

Methanol (CH_4O) – Acetonitrile (C_2H_3N)

[654]

X_2	t°C	λ = 10 cm		λ = 3,2 cm		λ = 0,408 cm		ε
		ε′	ε″	ε′	ε″	ε′	ε″	
90	10	21,1	14,3	10,2	9,05	5,16	3,24	35,8
	20	23,2	13,2	11,3	9,90	5,08	3,31	33,9
	30	24,3	11,8	12,2	10,5	5,06	3,42	32,0
	40	24,7	10,2	12,8	10,8	5,19	3,56	30,3
	50	24,6	8,41	13,2	10,7	5,45	3,85	28,5
70	10	29,1	11,7	16,8	12,5	6,18	5,42	36,4
	20	29,4	10,2	17,7	12,1	6,31	5,60	34,6
	30	29,2	8,70	18,4	11,6	6,46	5,80	32,7
	40	28,7	7,35	19,0	11,2	6,60	6,00	31,1
	50	27,9	6,30	19,4	10,7	6,74	6,19	29,4
50	10	33,6	8,32	25,9	11,8	7,53	8,09	37,5
	20	32,6	7,04	24,8	11,3	7,80	8,20	35,9
	30	31,5	5,80	24,1	10,7	8,08	8,35	34,2
	40	30,5	4,75	23,8	9,78	8,35	8,47	32,7
	50	29,4	3,95	24,0	8,88	8,63	10,6	31,3
10	10	37,6	3,16	36,5	8,60	10,3	13,4	38,0
	20	36,1	2,66	34,5	7,65	11,1	13,7	36,5
	30	34,7	2,25	33,1	6,71	11,9	13,9	34,9
	40	33,2	2,02	32,2	5,79	12,7	13,9	33,4
	50	31,7	1,74	31,5	4,85	13,4	13,8	32,0
0	10	38,1	2,69	36,7	7,86	11,1	15,3	38,3
	20	36,6	2,27	35,2	6,77	12,0	15,6	36,8
	30	35,0	1,90	34,0	5,84	12,9	15,5	35,2
	40	-	-	32,9	5,14	13,7	15,3	33,8
	50	-	-	31,7	4,53	14,3	15,0	32,4

Methanol (CH_4O) – Butanone (C_4H_8O)

[646], λ = 10 cm

t°C	ϕ_1 = 10		ϕ_1 = 30		ϕ_1 = 50		ϕ_1 = 70		ϕ_1 = 90	
	ε′	ε″	ε′	ε″	ε′	ε″	ε′	ε″	ε′	ε″
10	20,4	2,97	21,1	5,28	20,7	8,60	19,5	11,6	17,8	13,0
20	19,5	2,52	20,7	4,82	21,0	7,62	20,6	10,4	19,5	12,6
30	18,6	2,11	20,2	4,21	21,1	6,64	21,2	9,19	20,7	11,8
40	17,8	1,77	19,5	3,51	20,9	5,67	21,4	7,98	21,4	10,6
50	16,9	1,49	18,7	2,80	20,2	4,69	21,2	6,75	21,9	9,03

Methanol (CH_4O) – Benzene (C_6H_6)

[655]

X_1	t°C	λ = 10 cm		λ = 3,2 cm		λ = 1,8 cm		λ = 0,845 cm		λ = 0,408 cm		ε
		ε′	ε″	ε′	ε″	ε′	ε″	ε′	ε″	ε′	ε″	
10	10	3,16	0,58	-	-	-	-	2,68	0,216	2,57	0,19	-
	20	3,26	0,54	-	-	-	-	2,69	0,245	2,54	0,21	-
	30	3,33	0,49	-	-	-	-	-	-	2,54	0,22	-
	40	3,33	0,44	-	-	-	-	-	-	2,53	0,24	-
	50	3,25	0,33	-	-	-	-	-	-	2,57	0,27	-
20	10	4,01	1,46	3,22	0,73	3,11	0,54	-	-	2,73	0,35	6,46
	20	4,28	1,52	3,30	0,81	3,14	0,61	-	-	2,74	0,38	6,15
	30	4,50	1,45	3,36	0,89	3,19	0,67	-	-	2,75	0,42	5,77
	40	4,65	1,28	3,44	0,94	3,23	0,73	-	-	2,76	0,46	5,30
	50	4,77	1,04	-	-	-	-	-	-	2,77	0,50	5,06
30	10	4,79	2,60	3,61	1,25	3,43	0,88	3,20	0,65	2,97	0,51	10,1
	20	5,33	2,81	3,71	1,44	3,46	1,02	3,18	0,70	2,94	0,56	9,50
	30	5,85	2,76	3,86	1,67	3,50	1,14	3,19	0,84	2,95	0,59	8,90
	40	6,20	2,55	3,98	1,77	3,55	1,26	3,19	0,87	2,95	0,63	8,30
	50	6,38	2,20	-	-	-	-	3,20	0,95	3,01	0,68	7,87
50	10	6,70	5,42	4,44	2,57	3,95	1,66	3,66	1,16	3,39	0,88	17,7
	20	7,83	5,82	4,59	2,95	4,07	1,91	3,64	1,31	3,41	0,97	16,3
	30	8,76	5,71	4,80	3,30	4,20	2,17	3,66	1,42	3,38	1,00	15,4
	40	9,60	5,38	4,95	3,63	4,33	2,44	3,69	1,45	3,42	1,05	14,5
	50	10,3	4,77	-	-	-	-	3,76	1,41	3,43	1,11	13,9
70	10	9,47	8,84	5,54	4,05	4,86	2,70	4,20	1,78	3,81	1,30	25,0
	20	11,3	9,08	5,74	4,58	4,91	3,05	4,18	1,97	3,85	1,41	23,2
	30	12,3	8,71	6,12	5,22	4,96	3,49	4,18	2,08	3,84	1,49	22,0
	40	13,6	8,13	6,64	5,67	5,02	3,80	4,20	2,18	3,81	1,58	20,5
	50	14,3	7,31	-	-	-	-	4,20	2,25	3,84	1,64	19,5
90	10	13,6	12,6	6,90	5,94	5,71	3,97	5,00	2,48	4,34	1,91	32,0
	20	15,9	12,5	7,26	6,68	5,71	4,47	4,99	2,73	4,32	1,96	30,0
	30	17,2	11,9	7,90	7,46	5,90	4,99	4,96	2,88	4,29	2,06	28,2
	40	18,5	10,7	8,37	7,91	5,98	5,59	5,07	3,25	4,23	2,21	26,8
	50	19,5	9,46	-	-	-	-	5,20	3,26	4,24	2,54	25,5
100	10	16,6	14,9	7,45	7,28	6,13	4,58	5,34	3,00	4,59	2,26	35,5
	20	18,9	14,3	8,03	8,08	6,40	5,29	5,32	3,19	4,51	2,34	33,6
	30	20,8	13,5	8,80	8,66	6,49	5,85	5,33	3,52	4,55	2,47	31,4
	40	22,2	12,2	8,63	9,36	6,56	6,25	5,35	3,69	4,72	2,52	29,9
	50	23,0	10,0	-	-	-	-	5,55	4,23	4,75	2,57	27,9

1,2–Dichloroethane ($C_2H_4Cl_2$) – Tetrabutylammonium bromide ($C_{16}H_{36}NBr$)

[656], $\nu = 0{,}4–3$ GHz

t°C	N	ϵ	ϵ_∞	a	$10^{-12}\tau$
35	0,4	14,15	8,8	0,06	194
	0,3	13,85	9,25	0,03	186
	0,2	13,1	9,4	0,00	180
25	0,4	14,5	9,25	0,07	211
	0,3	14,45	9,6	0,07	217
	0,2	14,2	9,7	0,05	235
	0,1	13,1	9,9	0,06	225
	0,05	12,0	10,0	0,06	264
15	0,4	15,3	9,65	0,06	243
	0,3	14,9	9,8	0,07	247
	0,2	14,55	10,4	0,05	234
1	0,4	16,1	10,2	0,05	327
	0,3	15,8	10,5	0,05	306

Acetic acid ($C_2H_4O_2$) – Chlorobenzene (C_6H_5Cl)

[657]

λ, cm	ϕ_1	$t = 20$°C		$t = 30$°C		$t = 40$°C		$t = 50$°C	
		ϵ'	ϵ''	ϵ'	ϵ''	ϵ'	ϵ''	ϵ'	ϵ''
40	25	5,12	0,21	5,00	0,18	4,88	0,16	4,75	0,14
30		5,08	0,27	4,98	0,24	4,84	0,21	4,74	0,18
20		5,02	0,40	4,92	0,34	4,80	0,30	4,70	0,27
6,4		4,75	0,77	4,70	0,68	4,62	0,60	4,53	0,55
4,4		4,55	0,94	4,54	0,84	4,50	0,75	4,43	0,68
2,1		3,80	1,14	3,90	1,09	3,96	1,02	4,00	0,95
1,2		3,25	1,03	3,34	1,05	3,42	1,04	3,48	1,03
0,75		2,90	0,80	2,95	0,94	3,00	0,88	3,06	0,91
40	50	4,95	0,29	4,90	0,26	4,85	0,23	4,80	0,20
30		4,88	0,37	4,84	0,34	4,80	0,30	4,77	0,26
20		4,79	0,50	4,76	0,45	4,73	0,41	4,70	0,36
6,4		4,33	0,75	4,38	0,70	4,32	0,66	4,32	0,62
4,4		4,11	0,83	4,12	0,78	4,13	0,74	4,14	0,71
2,1		3,63	0,95	3,70	0,91	3,75	0,87	3,81	0,83
1,2		3,24	0,86	3,33	0,88	3,40	0,88	3,46	0,87
0,75		2,94	0,68	2,98	0,73	3,05	0,77	3,12	0,82
40	75	5,28	0,50	5,31	0,44	5,34	0,38	5,38	0,33
30		5,14	0,64	5,18	0,57	5,24	0,51	5,30	0,43
20		4,90	0,84	4,98	0,77	5,06	0,71	5,16	0,64
6,4		4,15	0,87	4,24	0,90	4,34	0,93	4,46	0,97
4,4		3,94	0,83	4,02	0,87	4,10	0,91	4,18	0,95
2,1		3,55	0,75	3,62	0,80	3,70	0,84	3,78	0,88
1,2		3,26	0,70	2,34	0,73	3,42	0,78	3,90	0,82
0,75		3,02	0,61	3,06	0,65	3,13	0,70	3,20	0,75

ϕ_1	t°C	ϵ'	ϵ_2	$\epsilon_{\infty 2}$	a	$10^{-12}\tau$	$10^{-12}\tau$	C_1
25	20	5,15	4,90	2,56	0,02	64,0	10,0	0,09
	50	4,78	4,55	2,51	0,02	40,0	7,2	0,10
50	20	5,03	4,43	2,56	0,06	76,0	9,3	0,24
	50	4,85	4,23	2,53	0,05	46,0	6,6	0,26
75	20	5,44	4,24	2,58	0,16	92,0	9,2	0,42
	50	5,44	4,20	2,54	0,13	53,0	6,4	0,42

Acetic acid ($C_2H_4O_2$) – Benzene (C_6H_6)

[658]

λ, cm	ϕ_1	$t = 20°C$		$t = 30°C$		$t = 40°C$		$t = 50°C$	
		ϵ'	ϵ''	ϵ'	ϵ''	ϵ'	ϵ''	ϵ'	ϵ''
20	20	2,48	0,044	2,46	0,040	2,44	0,035	2,43	0,030
10		2,45	0,062	2,44	0,057	2,43	0,052	2,42	0,046
6,4		2,42	0,074	2,41	0,069	2,40	0,064	2,40	0,058
4,4		2,40	0,078	2,39	0,074	2,38	0,069	2,38	0,064
2,1		2,36	0,080	2,34	0,076	2,35	0,072	2,34	0,068
1,2		2,33	0,074	2,32	0,072	2,31	0,070	2,31	0,067
40	60	3,52	0,25	3,35	0,23	3,63	0,21	3,70	0,20
30		3,46	0,28	3,51	0,27	3,60	0,26	3,66	0,24
20		3,35	0,34	3,42	0,34	3,49	0,33	3,56	0,32
10		3,15	0,35	3,20	0,37	3,26	0,38	3,33	0,40
6,4		3,04	0,33	3,09	0,34	3,14	0,39	3,20	0,40
4,4		2,96	0,30	3,00	0,31	3,05	0,35	3,10	0,37
2,1		2,85	0,28	2,89	0,30	2,93	0,34	2,97	0,34
1,2		2,74	0,26	2,77	0,28	2,81	0,32	2,86	0,33
0,75		2,64	0,23	2,67	0,25	2,70	0,28	2,74	0,31
40	90	5,05	0,67	5,18	0,59	5,32	0,52	5,47	0,47
30		4,87	0,81	5,02	0,71	5,19	0,65	5,36	0,60
20		4,54	0,93	4,70	0,92	4,88	0,89	5,09	0,85
10		4,04	0,86	4,19	0,90	4,35	0,95	4,54	1,00
6,4		3,80	0,78	3,91	0,84	4,04	0,91	4,20	0,98
4,4		3,60	0,73	3,70	0,79	3,83	0,86	3,97	0,92
2,1		3,31	0,62	3,39	0,66	3,47	0,72	3,57	0,77
1,2		3,15	0,52	3,21	0,56	3,29	0,62	3,37	0,68
0,75		3,00	0,45	3,05	0,50	3,12	0,54	3,20	0,59

Ethanol (C_2H_6O) – Tetrahydrofuran (C_4H_8O)

[645], $t = 25°C$

ν, GHz	$X_1 = 22,6$			$X_1 = 22,6$	
	ϵ'	ϵ''		ϵ'	ϵ''
0,207	9,37	0,13	1,5	9,28	0,60
0,304	9,30	0,17	3,00	9,00	1,06
0,450	9,30	0,42	4,00	8,78	1,25
0,600	9,31	0,28	8,52	7,81	1,48
0,900	9,26	0.33	-	-	-

Acetone (C_3H_6O) – Hexane (C_6H_{14})

[660], $t = 25°C$

ϕ_1	λ = 3,2 cm		λ = 7,86 mm		λ = 4,08 mm		ϵ
	ϵ'	ϵ''	ϵ'	ϵ''	ϵ'	ϵ''	
100	20,02	3,17	14,35	8,70	8,83	8,57	20,68
90	18,02	2,84	12,70	7,50	7,95	7,75	18,31
70	13,60	2,12	9,65	5,25	6,45	5,75	13,82
50	9,09	1,31	7,20	3,29	4,75	3,50	9,54
30	5,74	0,59	4,62	1,55	3,5	1,85	5,91
10	3,02	-	2,91	0,29	2,4	0,46	3,04

Propan-1-ol (C_3H_8O) – 1,4-Dioxan ($C_4H_8O_2$)

[661], $t = 25°C$

ν, GHz	X_1 = 58,1		X_1 = 79,0		X_1 = 85,6		X_1 = 94,0		X_1 = 100,0	
	ε′	ε″	ε′	ε″	ε′	ε″	ε′	ε″	ε′	ε″
0,356	8,19	0,85	12,63	2,79	13,83	4,02	14,90	6,12	14,41	7,97
0,515	8,20	1,13	11,89	3,67	12,63	4,98	12,63	6,94	11,64	8,43
0,697	7,94	1,51	10,96	4,28	10,54	5,64	9,79	6,85	8,92	7,71
1,00	7,51	1,93	9,20	4,63	9,15	5,53	8,23	6,36	7,06	6,39
1,50	6,81	2,21	7,47	4,43	7,19	5,03	6,21	5,21	5,53	4,93
2,11	6,22	2,32	6,57	4,17	6,05	4,38	5,52	4,43	4,67	3,92

Shows graphically dependence of τ, τ/η, ΔH, ΔS on X_1

Propan-1-ol (C_3H_8O) – Pyridine (C_5H_5N)

[661], $t = 25°C$

ν, GHz	X_1 = 57,7		X_1 = 71,9		X_1 = 85,5		X_1 = 90,5		X_1 = 91,7	
	ε′	ε″	ε′	ε″	ε′	ε″	ε′	ε″	ε′	ε″
0,356	16,75	1,76	17,44	2,83	17,48	4,78	16,52	6,06	16,86	6,06
0,515	16,27	2,29	16,95	3,77	15,51	6,08	14,75	6,84	14,51	7,12
0,697	15,93	3,10	15,99	4,72	13,58	6,77	12,69	7,25	12,66	7,60
1,00	14,84	4,14	14,24	5,69	11,30	6,76	9,94	6,85	9,58	6,82
1,50	13,18	4,69	12,32	6,10	8,73	6,16	7,93	5,93	7,94	5,96
2,11	12,01	5,22	10,27	5,87	7,23		6,38	4,99	6,05	4,86

ν, GHz	X_1 = 94,7		X_1 = 96,0		X_1 = 98,0		X_1 = 100,0.	
	ε′	ε″	ε′	ε″	ε′	ε″	ε′	ε″
0,356	16,07	6,82	15,81	7,14	15,37	7,63	14,41	7,97
0,515	13,38	7,56	13,12	7,82	12,37	8,14	11,64	8,43
0,697	10,93	7,76	10,62	7,85	9,71	7,72	8,92	7,71
1,00	8,55	6,78	9,35	7,16	7,57	6,55	7,06	6,39
1,50	6,64	5,57	6,19	5,30	5,97	5,15	5,55	4,93
2,11	5,48	4,55	5,34	4,41	4,97	4,12	4,67	3,92

Propan-1-ol (C_3H_8O) – Chlorobenzene (C_6H_5Cl)

[661], $t = 25°C$

ν, GHz	X_1 = 69,0		X_1 = 82,2		X_1 = 88,7		X_1 = 92,9		X_1 = 100,0	
	ε′	ε″	ε′	ε″	ε′	ε″	ε′	ε″	ε′	ε″
0,356	11,76	3,91	13,63	5,42	14,17	6,31	13,92	6,95	14,41	7,97
0,515	10,43	4,47	10,87	5,92	11,67	6,96	11,23	7,36	11,64	8,43
0,697	8,96	4,45	9,45	5,90	9,64	6,66	8,53	6,48	8,92	7,71
1,00	7,59	4,07	7,82	5,27	7,36	5,67	7,30	5,81	7,06	6,39
1,50	6,32	3,33	6,12	4,13	5,41	4,38	5,90	4,75	5,55	4,93
2,11	5,71	2,82	5,25	3,26	5,27	3,55	5,00	3,53	4,67	3,92

Graphs show relationships of τ, τ/η, ΔH, ΔS to X_1

Propan–1–ol (C_3H_8O) – Benzene (C_6H_6)

[661], $t = 25°C$

ν, GHz	$X_1 = 63,3$		$X_1 = 74,1$		$X_1 = 85,7$		$X_1 = 89,9$		$X_1 = 100,0$	
	ϵ'	ϵ''	ϵ'	ϵ''	ϵ'	ϵ''	ϵ'	ϵ''	ϵ'	ϵ''
0,356	9,61	3,35	11,30	4,83	12,75	6,10	13,33	6,79	14,41	7,97
0,515	8,43	3,82	9,56	5,33	10,76	6,72	11,11	7,22	11,64	8,43
0,697	7,47	4,07	8,08	5,25	8,98	6,59	8,48	6,54	8,92	7,71
1,00	6,07	3,67	6,45	4,63	6,75	5,42	6,79	5,70	7,06	6,39
1,50	4,87	3,07	5,16	3,77	5,55	4,37	5,36	4,48	5,55	4,93
2,11	4,24	2,47	4,45	3,07	4,52	3,42	4,41	3,55	4,67	3,92

Gives relationships of τ, τ/η, ΔH, ΔS to X_1

Propan–2–ol (C_3H_8O) – 1,4–Dioxan ($C_4H_8O_2$)

[663], $t = 25°C$

ν, MHz	$X_1 = 60,1$		$X_1 = 76,9$		$X_1 = 85,6$		$X_1 = 91,9$		$X_1 = 100$	
	ϵ'	ϵ''	ϵ'	ϵ''	ϵ'	ϵ''	ϵ'	ϵ''	ϵ'	ϵ''
0,356	7,85	0,91	11,15	2,39	12,73	3,80	13,38	5,44	13,77	7,58
0,515	7,59	1,15	10,43	3,08	11,47	4,67	11,50	6,05	10,28	7,81
0,697	7,33	1,43	9,62	3,72	9,85	5,24	9,66	6,32	8,44	7,31
1,00	6,90	1,76	8,18	3,83	8,13	5,08	7,18	5,56	6,18	5,76
1,50	6,04	2,03	6,60	3,77	6,24	4,52	5,54	4,64	4,57	4,23
2,11	5,50	2,21	5,66	3,54	5,23	3,83	4,70	3,76	4,07	3,36

Graphs show dependence of τ and τ/η on X_1

Propan–2–ol (C_3H_8O) – Pyridine (C_5H_5N)

[663], $t = 25°C$

ν, GHz	$X_1 = 55,0$		$X_1 = 75,2$		$X_1 = 82,2$		$X_1 = 91,2$		$X_1 = 100$	
	ϵ'	ϵ''	ϵ'	ϵ''	ϵ'	ϵ''	ϵ'	ϵ''	ϵ'	ϵ''
0,356	15,49	1,44	16,07	3,09	16,00	4,09	15,08	5,86	13,77	7,58
0,515	15,19	2,26	15,14	4,04	14,64	5,13	13,11	6,59	10,28	7,81
0,697	14,85	2,62	14,27	4,58	13,13	5,87	10,44	6,71	8,44	7,31
1,00	14,03	3,40	12,24	5,46	10,94	6,01	8,62	6,28	6,18	5,76
1,50	12,65	4,19	10,18	5,46	8,59	5,75	6,60	5,23	4,57	4,23
2,11	11,50	4,98	8,55	5,39	7,39	5,24	5,44	4,55	4,07	3,36

Shows dependence of τ and τ/η on X_1

Propan–2–ol (C_3H_8O) – Chlorobenzene (C_6H_5Cl)

[663], $t = 25°C$

ν, GHz	$X_1 = 62,5$		$X_1 = 81,4$		$X_1 = 89,4$		$X_1 = 93,7$		$X_1 = 100$	
	ϵ'	ϵ''	ϵ'	ϵ''	ϵ'	ϵ''	ϵ'	ϵ''	ϵ'	ϵ''
0,356	9,97	2,16	12,21	4,48	13,11	5,79	12,86	6,46	13,77	7,58
0,515	9,18	2,76	10,43	5,10	10,85	6,27	10,91	6,90	10,28	7,81
0,697	8,44	2,97	8,73	5,18	8,94	6,24	8,30	6,53	8,44	7,31
1,00	7,45	2,97	7,27	4,73	6,58	5,26	6,72	5,57	6,18	5,76
1,50	6,40	2,75	5,88	3,85	5,43	4,27	5,18	4,26	4,57	4,23
2,11	5,90	2,48	5,16	3,29	4,69	3,35	4,51	3,36	4,07	3,36

Graphs show dependence of τ and τ/η on X_1

Propan–2–ol (C_3H_8O) – Benzene (C_6H_6)

[663], $t = 25°C$

ν, GHz	$X_1 = 61,8$		$X_1 = 74,2$		$X_1 = 82,7$		$X_1 = 90,0$		$X_1 = 100$	
	ϵ'	ϵ''	ϵ'	ϵ''	ϵ'	ϵ''	ϵ'	ϵ''	ϵ'	ϵ''
0,356	8,76	2,34	10,58	3,76	11,96	5,01	12,56	6,01	13,77	7,58
0,515	7,80	2,83	9,19	4,40	10,05	5,51	10,36	6,44	10,28	7,81
0,697	6,99	3,24	7,76	4,66	8,25	5,60	8,46	6,36	8,44	7,31
1,00	5,72	3,21	6,25	4,22	6,39	4,87	6,85	5,49	6,18	5,76
1,50	4,74	2,70	4,85	3,47	5,01	3,90	4,75	4,16	4,57	4,23
2,11	4,19	2,35	4,23	2,93	4,10	2,95	4,25	3,39	4,07	3,36

Graphs show dependence of τ and τ/η on X_1

Butanone (C_4H_8O) – Heptane (C_7H_{16})

[664]

ϕ_1		$\lambda = 32$ mm		$\lambda = 7,96$ mm		$\lambda = 4,08$ mm	
	t°C	ϵ'	ϵ''	ϵ'	ϵ''	ϵ'	ϵ''
100	20	17,10	4,55	8,97	7,68	5,25	5,78
	40	15,78	3,51	9,57	7,05	5,75	5,86
	60	14,70	2,61	10,18	6,30	6,30	5,94
90	20	15,5	4,05	8,23	6,79	4,86	5,23
	40	14,1	3,03	8,83	6,38	5,33	5,35
	60	13,4	2,32	9,29	5,65	5,91	5,24
70	20	11,6	2,93	6,67	4,97	4,29	3,80
	40	10,8	2,24	7,12	4,50	4,64	3,91
	60	10,1	1,65	7,32	3,95	4,98	3,90
50	20	8,07	1,81	5,01	3,02	3,76	2,50
	40	7,54	1,35	5,16	2,69	3,95	2,54
	60	6,99	1,00	5,36	2,30	4,06	2,42
30	20	5,20	1,87	3,76	1,52	3,18	1,35
	40	4,87	0,64	3,80	1,36	3,23	1,32
	60	4,56	0,49	3,85	1,16	3,26	1,25
10	20	2,82	0,22	2,57	0,39	2,43	0,45
	40	1,73	0,15	2,51	0,33	2,42	0,42
	60	2,62	0,08	2,48	0,25	2,40	0,31

Tetrahydrofuran (C_4H_8O) – Benzene (C_6H_6)
[642], $\nu = 0{,}137–34$ GHz, $t = 25°C$

ν, GHz	Tetrahydrofuran		$N = 4{,}8$		$N = 1{,}8$	
	ϵ'	ϵ''	ϵ'	ϵ''	ϵ'	ϵ''
137	7,39	0,008	4,03	0,00	-	-
	7,34	0,011	4,01	0,003	2,90	0,005
260	7,38	0,024	4,04	0,006	2,90	0,013
	7,36	0,024	4,02	0,005	2,90	0,003
375	7,36	0,037	4,00	0,015	2,91	0,007
383	7,37	0,036	4,03	0,015	2,91	0,008
	7,34	0,038	4,01	0,014	2,90	0,007
625	7,36	0,058	4,00	0,023	2,91	0,008
1060	7,36	0,090	4,00	0,038	2,91	0,016
1666	7,35	0,131	4,01	0,047	2,92	0,019
2825	7,34	0,227	4,01	0,085	2,92	0,034
9347	7,20	0,73	3,94	0,265	2,86	0,094
34720	6,31	2,20	3,72	0,781	2,78	0,279

1,4–Dioxan ($C_4H_8O_2$) – Butan–1–ol ($C_4H_{10}O$)
[665], $\nu = 0{,}1$ MHz, $t = 20°C$

X_1	100,0	95,7	91,1	86,0	82,8	77,7	64,2	54,0
10^{-3} tan δ	11,2	16,0	16,4	19,0	20,0	17,0	14,4	12,6

1,4–Dioxan ($C_4H_8O_2$) – Hexan–1–ol ($C_6H_{14}O$)
[665], $\nu = 0{,}1$ MHz, $t = 20°C$

X_1	100,0	94,3	84,2	72,3	46,5
10^{-3} tan δ	7,0	15,6	15,0	12,4	1,8

Butan–1–ol ($C_4H_{10}O$) – Pyridine (C_5H_5N)
[661], $t = 25°C$

ν, MHz	$X_1 = 56{,}8$		$X_1 = 69{,}9$		$X_1 = 80{,}7$		$X_1 = 89{,}4$		$X_1 = 100$	
	ϵ'	ϵ''	ϵ'	ϵ''	ϵ'	ϵ''	ϵ'	ϵ''	ϵ'	ϵ''
0,356	14,59	1,77	14,52	2,90	13,96	4,18	12,62	5,69	9,44	6,84
0,515	14,10	2,44	13,67	3,74	12,56	5,00	10,35	5,91	7,35	6,17
0,697	13,51	3,10	12,70	4,40	11,14	5,43	8,88	5,93	5,85	5,24
1,00	12,53	3,79	10,87	4,84	9,15	5,51	7,17	5,11	4,79	3,99
1,50	10,91	4,33	9,00	4,89	7,18	4,69	5,67	4,20	4,01	2,95
2,11	9,90	4,64	8,08	4,67	6,20	4,26	5,00	3,48	3,70	2,11

Gives graph relating τ and X_1

Butan–1–ol ($C_4H_{10}O$) – Benzene (C_6H_6)

[661], $t = 25°C$

ν, GHz	X_1 =		X_1 =67,0		X_1 = 80,6		X_1 = 89,9		X_1 = 100,0	
	ε′	ε″	ε′	ε″	ε′	ε″	ε′	ε″	ε′	ε″
0,356	7,77	3,15	-	-	9,02	5,27	9,40	6,09	9,44	6,84
0,515	6,57	3,34	6,79	3,89	7,20	5,04	6,78	5,61	7,35	6,17
0,697	5,67	3,20	5,62	3,39	5,78	4,47	5,86	4,88	5,85	5,24
1,00	4,68	2,71	4,73	3,00	4,80	3,46	4,82	3,81	4,79	3,99
1,50	3,98	2,14	3,99	2,28	4,01	2,64	3,99	2,78	4,01	2,95
2,11	3,65	1,70	3,64	1,85	3,64	2,10	3,67	2,22	3,70	2,11

Gives graph of τ against X_1

X_1	49,8	59,5	67,8	78,7	81,0	89,3	94,7	96,6	100,0
10^{-3} tan δ	6,0	15,0	18,0	21,5	22,0	21,6	20,0	16,6	11,2

Isobutanol ($C_4H_{10}O$) – Pyridine (C_5H_5N)

[663], $t = 25°C$

ν, GHz	X_1 = 61,9		X_1 = 70,1		X_1 = 80,1		X_1 = 90,4		X_1 = 100,0	
	ε′	ε″	ε′	ε″	ε′	ε″	ε′	ε″	ε′	ε″
0,356	14,21	2,12	14,12	2,98	13,54	3,98	11,49	6,03	8,10	6,74
0,515	13,64	2,98	13,11	3,76	12,08	5,03	9,38	6,09	6,31	5,80
0,697	12,90	3,68	12,31	4,44	10,62	5,46	7,74	5,56	5,06	4,71
1,00	11,76	4,13	10,54	4,78	8,73	5,14	6,36	4,72	4,24	3,47
1,50	10,10	4,52	8,11	4,84	6,99	4,50	5,07	3,72	3,60	2,52
2,11	8,88	4,56	7,68	4,59	5,90	4,11	4,53	3,11	3,43	1,92

Gives graph of τ against X_1

Isobutanol ($C_4H_{10}O$) – Benzene (C_6H_6)

[663], $t = 25°C$

ν, MHz	X_1 = 60,9		X_1 = 66,4		X_1 = 79,9		X_1 = 88,8		X_1 = 100	
	ε′	ε″	ε′	ε″	ε′	ε″	ε′	ε″	ε′	ε″
0,356	7,42	2,96	7,86	3,65	8,43	5,14	8,36	6,06	8,10	6,74
0,515	6,33	3,10	6,51	3,71	6,67	4,83	6,69	5,35	6,31	5,80
0,697	5,30	2,97	5,44	3,31	5,39	4,21	5,47	4,55	5,06	4,71
1,00	4,52	2,46	4,49	2,78	4,46	3,26	4,42	3,47	4,24	3,37
1,50	3,85	2,01	3,89	2,19	3,81	2,51	3,71	2,56	3,60	2,52
2,11	3,52	1,62	3,53	1,74	3,45	1,96	3,44	1,96	3,43	1,92

Gives graph of τ against X_1

Butan–2–ol ($C_4H_{10}O$) – **Pyridine** (C_5H_5N)

[663], $t = 25°C$

ν, GHz	$X_1 = 61,1$		$X_1 = 69,9$		$X_1 = 80,9$		$X_1 = 90,0$		$X_1 = 100,0$	
	ϵ'	ϵ''	ϵ'	ϵ''	ϵ'	ϵ''	ϵ'	ϵ''	ϵ'	ϵ''
0,356	13,49	1,72	13,52	2,51	12,99	3,76	11,56	5,31	9,23	6,44
0,515	13,03	2,42	12,80	3,10	11,71	4,51	9,60	5,55	7,20	5,90
0,697	12,59	3,14	11,86	3,80	10,28	4,92	8,06	5,37	5,77	4,98
1,00	11,54	3,63	10,10	4,25	8,57	4,70	6,56	4,54	4,67	3,83
1,50	9,44	4,43	8,79	4,38	6,89	4,33	5,25	3,71	3,97	2,88
2,11	8,85	4,48	7,72	4,32	5,89	3,93	4,59	3,15	3,62	2,21

Gives graph of τ against X_1

Butan–2–ol ($C_4H_{10}O$) – **Benzene** (C_6H_6)

[633], $t = 25°C$

ν, GHz	$X_1 = 53,3$		$X_1 = 67,5$		$X_1 = 80,9$		$X_1 = 89,7$		$X_1 = 100,0$	
	ϵ'	ϵ''	ϵ'	ϵ''	ϵ'	ϵ''	ϵ'	ϵ''	ϵ'	ϵ''
0,356	6,04	1,31	7,86	2,52	8,95	4,25	9,23	5,34	9,23	6,44
0,515	5,78	1,51	6,87	2,74	7,42	4,41	7,37	5,36	7,20	5,90
0,697	5,21	1,62	5,93	2,96	5,96	4,02	5,85	4,47	5,77	4,98
1,00	4,80	1,60	4,99	2,57	5,01	3,42	4,65	3,64	4,67	3,83
1,50	4,09	1,53	3,98	2,06	4,10	2,71	3,92	2,80	3,97	2,88
2,11	3,77	1,35	3,73	1,83	3,68	2,10	3,52	2,20	3,62	2,21

Gives graph of τ against X_1

3–Pentanone ($C_5H_{10}O$) – **Tetrabutylammonium bromide** ($C_{16}H_{36}NBr$)

[656], $\nu = 0,4–3$ GHz

$t°C$	N	ϵ	ϵ_∞	a	$10\,\tau$
0,4	25	21,1	13,7	0,17	179
	35	19,8	13,85	0,08	162
	45	19,2	12,9	0,12	146
0,3	15	21,9	15,1	0,14	191
	25	20,9	14,4	0,13	177
	35	19,9	14,0	0,14	158
	45	19,0	13,2	0,16	141
0,2	15	20,9	15,7	0,10	190
	25	20,25	15,2	0,11	163
	45	18,3	14,3	0,07	138
0,1	25	19,2	15,8	0,12	210
0,05	25	18,2	16,1	0,19	161

Chlorobenzene (C_6H_5Cl) – Iodobenzene (C_6H_5I)

[669], $\phi = 1:1$

t°C	λ = 6,32 cm		λ = 4,01 cm		λ = 3,2 cm		ε
	ε′	ε″	ε′	ε″	ε′	ε″	
30	4,69	0,98	4,25	1,21	4,00	1,26	5,19
20	4,65	1,09	4,20	1,26	3,94	1,28	5,29
10	4,57	1,19	4,10	1,31	3,82	1,31	5,38
0	4,49	1,29	4,00	1,36	3,72	1,31	5,48
−10	4,41	1,36	3,90	1,37	3,61	1,29	5,59
−20	4,34	1,41	3,77	1,35	3,50	1,25	5,70
−30	4,23	1,45	3,64	1,32	3,39	1,18	5,82
−40	4,12	1,49	3,50	1,29	3,28	1,12	5,95

t°C	λ = 2,14 cm		λ = 1,18 cm		λ = 0,75 cm	
	ε′	ε″	ε′	ε″	ε′	ε″
30	3,55	1,23	3,00	0,96	2,79	0,73
20	3,48	1,21	2,98	0,91	2,79	0,69
10	3,39	1,19	2,96	0,86	2,78	0,65
0	3,32	1,14	2,94	0,81	2,78	0,61
−10	3,24	1,10	2,92	0,76	2,78	0,59
−20	3,17	1,03	2,90	0,70	2,78	0,55
−30	3,10	0,95	2,88	0,64	2,78	0,52
−40	3,04	0,89	2,86	0,58	2,78	0,49

Chlorobenzene (C_6H_5Cl) – Dibutyl phthalate ($C_{16}H_{22}O_4$)

[670], $\nu = 0{,}002–40$ GHz

ϕ	t°C	ϵ_1	$\epsilon_{\infty 3}$	C_1	C_2	$10^{-12}\tau_1$	$10^{-12}\tau_2$	a_3
1:1	20	6,33	2,39	0,55	0,35	85,0	12,5	0,15
	60	5,52	2,33	0,52	0,39	36,5	8,3	0,05
1:3	20	6,03	2,42	0,33	0,50	55,0	12,5	0,10
	60	5,23	2,35	0,30	0,63	28,0	8,3	0,04

Benzene (C_6H_6) – Hexan–1–ol ($C_6H_{14}O$)

[665], $\nu = 0{,}1$ MHz, $t = 20$°C

X_1	10,0	93,6	81,5	69,8	43,5
10^{-3} tan δ	7,0	11,2	15,0	11,8	2,3

Benzene (C_6H_6) – Tributylammonium picrate ($C_{18}H_{30}N_4O_7$)

[671], ν = 0,2–3,0 GHz

N	t°C	ϵ	ϵ_∞	$10^{-12}\tau$	N	t°C	ϵ	ϵ_∞	$10^{-12}\tau$
0,025	7	2,79	2,32	274	0,3	25	8,00	2,70	441
	25	2,71	2,30	223	0,33	7	10,0	2,80	679
	38	2,68	2,28	170		25	8,85	2,75	453
0,05	7	3,25	2,34	303		38	8,60	2,70	392
	25	3,15	2,32	238		50	8,25	2,70	331
	38	3,22	2,30	188	0,4	25	10,3	2,85	513
0,066	7	3,60	2,40	293	0,484	7	14,75	2,80	878
	25	3,50	2,40	246		25	13,0	2,85	594
	38	3,50	2,40	205		38	13,8	2,90	517
0,1	7	4,30	2,45	346		50	11,8	2,80	376
	25	4,26	2,48	286	0,55	25	14,0	2,85	625
	38	4,25	2,50	226	0,6	25	16,3	3,10	648
	50	4,12	2,36	173		38	15,2	3,00	534
0,133	7	4,75	2,50	367		50	14,5	3,00	442
	25	4,70	2,50	295	0,65	25	17,1	3,10	720
	38	4,66	2,50	232		38	16,0	3,05	540
0,166	25	5,50	2,68	315	0,7	25	18,1	3,15	744
	38	5,22	2,55	313	0,8	25	21,4	3,50	849
	50	5,10	2,55	260		38	20,6	3,35	669
0,2	25	6,25	2,70	364		50	19,4	3,05	525
	38	6,00	2,58	325		-	-	-	-

[672]

ν, GHz	N = 0,02		N = 0,05		N = 0,1	
	ϵ'	ϵ''	ϵ'	ϵ''	ϵ'	ϵ''
0,0082	2,61	0,0	3,12	0,0	4,03	0,0
0,0266	2,62	0,0	3,13	0,04	4,04	0,1
0,1062	2,61	0,041	3,10	0,12	3,97	0,32
0,140	2,60	0,064	3,07	0,18	3,90	0,444
0,265	2,56	0,112	2,97	0,294	3,64	0,672
0,375	2,54	0,138	2,91	0,349	3,44	0,733
0,385	2,53	0,125	2,88	0,342	3,38	0,755
0,625	2,47	0,158	2,73	0,385	3,07	0,743
1,060	2,39	0,149	2,54	0,338	2,73	0,623
1,666	2,35	0,118	2,43	0,266	2,56	0,479
2,825	2,31	0,085	2,36	0,180	2,44	0,331
9,334	2,28	0,03	2,30	0,07	2,34	0,12
37,72	2,31	0,01	2,34	0,03	2,35	0,05

ν, GHz	$N = 0,3$		$N = 0,6$		$N = 1,0$		$N = 1,4$	
	ϵ'	ϵ''	ϵ'	ϵ''	ϵ'	ϵ''	ϵ'	ϵ''
0,0082	8,47	0,993	16,4	6,50	27,8	16,2	37,5	22,1
0,0266	8,36	0,750	16,0	3,46	26,8	9,07	34,6	15,3
0,1062	7,80	1,80	13,6	5,43	18,7	12,4	16,8	17,1
0,140	7,37	2,21	12,1	6,42	15,1	12,9	12,5	15,9
0,265	5,99	2,79	8,31	6,48	8,87	10,2	7,17	10,3
0,375	5,08	2,71	6,42	5,52	6,65	7,89	5,74	7,46
0,385	5,01	2,70	6,30	5,63	6,53	8,08	5,62	7,59
0,625	3,95	2,26	4,58	4,08	4,80	5,40	4,58	4,92
1,060	3,25	1,62	3,65	2,73	3,92	3,48	4,02	3,14
1,666	2,92	1,18	3,24	1,90	3,55	2,38	3,75	2,17
2,825	2,71	0,778	2,99	1,24	3,30	1,54	3,52	1,43
9,334	2,49	0,295	2,70	0,482	2,96	0,623	3,21	0,620
34,72	2,46	0,118	2,63	0,204	2,84	0,285	3,08	0,318

Cyclohexanone ($C_6H_{10}O$) – Cyclohexanol ($C_6H_{12}O$)

[674]

ϕ_1	t°C	$\lambda = 0,42$ cm		$\lambda = 0,85$ cm		$\lambda = 3,2$ cm		$\lambda = 10$ cm	
		ϵ'	ϵ''	ϵ'	ϵ''	ϵ'	ϵ''	ϵ'	ϵ''
0,2	30	3,53	2,11	4,37	3,33	9,55	4,59	12,96	2,95
	40	3,55	2,24	4,58	3,44	9,73	4,34	12,79	2,55
	50	3,58	2,30	4,79	3,56	9,88	4,07	12,55	2,21
	60	3,61	2,47	5,01	3,63	9,96	3,74	12,12	1,83
	70	3,64	2,57	5,27	3,66	10,02	3,40	11,79	1,57
	80	3,67	2,67	5,50	3,66	9,92	3,10	11,44	1,32
	90	3,70	2,74	5,78	3,65	9,82	2,80	11,08	1,13
	100			6,05	3,61	9,70	2,53	10,70	0,97
	110			6,32	3,54	9,50	2,29	10,34	0,84
	120			6,59	3,45	9,33	2,04	9,96	0,76
	130			6,91	3,93	9,15	1,82	9,60	0,67
	140			7,22	3,19	9,00	1,63	9,24	0,60
0,4	30	3,19	1,57	4,00	2,47	7,40	3,64	10,38	3,23
	40	3,23	1,69	4,15	2,58	7,60	3,60	10,50	3,03
	50	3,35	1,79	4,30	2,71	7,85	3,52	10,63	2,73
	60	3,36	1,89	4,45	2,82	8,05	3,39	10,55	2,35
	70	3,38	1,96	4,60	2,90	8,20	3,23	10,45	1,93
	80	3,39	2,05	4,75	2,95	8,32	3,04	10,26	1,61
	90	3,40	2,12	4,90	2,99	3,36	2,80	10,00	1,34
	100			5,05	2,97	8,41	2,55	9,70	1,10
	110			5,20	2,88	8,37	2,29	9,48	0,97
	120			5,36	2,70	8,28	2,05	9,15	0,79
	130			5,50	2,77	8,15	1,81	8,81	0,68
	140			5,66	2,59	8,05	1,56	8,56	0,60
0,6	30	3,08	1,08	3,60	1,49	5,57	2,34	7,70	3,01
	40	3,18	1,22	3,71	1,57	5,77	2,47	8,15	3,10
	50	3,25	1,34	3,80	1,70	5,98	2,59	8,44	2,97
	60	3,29	1,46	3,89	1,82	6,22	2,69	8,76	2,77
	70	3,32	1,57	4,02	1,92	6,45	2,73	8,96	2,40

(continuation)

ϕ_1	t°C	λ = 0,42 cm		λ = 0,85 cm		λ = 3,2 cm		λ = 10 cm	
		ε′	ε″	ε′	ε″	ε′	ε″	ε′	ε″
	80	3,34	1,66	4,16	1,99	6,68	2,72	9,04	2,02
	90	3,36	1,74	4,32	2,07	6,83	2,64	8,95	1,69
	100	3,37	1,82	4,45	2,10	7,00	2,50	8,75	1,42
	110	3,37	1,88	4,56	2,13	7,09	2,32	8,55	1,20
	120	3,36	1,91	4,66	2,10	7,14	2,12	8,25	1,01
	130			4,75	2,05	7,08	1,82	7,93	0,84
	140			4,80	1,98	6,98	1,55	7,87	0,66
0,8	30	2,85	0,67	3,20	0,79	4,25	1,36	5,09	1,99
	40	2,88	0,74	3,27	0,86	4,35	1,53	5,57	2,39
	50	2,92	0,84	3,34	0,98	4,54	1,68	6,05	0,53
	60	2,96	0,93	3,40	1,12	4,75	1,81	6,50	2,72
	70	3,00	1,03	3,48	1,25	4,96	1,93	6,97	2,78
	80	3,04	1,12	3,60	1,36	5,18	2,06	7,39	2,57
	90	3,07	1,20	3,73	1,43	5,40	2,15	7,67	2,22
	100	3,11	1,29	3,85	1,51	5,63	2,16	7,74	1,79
	110	3,15	1,36	3,96	1,54	5,85	2,07	7,60	1,38
	120	3,19	1,41	4,04	1,55	6,06	1,90	7,37	1,00
	130	3,23	1,42	4,12	1,52	6,10	1,70	7,07	0,78
	140	3,27	1,38	4,09	1,45	5,95	1,47	6,72	0,62

Cyclohexane (C_6H_{12}) – Cyclohexanol ($C_6H_{12}O$)

[674]

λ, cm	t°C	ϕ_2 = 80		ϕ_2 = 60		ϕ_2 = 40		ϕ_2 = 20	
		ε′	ε″	ε′	ε″	ε′	ε″	ε′	ε″
10	20	2,80	0,60	-	-	-	-	-	-
	30	2,90	0,70	-	-	-	-	-	-
	40	3,09	0,87	-	-	-	-	-	-
	50	3,33	1,08	-	-	-	-	-	-
	60	3,64	1,36	-	-	-	-	-	-
	70	3,96	1,67	-	-	-	-	-	-
3,2	20	2,75	0,24	2,71	0,22	2,50	0,19	2,27	0,19
	30	2,79	0,39	2,73	0,24	2,52	0,21	2,28	0,20
	40	2,92	0,54	2,75	0,32	2,55	0,24	2,29	0,21
	50	3,03	0,60	2,83	0,36	2,61	0,27	2,31	0,22
	60	3,14	0,63	2,92	0,41	2,68	0,30	2,33	0,26
	70	3,28	0,65	3,04	0,45	2,76	0,34	2,34	0,30
0,4	20	2,63	0,21	2,61	0,18	2,24	0,14	2,24	0,14
	30	2,63	0,26	2,62	0,20	2,27	0,15	2,27	0,15
	40	2,63	0,31	2,63	0,22	2,28	0,18	2,28	0,18
	50	2,64	0,36	2,64	0,28	2,29	0,22	2,29	0,22
	60	2,65	0,41	2,65	0,31	2,30	0,25	2,30	0,25
	70	2,66	0,47	2,66	0,35	2,32	0,27	2,32	0,27

Cyclohexane (C_6H_{12}) – Hexan-1-ol ($C_6H_{14}O$)

[675], ν = 60–480 MHz

t°C	X_2	ϵ	ϵ_∞	ν_m, MHz	t°C	X_2	ϵ	ϵ_∞	ν_m, MHz
15	100,00	14,22	3,17	106	25	80,12	10,35	2,99	172
	89,88	12,78	3,08	104		70,38	8,82	2,94	175
	79,53	11,25	3,01	102		60,44	7,15	2,88	186
	69,74	9,61	2,93	105		50,26	5,47	2,77	209
	60,19	7,84	2,83	111		40,08	4,07	2,67	2,46
	49,98	5,97	2,74	125		32,58	3,29	2,51	308
	40,20	4,39	2,63	151	30	100,00	12,70	3,17	221
	30,58	3,24	2,47	205		90,20	11,37	3,08	219
20	100,00	13,66	3,16	135		80,59	10,04	3,06	216
	88,93	12,18	3,08	131		70,28	8,41	2,96	227
	79,76	10,81	3,00	132		59,62	6,69	2,88	244
	70,10	9,21	2,93	135		49,45	5,13	2,78	275
	59,92	7,40	2,83	145	35	100,00	12,18	3,22	277
	50,68	5,80	2,77	161		89,56	10,84	3,12	275
	40,15	4,21	2,64	194		80,25	9,50	3,07	280
	30,14	3,15	2,48	253		70,37	8,01	3,00	287
25	100,00	13,18	3,16	173		59,87	6,40	2,89	313
	89,73	11,75	3,03	171		49,70	4,94	2,81	3,49

Cyclohexane (C_6H_{12}) – Acetophenone (C_8H_8O)

[660]

X_2	t°C	λ = 20 cm		λ = 10 cm		λ = 3,2 cm		λ = 0,408 cm		ϵ
		ϵ'	ϵ''	ϵ'	ϵ''	ϵ'	ϵ''	ϵ'	ϵ''	
100	20	15,65	4,90	11,39	6,91	5,68	4,48	3,47	1,06	17,79
	40	15,15	3,65	12,82	5,64	6,73	5,42	3,42	1,26	16,27
	60	14,25	2,35	12,83	4,22	7,58	5,13	3,49	1,49	14,95
90	20	14,50	4,35	10,82	6,17	5,44	4,19	3,32	1,03	16,30
	40	14,00	3,10	12,10	4,75	6,37	4,72	3,31	1,23	14,80
	60	13,10	2,05	11,97	3,54	7,30	4,66	3,34	1,36	13,50
70	20	11,80	2,90	9,28	4,50	4,89	3,43	3,03	0,83	12,82
	40	11,10	2,15	9,91	3,23	5,69	3,58	3,02	0,98	11,50
	60	11,30	1,45	9,53	2,49	6,28	5,21	3,05	1,13	10,60
50	20	8,52	1,77	11,80	2,90	4,37	2,44	2,76	0,68	9,17
	40	8,05	1,17	11,10	2,15	4,85	2,45	2,76	0,77	8,32
	60	7,55	0,86	11,30	1,43	5,21	2,24	2,81	0,85	7,72
30	20	5,33	0,75	8,52	1,77	3,58	1,34	2,47	0,42	5,58
	40	5,19	0,54	8,05	1,17	3,78	1,26	2,45	0,50	5,30
	60	5,0	0,35	7,55	0,86	3,97	1,09	2,44	0,55	5,07
10	20	-	-	-	-	2,61	0,40	2,20	0,18	3,05
	40	-	-	-	-	2,64	0,33	2,17	0,19	2,93
	60	-	-	-	-	2,72	0,27	2,11	0,22	2,80

Cyclohexane (C_6H_{12}) – Octan–1–ol ($C_8H_{18}O$)

[675], $\nu = 60–480$ MHz

t°C	X_2	ϵ	ϵ_∞	ν_m, MHz	t°C	X_2	ϵ	ϵ_∞	ν_m, MHz
20	100,00	10,55	3,03	89	25	50,66	4,59	2,71	1,75
	99,91	9,69	2,98	90		40,49	3,60	2,60	218
	80,93	8,61	2,92	94	30	100,00	9,70	3,04	153
	70,07	7,23	2,85	102		91,45	8,93	3,01	155
	60,32	6,02	2,78	112		80,22	7,76	2,94	164
	50,69	4,78	2,70	134		69,69	6,57	2,90	178
	40,41	3,68	2,58	172		59,75	5,42	2,82	197
	30,08	2,93	2,44	235		49,57	4,32	2,73	233
25	100,00	10,14	3,04	116	35	100,00	9,32	3,06	192
	90,77	9,26	2,99	119		89,71	8,36	3,03	202
	79,96	8,12	2,94	124		80,09	7,36	2,93	216
	70,07	6,95	2,88	133		70,54	6,35	2,88	230
	60,70	5,79	2,81	149		59,83	5,20	2,83	254

ν_m – relaxation frequency

Hexane (C_6H_{14}) – Hexan–1–ol ($C_6H_{14}O$)

[665], $\nu = 0{,}1$ MHz, $t = 20$°C

X_1	100,00	94,3	83,3	71,7	45,8
10^{-3} tan δ	7,0	10,4	12,6	13,0	2,0

***p*–Xylene (C_8H_{10}) – Tributylammonium picrate ($C_{18}H_{30}N_4O_7$)**

[676], $\nu = 0{,}3–3{,}5$ GHz

	t°C	ϵ	ϵ_∞	$10^{-12}\tau$		t°C	ϵ	ϵ_∞	$10^{-12}\tau$
0,05	12,5	3,10	2,29	304	0,2	12,5	5,23	2,44	403
	25	3,03	2,28	217		25	5,66	2,48	416
	44	3,01	2,26	187		40	5,17	2,38	288
	67	3,45	2,32	172		67	5,13	2,40	206
0,075	12,5	3,59	2,41	350	0,3	12,5	6,62	2,48	497
	25	3,58	2,33	259		25	6.90	2,55	451
	44	3,43	2,31	201		40	6,54	2,51	325
	67	4,12	2,38	198		67	6,67	2,48	255
0,125	12,5	4,17	2,39	387	0,5	12,5	6,85	2,54	522
	25	4,14	2,34	296		25	9,18	2,57	530
	44	4,25	2,39	264		44	10,11	2,73	433
	67	4,76	2,42	217		67	9,71	2,71	323

Fluorenone ($C_{13}H_8O$) – *o*–Terphenyl ($C_{18}H_{14}$)

[677], $\nu = 1 – 10^5$ Hz

ω_1	T°Π	ϵ''_m	log ν_m (Hz)	ω_1	T°K	ϵ''_m	log ν_m (Hz)
5,10	253,7	0,140	0,975	15,06	250,0	0,439	0,575
	258,7	0,140	2,150		251,2	0,444	0,925
8,02	253,7	0,228	0,950		253,6	0,434	1,425
	258,7	0,211	2,00	22,48	248,3	0,710	0,850
	261,7	0,214	2,650		249,9	0,696	1,175
	-	-	-	-	251,2	0,684	1,375

Dibutyl phthalate ($C_{16}H_{22}O$) – *o*–Terphenyl ($C_{18}H_{14}$)

[677], $\nu = 1 – 10^5$ Hz

ω_1	T°K	ϵ''_m	log ν_m (Hz)	ω_1	T°K	ϵ''_m	log ν_m (Hz)
4,80	249,0	0,064	0,400	49,82	224,0	0,515	2,725
	252,2	0,064	1,15		228,2	0,534	3,300
9,65	245,2	0,114	0,675	74,97	206,8	1,126	3,050
	248,4	0,115	1,275		210,2	1,172	3,550
19,52	240,1	0,228	1,400	100,0	198,8	2,051	3,875
	242,2	0,221	1,750		202,95	2,051	4,400
29,92	235,6	0,348	2,100	-	-	-	-
	239,3	0,330	2,825	-	-	-	-

Supplementary information on the dielectric dispersion parameters of non-aqueous solutions of inorganic and organic compounds.

Lithium (Li) – Ammonia (NH_3) [639], ν = 1 – 10 GHz, t = 35°C, x = 0.1 – 1.5.

Lithium bromide (LiBr) – Dimethyl carbonate ($C_3H_3O_3$) [640], $\nu = 0.45 –$ 67 GHz, $t =$ 25°C.

Lithium perchlorate ($LiC10_4$) – Dimethyl carbonate ($C_3H_3O_3$) [640], $\nu = 0.45$ –67 GHz, $t =$ 25°C.

Lithium perchlorate ($LiC10_4$) – 1,2-Dimethoxyethane ($C_3H_8O_2$) [641], $\nu = 0.1$ – 35 GHz, $t = 25$°C, $N = 0 – 0.5$.

Lithium perchlorate ($LiC10_4$) – 1,3-Dioxan ($C_4H_8O_2$) [641], ν = 0.1 – 35 GHz, $t = 25$°C, $N = 0 – 1$.

Lithium thiocyanate (LiSCN) – Dimethyl carbonate ($C_3H_3O_3$) [640], $\nu = 0.45$ – 67 GHz, $t = 25$°C.

Sodium (Na) – Ammonia (NH_3) [639], ν = 1 – 10 GHz, t = 35°C, $x = 0.1 – 1.5$.

Carbon tetrachloride (CCl_4) – 1-Bromopropane (C_3H_7Br) [647], $\lambda = 3.2$ cm, $t = 20°C, 50°C$.

Carbon tetrachloride (CCl_4) – *N*-Methylacetamide (C_3H_7NO) [648], $\nu = 0.45 - 1$ GHz, $t = 30°C, N = 0.1 - 15$.

Carbon tetrachloride (CCl_4) – 1-Bromobutane (C_4H_9Br) [647], $\nu = 3.2$ cm, $t = 20°C$.

Carbon tetrachloride (CCl_4) – 1-Bromopentane ($C_5H_{11}Br$) [647], $\nu = 3.2$ cm, $t = 20°C, 50°C$.

Carbon tetrachloride (CCl_4) – 1-Bromohexane ($C_6H_{13}Br$) [647], $\nu = 3.2$ cm, $t = 20°C, 50°C$.

Carbon tetrachloride (CCl_4) – 1-Bromoheptane ($C_7H_{15}Br$) [647], $\nu = 3.2$ cm, $t = 20°C, 50°C$.

Carbon tetrachloride (CCl_4) – Methyl benzoate ($C_8H_8O_2$) [649], $\nu = 3.6$ and 9.17 GHz, $t = 275 - 323°K, x = 50 - 100$.

Carbon tetrachloride (CCl_4) – 1-Bromooctane ($C_8H_{17}Br$) [647], $\lambda = 3.2$ cm, $t = 20°$ C, 50°C.

Carbon tetrachloride (CCl_4) – Octan-1-ol ($C_8H_{18}O$) [650], $\nu = 2 - 400$ MHz, $t = 20 - 60°C$.

Carbon tetrachloride (CCl_4) – Ethyl benzoate ($C_9H_{10}O_2$) [649], $\nu = 3.6$ and 9.17 GHz, $t = 275 - 323°K, x = 50 - 100$.

Carbon tetrachloride (CCl_4) – 1-Bromononane ($C_9H_{19}Br$) [647], $\lambda = 3.2$ cm, $t = 20°C, 50°C$.

Carbon tetrachloride (CCl_4) – 1-Bromodecane ($C_{10}H_{21}Br$) [647], $\lambda = 3.2$ cm, $t = 20°C, 50°C$.

Carbon tetrachloride (CCl_4) – Tetrabutylammonium bromide ($C_{16}H_{36}NBr$) [651], $\nu = 8 - 3500$ MHz, $N = 0 - 1.2$.

Chloroform ($CHCl_3$) –Lecithin ($C_{42}H_{84}O_9NP$) [653], $\nu = 0.5 - 250$ MHz, $t = 25°C$.

Methanol (CH_4O) –Benzene (C_6H_6) [649], $\nu = 3.6$ and 9.17 GHz, $t = 275 - 323°K, x = 50 - 100$.

Methanol (CH_4O) –Lecithin ($C_{42}H_{84}O_9NP$) [653], $\nu = 0.5 - 250$ MHz, $t = 25°C$.

Acetone (C_3H_6O) – Cyclopentanone (C_5H_8O) [659], $\lambda = 8.15$, 3,2, 1.75 cm, $t = -40$ to $40°C, x = 0 - 100$.

Propan-2-ol (C_3H_8O) – 1,4-Diocan ($C_4H_8O_2$) [662], $\nu = 350 - 2100$ MHz, $t = 25°C, N = 1 - 6$.

Propan-1-ol (C_3H_8O) – Pyridine (C_5H_5N) [662], $\nu = 350 - 2100$ MHz, $t = 25°C, x = 8.3 - 42$.

Propan-1-ol (C_3H_8O) – Chlorobenzene (C_6H_5Cl) [662], $\nu = 350 - 2100$ MHz, $t = 25°C, N = 1 - 6$.

Propan-1-ol (C_3H_8O) – Benzene (C_6H_6) [662], ν = 350 – 2100 MHz, t = 25°C, N = 6 – 12.

Propan-2-ol (C_3H_8O) – 1,4-Dioxan ($C_4H_8O_2$) [662], ν = 350 – 2100 MHz, t = 25°C, N = 6 – 12.

Propan-2-ol (C_3H_8O) – Pyridine (C_5H_5N) [662], ν = 350 – 2100 MHz, t = 25°C, N = 6 – 12.

Propan-2-ol (C_3H_8O) – Benzene (C_6H_6) [662], ν = 350 – 2100 MHz, t = 25°C, N = 6 – 12.

tert-Butyl alcohol ($C_4H_{10}O$) – Pyridine (C_5H_5N) [666, 667], ν = 0.35 – 2.1 GHz, t = 25 – 40°C, x = 70 – 100.

tert-Butyl alcohol ($C_4H_{10}O$) – Benzene (C_6H_6) [666, 667], ν = 0.35 – 2.1 GHz, t = 25 – 40°C, x = 70 – 100.

Isopentyl bromide ($C_5H_{11}Br$) – Chlorobenzene (C_6H_5Cl) [668], ν = 1 MHz – 50 GHz, t = 16°C.

Isopentyl bromide ($C_5H_{11}Br$) – Hexane (C_6H_{14}) [668], ν = 1 MHz – 50 GHz, t = 25°C.

Isopentyl bromide ($C_5H_{11}Br$) – 1-Bromooctane ($C_8H_{17}Br$) [668], ν = 1 MHz – 50 GHz, t = 16°C.

Bromobenzene (C_6H_5Br) – Chlorobenzene (C_6H_5Cl) [668], ν = 1 MHz – 50 GHz, t = 16°C.

Chlorobenzene (C_6H_5Cl) – Hexane (C_6H_{14}) [668], ν = 1 MHz – 50 GHz, t = 16°C.

Nitrobenzene ($C_6H_5NO_2$) – Benzene (C_6H_6) [649], ν = 3.6 and 9.17 GHz, t = 275 – 323°C, x = 50 – 100.

Benzene (C_6H_6) – *N,N*-Diethylaniline ($C_{10}H_{15}N$) [649], ν = 3.6 and 9.17 GHz, t = 275 – 323°C, x = 50 – 100.

Benzene (C_6H_6) – Tetrabutylammonium bromide ($C_{16}H_{36}NBr$) [651], ν = 8– 3500 MHz, N = 0 – 0.6.

Benzene (C_6H_6) – Tributylammonium picrate ($C_{18}H_{30}N_4O_7$) [651], ν = 9 – 3500 MHz, N = 1.4.

Aniline (C_6H_7N) – Cyclohexane (C_6H_{12}) [673], ν = 5 kHz, x = 47.5 – 40.8, t = 29.5 – 32.0°C.

Hexane (C_6H_6) – 1-Bromooctane ($C_8H_{17}Br$) [668], ν = 1 MHz – 50 GHz, t = 16°C, x = 0 – 100.

o-Toluidine (C_7H_9N) – Heptane (C_7H_{16}) [649], ν = 3.6 and 9.17 GHz, t = 275 – 323°K, x = 50 – 100.

Heptane (C_7H_{16}) – Methyl benzoate ($C_8H_8O_2$) [649], ν = 3.6 and 9.17 GHz, t = 275 – 323°K, x = 50 – 100.

Heptane (C_7H_{16}) – Diethylamine ($C_{10}H_{15}N$) [649], ν = 3.6 and 9.17 GHz, t = 275 – 323°K, x = 50 – 100.

Decahydronaphthalene ($C_{10}H_{18}$) – Dodecan-1-ol ($C_{12}H_{26}O$) [650], $\nu = 2 - 400$ MHz, $t = 20 - 60°C$.

o–Terphenyl ($C_{18}H_{14}$) – Tritolyl phosphate ($C_{21}H_{21}O_4P$) [678], $\nu = 10^{-2} - 10^5$ Hz, $t = 214 - 245°K$.

§3. Permittivity of aqueous solutions of inorganic and organic compounds

Water – Acetonitrile (C_2H_3N)

[681], $t = 25°C$

X_2	100	84,8	68,3	50,8	29,5	9,71	0
ϵ	35,94	38,58	41,98	46,90	56,41	70,92	78,45

Water – Sodium Acetate ($C_3H_3O_2Na_1$)

[685], $\nu = 500–100$ MHz, $t = 20°C$

N	5,0	4,0	3,0	2,0	1,0	0,5	0,25
ϵ	45,0	50,0	55,0	61,0	66,5	71,5	74,0

Water –1,2–Propandiol ($C_3H_8O_2$)

[682], $\nu = 1,8$ MHz

W_2	ϵ at t°C							
	15	20	25	30	35	40	45	50
100	31,05	30,00	29,02	27,96	27,19	26,40	25,67	24,97
90	37,33	36,21	35,11	34,04	33,18	32,25	31,39	30,54
70	49,30	47,92	46,68	45,29	44,11	42,81	41,70	40,60
50	60,51	58,77	57,12	55,65	54,17	52,78	51,55	50,26
30	70,17	68,36	66,63	64,66	63,45	61,81	60,19	58,97
10	78,41	76,27	74,52	72,72	71,17	69,41	67,38	65,97
0	82,07	80,23	78,21	76,40	74,84	73,18	71,49	69,80

Water – Butan–2–ol ($C_4H_{10}O$)

[683], $\nu = 8$ MHz

X_2	ϵ at t°C						
	20	25	35	45	55	65	75
100,00	20,13	16,73	-	-	-	-	-
89,404	16,44	15,67	14,22	12,92	11,72	10,58	9,46
81,011	16,39	15,64	14,33	13,18	12,08	11,07	10,03
60,872	18,28	17,63	16,33	15,15	14,06	13,04	12,10
50,947	20,28	19,63	18,33	17,14	16,01	14,92	13,89
40,283	24,07	23,38	21,83	20,36	19,06	18,03	17,17
31,64	28,28	27,47	25,78	-	-	-	-
29,581	29,72	29,55	-	-	-	-	-
5,004	66,80	65,02	-	-	-	-	-
4,051	69,60	67,72	64,02	-	-	-	-
3,926	69,80	68,00	64,43	61,10	-	-	-
3,187	72,00	70,00	66,37	63,02	59,79	57,08	54,77
0,00	80,20	78,40	74,90	71,55	68,35	65,27	62,34

Water – *tert*–Butyl alcohol ($C_4H_{10}O$)

[683], $\nu = 8$ MHz

X_2	ϵ at t°C							
	10	20	25	30	40	50	60	70
100,00	-	-	12,40	11,25	9,43	8,13	7,18	6,53
89,64	13,79	12,26	11,54	10,87	9,66	8,65	7,89	7,37
79,389	13,71	12,64	12,13	11,63	10,64	9,73	8,96	8,33
70,23	14,60	13,43	12,92	12,44	12,59	10,65	9,90	9,37
59,755	16,35	16,23	14,71	14,22	13,36	12,64	12,10	-
49,677	19,23	18,10	17,56	17,03	16,05	15,14	14,33	13,65
40,157	22,79	21,67	21,13	20,59	19,53	18,52	17,50	16,52
29,642	29,56	27,96	27,17	26,42	24,91	23,44	22,09	20,83
21,134	38,37	36,19	35,15	34,13	32,22	30,54	29,03	27,60
10,007	58,34	54,87	53,21	51,57	48,43	45,57	42,93	40,53
5,502	70,04	66,20	64,34	62,57	59,17	56,08	53,13	50,33
0,00	83,80	80,20	78,40	76,65	73,20	69,90	66,88	63,77

[684], $t = 25$°C

X_2	ϵ	X_2	ϵ	X_2	ϵ	X_2	ϵ
50,00	17,24	20,06	36,79	5,00	66,06	0,00	78,40
30,00	26,76	15,00	43,09	4,01	68,43	-	-
23,87	32,18	10,01	53.46	3,00	70,84	-	-

Supplementary information on the permittivity of aqueous solutions of inorganic and organic substances.

Water – Formamide (CH_3NO) [679], ν = 10 MHz, t = 25°C, x = 0 – 100.
Water – Methanol (CH_4O) [680], t = 15 – 35°C, ϕ = 0 – 80.
Water – Acetamide (C_2H_5NO) [679], ν = 10 MHz, t = 25°C, x = 0 – 100.
Water – Methylformamide (C_2H_5NO) [679], ν = 10 MHz, t = 25°C, x = 0 – 100.
Water – *N, N*-Dimethylformamide (C_3H_7NO) [679], ν = 10 MHz, t = 25°C, x = 0 – 100.
Water – *N*-Methylacetamide (C_3H_7NO) [679], ν = MHz, t = 25°C, x = 0 – 100.
Water – Propionamide (C_3H_7NO) [679], ν = 10 MHz, t = 25°C, x = 0 – 100.
Water – *N*-Ethylformamide (C_3H_7NO) [679], ν = 10 MHz, t = 25°C, x = 0 – 100.
Water – 1,4-Dioxan ($C_4H_8O_2$) [680], t = 25°C, ϕ = 10 – 90.
Water – *N,N*-Dimethylacetamide (C_4H_9NO) [679], ν = 10 MHz, t = 25°C, x = 0 – 100.
Water – Isopropylformamide (C_4H_9NO) [679], ν = 10 MHz, t = 25°C, x = 0 – 100.
Water – *N*-Ethylacetamide (C_4H_9NO) [679], ν = 10 MHz, t = 25°C, x = 0 – 100.
Water – *N,N*-Diethylformamide ($C_5H_{11}NO$) [679], ν = 10 MHz, t = 25°C, x = 0 – 100.
Water – *N,N*-Diethylacetamide ($C_6H_{13}NO$) [679], ν = 10 MHz, t = 25°C, x = 0 – 100.
Water – *N,N*-Dipropylacetamide ($C_8H_{17}NO$) [679], ν = 10 MHz, t = 25°C, x = 0 – 100.

§4. Dielectric dispersions parameters of aqueous solutions of inorganic and organic compaounds

Water – Potassium chloride (KCl)

[637], $\nu = 9370$ MHz

X_2	ϵ at t°C					
	4,6	8,5	15	25	35	50
0,00	49,3	53,4	58,6	63,6	65,2	64,9
0,18	49,2	53,1	58,3	62,7	64,1	63,5
0,45	49,2	52,6	57,9	61,6	62,7	62,1
0,89	48,8	52,1	56,8	60,0	60,7	59,8
1,33	48,7	51,7	55,5	58,5	59,0	58,0
1,77	48,2	51,0	54,6	57,4	57,4	56,8
3,48	46,3	48,4	50,7	52,6	52,9	52,1

Water – Lithium chloride (LiCl)

[637], $\nu = 9370$ MHz

X_2	ϵ at t° C					
	4,6	8,5	15	25	35	50
0,00	49,4	53,3	58,5	63,7	65,4	64,9
0,18	48,9	52,3	57,9	62,8	63,8	63,4
0,45	48,1	51,4	56,7	61,2	62,0	61,8
0,88	46,8	49,8	54,7	58,2	59,4	58,7
1,32	45,5	48,2	52,7	56,2	56,9	56,5
1,74	44,2	46,7	51,0	54,0	55,1	54,8
3,36	39,7	41,8	45,0	47,0	48,7	48,0

Water – Sodium chloride (NaCl)

[637], $\nu = 9370$ MHz

X_2	ϵ at t° C					
	4,6	8,5	15	25	35	50
0,00	49,3	53,3	58,8	63,6	65,2	64,9
0,18	49,1	53,1	58,2	62,8	63,7	63,4
0,45	48,6	52,5	57,4	61,1	62,2	61,8
0,89	47,8	51,3	56,0	58,8	59,6	59,2
1,33	47,2	50,2	54,3	57,2	57,7	57,1
1,77	46,0	48,9	52,7	55,0	55,7	55,2
3,48	42,5	44,9	47,4	49,0	49,9	49,0

Water – Acetonitrile (C_2H_3N)

[689]

X_2	t°C	$\nu' = 3$ GHz		$\nu = 9{,}37$ GHz		$\nu = 14$ GHz		10^{-12}T	a
		ϵ'	ϵ''	ϵ'	ϵ''	ϵ'	ϵ''		
0	5	79,7	20,8	50,8	9,8	34,9	38,9	14,8	0,02
	15	78,9	14,8	60,2	34,4	45,5	38,4	10,7	0,015
	25	76,6	11,1	64,4	28,6	53,2	34,4	8,25	0,01
	35	73,8	8,5	65,9	23,3	58,4	30,3	6,5	0,005
	50	69,5	5,9	65,2	17,2	60,4	23,4	4,8	0
5	5	75,3	20,9	46,1	37,9	31,2	36,0	15,9	0,055
	15	75,6	15,3	55,6	34,2	41,5	36,5	12,0	0,04
	25	37,0	11,6	59,5	28,3	47,9	33,9	9,0	0,025
	35	70,5	8,9	61,2	23,4	52,4	29,8	7,3	0,015
	50	66,2	5,8	61,5	16,9	56,5	22,6	5,0	0,005
10	5	71,1	20,4	42,5	36,1	28,6	33,8	16,4	0,065
	15	71,0	14,9	51,4	32,5	38,0	34,4	12,3	0,050
	25	67,1	10,9	54,2	26,9	43,3	31,3	9,5	0,035
	35	66,4	8,6	57,6	23,0	48,8	28,7	7,4	0,025
	50	62,4	5,7	58,0	16,4	52,9	22,2	5,2	0,01
20	5	65,3	16,5	42,4	32,1	29,5	31,6	14,5	0,048
	15	63,6	11,9	48,1	27,6	37,2	30,5	10,7	0,028
	25	60,6	9,0	50,5	23,0	41,4	27,5	8,5	0,017
	35	59,1	6,9	52,5	19,0	45,7	24,3	6,7	0,005
	50	56,4	4,8	52,4	13,8	48,8	18,9	4,8	0
30	5	59,3	13,8	40,6	28,3	29,1	28,9	13,3	0,027
	15	56,3	9,9	44,4	23,7	35,7	26,9	10,0	0,01
	25	54,9	7,6	46,7	19,8	39,1	24,3	7,94	0
	35	52,7	5,9	47,3	16,3	41,5	21,2	6,4	0
	50	51,0	4,1	48,1	12,1	44,6	16,7	4,6	0
40	5	54,7	11,1	40,2	24,7	30,0	26,5	11,6	0,01
	15	51,8	8,1	42,5	20,4	34,8	24,1	8,9	0
	25	50,1	6,4	43,7	17,1	37,3	21,5	7,25	0
	35	48,4	5,1	44,0	14,3	39,2	18,8	8,0	0
	50	46,7	3,6	44,3	10,6	41,3	14,7	4,4	0
60	5	45,8	6,8	38,2	17,4	31,3	20,9	8,5	0
	15	45,4	5,5	40,0	15,1	34,1	19,1	6,9	0
	25	43,5	5,1	39,6	12,7	35,3	16,8	5,9	0
	35	42,1	3,6	39,4	10,5	36,2	14,4	4,9	0
	50	40,4	2,8	38,7	8,3	36,6	11,6	3,9	0
80	5	43,2	4,98	38,5	13,8	33,6	17,8	6,5	0
	15	40,8	4,1	37,4	11,6	33,6	15,4	5,6	0
	25	38,9	3,4	36,4	9,8	33,5	13,4	4,9	0
	35	37,6	2,9	35,6	8,5	33,3	11,9	4,4	0
	50	36,3	2,3	34,8	6,9	33,2	9,8	3,6	0
100	5	39,4	3,76	36,4	10,8	33,0	14,5	5,3	0
	15	37,3	3,0	35,3	8,9	32,7	12,2	4,5	0
	25	36,0	2,5	34,5	7,5	32,6	10,6	3,9	0
	35	34,6	2,7	33,4	6,7	31,8	9,5	3,6	0
	50	33,2	2,0	32,1	6,1	30,7	8,7	3,4	0

X_2	0	5	10	20	30	40	60	80	100
ΔF	2,31	2,36	2,40	2,33	2,27	2,20	2,08	1,99	1,88
ΔH	3,95	3,96	3,96	3,80	3,60	3,25	3,50	1,75	1,20
TΔS	1,64	1,60	1,66	1,47	1,33	1,05	0,42	0,24	0,68

Water – Acetamide (C_2H_5NO)

[690]

ν, GHz	X_2	$t = 5°C$		$t = 25°C$		$t = 50°C$	
		ϵ'	ϵ''	ϵ'	ϵ''	ϵ'	ϵ''
9,37	0	52	39,5	63,6	28,5	65,0	17
	3	42	37,3	57	32	62,9	21,1
	4,66	39,5	36	53	33	61	24,1
	10	35,5	35	45	35,5	54,4	28,9
	15	26,3	30,5	35	33	47,5	31,1
	20	17	23	25	29,5	40,1	31,3
	30	11	14	18	24	28	31,4
14	0	35	38,4	53,1	34,7	60,6	23,5
	3	28,7	35,5	45,9	36,3	55,9	27,9
	4,66	26	34	40,9	35,7	52,2	29,6
	10	22	31,9	34,6	33,5	43	32,5
	15	17	25	25	29,5	36	31,5
	20	10,7	17,3	17,7	23,1	28,5	30,7
	30	7,3	9,4	10,8	16,3	16,9	24

X_2	t°C	a	$10^{-11}\tau$	ΔF	ΔH	$T\Delta S$	ΔS
0	5	0,01	1,58	2,48	4,2	1,72	6,18
	25	0,01	0,88	2,36	4,16	1,80	6,04
	50	0,00	0,47	2,21	4,11	1,90	5,88
3	5	0,04	1,83	2,56	3,65	1,09	3,92
	25	0,04	1,08	2,48	3,61	1,13	3,79
	50	0,02	0,63	2,39	3,56	1,17	3,62
4,66	5	0,06	2,06	2,64	3,45	0,81	2,92
	25	0,05	1,24	2,56	3,41	0,85	2,85
	50	0,04	0,74	2,5	3,36	0,86	2,66
10	5	0,09	2,5	2,73	3,04	0,32	1,15
	25	0,08	1,62	2,71	3,01	0,30	1,04
	50	0,07	1,0	2,61	2,96	0,27	0,84
15	5	0,1	3,6	2,92	3,15	0,23	0,83
	25	0,09	2,35	2,92	3,11	0,19	0,64
	50	0,08	1,4	2,89	3,06	0,17	0,53
20	5	0,11	6,15	3,22	4,05	0,83	2,98
	25	0,1	3,52	3,17	4,01	0,84	2,82
	50	0,09	1,91	3,1	3,96	0,86	2,66
30	5	0,12	12,4	3,61	5,05	1,44	5,18
	25	0,11	6,33	3,52	5,01	1,49	5,0
	50	0,09	3,0	3,38	4,96	1,58	4,89

Water – *N*–Methylformamide (C_2H_5NO)

[691]

X	t°C	λ = 10 *cm*		λ = 3,2 *cm*		λ = 0,81 *cm*		λ = 0,416 *cm*		ε
		ε′	ε″	ε′	ε″	ε′	ε″	ε′	ε″	
5	80	62,5	6,82	60,1	12,8	35,0	28,1	19,5	24,5	63,5
	60	68,9	9,73	62,8	20,5	28,3	30,2	13,0	21,2	69,5
	40	74,8	12,7	61,6	28,8	20,2	28,1	8,90	17,8	77,0
	20	78,5	20,6	53,0	38,2	13,3	22,0	7,25	13,3	84,1
	10	77,4	28,5	45,0	40,8	10,7	18,7	6,72	11,0	88,2
	3	76,6	31,5	37,3	39,0	8,90	16,5	6,43	9,22	90,8
10	80	65,9	10,1	60,2	19,8	27,4	29,1	13,3	20,5	67,0
	60	70,4	13,0	58,4	28,1	22,0	27,6	9,62	17,6	72,5
	40	75,6	18,5	55,2	35,2	15,5	24,2	7,48	14,6	80,0
	20	76,3	28,3	44,1	39,8	10,1	18,5	6,82	10,5	87,5
	10	74,2	37,4	36,0	39,2	8,73	15,0	6,45	8,95	92,0
	3	70,6	42,3	30,2	34,6	7,50	12,7	7,63	6,15	95,0
20	80	68,5	14,4	54,2	28,2	19,3	27,2	9,02	15,5	71,0
	60	73,0	20,4	49,8	35,1	15,2	22,5	7,15	13,4	78,0
	40	75,2	29,8	41,2	37,9	10,9	18,7	6,65	10,4	85,5
	20	68,5	42,2	26,8	36,0	8,07	13,4	6,38	7,52	94,0
	10	60,2	48,3	20,1	30,3	7,43	10,8	6,25	6,43	99,0
	3	53,0	51,6	16,1	28,2	6,95	8,92	6,15	5,40	102
40	80	71,2	24,2	44,4	35,2	12,3	18,6	6,40	10,5	81
	60	74,0	32,2	35,2	37,6	9,60	15,7	5,90	8,71	89
	40	69,3	44,6	24,8	35,2	7,62	12,6	5,82	6,88	100
	20	54,2	52,3	15,3	28,2	6,80	9,31	5,75	5,43	111
	10	42,3	53,5	12,5	23,8	6,60	7,80	5,73	4,62	117
	3	35,0	52,5	11,2	21,5	6,45	7,05	5,70	4,09	122
60	80	72,2	33,0	35,1	39,2	8,43	14,8	5,52	8,32	93
	60	72,0	44,8	25,8	36,6	7,20	12,3	5,35	6,92	102
	40	61,3	53,2	15,2	32,0	6,45	10,1	5,30	5,68	117
	20	44,1	54,2	11,1	23,8	6,12	7,88	5,28	4,43	132
	10	31,2	53,2	10,5	20,2	6,05	6,92	5,26	3,92	140
	3	25,3	51,8	8,8	17,5	5,89	6,28	5,25	3,49	147
80	80	74,0	43,2	25,9	39,0	6,30	13,0	5,02	7,13	105
	60	67,6	55,7	18,8	35,6	5,70	10,9	4,97	6,08	117
	40	52,7	60,2	11,5	29,3	5,60	8,85	4,95	5,05	136
	20	35,5	59,0	9,20	22,5	5,55	7,21	4,90	4,02	156
	10	24,8	54,2	8,12	19,5	5,50	6,42	4,87	3,53	167
	3	18,2	49,3	7,02	16,3	5,45	5,80	4,85	3,25	175

Water – Dimethyl sulphoxide (C_2H_6OS)

[698]

X_2	t°K	ν = 2,856 GHz		ν = 9,368 GHz		ν = 37,034 GHz		ν = 79,310 GHz		ϵ
		ϵ'	ϵ''	ϵ'	ϵ''	ϵ'	ϵ''	ϵ'	ϵ''	
5	273,2	62,0	35,0	26,0	30,2	8,00	12,0	6,50	6,85	88,1
	293,2	69,7	21,7	45,1	34,5	12,0	20,0	7,90	10,0	70,3
	313,2	69,5	12,5	54,5	28,0	18,2	24,7	9,15	14,5	72,6
	333,2	65,0	7,60	58,3	20,0	25,5	26,7	10,7	18,2	65,9
	353,2	59,8	4,20	54,1	14,0	31,5	25,8	12,3	19,0	60,0
10	263,2	30,0	34,0	12,0	18,0	6,30	5,85	5,62	3,50	91,9
	273,2	47,6	39,5	16,1	25,0	7,00	8,45	5,95	4,75	87,1
	293,2	61,6	29,5	30,0	32,6	8,80	13,5	6,80	7,15	78,3
	313,2	66,2	17,7	43,5	29,7	12,0	18,7	7,80	9,51	71,7
	333,2	63,0	10,5	50,3	24,3	16,8	22,7	8,40	12,0	65,1
	353,2	58,8	6,90	50,3	19,0	22,7	23,9	10,1	13,5	59,3
20	243,2	9,00	9,90	6,57	4,34	5,30	1,70	4,85	1,25	98,0
	253,2	13,9	16,9	7,40	7,00	5,65	2,46	5,05	1,75	92,0
	263,2	19,5	24,5	8,35	10,2	5,90	3,75	5,20	2,25	88,5
	273,2	27,4	30,7	9,65	14,0	6,28	5,10	5,40	2,80	84,1
	293,2	47,8	31,5	16,2	23,2	7,10	8,00	5,80	4,15	75,9
	313,2	57,7	22,7	28,0	28,0	8,30	11,8	6,25	5,85	69,3
	333,2	59,2	14,6	37,3	27,3	10,2	15,7	6,90	7,65	63,0
	353,2	55,8	9,60	42,7	23,0	13,7	18,9	7,90	10,1	57,6
30	243,2	8,70	8,50	6,30	3,60	5,30	1,50	4,97	1,12	93,4
	253,2	12,4	15,0	7,20	5,40	5,55	2,20	5,14	1,45	88,9
	263,2	16,0	20,5	8,00	8,00	5,85	2,80	5,20	1,90	84,5
	273,2	20,7	27,4	9,00	11,4	6,20	4,00	5,30	2,45	80,4
	293,2	39,2	31,1	13,8	18,3	6,55	6,35	5,42	3,60	72,8
	313,2	50,6	24,6	21,7	24,5	7,20	9,25	5,55	5,00	66,4
	333,2	54,5	16,9	29,2	26,4	8,50	12,2	5,85	6,60	60,7
	353,2	52,4	10,9	36,8	23,8	10,5	15,5	6,40	8,75	55,7
35	243,2	8,80	8,80	6,50	3,50	5,35	1,50	5,02	1,15	90,9
	253,2	12,5	14,5	7,40	5,10	5,60	2,25	5,18	1,50	86,4
	263,2	15,9	19,5	8,20	7,60	5,95	2,90	5,25	1,30	82,3
	273,2	20,0	25,6	9,00	11,0	6,30	4,00	5,35	2,45	78,3
	293,2	37,3	30,3	13,4	17,2	6,60	6,20	5,45	3,50	71,0
	313,2	48,4	24,6	20,8	23,1	7,20	8,75	5,55	4,90	65,1
	333,2	52,1	17,2	27,2	26,0	8,20	11,5	5,77	6,45	59,4
	353,2	50,8	11,2	34,9	23,6	10,0	14,5	6,15	8,25	54,6
40	243,2	9,80	8,20	7,20	3,70	5,50	1,70	5,05	1,25	88,5
	253,2	13,0	14,5	7,70	5,30	5,80	3,35	5,20	1,65	83,9
	263,2	16,8	19,5	8,50	7,60	6,10	3,10	5,30	2,02	80,0
	273,2	20,4	25,0	9,30	11,1	6,35	4,20	5,40	2,55	76,2
	313,2	47,1	23,9	20,2	22,0	7,30	8,65	5,60	5,00	63,4
	333,2	51,4	16,7	26,5	25,4	8,15	11,1	5,72	6,35	58,1
	353,2	49,5	11,1	34,0	23,3	9,90	13,8	6,00	7,80	53,4
50	253,2	14,6	15,3	8,60	6,80	6,00	2,65	5,35	2,00	78,7
	263,2	19,5	20,7	9,40	9,00	6,20	3,55	5,40	2,30	74,8
	273,2	24,0	25,2	10,2	12,0	6,50	4,66	5,45	2,85	71,5
	313,2	45,7	21,8	20,1	21,4	7,50	8,80	5,60	5,15	59,8
	293,2	36,6	28,1	14,2	17,8	6,8	6,75	5,50	3,85	65,0
	333,2	49,4	15,4	27,1	23,7	8,35	11,1	5,70	6,42	55,4

Continuation

	353,2	48,0	10,5	33,2	22,0	9,80	13,5	5,90	7,40	51,0
60	263,2	22,2	22,0	10,7	10,6	6,40	3,95	5,41	2,60	69,5
	273,2	27,4	27,0	11,6	13,2	6,70	5,20	5,42	3,20	67,0
	293,2	39,0	26,1	15,5	18,7	7,00	7,30	5,45	4,25	61,0
	313,2	45,2	19,9	21,2	21,7	7,70	9,30	5,55	5,36	56,5
	333,2	47,5	13,8	27,3	22,6	8,70	11,3	5,65	6,55	52,5
	353,2	46,4	9,80	32,8	20,4	9,80	13,6	5,92	7,45	48,6
80	293,2	42,2	20,2	19,9	20,4	7,20	8,60	5,30	4,87	53,6
	313,2	45,3	15,9	26,3	21,3	8,10	10,3	5,42	5,83	50,3
	333,2	46,0	10,8	30,0	19,5	9,40	12,2	5,55	6,95	47,0
	353,2	43,9	7,60	32,6	16,5	10,6	14,0	5,85	7,75	44,0

$t°K$	$t°K$	$10^{-11}\tau\beta$	β	$10^{-11}\beta\tau\beta$	ϵ	$\epsilon\infty$
5	273,2	5,03	0,746	3,75	88,3	2,79
	293,2	2,14	0,793	1,70	79,4	2,88
	313,2	1,23	0,818	1,00	72,7	2,93
	333,2	0,770	0,833	0,649	66,5	1,80
	353,2	0,581	0,821	0,477	60,0	1,80
10	263,2	10,9	0,799	8,69	91,7	4,08
	273,2	6,92	0,794	5,50	87,4	3,93
	293,2	3,20	0,820	2,63	78,4	3,80
	313,2	11,95	0,794	1,55	71,6	2,98
	333,2	1,23	0,801	0,985	65,0	2,45
	353,2	0,88	0,801	0,705	59,2	2,40
20	243,2	87,2	0,838	64,4	97,8	4,35
	253,2	40,0	0,758	30,0	92,8	4,38
	263,2	23,0	0,767	17,6	88,5	4,28
	273,2	13,5	0,798	10,8	83,9	4,08
	293,2	5,73	0,822	4,71	75,6	4,17
	313,2	2,90	0,852	2,47	69,3	4,10
	333,2	1,76	0,881	1,55	62,9	4,24
	353,2	1,16	0,893	1,04	57,6	2,29
30	243,2	113	0,712	80,2	93,2	4,39
	253,2	50,5	0,743	37,5	88,9	4,52
	263,2	29,0	0,756	21,9	84,4	4,34
	273,2	17,8	0,775	13,8	80,5	4,30
	293,2	7,60	0,791	6,01	72,8	4,11
	313,2	3,70	0,839	3,10	66,4	3,40
	333,2	2,30	0,860	1,98	60,7	3,60
	353,2	1,45	0,878	1,27	55,7	3,48
35	243,2	112	0,710	79,8	90,6	4,59
	253,2	47,9	0,746	35,8	86,2	4,59
	263,2	29,0	0,755	21,9	82,3	4,40
	273,2	17,8	0,765	13,6	78,3	4,35
	293,2	7,60	0,800	6,08	71,0	4,10
	313,2	4,00	0,820	3,28	65,1	4,00
	333,2	2,46	0,847	2,09	59,5	3,72
	353,2	1,56	0,866	1,35	54,5	3,38
40	243,2	103	0,691	71,6	88,2	4,62
	253,2	45,0	0,740	33,3	83,9	4,70
	263,2	27,0	0,750	20,2	80,0	4,60
	273,2	18,0	0,760	13,7	76,2	4,60

Continuation

	293,2	7,79	0,781	6,08	69,1	4,15
	313,2	4,10	0,806	3,30	63,1	3,81
	333,2	2,53	0,832	2,11	58,1	3,57
	353,2	1,60	0,867	1,39	53,3	3,52
50	353,2	36,8	0,710	26,1	77,0	4,60
	263,2	20,7	0,739	15,3	74,6	4,49
	273,2	14,0	0,752	10,5	71,3	4,27
	293,2	6,80	0,772	5,25	65,0	3,86
	313,2	3,90	0,786	3,07	59,7	3,61
	333,2	2,39	0,817	1,96	55,3	3,30
	353,2	1,59	0,845	1,35	50,9	3,13
60	263,2	16,8	0,719	12,1	69,4	4,22
	273,2	11,5	0,723	8,31	67,2	3,92
	293,2	5,59	0,770	4,31	61,0	3,60
	313,2	3,54	0,785	2,78	57,2	3,39
	333,2	2,18	0,814	1,77	52,6	3,13
	353,2	1,57	0,834	1,31	48,5	3,00
80	293,2	3,78	0,774	2,92	55,0	3,16
	313,2	2,59	0,795	2,06	52,0	3,05
	333,2	1,61	0,868	1,40	47,8	3,41
	353,2	1,14	0,913	1,04	43,8	3,40

Water – Dimethylacetamide (C_4H_9NO)

[693]

ν, GHz	t°C	$X_2 = 0$		$X_2 = 10$		$X_2 = 20$		$X_2 = 30$	
		ϵ'	ϵ''	ϵ'	ϵ''	ϵ'	ϵ''	ϵ'	ϵ''
9,37	5	50,8	39,8	16,3	23,4	9,6	12,3	7,6	8,6
	15	60,2	34,4	20,6	24,8	12,0	14,5	9,8	10,8
	25	64,4	28,6	26,0	27,2	16,2	18,4	13,2	15,0
	35	65,9	23,3	34,7	28,0	21,4	20,8	18,0	17,8
	50	65,2	17,2	48,0	24,0	35,0	21,8	29,6	20,0
14	5	34,9	38,9	9,6	12,8	6,3	6,0	6,1	4,6
	15	46,0	38,1	13,4	17,4	8,2	9,0	7,5	6,8
	25	53,2	34,6	18,0	22,0	10,8	15,0	9,0	11,0
	35	57,6	30,2	21,6	25,0	13,6	17,0	10,8	14,1
	50	60,1	23,5	30,0	28,0	18,1	20,6	14,2	16,3

ν, GHz	t°C	$X_2 = 40$		$X_2 = 60$		$X_2 = 80$		$X_2 = 100$	
		ϵ'	ϵ''	ϵ'	ϵ''	ϵ'	ϵ''	ϵ'	ϵ''
9,37	5	8,2	8,6	10,6	11,4	12,6	14,9	17,8	18,5
	15	9,8	10,7	12,6	14,0	16,7	16,0	21,1	17,2
	25	13,0	14,0	16,0	16,5	19,9	17,0	24,4	16,1
	35	16,6	16,2	19,4	16,9	23,4	17,0	26,8	14,0
	50	27,0	19,0	27,1	16,9	26,8	15,0	28,0	10,8
14	5	56,6	4,6	7,2	6,7	8,2	11,0	10,9	15,0
	15	7,4	6,8	9,0	9,4	11,0	12,7	14,0	15,3
	25	8,7	9,0	10,4	12,0	13,0	14,4	16,2	16,1
	35	10,8	11,6	12,1	14,2	15,0	15,2	18,8	15,0
	50	13,4	15,0	15,6	16,3	18,7	15,3	22,2	14,1

Continuation

X_2	0	10	20	30	40	60	80	100
ΔF	2,31	3,0	3,3	3,4	3,3	3,1	2,8	2,6
ΔH	3,95	4,8	5,2	5,4	4,9	4,0	3,2	2,9
$T\Delta S$	1,64	1,8	1,9	2,0	1,6	0,9	0,4	0,3

Water – Diethylformamide ($C_5H_{11}NO$)

[693]

ν, GHz	t°C	$X_2 = 0$		$X_2 = 10$		$X_2 = 20$		$X_2 = 30$	
		ϵ'	ϵ''	ϵ'	ϵ''	ϵ'	ϵ''	ϵ'	ϵ''
9,37	5	50,8	39,8	16,0	20,0	8,6	10,4	7,1	8,4
	15	60,2	34,4	20,0	23,1	10,1	13,0	8,1	10,3
	25	64,4	28,6	24,1	24,6	11,6	16,9	9,0	11,7
	35	65,9	23,3	27,9	25,1	13,6	16,2	10,4	12,6
	50	65,2	17,2	33,2	24,0	17,1	17,3	12,3	15,0
14	15	46,0	38,1	12,4	14,9	7,2	9,2	6,4	7,0
	25	53,2	34,6	16,2	8,1	9,0	10,9	7,2	8,0
	35	57,6	30,2	19,0	19,9	11,1	12,6	8,4	9,8
	50	60,1	23,5	24,3	22,9	13,3	14,4	10,6	11,8

ν, GHz	t°C	$X_2 = 40$		$X_2 = 60$		$X_2 = 80$		$X_2 = 100$	
		ϵ'	ϵ''	ϵ'	ϵ''	ϵ'	ϵ''	ϵ'	ϵ''
9,37	5	7,0	8,3	7,8	8,4	8,4	10,2	9,6	11,4
	15	7,4	8,4	7,7	8,8	8,9	10,4	10,0	11,2
	25	8,4	9,8	8,8	10,0	10,0	10,8	11,0	11,0
	35	9,4	11,6	9,8	11,2	10,6	11,0	11,8	10,8
	50	11,0	13,4	11,4	12,4	12,2	11,1	12,6	10,5
14	15	5,9	6,2	6,1	6,4	6,3	7,8	7,0	8,8
	25	6,6	6,6	6,9	7,2	7,2	7,9	8,1	9,0
	35	7,6	8,2	8,0	8,4	8,4	8,8	8,8	9,2
	50	9,2	10,6	8,6	10,0	9,0	8,8	9,8	9,5

X_2	0	10	20	30	40	60	80	100
ΔF	2,31	3,0	3,3	3,45	3,34	3,3	3,1	3,0
ΔH	3,95	4,6	4,9	5,0	4,8	4,2	3,6	3,1
$T\Delta S$	1,64	1,6	1,6	1,55	1,4	0,9	0,5	0,1

Water – Diethylacetamide ($C_6H_{13}NO$)

[693]

ν, GHz	t°C	$X_2 = 0$		$X_2 = 10$		$X_2 = 20$		$X_2 = 30$	
		ϵ'	ϵ''	ϵ'	ϵ''	ϵ'	ϵ''	ϵ'	ϵ''
9,37	15	60,2	34,4	15,2	20,8	8,8	10,8	7,2	6,0
	25	64,4	28,6	18,0	22,9	10,1	12,7	8,1	7,9
	35	65,9	23,3	20,1	23,4	11,0	13,6	8,5	8,8
	50	65,2	17,2	23,3	23,6	13,0	15,6	9,8	10,8
14	15	46,0	38,1	10,1	14,1	7,1	6,9	5,4	4,0
	25	53,2	34,6	14,0	16,0	8,8	9,0	6,4	5,1
	35	57,6	30,2	17,1	17,0	10,6	9,4	7,4	6,1
	50	60,1	23,5	20,4	17,4	13,0	10,8	8,4	7,4

ν, GHz	t°C	$X_2 = 40$		$X_2 = 60$		$X_2 = 80$		$X_2 = 100$	
		ϵ'	ϵ''	ϵ'	ϵ''	ϵ'	ϵ''	ϵ'	ϵ''
9,37	15	6,4	6,0	6,3	6,7	7,7	6,9	8,2	7,9
	25	6,8	7,2	6,4	7,2	7,6	7,6	8,3	8,2
	35	7,4	7,8	6,9	7,6	7,5	8,0	8,4	8,4
	50	7,8	8,8	7,3	8,0	7,4	9,0	8,6	9,4
14	15	5,2	3,0	4,4	3,4	4,9	3,8	5,0	4,0
	25	15,6	4,0	4,8	4,2	5,3	4,4	5,4	4,6
	35	6,0	5,0	5,2	4,5	5,6	4,8	5,8	6,0
	50	7,0	6,4	6,0	5,4	6,2	5,7	6,4	6,0

X_2	0	10	20	30	40	60	80	100
ΔF	2,31	3,2	3,6	3,8	3,7	3,6	3,4	3,3
ΔH	3,95	4,9	5,4	5,7	5,3	4,6	3,9	3,3
$T\Delta S$	1,64	1,7	1,8	1,9	1,6	1,0	0,5	0,0

Water – 1,6–Diaminohexane ($C_6H_{16}N_2$)

[638], $\nu = 1{,}8$–40 GHz, $t = 25$°C

N	ϕ	ϵ	$\epsilon\infty$	$10^{-12}\tau$	a
0	0	78,5	5,2	8,25	0,007
0,3450	4,044	76,2	4,4	9,24	0,032
0,6701	7,303	73,5	4,5	10,46	0,045
1,00	10,44	70,5	4,8	12,04	0,053
2,00	20,75	63,3	4,7	18,49	0,085
3,00	30,65	55,4	5,0	31,32	0,098

Supplementary information on the dielectric dispersion parameters of aqueous solutions of inorganic and organic substances.

Water – Caesium fluoride (CsF) [686], $\nu = 8 - 38$ GHz, $t = 25°$C, $N = 0 - 8$.
Water – Caesium iodide (CsI) [687], $\nu = 9370$ GHz, $t = 1 - 50°$C.
Water – Potassium chloride (KCl) [687], $\nu = 9370$ GHz, $t = 1 - 50°$C.
Water – Potassium fluoride (KF) [686], $\nu = 8 - 38$ GHz, $t = 25°$C, N = 0.5 – 8.
Water – Potassium iodide (KI) [686], $\nu = 8 - 38$ GHz, $t = 25°$C, N = 0.5 – 8; [687], $\nu = 9370$ GHz, $t = 1 - 50°$C.
Water – Lithium chloride (LiCl) [687], $\nu = 9370$ GHz, $t = 1 - 50°$C.
Water – Lithium iodide (LiI) [687], $\nu = 9370$ GHz, $t = 1 - 50°$C.
Water – Sodium chloride (NaCl) [687], $\nu = 9370$ GHz, $t = 1 - 50°$C.
Water – Sodium iodide (NaI) [687], $\nu = 9370$ GHz, $t = 1 - 50°$C.
Water – Formamide (CH_3NO) [688], $\nu = 9.4$ GHz, 14 GHz, $t = 5 - 80°$C, $x = 0 - 100$.
Water – Glycine ($C_2H_5NO_2$) [692], $\nu = 3 - 35$ GHz, $t = 5 - 90°$C.
Water – 4-Aminobutyric acid ($C_4H_9NO_2$) [694], $\nu = 2 - 35$ GHz, $t = 5 - 90°$C, $N = 1 - 5$.
Water – Butylammonium chloride ($C_4H_{12}NCl$) [695], $\nu = 5 - 37$ GHz, $t = 25°$C, $N = 1$.
Water – 6-Aminohexanoic acid ($C_6H_{13}NO_2$) [696], $\nu = 3 - 35$ GHz, $t = 5 - 80°$C, $N = 1 - 5$.
Water – Hexylammonium chloride ($C_6H_{16}NCl$) [695], $\nu = 5 - 37$ GHz, $t = 25°$C, $N = 0.75$.
Water – Triethylenediamine ($C_6H_{16}N_2$) [697], $\nu = 17.6$ GHz, $t = 2 - 20°$C.
Water – Heptylammonium bromide ($C_7H_{18}NBr$) [695], $\nu = 5 - 37$ GHz, $t = 25°$C, $N = 0.4$.
Water – Octylammonium chloride ($C_8H_{20}NCl$) [695], $\nu = 5 - 37$ GHz, $t = 25°$C, $N = 0.07 - 0.15$.

LITERATURE TO CHAPTER V

613. Pan W. P., Mady M. H., Miller R. C., AIChe Jour. **21**, 2, 283 (1975).
614. Nath J., Singh B., Indian J. Chem. **16A**, 7, 620 (1978).
615. Ratzsch M. T., Wohlfarth C., Credo U., Jarmuschwitsch A., Nehmer U., Z. Phys. Chem. **257**, 1, 161 (1976).
616. Sabesan R., Varadharajan R., Indian J. Pure Appl. Phys. **15**, 8, 538 (1977).
617. Katz M., Lobo P. W., Solimo H., An. Acoc. quim. Argentina **62**, 4, 171 (1974).
618. Nowak J., Acta Phys. Polonica **41A**, 5, 617 (1972).
619. Campbell C., Brink G., Glasser L., J. Phys. Chem. **79**, 6, 610 (1975).
620. Ratzsch M. T., Rickelt E., Rosner H., Z. Phys. Chem. **256**, 2, 349 (1975).
621. Krasnoperova A. P., Asheko A. A., Zhur. strukt. khim. **18**, 5, 960 (1977).
622. Ratzsch M. T., Wohlfarth C., Claudius M., J. Prakt. Chem. **319**, 3, 353 (1977).
623. Dhillon M. S., Chugh H. S., J. Chem. Eng. Data **23**, 4, 263 (1978).
624. Kenttamaa J., Jarvi P., Lindberg J. J., Suomen Kemistilehti **33B**, 3, 101 (1960).
625. Singh B., Vij J. K., Bull. Chem. Soc. Jap. **49**, 7, 1824 (1976).
626. Jadzyn J., Malecki J., Jadzyn C., J. Phys. Chem. **82**, 19, 2128 (1978).
627. Malecki J., Nowak J., Acta Phys. Polonica **55A**, 1, 55 (1979).
628. Malecki J., Nowak J., Kulek T., Acta Phys. Polonica **52A**, 1, 171 (1977).
629. Horig H. J., Michel W., Bittrich H. J., Wis. Z. Techn. Hochsch. Chemie, Leuna-Merseburg 8, 4, 298 (1966).
630. Rivail J. L., Thiebaut J. M., J. Chem. Soc. Faraday Trans. **2**, 3, 430 (1974).
631. Hollecker M., Goblon J., Thiebaut J. M., Chem. Phys. **11**, 1, 99 (1975).
632. Ratzsch M. T., Rickelt E., Rosner H., Z. Phys. Chemie **255**, 5, 933 (1974).
633. Szurkowski B., Poluch A., Acta Physica Polonica **47A**, 3, 353 (1975).
634. Givon M., Pelan I., Phys. Lett., **48A**, 1, 1 (1974).
635. Glushkova N. V., Shakhparonov M. I., Vestnik Moskovsk. Gosudar. Univers., deppon. VINITI No. 4008–76.
636. Sliwinska-Bartkowiak M., Hilczer T., Acta Phys. Polonica **45A**, 6, 915 (1974).
637. Samoylov O. Y., Yastremsky P. S., Zhur. strukt. khim. **12**, 3, 379 (1971).
638. Kaatze U., Wen W. Y., J. Phys. Chem. **82**, 1, 109 (1978).
639. Breitschwerdt K. G., Radscheit H., Phys. Lett. **50A**, 6, 423 (1975).
640. Saar D., Brauner J., Farber H., Petrucci S., J. Phys. Chem. **82**, 5, 545 (1978).
641. Beuzelin P., Cacnet H., Cyrot A., Lestrade C., J. Chim. Phys. Biol. **74**, 11, 1131 (1977).
642. Badiali J. P., Cacnet H., Cyrot A., Lestrade C., J. Chem. Soc. Faraday Trans. **2**, 9, 1339 (1973).
643. Badiali J. P., Cacnet H., Lestrade C., Electrochem. Acta **16**, 731 (1971).
644. Breitschwerdt K. G., Radscheit H., Berich. Bunsen. Phys. Chem. **75**, 7, 644 (1971).

645. Farber H.,Petrucci S., J. Phys. Chem. **80**, 3, 327 (1976).
646. Kazantseva S. I., Shakhparonov M. I., Fizika i fiziko-khimiya zhidkostei, Izd. Mosk. Gosud. Univers. vyp. 3, 52 (1976).
647. Rourret D., Maham Ya., Sempere R., Regnier J. F., J. Chim. Phys. Biol. **74**, 9, 908 (1977).
648. Omar M. M., J. Chem. Soc. Faraday Trans. **1**, 74, 1, 115 (1978).
649. Sarma B. S., Venkateswara R. V., Austr. J. Phys. **27**, 1, 87 (1974).
650. Hanna F. F., Hakim I. K., Z. Naturforsch. **27**a, 8–9, 1363 (1972).
651. Cacnet H., Lestrade C., Bull. Soc. Chim. Belg. **85**, 7, 481 (1976).
652. Kazantseva S. I., Levin V. V., Zhur. fiz. khim. **47**, 7, 1842 (1973).
653. Pennock B. E., Goldman E. E., Chacko G. K., Chock S., J. Phys. Chem. **77**, 20, 2383 (1973).
654. Kazantseva S. I., Levin V. V., Zhur. strukt. khim. **14**, 3, 552 (1973).
655. Kazantseva S. I., Levin V. V., Zhur. strukt. khim. **47**, 3, 731 (1973).
656. Cavell E. A. S., Knight P. C., J. Chem. Soc. Faraday Trans. **2**, 68, 5, 765 (1972).
657. Zul'fugarzade K. E., Guliev L. A., Imanov L. M., Izv. AN Azerb. SSR, ser. fiz.-tekh. **3**, 21 (1973).
658. Zul'fugarzade K. E., Guliev L. A., Imanov L. M., Izv. AN Azerb. SSR, ser. fiz.-tekh. **3**, 15 (1973).
659. Useinova S. M., Kasimov R. M., Uch. zap. Azerb. Gos. Univers., ser. fiz.-mat. **4**, 76 (1972).
660. Kalacheva E. I., Nauch. trudy Mosk. lesotekhn. inst. vyp. 113 (1979).
661. Koshii T., Aril E., Nakamura M., Takahashi H., Bull. Chem. Soc. Jap. **47**, 3, 618 (1974).
662. Aril E., Nakamura M., Takahashi H., Higasi K., Chem. Lett. **5**, 533 (1973).
663. Koshii T., Aril E., Nakamura M., Bull. Chem. Soc. Jap. **47**, 3, 623 (1974).
664. Kalacheva E. I., Nauch. trudy Mosk. lesotekhn. inst. vyp. 103, 182 (1978).
665. Furmanski W., Stralkowski I., Skulski L., Bull. Acad. Sci. **23**, 10, 875 (1975).
666. Sato H., Koshii T., Takahashi H., Higasi K., Chem. Lett. **6**, 579 (1974).
667. Sato H., Koshii T., Takahashi H., Higasi K., Chem. Lett. **5**, 491 (1975).
668. Baba K., Kamiyoshi K., J. Phys. Chem. **81**, 19, 1872 (1977).
669. Zul'fugarzade K. E., Gadzhiev G. A., Imanov L. M., Izv. AN Azerb. SSR, ser. fiz.-tekh. **1**, 13 (1969).
670. Zul'fugarzade K. E., Turakulov K. T., Imanov L. M., Zhur. strukt. khim. **15**, 1, 135 (1974).
671. Cavell E. A. S., Sheikh M. A., J. Chem. Soc. Faraday Trans. **2**, 3, 315 (1973).
672. Badiali J. P., Cacnet H., Cyrot A., Lanstrade C., J. Chem. Soc. Faraday Trans. **2**, 72, 7, 1231 (1976).
673. Lubezky I., McIntosh R., Canadian J. Chem. **52**, 18, 3176 (1974).
674. Glushkova N. V., Shakhparonov M. I., Vestnik Moskov. Gosud. Univers., deppon. VINITI No. 4008–76.
675. Komooka H., Bull. Chem. Soc. Jap., **15**, 6, 1696 (1972).
676. Cavell E. A. S., Sheikh M. A., J. Chem. Soc. Faraday Trans. **2**, 71, 3, 474 (1975).
677. Shear M. F., William G., J. Chem. Soc. Faraday Trans. **2**, 69, 4, 608 (1973).
678. Beevers M. S., Crossley J., Garrington D. C., William G., J. Chem. Soc. Faraday Trans. **2**, 74, 4, 458 (1977).

679. Rohdewold P., Moldner M., J. Phys. Chem. **77**, 3, 373 (1973).
680. Jannakoudakis D., Papanastasiou G., Mavridis P. G., J. Chim. Phys. Biol. **73**, 2, 156 (1976).
681. Luhrs C., Schwitzgebel G., Berich. Bunsen. Phys. Chem. **83**, 6, 623 (1979).
682. Verbeeck R. M. H., Thun H. P., Verbeeck F., Bull. Soc. Chim. Belg. **86**, 3, 125 (1977).
683. Winkelmann J., Z. Phys. Chem. **255**, 6, 1109 (1974).
684. Broadwater T. L., Kay R., J. Phys. Chem. **74**, 21, 3802 (1970).
685. Ermakov V. I. Shcherbakov V. V., Khubetsov S. B., Elektrokhimiya **12**, 1, 133 (1976).
686. Kruger J., Scholmeyer E., Barthel J., Z. Naturforsch. **30a**, 11, 1476 (1975).
687. Samoylov O. Y., Yastremsky P. S., Tarasov A. P., Zhur. strukt. khim. **14**, 4, 600 (1973).
688. Yastremsky P. S., Verstakov E. S., Kessler Y. M., Mishustin A. I., Yemelin V. P., Bobrinov Y. M., Zhur. fiz. khim. **49**, 11, 2950 (1975).
689. Boroda Y. P., Candidate's Dissertation, Kharkov State University, 1979.
690. Goncharov V. S., Lyashchenko A. K., Yastremsky P. S., Zhur. strukt. khim. **17**, 4, 662 (1976).
691. Karamyan G. G., Shakhparonov M. I., Fizika i fiziko-khimiya zhidkostei, Izd. Mosk. Gosud. Univers. vyp. 3 (1979).
692. Bottreau A. M., Delbos G., Marzat C., Lacroix Y., Salefran J. L., Dutuit Y., Comp. Rend. Acad. Sci. Paris **276B**, 373 (1973).
693. Boroda Y. P., Verstakov E. S., Yastremsky P. S., Kessler Y. M., in a review Termodinamika i stroenie rostvorov, Ivanovo, 85, 1978.
694. Bottreau A. M., Delbos G., Marzat C., Salefran J. L., Moreau J. M., Comp. Rend. Acad. Sci. Paris **278B**, 17, 676 (1974).
695. Kaatze U., Limberg C. H., Pottel R., Berich. Bunsen. Phys. Chem. **78**, 6, 555 (1974).
696. Bottreau A. M., Delbos G., Dutuit Y., Marzat C., Salefran J. L., Comp. Rend. Acad. Sci. Paris **277B**, 639 (1973).
697. Pottel R., Asselborn E., Berich. Buns. Phys. Chem. **83**, 1, 29 (1979).
698. Galiyarova N. M., Shakhparonov I. M., Vestnik Mosk. Gosud. Univers., dep. VINITI No. 3613 (1976).

INDEX TO CHAPTER V